STATISTICS AND PROBABILITY

THE SCHOOL MATHEMATICS PROJECT

STATISTICS
AND
PROBABILITY

J. H. DURRAN
Winchester College

CAMBRIDGE
AT THE UNIVERSITY PRESS
1970

Published by the Syndics of the Cambridge University Press
Bentley House, 200 Euston Road, London N.W.1
American Branch: 32 East 57th Street, New York, N.Y.10022

© Cambridge University Press 1970

Library of Congress Catalogue Card Number: 70–96086

Standard Book Number: 521 06933 5

Printed in Great Britain
at the University Printing House, Cambridge
(Brooke Crutchley, University Printer)

TO

A. J. MORNARD

THIS VOLUME IS GRATEFULLY DEDICATED
BY HIS FORMER PUPIL

CONTENTS

CONTENTS

FOREWORD

By Professor H. E. Daniels, University of Birmingham

In recent years Probability and Statistics have been increasingly introduced into the mathematical curriculum of sixth forms. This is partly a consequence of the movement to revitalize school mathematics—the manipulations of set theory have their most natural application in probability theory—but, more significantly, it is a recognition of the fact that these subjects form one of the most important areas of applicable mathematics in the present-day world and should no longer be ignored at school level.

The need has therefore arisen for books which are suitable for use in schools. A number of these are already available and some of them are admirable. With the occasional exception, however, the pattern of such books generally tends to follow rather closely that of the typical elementary university text.

When I first heard Mr Durran expound his ideas on the teaching of probability and statistics for the School Mathematics Project to a fascinated audience, I realized that here was an exceptionally gifted and original teacher who could approach the problem of presenting the subject in the classroom with a fresh mind. The result of his efforts is a book which must surely excite the enthusiasm of all those who teach and are taught from it. Not everyone will agree with some of the innovations of nomenclature and presentation, and it must be said that this is a book for the bright student who is not afraid of hard work. But I believe it will come to be regarded as a classic of its kind.

PREFACE

Events have causes and occasions. The causes of the writing of this book were two questions.

The first of these questions, '..., but why?', was believed by his pupils to be always on the lips of their mathematics master Mr A. J. Mornard; the author was fortunate enough to be his pupil from 1943 to 1947. The second question, 'What *is* a distribution?' was put to the author by Mr R. M. N. Montgomery at about the time of the occasion for the book. This occasion was the invitation to provide the probability and statistics chapters for the A-level books being written for the School Mathematics Project.

A series of lectures on 'The Mathematical Study of Random Phenomena' given by Dr D. G. Kendall, and on 'Monte Carlo Methods' by Dr J. M. Hammersley at The Oxford Mathematics Conference of 1957 had interested me in probability and statistics, and I eagerly accepted the invitation. This book grew from the attempt to meet the SMP's requirements. As time passed, however, I felt more and more compelled to write a book which would try to answer those two questions—questions too seldom asked in statistics: 'Why?' and 'What is a ...?'

Many books set out to teach statistical techniques and methods. What would be the justification for my adding to their number? None. This book attempts a different job. It is intended for those who wish to see the subject of statistics from the sort of standpoint from which mathematical students have been accustomed to study any other branch of applied mathematics, at school or as undergraduates. Many introductory treatments concentrate on 'How?', and that does not seem right after the very early stages of a subject. Misdirection of effort in the study of statistics is a quite sufficient reason for the distaste expressed by many competent mathematical pupils (and their teachers) at what they think is 'Statistics'.

Statistics is regarded here as a branch of applied mathematics, and that phrase needs to be read both as *applied* mathematics and as applied *mathematics*. Mechanics in its early stages has often become Classical Mechanics and is not applied mathematics any more: it is a closed system based on a few axioms. Statistics, however, gives scope for the formulation of potentially provable results from the examination of what is in fact happening. Proofs may then follow. That is what the present work tries to show.

The order of chapters may call for explanation. After a brief introduction to some of the typical problems with which statistics is concerned (Chapter 1), an extended discussion is made in Chapter 2 of the frequency-based approach to probability and an attempt is made to put it on a sufficiently sound footing for mathematical progress to be made. I do not now believe that the classical definition of probability makes a proper starting point; but I do believe that the true place of 'equiprobable cases' can nowadays be seen to be in the field of model building, where they have an indispensable role. Chapter 3 gives a prolonged development (fuller than in many statistics books) of the manipulation of probabilities and ends with an exercise showing a wide range of applications.

Chapter 4 attempts a fresh look at descriptive statistics, and keeps their merely descriptive uses separate from any possibly predictive ones. Chapter 5 makes clear that when data about values of a discrete variable are grouped all that happens is a loss of some information which we could otherwise have had. We do not by what is a manipulative convenience acquire a theory of estimation: in problems of estimation our uncertainty differs in origin from the uncertainty introduced by the mere approximations in grouped data. This chapter prepares the ground for the discussion in a later volume of continuous variables.

Chapter 6 looks at the descriptive employment of the standard deviation and introduces Chebyshev's Inequality. In Chapters 7 and 8 we draw together all the work so far, and introduce one of the key ideas of this treatment: model building. We develop the models due to Bernouilli and Poisson for the results of repeated trials. In Chapter 9 there evolves the concept of a random variable which has begun to emerge in the previous two chapters, and we apply to it the ideas about descriptive statistics from Chapter 4.

In Chapter 10 we return to the study of sequences of trials, but under wider conditions (using Markov's model). In Chapter 11 the results about random variables from Chapter 9 are applied to samples, and a whole apparatus is constructed for a fairly comprehensive discussion of point-estimation in Chapter 12.

Throughout the book the exercises are used for teaching new material. Questions designed to teach material not in the main text are marked T; questions designed to keep the reader looking forward to develop for himself material treated later in the text are marked F. Questions marked R are designed to revise important material introduced in the preceding sections. Almost all the 500 questions (divided into about 50 exercises) were specially composed for the book (though a small proportion are re-used in the SMP Advanced Mathematics texts), and extended solutions are given to odd-numbered ones.

It will soon be noticed that the book deals only with discrete random variables, but it is important to realize that this limitation does not prevent most of the major ideas of the subject from being introduced. I would go further and say that the limitation, in a first course, to discrete variables is

xii

actually beneficial: the statistical wood is not obscured by nearby manipulative undergrowth.

A person who has studied this book will not have been prepared for a specific examination, and yet he should find himself able to complete the syllabus for any statistics examination he is likely to meet at this level. Meanwhile in carrying things further, he should read only those statistics books whose *mathematics* is as far advanced as he can cope with. The danger with anything less is that he may be fobbed off with recipes.

This book would never have seen the light of day without the early encouragement of Mr L. E. Ellis and Dr A. G. Howson and the constant support of Mr A. J. Collier and Mr T. A. Jones. Mr J. C. Manisty has for the last five years been sharing house with an abstracted (and at times distracted) author; over and over again his nicety of mathematical judgement, and his wisdom, have been richly poured out; he has always seemed quite tireless; and I am profoundly grateful. My debt to Professor H. E Daniels and Dr V. D. Barnett, both of Birmingham University, who read parts of the manuscript, is enormous. The changes they suggested in the treatment of probability caused me to look at the whole book again with fresh eyes. Professor M. H. Quenouille of Southampton University was most generous with exercise material. As the book bulged further and further from the niche planned for it in the SMP A-level texts and disaster for the author loomed ahead, Dr B. Thwaites (Director of the School Mathematics Project) offered to take it on as a separate SMP production. The Cambridge University Press kindly undertook publication and their patience and helpfulness have been beyond words. I am grateful also to the then Headmaster of Winchester College, Sir Desmond Lee, who relieved me of nearly two thirds of my teaching during the summer term of 1967, and to Mr A. H. Brodhurst and the members of Kingsgate House, who have had short measure from their house tutor for some years now.

J. H. DURRAN

Winchester
September 1969

1

GENERAL OUTLINES

Statistics is the art of designing experiments and extracting information from them. Sometimes the information is hidden in large masses of figures; such collections of figures as accumulate, for instance, when a series of scientific measurements is made or a biological or social survey is carried out: but the better the design of the experiment the fewer the measurements can be.

1. REPETITIONS OF EXPERIMENTS

Let us be quite clear why we carry out a series of experiments at all, and not merely, as so often in elementary science courses, a single experiment, or perhaps an experiment and a check on it. Imagine the following situation:

An industrial process designed to produce a certain chemical can be run at various temperatures. An attempt is being made to determine the temperature which gives the maximum yield, and to find out how important it is to get the temperature correctly adjusted. Two workers each carry out an experiment using the same amount of material at each of four temperatures. The yields in grams are as shown in Table 1.1.

Table 1.1

Temperature/°C	460	480	500	520
Yield (1st worker)/gm	5·27	5·37	5·42	5·41
Yield (2nd worker)/gm	5·04	5·82	5·50	5·36

Each worker draws a graph of his results: see Figures 1.1 and 1.2.

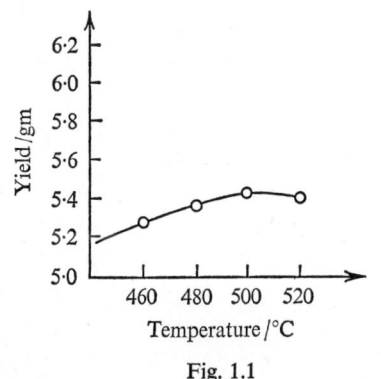

Fig. 1.1

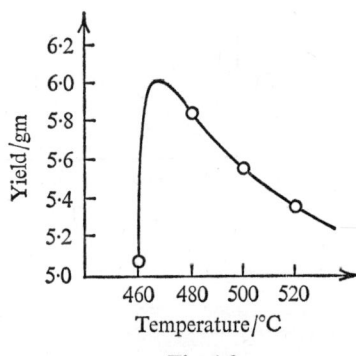

Fig. 1.2

1

Their reports are as shown in Table 1.2.

Table 1.2

	Best temperature	Critical?	Max. yield
1st worker	about 500 °C	No	about 5·4 gm
2nd worker	about 470 °C	Yes	about 6 gm

Now a full table of five runs might well look like Table 1.3, giving the general picture shown in Figure 1.3, in which a pattern begins to emerge.

Table 1.3

	Temp/°C			
Run	460	480	500	520
1st run	5·27	5·37	5·42	5·41
2nd run	5·04	5·82	5·50	5·36
3rd run	4·87	5·59	5·31	5·35
4th run	4·97	5·69	5·58	5·12
5th run	5·15	5·79	5·47	5·14

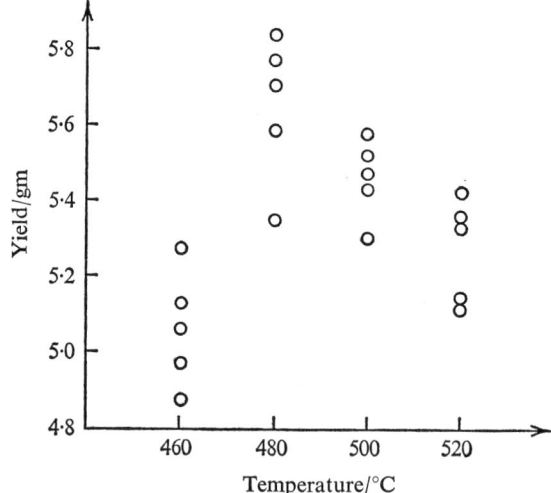

Fig. 1.3

2. ESTIMATION

From this pattern a statistician would not merely estimate a best temperature and a likely yield, but would want to estimate the degree of accuracy
to be expected in view of the fact that five runs have now been made; he
would assess the likely errors more closely than if only the two runs had
been made; and if twenty runs were to be made he would give closer
estimates still.

One simple way he might begin to tackle the complications of this
pattern would be to select the middle value of the five yields at each
temperature, as a sort of average, and get the graph, Figure 1.4.

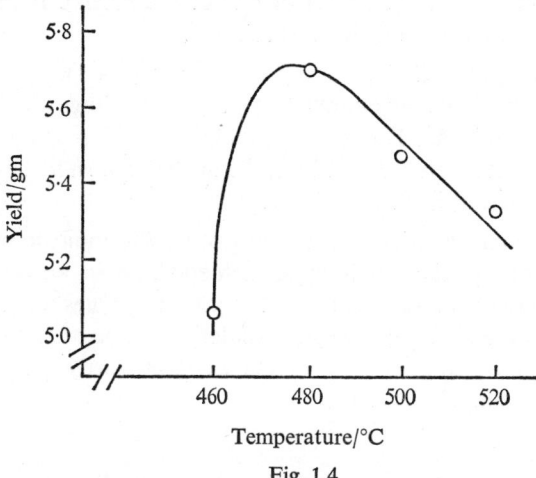

Fig. 1.4

This would suggest:

best temperature: 470° to 485°
critical?   better to err on the high side of temperature
maximum yield: 5·6 to 5·7 gm.

A comparison with the whole set of readings suggests that the best
temperature is likely to have been fairly well estimated but that the maximum yield is rather variable; presumably some other factor (perhaps
degree of agitation?) is important.

3. ORDER-STATISTICS

To simplify the numerical picture presented by the problem in Section 2,
we suggested taking at each temperature the middle reading in order of
size. This is a special case of a useful procedure in which we do not use the

actual values of the readings, but merely list them in order of size, label the smallest with the number 1 and the greatest (say of N of them) with the number N, and use the labels in the calculations, decoding them at the end. If several readings are equal, then they can be labelled in the way suggested by Table 1.4.

Table 1.4

Readings	1	1	2	3·5	3·5	3·5	4	4	4	4	5	5	6	7
Order	1	2	3	4	5	6	7	8	9	10	11	12	13	14

The 'middle' reading (or an effectively equivalent number if a middle reading does not exist) is called the *median* (and it follows that the median plays the role of a reading and not of one of the labels). If there are N readings the median can be defined as follows:

N *odd:* the median is the reading with label $\frac{1}{2}(N+1)$.

N *even:* the median is any number between the readings with labels $\frac{1}{2}N$ and $\frac{1}{2}N+1$. If these readings are different numbers and not merely repetitions, the median is *strictly* between them and is indeterminate.

We also define *quartiles* in the same sort of way to indicate readings or numbers 'a quarter of the way in from each end'. Their precise definition is left as an exercise for the reader in Exercise 1 A, Question 5. Modern developments in the theory of statistics, under the name of *order-statistics*, are concerned more and more with the extraordinarily general results developing these simple ideas.

4. CURVES OF BEST FIT

One problem that was skated lightly over in drawing Figures 1.1, 1.2 and 1.4 was where to draw the curve to give the best impression of the data, or even whether the data merit a curve being drawn at all.

The diagrams Figures 1.5–1.8, represent some typical situations, where two variables are being measured for each of a series of individual items or events. Each dot represents an individual and the diagrams are often called *scatter-diagrams*.

The statistician's first problem here is to test whether the data appear to be numerically related (or not) in the sense that changes in a given direction along one axis tend to occur with changes in a definite direction along the other (or not).

The data of Table 1.3 (the chemical process) might well fail under too simple a test of this sort since an increase in one direction produces an increase to start with and a decrease later, but even a simple test would show what is called *correlation* in Figures 1.6, 1.7 and 1.8. Any proposed test of correlation would be required to show none in Figure 1.5 of course. Next, Figure 1.8 reminds us that correlation is a merely numerical matter and cannot prove causation. To take another example, no one doubts the correlation of risk of lung cancer with intensity of cigarette-smoking—the quarrels are over the existence or not of causation; both might be due to some other factor, such as stress.

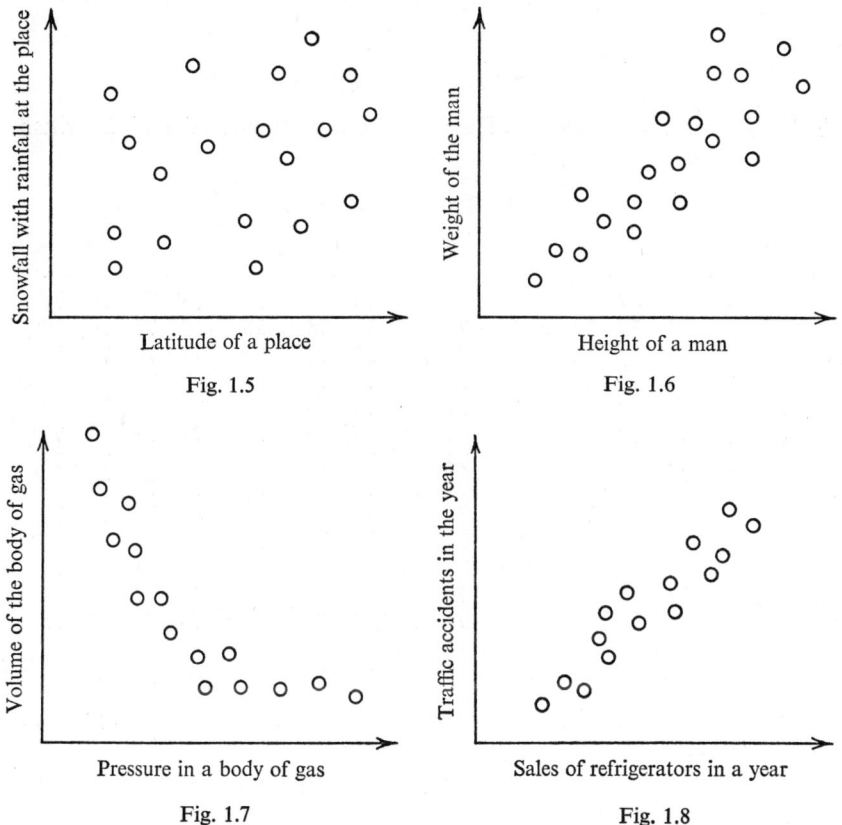

Fig. 1.5

Fig. 1.6

Fig. 1.7

Fig. 1.8

Once correlation is identified convincingly (and 'how do we convince?' is another problem for statisticians) then the problem of choosing a satisfactory curve to show the trend of the relationship is one of *regression*. The chosen curve is called a *regression curve*; its equation may represent an idealized state of association, from which the real values depart both because of the variability of the individuals in nature, and also because of the inevitability of approximations and errors in each single measurement. In the case of Figure 1.6 we do not believe that the regression line is more than a guide; the scatter is inevitable in nature and is not all due to mere

5

difficulties of measurement; the attribution of parts of a scatter to two or more causes is a problem in the *analysis of variance*. In a case like Figure 1.6 the regression line must be used cautiously: in Figure 1.7, however, physicists have long believed that the regression curve represents a law of nature which would be seen to be followed even more closely as measurement techniques became more refined; the equation of the regression curve can be used for highly accurate prediction expressed by Boyle's Law.

5. TESTING HYPOTHESES

Table 1.5 shows the scores of the first four rounds of the 1962 British Open Golf Championships.

Table 1.5

Player	Scores				Player	Scores				Player	Scores			
1	71	69	67	69	14	76	73	76	71	27	73	78	74	76
2	71	71	70	70	15	75	73	73	75	28	69	77	76	79
3	75	71	74	69	16	74	78	73	72	29	77	75	77	73
4	75	70	72	72	17	78	73	76	70	30	78	74	72	79
5	75	70	70	75	18	77	73	72	75	31	72	75	81	75
6	76	73	72	71	19	74	75	75	73	32	74	73	79	77
7	70	77	75	70	20	74	73	79	71	33	72	75	79	78
8	77	70	71	75	21	73	72	76	76	34	74	78	79	74
9	77	69	73	73	22	76	75	75	72	35	80	72	74	79
10	77	74	75	68	23	72	79	74	74	36	76	74	81	76
11	75	70	74	76	24	78	74	76	72	37	76	75	81	76
12	72	79	73	72	25	76	75	78	71	38	76	72	79	81
13	74	75	72	75	26	77	73	74	76	39	76	74	81	78

In this case it is the table which suggests certain questions:

(i) Are the scores significantly lower for one round than for another? Systematic differences might be produced by growing familiarity with the course (tending to lower scores), by increasing tiredness and strain (tending to increase scores) or by weather (tending to alter scores in an unpredictable direction).

(ii) Are the different scores scattered more or less at random among the different players, so that in a sense the leader takes his position rather by chance?

(iii) Is there any evidence that the thirty-nine players were selected from some larger number by the very scores recorded, so that the list is truncated?

(iv) Is there any tendency for the scores to be basically low but with a scatter of higher ones (perhaps due to particular difficulties in a round) or are the scores more or less evenly scattered on both sides of an average?

(v) Do the players at the extremes of the table play more consistently (whether well or badly) than those in the middle, or are the better players the most consistent ones?

These questions are really about *hypotheses* which we wish to *test* against the data. This is another important type of statistical problem.

6. FREQUENCY TABLES

We might examine the data of Table 1.5 by constructing tables such as Table 1.6 or Table 1.7.

The number of occurrences of each score for each category is called the *frequency* of that occurrence, so that Tables 1.6 and 1.7 really provide us

Table 1.6

Score	Round 1	Round 2	Round 3	Round 4	All rounds
	Number of occurrences in				
67			1		1
68				1	1
69	1	2		2	5
70	1	4	2	3	10
71	2	2	1	4	9
72	4	3	5	5	17
73	2	8	4	3	17
74	6	5	6	2	19
75	5	8	4	6	23
76	8		5	6	19
77	6	2	1	1	10
78	3	3	1	2	9
79		2	5	3	10
80	1				1
81			4	1	5

Table 1.7

Scores in Round 2	1st thirteen	2nd thirteen	3rd thirteen
	Number of occurrences for		
69	2		
70	4		
71	2		
72		1	2
73	1	6	1
74	1	1	3
75	1	3	4
76			
77	1		1
78		1	2
79	1	1	

with eight separate *frequency tables*. The frequencies are not all independent, of course, because the frequencies in the last column of Table 1.6 equal the totals of those in the first four columns; and the frequencies of Table 1.7 sum to those of Round 2 in Table 1.6. A skilful presentation of data may be very helpful in a statistical exercise.

7. MODELS AND SIMULATION

Another problem for statistical analysis might arise as follows:

An attempt is being made to improve the appointment system at a clinic where there is a single specialist to see a certain type of patient. Table 1.8 shows the information available from part of the records of the clinic.

Table 1.8

Patient number	Time booked	Time of arrival	Time of seeing doctor	Time of departure
16	11·20	10·55	11·07	11·18
17	11·25	11·41		
18	11·30	11·10	11·18	11·30
19	11·35	11·18	11·30	11·32
20	11·40	11·34		
21	11·45	11·31	11·32	11·33
(20)			11·34	11·38
(17)			11·41	11·44
22	11·50	11·45	11·45	11·46
23	11·55	11·54		
24	12·00	11·45	11·46	11·56
(23)			11·56	11·59
25	12·05	Never came		
26	12·10	11·54	11·59	12·07

An analysis of these figures might be designed to answer such questions as:

(i) How much time does the doctor spend doing nothing when he could be treating patients, and could this time be cut down by booking patients more closely; or by having more patients booked at the start of the session so that there is a line of patients ready?

(ii) How long do patients wait in the line (or 'queue', to use the technical term), and could this be cut down by booking patients less closely; or would that start to be inefficient over the employment of the doctor?

To answer these questions we would probably want to know:

(iii) Does the length of the line, or queue, settle down to a stable value, does it fluctuate fairly violently, or does it increase fairly steadily?

And that would drive us to the following, among other, questions:

(iv) How long are the consultations?

(v) How varied are their lengths?

Once we knew the answers to these questions we could study the queue theoretically; or could construct an abstract system which would behave like the queue, by drawing suitably marked cards from one mixed hatful, each card representing a patient and having his length of consultation written on it, and a card from another hatful to indicate how early or late the patient arrives, (and of course much more sophisticated *models* than this could be set up). We could then rapidly study the effects in the model of making changes in the various quantities represented. By using a computer many thousands of consultations could be simulated under each of various conditions and general conclusions drawn. If the statistical problem is studied theoretically, we are said to be using a *mathematical model*; and if the process is reversed to solve a purely mathematical problem by a suitable statistical model, we are said (in a surprisingly light-hearted way) to be using a *Monte Carlo method.* In attempting a Monte Carlo solution of a mathematical problem we probably supply the random element via a computer, to speed things up.

8. PURE AND APPLIED MATHEMATICS

A statistician, like any other applied mathematician, requires various skills:

(i) An imaginative power enabling him to see the possible mathematical models that may express the essence of a physical situation or process, and so would enable the process to be concisely represented;

(ii) a more purely mathematical skill that will enable him to handle the models either by analysis or by simulated experiment and so predict any of their consequences;

(iii) a thoroughly commonsense skill that will enable him to select those consequences that are crucial as tests, or will lead to physically meaningful conclusions.

9. DESIGN OF EXPERIMENTS

Too often, in science or economics, lists of measurements are regarded literally as *data* (that is *given things*) that we happen to be lucky enough or persevering enough to have recorded; they are then merely brought to a statistician for interpretation. The statistician's skills may tell him that the underlying mathematical model that would suggest the collection of such types of measurement is inappropriate, or that it has been incorrectly handled, so that in either case the data may be blonde, or even beautiful, but they are dumb. Very often, however, and perhaps even more sadly, the model is appropriate and the deductions are sound as far as they go, but the tests proposed are not at all crucial and so the experiment narrowly fails to be worthwhile and is again a waste of time and money.

9

It is now a commonplace that such information as 'Of 100 patients treated with the drug 89 recovered fully in 7 days' is by itself meaningless, as it *might* turn out that 'Of 100 patients NOT treated with the drug 89 recovered fully in 7 days'. But what are we to make of an experiment showing the results in Table 1.9 when a further analysis shows us the spreads of weights in Figure 1.9?

Table 1.9

(Average weights of ten in each category; 1 lb = 0·454 kg)

	before	after
children with tonic	86 lb	88 lb
children without tonic	86 lb	88 lb

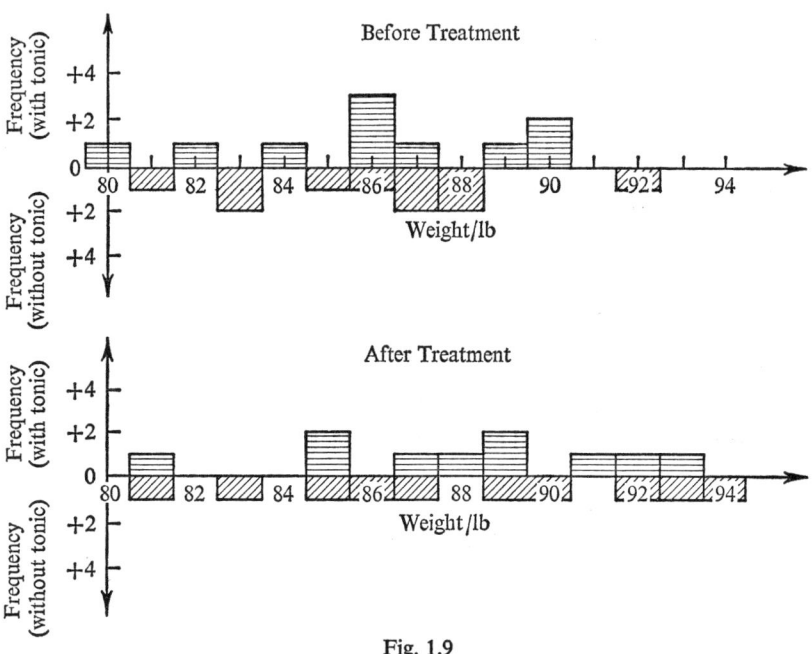

Fig. 1.9

These are four *frequency diagrams*, two of them having their frequency axes, slightly unconventionally, downwards. A study of these diagrams tells us something that we could not know from the bald averages, namely that the weights seem more dispersed after treatment than they were before. It is, however, the next analysis, using *paired comparisons*, that tells us all we can discover from the experiment, less than which is, *in this case*, almost useless.

10

Table 1.10

With tonic	Before	Weight/lb After	Gain	Without tonic	Before	Weight/lb After	Gain
child 1	80	81	+1	child 11	81	87	+6
2	82	85	+3	12	83	81	−2
3	84	85	+1	13	83	89	+6
4	86	89	+3	14	85	83	−2
5	86	87	+1	15	86	92	+6
6	86	89	+3	16	87	85	−2
7	87	88	+1	17	87	93	+6
8	89	92	+3	18	88	86	−2
9	90	91	+1	19	88	94	+6
10	90	93	+3	20	92	90	−2

In specific work in any field—politics, economics, psychology, biology, chemistry, physics—the statistical analysis should precede the experiment or survey and the experimenter or statistician be given a chance to use his skills in the very *design of the experiment*, so that the right questions are answered by the measurements.

It is the full employment of both pure and applied mathematical skills that makes statistics such a fascinating subject to those who like that sort of thing.

10. DESCRIPTIVE STATISTICS

When attempting to draw conclusions from lists of data such as those of Tables 1.3 and 1.5 we soon found that the information they contained was not sufficiently condensed for us to be able to carry away anything firm. Arranging the numbers in order was one first step towards a simplification.

For other types of data such as cricket scores, census returns, political poll answers, stock exchange prices, car production figures or numbers of spots on butterfly wings, we would start calculating numbers like greatest score, number of children per family, percentage of voters backing each candidate, share index, rise in output, average numbers of spots; and we would then use these numbers instead of all the data when comparing two collections of similar data, such as the car production figures for two similar factories or for the same factory at different times.

In the words of R. A. Fisher 'A quantity of data which by its mere bulk may be incapable of entering the mind is to be replaced by relatively few quantities which shall adequately represent the whole, or which, in other words, shall contain as much as possible, ideally the whole, of the relevant information contained in the original data.'

Any quantity calculated in this fashion or indeed obtained in any way

11

for this purpose from a collection of data is called a *statistic*. The actual numbers obtained are the *values* of the statistic. The branch of study which devises statistics and calculates their values is called *descriptive statistics*.

11. SUMMARY

The three principal types of problem we shall deal with in this study are of the following sorts:

(i) Knowing the percentage, in a 'sample', of people intending to vote for a particular candidate, what can we give as a range of values within which we can safely estimate lies the percentage of people in the whole constituency who intend to vote for that candidate? and what do we mean by safely? This is a problem of *estimation*.

(ii) Knowing the results at this particular hospital of using one drug on one group of patients and another drug (or even no drug, because their pills may have been deliberately inert) on another group, have we any evidence that the first treatment is more effective? Here we are *testing a hypothesis*.

(iii) Knowing the details of traffic flow, such as how long during the day the traffic is flowing at any particular volume rate, and how many vehicles are travelling at given speeds and how long the lights are at 'red', can we construct a mathematical system of equations which will predict the range of the observed phenomena, *and their variability*, so that by varying the constants in our equations we can study the effects of various changes in traffic control and road design? This is the problem of building a *model* and *simulating* from it the conditions of real life. A mathematical model which has an element of chance built in is called *probabilistic* or *stochastic*; mathematical models without such an element are called *deterministic*.

These three problems by no means exhaust the field of statistics, but their study will leave the reader well poised to read further on his own.

12. PROBABILITY THEORY

Any study of statistics which is going to be more than a list of recipes must be firmly based on a knowledge of *probability theory* and the reader will find that a great deal of time is taken at the outset of this course to put this theory at his command; the next two chapters are devoted to it, the first in a fairly general way and the second in more detail.

13. THE EXERCISES

Throughout the course the exercises form an integral part of the text; many of the ideas are developed in the exercises; so that a later chapter often refers back to the results of an earlier exercise, either for some piece of theory developed there by the reader, or for some experimental results obtained there by the reader, or simply to carry some question a stage further using answers already obtained. *The reader should keep and file the solutions to all the exercises he does.*

Questions marked 'R' are of a fairly Revisionary nature, developing an understanding of what has just been covered in the text. An adequate number of these should be done as each exercise is reached.

Questions marked 'T' aim to Teach the reader something which is not covered or emphasized in the text. As many as possible of these should be covered. They are not necessarily difficult.

Questions marked 'F' look Forward to matter not yet reached in the text and often enable the reader to discover some key matter for himself and thus to learn it in a more stimulating way.

Questions marked with a star may be of any category but are considered to be distinctly above the average level of difficulty of that particular exercise at that stage. *The reader may come back to them later and find them quite straightforward.*

Exercise 1 A

1T. (*a*) Take a book with large successions of leaves of the same paper (that is, not too broken up with illustrations on different paper) and measure the thicknesses of bunches of leaves, noting the number of leaves involved each time. Discuss the relative merits of the following methods of using the data to determine the average thickness of a leaf:

(*A*) Find the average thickness in each bunch and average the averages;

(*B*) Add all the thicknesses of bunches and divide by the total number of leaves;

(*C*) Plot the thickness of a bunch against the number of leaves and find the gradient of the graph.

(*b*) Redesign the experiment in such a way that (*A*) is an appropriate way of handling the data. What is then the status of (*B*)?

(*c*) Outline some of the experimental difficulties and suggest precautions.

(*d*) Suggest an equivalent biological experiment in which only method (*C*) would be appropriate.

2T. *Experiment on the Judgement of Ranked Size.* [It is particularly important to keep the results of experiments as they are often referred to later and take time to repeat.]

Material Required: Ten plain postcards, each having one line drawn on it; the lines to be of length 5·0 cm, 5·3 cm, 5·6 cm, ..., 7·4, 7·7 cm. One sheet of paper with the ten lines drawn side by side and ranked 1, 2, ..., 10 from the shortest to the longest.

Procedure: There are four possible procedures:

(*a*) The cards are shuffled and presented by the experimenter singly, once each, to the subject. The subject writes down the guessed rank after each showing of

13

a card. He is not allowed to alter earlier guesses, although he knows there can be no repetitions. Disadvantage: The later guesses are affected by the order of showing.

(*b*) The cards are presented as before. The subject makes any notes he likes (for instance: guessed rank; or 5+ ; or 6+ ; and so on). When all the cards have been shown the subject produces his final list of guessed ranks. Disadvantage: the later cards are better remembered for reassessment later.

(*c*) [This method avoids the disadvantages of (*a*) and (*b*).] The cards have a code number on the back which the subject is not allowed to see. The cards are shuffled and cut between each showing and any number of showings in succession, with all manner of possible repetitions, are made to the subject. Disadvantages: there may be a learning effect *but*

(i) This could provide an experiment for research into such an effect, and in this case it might be helpful to have a large number of subjects some of whom are told the true rank after each showing and some of whom are not;

(ii) Any learning effect might be reduced by telling the subject the answers after each card of the first dozen or so showings, and only then starting to record his results.

(*d*) The cards are presented simultaneously in a pattern, as in Figure 1.10, and the subject may move around them before producing his guessed ranks in a corresponding pattern such as

$$\begin{array}{ccccc} 1 & 3 & 5 & 7 & 2 \\ 4 & 8 & 10 & 9 & 6 \end{array}$$

Each procedure may also be altered by whether or not the master sheet is made available for the subject to see (at a reasonable distance).

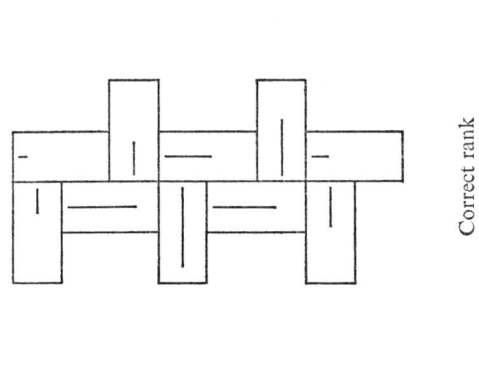

Fig. 1.10

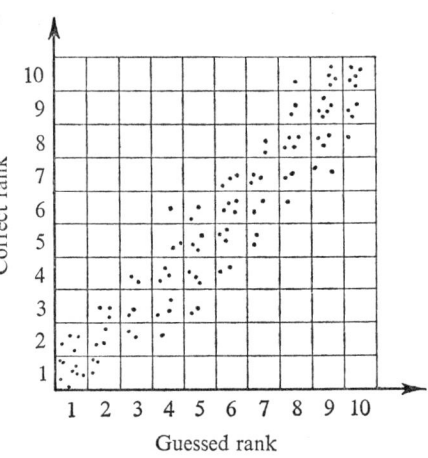

Fig. 1.11

Analysis: The data for each subject can be entered in a table as shown in Figure 1.11.

This table can be used in two ways:

(i) Given the correct rank we can predict the way the subject will guess and how often he is how far out.

14

(ii) Given the subject's guessed rank we can see whether his guesses have a bias towards the small sizes or towards the extreme sizes or towards the central sizes and so on, and could determine how best to use his guess to *estimate* the correct rank.

N∫ **3.** *Experiment on two-point acuity*

Material required: Six geometrical compasses with fairly blunt points, effective blindfold.

Procedure: The subject is blindfolded and sometimes one point, sometimes two points (simultaneously), of a compass are pressed on his forearm, about midway between the hand and elbow, on the inside. The subject must say whether one or two points have been presented.

Let the distance between the points of the compass be d mm, then in the first stage the compass is presented with one point or with $d = 3, 6, 9, \ldots$ until the first correct two-point judgement is made. Include a few repetitions at some distances, but there is no need to decrease d in the sequence provided one-point presentations are intermixed. Then with d about 20 larger than the last value start again, decreasing the distance by 3 mm or zero at a time (and including one-point presentations) until the first incorrect one-point judgement is made. Use the informations from these runs to select six equally-spaced values of d which just overlap the region where the response is inaccurate.

With the six distances selected set one compass at each distance and present them in random order, including ample one-point presentations. Record all the presentations and their responses. Repeat until all six distances have been used five times each. The subject should be prevented from giving an undecided response but discouraged from guessing. Care must be taken over the equality of pressure on the two points, and saturation of the region with sensation needs to be avoided! The effect of presenting the points *across* the arm or *along* the arm can be investigated in separate experiments. Any distinction between left and right arms can also be investigated.

Analysis: Calculate the proportion of two-point responses at each value of d and plot the proportion against d; discuss the concept of a minimum separation at which two points can be distinguished and the problem of specifying it.

4. A useful method of characterizing a collection of objects which have no numerical basis for ordering them is to pick out the commonest one. It is called the *mode*. The idea is equally applicable to items that do have a numerical order, such as the sizes of shoes (not the shoes themselves) in the stock of a shoe shop.

(i) By using dictionaries estimate the modal initial letter of the English vocabulary; of the French vocabulary. Compare these modes with those of the initial letters in extensive passages of English, French. Can any of these modes be deduced from a knowledge of others of them?

(ii) Tabulate and compare the modal ages of marriage of females, of males, and of all persons in three separate regions of your choice, and *overall* in the three regions; (for instance in England and Wales, in Scotland, in Northern Ireland and in the U.K.).

Can any of these modes be deduced from a knowledge of others of them?

5F. In the schematic diagrams (Figure 1.12) arrows represent readings arranged in order of size; and the median, b, and the quartiles q_1 and q_3, called 1st and 3rd, are indicated.

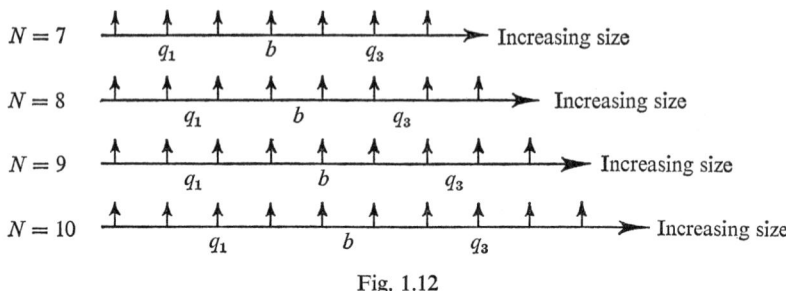

Fig. 1.12

Construct rules, similar to those in Section 3, to define precisely the quartiles of a collection of unequal readings in each of the cases $N \equiv 0, 1, 2, 3$ [mod 4], (the last case may be found the easiest). What alterations would you make if some of the readings were repeated?

6T. Draw a single diagram, with axes as in Figure 1.13, on which the median and quartile scores of each round of the Golf Tournament of Table 1.5 can be shown, and analyse the result in terms of the problems posed in Section 6. A more refined method is given in Question 8.

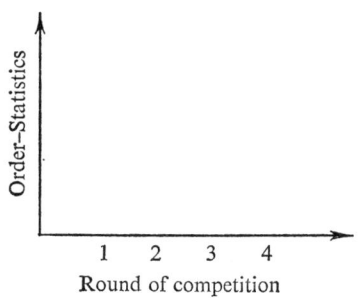

Fig. 1.13

7. The readings of Table 1.4 could have *ranks* allotted to them as follows:

Readings	1	1	2	3·5	3·5	3·5	4	4	4	4	5	5	6	7
Ranks	$13\frac{1}{2}$	$13\frac{1}{2}$	12	10	10	10	$6\frac{1}{2}$	$6\frac{1}{2}$	$6\frac{1}{2}$	$6\frac{1}{2}$	$3\frac{1}{2}$	$3\frac{1}{2}$	2	1

the rule for equal readings being that the rank-total is the same as the total of the ranks that would have been allotted, had the readings been just perceptibly different from each other.

Can you in general define the median in terms of ranks?

16

NS

8 T. M. H. Quenouille: *Rapid Statistical Calculations* (Griffin, 1959) gives the following rough and ready test for the hypothesis that there is no significant difference between two large-sized groups of readings (N.B. *not* groups of large-sized readings). If there are twenty-five or more readings in each group and if the larger group contains not more than one third as many readings again as the smaller group then the sizes of the groups are suitable and we proceed as follows:

Procedure. (i) Consider the two groups together. Let the group with the largest reading (*not* necessarily *more* readings) be called the 'top-group' and the group with the smallest reading be called the 'bottom-group'. *These may both be the same group.*

(ii) Count how many readings of the top-group exceed the largest reading of the other group.

(iii) Count how many readings of the bottom-group are less than the least reading of the other group.

(iv) Add these two values and call the result *T*.

Consequences: Thought shows that if *T* is large enough, then *either:* one group may be considered to be composed of genuinely larger numbers than occur in the other group, *or:* one group is more scattered than the other group. In either case the hypothesis that there is no difference may be rejected. The question is: 'Where do we draw the line for size of *T*'? and the answer is as follows:

If the groups are really drawn 'at random' from two groups of roughly equivalent readings, then *T* would exceed 9 'by chance' only about 1 time in 20; and would exceed 12 'by chance' only about 1 time in 100. (More precise meanings will be given later to 'at random' and 'by chance'; at the moment their obvious meanings are enough.)

Conclusion: For example: if *T* is 10 we are at liberty to say: *either* 'The readings are really drawn "at random" from two groups of equal-sized readings but we have witnessed a fairly unlucky coincidence', *or* 'We do not believe in bad luck and would prefer the hypothesis that the readings are *not* drawn at random from groups of equivalent readings'. On the other hand if *T* is 14 we have the same choice but for 'fairly unlucky' we might substitute 'very unlucky'.

If *T* is less than 9 the readings could easily be drawn from two groups of equivalent readings, and we would *not* reject the hypothesis that they were.

Which conclusion we draw depends on us—mathematics has nothing more to say (unless we can quantify in some way the seriousness to us of an incorrect decision)—but it is very important to formulate the *decision rule* for each possible range of values of *T* before calculating *T* from the data.

The proofs of these easily-applied results are obtainable from Exercise 7 *E*, Question 3 but the results themselves provide now a glimpse into more advanced statistics, and the reader should apply them to certain of the problems suggested about the golf data in Table 1.5.

9 T. Two chains were examined and their links individually broken in succession to determine their breaking strains. The measurements were as follows. (All units are kg.)

First chain	7000,	7200,	6800,	7400,	7100,	6900
Second chain	7100,	7400,	6900,	7500,	7300,	6700

Find the mean breaking strain of the links of each chain. Which chain was the stronger?

10 T. The following rainfalls occur in the seaside resorts of Puddling Regis and Llandrwnch. (All units are inches except for those of the February rainfall at Llandrwnch which are feet.)

	Jan.	Feb.	Mar.	Apr.	May	June	Jul.	Aug.	Sep.	Oct.	Nov.	Dec.
P.R.	4	4	3	3	2	2	2	2	3	3	3	4
Ll.	4	4	4	2	1	0	1	1	2	2	4	4

Draw up the frequency tables. What are the medians and modes from these sets of data? Which town has the higher 'average' monthly rainfall? Which gives the better holiday prospects? [For the definition of *mode* see Question 4.]

11 F. The ages of people on a holiday tour are given by the following table.

Age in years	Frequency
under 13	6
13	8
14	7
15	6
over 16	8

It is known that some of the people (included in the table) were adult guides. Select two statistics to describe an 'average' age. Give a measure of the spread of ages. (This is an example of a problem where very little detail is known about the form of the frequencies. The statistics useful are the order-statistics of Section 3.)

$N\int$ **12.** The following are two actual quotations from newspapers, describing a protest march, and traffic conditions during the Easter holiday 1965.

(i) 'A survey carried out on "The March" showed that the average age was 24 and the commonest age 18'.
(ii) 'JAMS FOR 200 MILES ON THE WAY HOME'.

Each of these summarizes some data obtained by reporters. For each set of data draw up two possible frequency tables making them as different as possible, but consistent with the quotations.

2

PROBABILITY—OUTLINES

1. INTRODUCTION

In order to carry out certain statistical experiments, which will be described later, the author bought six boxes of drawing pins, contents advertised as 36. [Any American reader will, throughout this chapter, need to translate 'drawing pin' into 'thumb tack'; over some words and phrases it is impossible to please all readers simultaneously.] The boxes contained 36, 39, 33, 34, 36, 37 pins and thus clearly did not contain 36 each. There is, however, a lesson to be drawn here already, namely that there is a variety in any series of measurements, due in this case to imperfections or even errors in the machinery which did the counting. This is a commonplace.

A closer look at the boxes, however, reminds us of a more important lesson, for the label said 'AVERAGE CONTENTS 36'. Our acceptance that this phrase—average contents—means anything at all when printed on a single box shows that we believe that within the variety there is a streak of constancy which enables us to say *something* about each box. The number of pins in a box may alter from box to box but it will be located near, and the collection of such numbers will be scattered slightly about, 36. This is a different view of average from that taken by elementary arithmetic. In elementary arithmetic average is purely descriptive. If large-sized half-used boxes of drawing pins had their contents counted an average could be calculated *after* the counting; even if they contained 1, 17, 72, 3, 100, 23 pins then one description of this state of affairs would be that the average content was $\frac{216}{6}$, that is 36 exactly. The only suitable comment about this average is 'so what?' The fact that, after the counting, the actual average content of the six *bought* boxes was $35\frac{5}{6}$, and not 36, in no way makes the statement printed on those boxes less useful. For the statement printed on the boxes does not purport to tell us the actual average content of a particular collection of bought boxes calculated after the boxes are bought. What it is intended to convey is however quite clear: it is that the manufacturers have set the machinery in such a way that frequent repetitions will produce an approximately constant number of drawing pins in a box, and that the number approximated to is 36. Our belief that this is a reasonable sort of statement for the manufacturers to be able to make is so strong that if during a visit to the factory we noticed a marked change taking place in the number of drawing pins per box coming off the production line, we would suspect an alteration in the machinery and would not merely shrug our shoulders. It is in the bland acceptance of obvious anomalies that there lies, for instance, the humour of early silent films.

This constancy within variety which occurs when a process is repeated time and time again is, of course, an experimental fact and is no more deducible from axioms of mathematics than any other fact of the real world. It was to demonstrate the constancy for another process that the drawing pins were bought; that the labelling of the boxes also implied a constancy only shows the widely different types of situation where such constancy can be observed by those who keep their eyes open.

2. AN EXPERIMENT

2.1 The experiment. The experiment or trial or process envisaged in the purchase of the drawing pins was to take the lid (22 cm × 24 cm) of a biscuit tin; put a boxful of drawing pins into a teacup and scatter them lightly over the level lid like salt on a plate of soup.

The result concerned in this experiment was numerical and was simply the number of drawing pins that rested point-down when a boxful was scattered *in the way described*. The simile about salt is not mere verbiage; a different process would be being carried out if the pins were dumped violently in a heap, and a different one again if they were slid gently into a long strip.

(Whether we felt that such apparently minor alterations in the conditions made an experiment *effectively* different would, in practice, depend on whether we regarded the numbers obtained from the experiment as seriously different. This test of a process by its outcome betrays once more our conviction that repeating the conditions of processes does tend to produce repetitions of their outcomes.)

Our purposes in the chosen experiment are:

(i) to illustrate that the 'approximate constancy' can occur in a much wider variety of situations than might be suspected. (No one doubts that train timetables, for instance, are useful, but we should be rash to extrapolate from train timetables to teacupfuls of drawing pins.)

and (ii) to discuss a method of measuring the result of the constancy so that we can calculate with it, and ultimately predict with sufficient confidence to take decisions.

2.2 The results. In effect six parallel experiments were run by throwing, quite separately, six different kinds of pin.

Five of the six boxes contained light pins with plastic tops, of different colours: one box each of Green, Red, Purple, Blue, Yellow; the sixth box contained heavy brass pins. The results of the first four throws of each box were as in Table 2.1.

Table 2.1

Experiment	Box of	Numbers of pins landing point-down			
1	36 Green pins	12,	18,	10,	11
2	39 Red	23,	17,	20,	21
3	33 Purple	18,	15,	21,	12
4	34 Blue	20,	19,	20,	14
5	36 Yellow	14,	23,	11,	17
6	37 Brass	15,	18,	14,	11

Since we do not expect the *numbers* to be the same for each type of pin regardless of whether there were 39 or 33 pins in a box, and since constancy is only expected to emerge in long runs, it seems sensible to do two things:

(i) to calculate cumulative totals of 'point-down' and (ii) to reduce the cumulative totals of 'point-down' to percentages of the cumulative totals of that type thrown. These things are done in Table 2.2, and the results are shown in Figure 2.1.

Table 2.2. Cumulative totals and proportions

Experiment	Type					
1	Green	(totals)	12	30	40	51
		(per cent)	33	42	37	35
2	Red	(totals)	23	40	60	81
		(per cent)	59	51	51	52
3	Purple	(totals)	18	33	54	66
		(per cent)	54	50	55	50
4	Blue	(totals)	20	39	59	73
		(per cent)	59	57	58	57
5	Yellow	(totals)	14	37	48	65
		(per cent)	39	51	44	45
6	Brass	(totals)	15	33	47	58
		(per cent)	41	45	42	39

Nothing very striking has emerged yet, but then the runs have not yet been very long—no more than 160 of any type of drawing pin have been thrown. If the cumulative effects are being studied, it seems helpful to add at each stage a number of throws equal to that *already* thrown, otherwise succeeding steps do not have equal effects on the final result and the results are not so easy to take in at a glance.

21

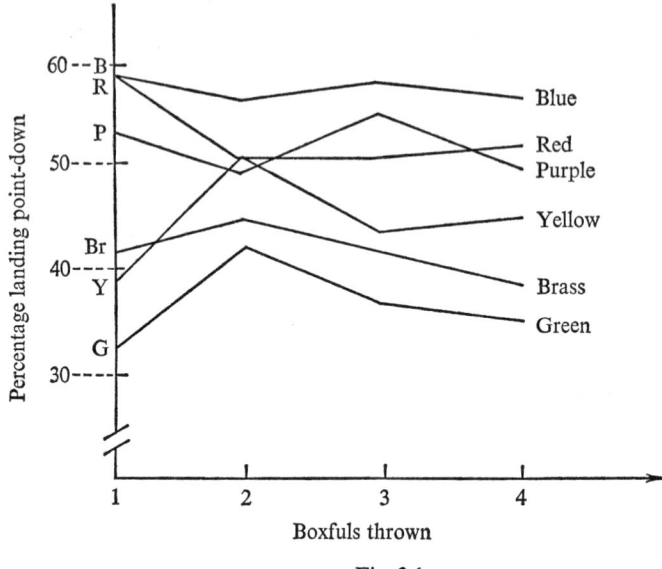

Fig. 2.1

The last two columns of Table 2.3 show the results of four more, and then eight more, throws (each of a boxful) for each type of pin. Constancy of proportion would be reflected in a doubling of cumulative totals at each stage. The proportions are now shown as decimal fractions not as percentage fractions. These results are represented in Figure 2.2.

Table 2.3. Cumulative totals and proportions since start

		Number of boxfuls				
Experiment	Type	1	2	4	8	16
1	Green	12	30	51	123	244
		0·333	0·417	0·354	0·417	0·424
2	Red	23	40	81	142	274
		0·590	0·513	0·519	0·456	0·439
3	Purple	18	33	66	112	230
		0·545	0·500	0·500	0·424	0·436
4	Blue	20	39	73	126	241
		0·588	0·573	0·568	0·463	0·443
5	Yellow	14	37	65	128	241
		0·389	0·514	0·451	0·444	0·418
6	Brass	15	33	58	101	186
		0·406	0·446	0·392	0·342	0·315

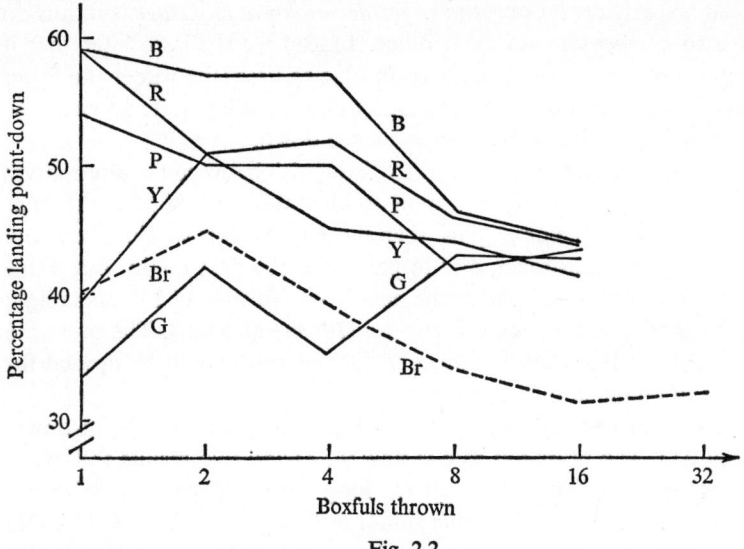

Fig. 2.2

2.3 Conclusions. Two conclusions seem to emerge:

(i) If we fix our attention on a particular experiment, then the graph indicating the proportion in that experiment tends to level off as the number of throws gets larger. The reader is at liberty to doubt this conclusion, of course. He should attempt similar experiments himself and carry them on much longer, or he should merely reflect on the experience of insurance companies. Those companies which accept that results of this type describe the accidents and events of ordinary life make large profits; those that think they can circumvent them collapse.

(ii) The coloured pins behave fairly similarly whatever their colour, or at least behave in a distinguishably different way from the brass pins. It would be pertinent of the reader to say 'You talk of long runs; another 16 boxfuls might restore the brass pins to the fold'. But consider: to provide even a 0·41 rate over the first 32 boxfuls would require a rate of 0·51 over the next 16 boxfuls and this rate is higher than that in any single boxful of brass pins yet thrown and would have to be maintained as an average for 16 consecutive boxfuls. In fact the reader is being unnecessarily strict in expecting the brass pins to be restored to the fold *in this way*. Suppose that the construction of brass pins really was such that about 0·43 was their usual ratio in this experiment. All that is required for an acceptance of this value as a basis for proceeding is that reasonably long runs should show it. There is no need for the next 16 boxfuls to be punished for the last 16 boxfuls being so remiss. If the brass pins promise to be good, then all will eventually be forgiven by a long run of 0·43 *swamping* what will be seen to

23

have been an early deviation. *And so for deviations at any stage.* Boxfuls do not have to *compensate* for each other. Indeed we shall see later that if wildly aberrant runs did not occur from time to time (the wilder the fewer of course), then we would begin to suspect that (possibly quite unwittingly) the experimenter was damping the sequence down.

We could provide a numerical test to distinguish between the alternatives as follows: if the present rate is *not* an accidental feature of this particular run but is typical of the brass pins, then most 16-boxfuls would give about 185 pins landing point-down, and in particular the next might. But if the usual rate was about 0·43 then in the next 16-boxfuls about 255 pins might land point-down, *and the overall rate would then reach* 0·37. The brass pin graph would not have levelled off, and further runs would be needed for a decision to be made by a sceptic.

In fact the next 16-boxfuls of brass pins thrown gave 193 pins point-down, instead of 255. The brass-pin rate is now 0·320 and two results follow.

(i) It becomes harder to envisage the long term rate reaching 0·43.

(ii) It becomes easier to imagine that a number near 0·31 or 0·32 could be used to provide a measure of one aspect of the behaviour of brass pins in this experiment.

3. GENERALIZATION

With all this in mind let us make a general statement: Suppose that an arbitrary trial or process or experiment has been set up and that we have described it as carefully as we can so that we or others can repeat its conditions. As a result of this experiment various events can take place. Then, even if it is impossible before each repetition of the experiment to know which event will take place next, nevertheless for each of the possible events we can obtain a number k such that if we intend to carry out a large number N of repetitions of the experiment then we can predict that in nearly kN of them that event will occur.

We note the following: (i) This is a statement about the real world. It does not claim to be a statement of mathematics but does claim to represent observed facts.

(ii) Although k depends only on the definition of the experiment and the event, these definitions are not in general *sufficient* for us to be able to infer a suitable numerical value for k.

(iii) One method of obtaining a numerical value for k corresponding to a given event is to carry out, preliminarily, a large number of repetitions of the experiment and then to take k as the fraction

$$\frac{\text{Number of occurrences of the event so far}}{\text{Number of repetitions of the experiment so far}}.$$

As with all measurements in science, different experimenters, or the same experimenter at different times, will obtain different numerical values. The following for instance are the results of attempts between 1895 and 1961 to determine the ratio of the mean radius of the earth's orbit to a carefully defined length of 10^6 miles (1 mile = 1·61 km):

$$93·28 \quad 92·83 \quad 92·91 \quad 92·87 \quad 93·00 \quad 92·91 \quad 92·84 \quad 92·977$$
$$92·9148 \quad 92·874 \quad 92·876 \quad 92·9251 \quad 92·960 \quad 92·956 \quad 92·813$$

Whether there exists a 'true' value of this ratio is an interesting talking point but for the purposes of astronomy is irrelevant. If there did exist a 'true' value, it is clear that we could never know whether we had found it. What we do know is that the measurements above are all ones which have some claim to usefulness in view of the care taken in making them.

In the case of the experiment defined in Section 2.1 we might feel that for a plastic drawing pin landing point-down we could take $k = 0·43$, and that for a brass pin landing point-down we could take $k = 0·32$. The meaning to be attached, for instance, to the first of these two statements is that we would expect that in any 1000 throws, say, of a single plastic-topped pin carried out in the way described the pin would land point-down about 430 times. 'About' is vague, and at the moment deliberately so; we have not yet got beyond the vague stage.

Someone else examining our data might decide that $k = 0·44$ would be a better value to choose. We have both made 'estimates' based on the data and even 1000 throws would scarcely be enough to distinguish between our expectation of 430 and his of 440. Each value might prove satisfactory for most purposes.

What we have just said must not be taken to imply that no distinction can ever be made between nearly equal estimates. Imagine a situation where someone else maintains that he has drawing pins of the same manufacture as ours. He has carried out the by now standard experiment and obtained 0·42 for the value of k—not noticeably different, he maintains. He proposes that each of us should throw 1000 pins, because he knows full well that we might each easily get about 425 point-down, and his claim will then be substantiated. We could propose a more sensitive test: a million throws give a predicted 430,000 for us and 420,000 for him. Now these numbers are proportionately no more different than 430 and 420 or even 43 and 42 but we shall see later that a difference of this magnitude in numbers of this magnitude is almost inconceivably unlikely to arise by chance and might indicate that we ought to look for a real underlying difference—though whether in the pins, or the cups, or the trays or the manner of throwing or some other detail, we could not yet say. And since in a long enough run the inconceivably unlikely will occur *sometime* we might say 'This is that moment', shrug our shoulders and ignore the indication to look further. On the other hand, Lord Rayleigh started himself and Ramsay on the investigation which led to the isolation of argon, and the other 'noble' gases, by using the difference between 2·3103, 2·3102, 2·3100 gm (as he measured them) for the masses of a fixed volume of nitrogen obtained by various means from the atmosphere, and 2·3001, 2·2990, 2·2987, 2·2987, 2·2985 gm for nitrogen obtained from chemical compounds.

4. MODELS

In order to apply mathematics to the study of some observed regularity of measurements in the real world a mathematician sets up an idealized system. The structure of the system is based on the qualitative behaviour of a relevant part of the real world. In order to handle some cluster of measurements which make up part of the observed regularity of the real world the mathematician introduces into his system a number whose single value adequately represents the members of the cluster; and similarly for other clusters, possibly. This does not imply that there were true values in the real world to which nature only succeeded in making approximations, or that the qualitative behaviour of the real world was an attempt by nature to exhibit the mathematician's structure; quite the reverse, the idealized system is only called a *model* of the real world.

The mathematician follows the logical consequences of the rules that govern his model and compares the results with the supposedly corresponding results from the real world. It is important, however, to realize that during the actual handling of the mathematics it is the *rules* of the model which are the last authority. The rules of operation in a mathematical field are called *axioms*.

5. PROBABILITY

5.1   Axiom I. Consider a real experiment 'E' and let A be one of the events which could occur as a result of 'E'. For instance, 'E' might consist of the author, with all his imperfections, sitting with his crudely made plastic-topped drawing pins casting them onto a tray from a teacup and trying to repeat the conditions as closely as possible. The event A would occur when a drawing pin rested point-down.

In a large number N of repetitions of the experiment let the number of times that A occurs (called the *frequency* of A) be written fr (A), then $0 \leqslant$ fr $(A) \leqslant N$ and it follows that $0 \leqslant \dfrac{\text{fr}(A)}{N} \leqslant 1$. We have seen that we can obtain a number k such that for large N we have fr $(A)/N$ nearly equal to k.

In the model we consider an idealized experiment E corresponding to 'E'. E, in our example, would be concerned with perfectly equal drawing

26

pins being thrown onto an absolutely level uniform tray with a predetermined wrist action and degree of violence. It is of course important
that the idealized experiment is not made so uniform that the outcome is
actually determined. To put it loosely the object is to study the workings
of chance. We try to decide which variations in the conditions are ones we
consider would actually change the experiment, and those variations we
try to eliminate. The outcomes are then affected by chance to the extent
in which we are interested. The purpose of the investigation is to predict
the pattern of behaviour of some system over which neither we nor anyone
else will have complete control. We know the sort of thing we are talking
about in the simple case of firing, with all our concentration and un-
hampered by wasps, a succession of carefully made bullets through an
accurate rifle at a stationary target. We do not even then find that all the
bullets go through the same hole. Again, if we carefully repeat the conditions of flipping a true coin, then it does not always land the same way up;
and in this case we probably believe we can be much more free over the
conditions and might not insist on a re-throw if the coin rolled off the
carpet onto the floor, for we might feel the carpet-floor variation
insignificant.

We will write A for the event whether the experiment is real or idealized
since we describe it in the same words. If then we have an idealized experiment E of which an event A is a possible outcome we will postulate the
existence of a number (corresponding to k in the case of a real experiment)
associated with A and we will call it the probability that the event A occurs
as the result of the experiment E and we will write it pr $(A:E)$. In cases
where it is clear what experiment we are concerned with we will immediately shorten the name of the number to the probability of A and will
write it as pr (A).

In order to preserve the correspondence between probability and k we
will take:

Axiom I: $\qquad\qquad 0 \leqslant \mathrm{pr}\,(A) \leqslant 1.$

It may help the reader to consider the way in which this situation resembles
that in the study of mechanics where there seem to exist 'rigid bodies' in
the real world with masses; but where a closer examination reminds us
that molecules may come and go by adhering and evaporating and no
body is perfectly rigid, so that before being able to introduce mathematics
we assume the existence of ideal bodies with which to associate numbers
which measure masses. If we have a closed Newtonian mechanical system,
say of two trolleys with men on them throwing bricks at each other (this
seems a common mode of propulsion in mechanics questions), then we
have a total mass represented by a number M in the system and the mass
m of any body in the system must satisfy $0 \leqslant m \leqslant M$.

It is important to realize that, although the nouns 'possibility' and 'probability' in English have been constructed in similar ways from the adjectives 'possible' and 'probable', their technical usages here are quite different: a *possibility* in this context is something which might happen; a *probability* is a number between 0 and 1. (Remember that a probability is not a ratio; we shall add, subtract and multiply probabilities; a probability might have the value 0·6 or $\frac{3}{5}$, but we would *not* write this as 3:5.) In ordinary English a possibility suggests something that might happen but is unlikely, whereas a probability suggests something that might happen and is more likely. No such meanings are relevant here.

5.2 Axiom II. To get any further we need to have a more complicated experimental situation in mind. Table 2.4 shows the actual results of the single experiment throwing about half of a *mixed teacupful* of drawing pins onto a larger wooden tray and counting those of various sorts that fell point-down. In this case there were white plastic-topped, coloured plastic-topped and brass-topped, and any of these might have landed 'down' or 'up', giving six possible events.

Table 2.4. Frequencies of various events

	Point-Down	Point-Up
White	104	109
Coloured	62	64
Brass	38	56

The total number of throws is obtained as

$$104 + 62 + 38 + 109 + 64 + 56 = 433.$$

In general if the events A_1, A_2, A_3, ..., A_n form a set of events of which *at most one can occur* at a time (so that they are called *exclusive*) and such that *at least one must occur* (so that they are called *exhaustive*) then

$$\text{fr}(A_1) + \text{fr}(A_2) + \ldots + \text{fr}(A_n) = N$$

or

$$\frac{\text{fr}(A_1)}{N} + \frac{\text{fr}(A_2)}{N} + \ldots + \frac{\text{fr}(A_n)}{N} = 1.$$

To make the probabilities correspond to the various numbers k_1, k_2, ..., k_n to which these frequency ratios are nearly equal for large N, we take:

Axiom II: The events A_1, A_2, ..., A_n are exclusive, exhaustive

$$\Rightarrow \text{pr}(A_1) + \text{pr}(A_2) + \ldots + \text{pr}(A_n) = 1.$$

The corresponding axiom for the system of numbers measuring masses in mechanics should be obvious to the reader. He will not however expect to find corresponding axioms everywhere, for if he has really understood the role of axioms he will see that such a correspondence would suggest a greater similarity between the studies of probability and Newtonian mechanics than seems to exist.

We will further say: A is certain $\Leftrightarrow$ only A can occur
$$\Leftrightarrow A \text{ is exhaustive}$$
$$\Rightarrow \text{pr}(A) = 1.$$

This amounts to a rule for interpreting the word *certain* in terms of probabilities.

And also we will say: A is impossible $\Leftrightarrow$ (not-A) is certain
$$\Rightarrow \text{pr}(\text{not-}A) = 1.$$

This amounts to a rule for interpreting the word *impossible* in terms of probabilities.

We can obtain a very important corollary of Axiom II by applying it to the exclusive, exhaustive set of events consisting of A and (not-A); for we then have:
$$\text{pr}(A) + \text{pr}(\text{not-}A) = 1.$$

Our first result from this corollary is obtained by applying our rule for interpreting 'A is impossible' (namely pr (not-A) = 1).

We obtain A is impossible $\Rightarrow$ pr $(A) = 0$.

Notice that we do not have pr $(A) = 1 \Rightarrow A$ is certain. We will interpret pr $(A) = 1$ to mean that $\dfrac{\text{fr}(A)}{N}$ is as near 1 as we need to consider, for large N. This becomes of importance in later work when we consider idealized experiments with infinitely many possible outcomes; experiments involving measurements, in fact.

Notice also that pr $(A) = 0 \not\Rightarrow A$ is impossible; we will interpret pr $(A) = 0$ to mean that fr $(A)/N$ is as near 0 as we need to consider, for large N.

As an application of these remarks we can see that if the trial consisted of spinning a coin from about a foot above a table we might well take in our calculations pr (coin stands on edge) = 0; the real coin *can* stand on its edge but we will ignore the possibility of this happening in our experiment. [On the other hand, we must remember that 'the probability of an event' only acquired its meaning in the context of a particular experiment (as the notation pr $(A:E)$ would show), and if the experiment is to spin a coin from about five feet above a muddy football pitch and 'standing on its edge' is carefully defined, we might well take pr (coin stands on edge) $\neq 0$.]

5.3   **Axiom III.**   To return to our results in Table 2.4 we see that a drawing pin fell point-down 204 times, obtainable from $204 = 104 + 62 + 38$.

Using an obvious code from the initial letters we could write this as:
$$\text{fr}\,(D) = \text{fr}\,(D \text{ and } W) + \text{fr}\,(D \text{ and } C) + \text{fr}\,(D \text{ and } B)$$
and derive
$$\frac{\text{fr}\,(D)}{N} = \frac{\text{fr}\,(D \text{ and } W)}{N} + \frac{\text{fr}\,(D \text{ and } C)}{N} + \frac{\text{fr}\,(D \text{ and } B)}{N}.$$
Now the events on the right interest us at this stage because they form an *exclusive* set of events such that one of them must occur whenever D occurs. They are not *exhaustive* since the experiment can be carried out without any of them occurring, but they could be described as exhausting D, since
$$D = (D \text{ and } W) \text{ or } (D \text{ and } C) \text{ or } (D \text{ and } B).$$
We frame our third axiom as follows:

Axiom III: A_1, A_2, ..., A_n form an exclusive set of events
$$\Rightarrow \text{pr}\,(A_1 \text{ or } A_2 \quad \ldots \quad \text{or } A_n) = \text{pr}\,(A_1) + \text{pr}\,(A_2) + \quad \ldots \quad + \text{pr}\,(A_n).$$
A *much used* corollary of Axiom III is obtained by applying it to $(A \text{ and } B)$, $(A \text{ and not-}B)$ as two exclusive events. Noting that
$$(A \text{ and } B) \quad \text{or} \quad (A \text{ and not-}B) = A,$$
we get $\qquad \text{pr}\,(A) = \text{pr}\,(A \text{ and } B) + \text{pr}\,(A \text{ and not-}B).$

5.4 Consistency. The plot of Gilbert and Sullivan's light opera *The Mikado* depends on the inconsistency of the laws of the state of Titipu with each other. It is very important when framing a set of axioms to see that they are consistent. This is usually done by demonstrating that a system already agreed to be consistent can be found which satisfies the axioms. If a system which satisfies the axioms can be found but cannot be agreed to be itself consistent, then at least we have shown that if the system is later agreed to be consistent then the axioms are consistent. Temporary failure to find a suitable system does not, of course, prove the axioms inconsistent.

The system we will choose for what is called a *realization of the axioms* consists of a set of points and its subsets.

We set up the system as follows: Figure 2.3 represents a set $\mathscr{E}$ consisting of subsets $\mathscr{D}$, $\mathscr{U}$, $\mathscr{W}$, $\mathscr{C}$, $\mathscr{B}$, related as shown (the crosses representing the only points in $\mathscr{E}$).

Set $\mathscr{E}$		Experiment E
sets $\mathscr{D}$, $\mathscr{U}$, $\mathscr{W}$, $\mathscr{C}$, $\mathscr{B}$	corresponds to	events D, U, W, C, B
$\cap$		and
$\cup$		or
$\mathscr{D} \cap \mathscr{W}$		D and W
$\mathscr{D} \cup \mathscr{W}$		D or W
$\mathscr{D}'$, (the complement of $\mathscr{D}$)		D', (not-D)
exclusive sets		exclusive events
exhaustive sets		exhaustive events

30

Then for each event A the quantity $\pi(\mathscr{A}) = \dfrac{n(\mathscr{A})}{n(\mathscr{E})}$ will correspond to pr $(A:E)$, and

I   For each $\mathscr{A}$ we have $0 \leqslant \pi(\mathscr{A}) \leqslant 1$; so the numbers $\pi(\mathscr{A})$ satisfy Axiom I.

II   The sets $\mathscr{W}$, $\mathscr{C}$, $\mathscr{B}$ are exclusive and exhaustive, and

$$\pi(\mathscr{W}) + \pi(\mathscr{C}) + \pi(\mathscr{B}) = 1,$$

also $\mathscr{D}$, $\mathscr{U}$ are exclusive and exhaustive, and

$$\pi(\mathscr{D}) + \pi(\mathscr{U}) = 1;$$

so the numbers $\pi(\mathscr{A})$ satisfy Axiom II.

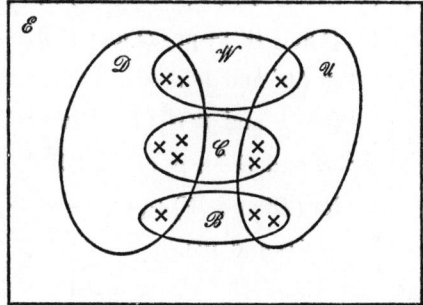

Fig. 2.3

III   The sets $(\mathscr{D} \cap \mathscr{W})$, $(\mathscr{D} \cap \mathscr{C})$, $(\mathscr{D} \cap \mathscr{B})$ are exclusive,

but $\qquad \mathscr{D} = (\mathscr{D} \cap \mathscr{W}) \cup (\mathscr{D} \cap \mathscr{C}) \cup (\mathscr{D} \cap \mathscr{B})$

and $\qquad \pi(\mathscr{D}) = \pi(\mathscr{D} \cap \mathscr{W}) + \pi(\mathscr{D} \cap \mathscr{C}) + \pi(\mathscr{D} \cap \mathscr{B})$

(and similarly for other sets related in this way); so the numbers $\pi(\mathscr{A})$ satisfy Axiom III.

The numbers $\pi(\mathscr{A})$ thus form a set of numbers satisfying all the axioms and, belonging to elementary arithmetic, are usually taken to be consistent. Another useful realization of the axioms can be obtained by using the areas of various regions of the plane, as will emerge later.

5.5   Development. We next show how we might start to develop the theory of probability from these axioms. We will use the properties of the relative frequencies to guide us as to what results are likely to be provable.

We return to Figure 2.3 and notice that for the events D and W

$$\text{fr}\,(D \text{ or } W) = \text{fr}\,(D) + \text{fr}\,(W) - \text{fr}\,(D \text{ and } W)$$

If we divide this through by N the resulting equation suggests that we should have:

$$\text{pr}\,(D \text{ or } W) = \text{pr}\,(D) + \text{pr}\,(W) - \text{pr}\,(D \text{ and } W).$$

31

It is interesting to see that this can be derived from the axioms and so does not need separate axiomatic status. The dry bones of the axioms are beginning to take on life.

Theorem.
$$\text{pr}\,(D \text{ or } W) = \text{pr}\,(D) + \text{pr}\,(W) - \text{pr}\,(D \text{ and } W).$$

Proof. [Remember, we will now write not-A as A'.]

$(D \text{ and } W')$, $(D \text{ and } W)$, $(D' \text{ and } W)$ form a set of exclusive events such that
$$(D \text{ or } W) = (D \text{ and } W') \text{ or } (D \text{ and } W) \text{ or } (D' \text{ and } W).$$

It follows, from *Axiom III*, that

$$\begin{aligned}
\text{pr}\,(D \text{ or } W) &= \text{pr}\,(D \text{ and } W') + \text{pr}\,(D \text{ and } W) + \text{pr}\,(D' \text{ and } W)\\
&= [\text{pr}\,(D \text{ and } W') + \text{pr}\,(D \text{ and } W)]\\
&\quad + \text{pr}\,(D' \text{ and } W)\\
&\quad + [\text{pr}\,(D \text{ and } W) - \text{pr}\,(D \text{ and } W)]\\
&= [\text{pr}\,(D \text{ and } W') + \text{pr}\,(D \text{ and } W)]\\
&\quad + [\text{pr}\,(D' \text{ and } W) + \text{pr}\,(D \text{ and } W)]\\
&\quad - \text{pr}\,(D \text{ and } W).
\end{aligned}$$

Now by the corollary of Axiom III
$$\text{pr}\,(D \text{ and } W') + \text{pr}\,(D \text{ and } W) = \text{pr}\,(D)$$
and
$$\text{pr}\,(D' \text{ and } W) + \text{pr}\,(D \text{ and } W) = \text{pr}\,(W)$$
so we have finally
$$\text{pr}\,(D \text{ or } W) = \text{pr}\,(D) + \text{pr}\,(W) - \text{pr}\,(D \text{ and } W). \quad \text{(Q.E.D.)}$$

Anyone who turned over to Figure 2.3 at the first step of this argument will have had a much easier time following it. The purpose of setting the proof out in this way has not been to convince the reader of the truth of the theorem (proofs rarely do that); it has been to demonstrate that such things can be deduced from the axioms if we wish to develop the subject along 'purely' mathematical lines.

6. ALLOTMENT OF VALUES

These three axioms are all that we require to develop the *theory* of probability. But notice that just as the definition of an event in the setting of a real experiment is in general not sufficient for us to be able to find a suitable value for k but that we have in general actually to carry out the experiment a large number of times to give ourselves an idea of what value to select for k, so the definition of an event in the setting of an ideal experiment is in general not sufficient to give the numerical value of the probability of the event. It follows that at the start of an investigation into a real process, after we have set up a model to allow us to use mathematics at all, we shall usually find that the model does not yet include numerical values for probabilities and that there are still many values which the probabilities might take. Any set of numbers associated with a set of ideal events in such a way that they satisfy the axioms of probability will be called a set of *permissible probabilities*.

Now the model could only be called a good model or a suitable model of the real process if its behaviour closely matched that of the real process. For this to happen we shall need to allot values to the probabilities in such a way that the probabilities equal the various values of k, so that by handling the probabilities mathematically we can predict the long term behaviour of the process. To put this more precisely: If A is an event defined in terms of a real and a corresponding ideal experiment then we will try to choose the value of pr (A) so that in any subsequent large number N of repetitions of the real experiment fr $(A)/N$ will be nearly equal to pr (A).

We will say that *suitable probabilities* are permissible probabilities chosen in this way. It is with suitable probabilities that we will in future be concerned; the study of merely permissible probabilities is a branch of pure mathematics forming part of Measure Theory.

7. SYMMETRY

We have remarked already that in general the definitions of real experiments and of events A are not sufficient for us to infer a value of k to which fr $(A)/N$ will approximate in a large number N of repetitions of the real experiment. There are, however, special cases where we can feel sure of what value to take for k from the very way the experiment is described. If we take a die, very accurately cut from uniform material and with minimally influential marks to distinguish the faces, and throw it with spin onto a table, we expect that every face will come up with approximately the same long-run frequency ratio as every other face, since there seems to be no reason for one face to be 'preferred' before another. Experience will show that dice made and treated like this behave like this, and we describe such dice as 'fair'. We now have only to include the word 'fair' in our description of the dice in an experiment to indicate immediately what is the appropriate value of k for each face. It is obviously $\frac{1}{6}$.

This does not apply, of course, only to standard dice; we might be using discs (for example, coins), tetrahedra, triangularly-cross-sectioned rods, and so on, and could choose the appropriate value of k for each.

That the meaning of the word 'fair' can be understood like this is best seen by considering the results of some throws of a die (see Table 2.5).

If the reader feels some hesitation about calling the particular die concerned 'fair' it is probably because 0·140 (for 'fives') seems rather low. This is itself an indication that a more equal treatment was expected. Of real dice there are of course only degrees of fairness and if we analyse what we mean by this we can see that we have the notion of an idealized die which could be involved in idealized experiments in which we would allot equal probabilities to the occurrences of the various faces. It is clear the probabilities will be $\frac{1}{6}$, but it is important to see that the axioms of probability allow us to obtain the probabilities directly from our assumption that

pr (one) = pr (two) = pr (three) = pr (four) = pr (five) = pr (six).

Table 2.5. Results of throwing a die

Face	Frequency and relative frequency in	
	1st 150 throws	1st 300 throws
One	26	54
	0·173	0·180
Two	31	53
	0·207	0·177
Three	26	50
	0·173	0·167
Four	24	48
	0·160	0·160
Five	20	42
	0·133	0·140
Six	23	53
	0·153	0·177
Three or six	49	103
	0·327	0·341

The method is simple: Write each probability as x, then, since the occurrences of the faces form an exclusive exhaustive set of events, Axiom II gives $6x = 1$; from which $x = \frac{1}{6}$. We notice in passing that this value satisfies Axiom I. From Axiom III we can then deduce such other numerical values as pr (three or six) $= \frac{1}{6} + \frac{1}{6} = \frac{1}{3}$. This agrees well with the value for the relative frequency of (three or six) in Table 2.5, namely 0·341, so our model fits facts outside those of its original construction.

The method can be generalized: if we have an idealized experiment with N possible outcomes and they are all symmetrical in the sense that in a real experiment corresponding to the ideal one we should obtain nearly equal frequencies in the long run, then we will allot the value $1/N$ to the probabilities of each of the possible outcomes of the idealized experiment.

It then follows from Axiom III that if an event A occurs whenever the outcome belongs to a subset $\mathcal{A}$ of these symmetrical outcomes and if $n(\mathcal{A}) = r$ then we must allot a numerical value to the probability of A as follows:

$$\text{pr}\,(A) = r/N.$$

This allotment of values to probabilities was taken by Laplace as his definition of probabilities. We shall refer to it as an *allotment on grounds of symmetry*.

8. DIFFICULTIES

If two events arising from an experiment are such that the long-run frequency-ratios associated with them are equal enough for us to wish to take the relevant values of k as equal, then we may call the events *equally-likely*. In a corresponding idealized experiment we may well detect some symmetry which would confirm our views and make us take the corresponding probabilities as equal. Sometimes, however, there is more than one way of reading symmetry into an idealized experiment; so that it is dangerous to allot equal probabilities on the ground of symmetry *alone* unless experience has already convinced us that we have detected the appropriate symmetry. Detecting the appropriate symmetry (if there is symmetry at all) is sometimes difficult, but the reader will gradually build up a body of experience and in the case of dice and coins he probably already has correct ideas.

The possible dangers are best seen by examining a few situations; the linking factor here is that these are situations not entirely under our control; in the next section we shall discuss the possibility of so arranging the situation that difficulties of the sort discussed here are removed. This is one of the objects of the 'Design of Experiments'.

8.1 Situation 1. Suppose we examine the first two children of families of two or more children. There are three possibilities: two boys; one of each; two girls. Do we on grounds of symmetry allot to each possibility a probability of $\frac{1}{3}$? Alternatively we might say there are two possibilities: both the same; one of each. We should then be tempted to allot probabilities of $\frac{1}{2}$ to each of these events; and we see that the different probabilities allotted by the symmetries to 'one of each' force us to go beyond the mere apparent symmetry and to return to the frequencies of the real events for which we are trying to make a model. In applied mathematics *the facts* are sacred.

In 130 such families there were

	2 boys	1 of each	2 girls
Numbers	37	59	34
Frequency ratio	0·285	0·454	0·262

These figures do not suggest that

$$0{\cdot}333 \qquad 0{\cdot}333 \qquad 0{\cdot}333$$

would be good values to take for k to allow prediction of frequencies in further long runs and so do not suggest that

$$\tfrac{1}{3} \qquad \tfrac{1}{3} \qquad \tfrac{1}{3}$$

(which is a *permissible* allotment of probabilities) would be a *suitable* allotment. The first symmetry that we suggested does not seem to be one that would be helpful in building a model. The same figures rearranged give:

	Both the same	1 of each
Numbers	71	59
Frequency-ratio	0·546	0·454

and these figures hardly seem conclusive either for or against using a model with the second suggested symmetry.

If we take families consisting of exactly two children we have slightly changed the conditions of the experiment. Of 62 such families we have the following details:

	2 boys	1 of each	2 girls
Numbers	12	42	8
Frequency-ratio	0·194	0·676	0·129

or alternatively

	Both the same	1 of each
Numbers	20	42
Frequency-ratio	0·324	0·676

These figures seem conclusively against *both* the suggested allotments of probability by symmetry.

We return to the first experiment and take a larger sample of families of two or more children to obtain:

	2 boys	1 of each	2 girls
Numbers	281	594	275
Frequency-ratios	0·244	0·516	0·239

If we are hunting for simple numbers we might propose to allot the probabilities

$$\tfrac{1}{4} \qquad \tfrac{1}{2} \qquad \tfrac{1}{4},$$

though we have not yet devised the rest of the model.

We could link this choice of probabilities to the model which is suggested by the following tabulation:

		2nd born child	
		Boy	Girl
1st born child $\begin{cases} \text{boy} \\ \text{girl} \end{cases}$		Results in 2 boys Results in 1 of each	Results in 1 of each Results in 2 girls

This model could be made plausible by the claim that since we can detect which child of two is the first born we have at least achieved a proper symmetry in the analysis (provided we assume that boy/girl is a proper symmetry at each birth).

The crux of the matter remains however that this analysis gives good answers for the frequencies; the rabbit of symmetry was put into the hat afterwards. It is obvious that the symmetry we have supplied is one that would lead to the second symmetry originally suggested (namely: both the same/one of each), but there seems no obvious way of justifying that symmetry *directly*.

The model just suggested *is* a good one in the sense that by an extension to families of three or more children we would allot probabilities for the first three children to be born as follows (the methods of calculation for this sort of problem are explained in Chapter 7):

3 boys	2 boys, 1 girl	2 girls, 1 boy	3 girls
$\frac{1}{8}$	$\frac{3}{8}$	$\frac{3}{8}$	$\frac{1}{8}$

and these are in good agreement with long run frequencies of real families. It is after all no test of a theory that it fits the facts it was designed to fit; it is its powers of prediction that really matter.

We can explain the failure of the model to fit families of *exactly* two children (as opposed to *two or more children*) by the remark that many couples with their first two children of the same sex attempt to restore the balance, with a third child, but that many couples with their first two children of opposite sexes are content with their lot. This would produce a preponderance of 'one of each' families of exactly two children.

8.2   Situation 2. The assumed symmetry at *each birth* of boy and girl is a good enough model for many purposes in that it fits the long-run frequencies well, but for more definite calculations the long-run frequencies in recent years show that better answers are obtainable from a model in which pr (child is a boy) = 0·516 and pr (child is a girl) = 0·484. The probabilities *at birth* are not the best values to take at later ages since the death-rates of boys and girls are different. Even the frequency-ratios of

37

sexes at birth seem to change significantly with time. This only illustrates the fact there is not one correct model and many incorrect models, but rather various suitable models and many unsuitable ones.

8.3 Situation 3. The reader can test his powers in this field by devising a mathematical model for the throwing of two indistinguishable coins, and noting that here we are without the immediate distinction 'first-born' and 'second-born'. Various models would lead to:

(i) pr (2 heads) $= \frac{1}{2}$; pr (not-2-heads) $= \frac{1}{2}$;

(ii) pr (2 heads) $= \frac{1}{3}$;

pr (not-2-heads) $=$ pr (1 of each) $+$ pr (2 tails)

$$= \frac{1}{3} + \frac{1}{3};$$

(iii) pr (2 heads) $= \frac{1}{4}$;

pr (not-2-heads) $=$ pr (1 of each) $+$ pr (2 tails)

$$= \frac{1}{2} + \frac{1}{4}.$$

A long run of throws of a pair of coins, say 400 throws, would safely produce telling evidence as to which allotment of probabilities, and hence which model, was most suitable. The reader should carry out such a run and decide on a model before reading on.

8.4 Situation 4. A more sophisticated example follows. We are to put objects into boxes under various rules as below.

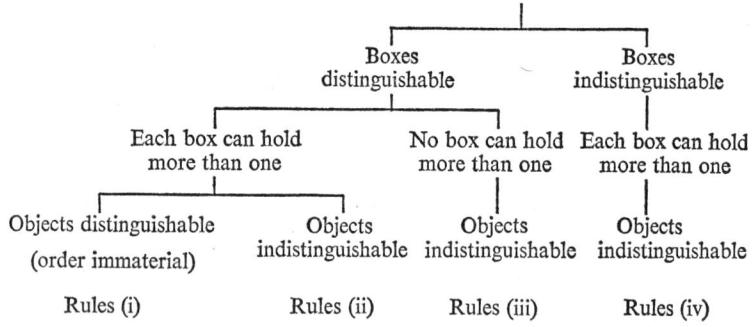

In Table 2.6 each row contains a possible outcome of the experiment of placing two such objects in three such boxes. Distinguishable objects are called A, B; indistinguishable ones merely X.

Experience shows that if (a) the objects are, say, ping-pong balls and (b) the boxes can hold more than one of them and (c) the ping-pong balls are scattered into the boxes in a way which treats the boxes symmetrically in the long run, *then even if we cannot in practice tell the ping-pong balls apart* it is the symmetries from model (i) and NOT those from model (ii) that suggest probabilities agreeing with the long-run frequency ratios. Thus the probability of getting the arrangement $X|X|-$ (for example) with ping-pong balls is $\frac{2}{9}$, not $\frac{1}{6}$. Once this is seen to fit the facts we say that the ping-pong balls *could* in principle be distinguished by a tiny mark which did not affect their behaviour and rules (i) would then be appropriate and would lead to model (i) for the correct choice of symmetries, so that $X|X|-$ conceals two possibilities namely $A|B|-$ and $B|A|-$. *This* is *hindsight*.

Table 2.6

Model (i) from rules (i)	Model (ii) from rules (ii)	Model (iii) from rules (iii)	Model (iv) from rules (iv)
$AB\|-\|-$	$XX\|-\|-$		$XX\|-\|-$
$-\|AB\|-$	$-\|XX\|-$		
$-\|-\|AB$	$-\|-\|XX$		
$A\|B\|-$	$X\|X\|-$	$X\|X\|-$	$X\|X\|-$
$B\|A\|-$			
$A\|-\|B$	$X\|-\|X$	$X\|-\|X$	
$B\|-\|A$			
$-\|A\|B$	$-\|X\|X$	$-\|X\|X$	
$-\|B\|A$			

It is important for the reader to remember model (i), because for large every-day objects like coins, germs or elephants it is the one which gives suitable probabilities.

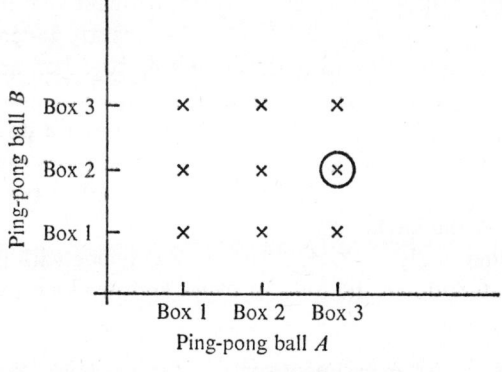

Fig. 2.4

It can be diagrammatically represented as in Figure 2.4. Each cross represents a possible configuration of balls and boxes (for instance the circled one corresponds to: $-|B|A$) and the configurations are equiprobable. We return to such diagrams in the next chapter.

The question arises 'Do models (ii) and (iii) serve any purpose as models for *real* experiments?' The answer is 'Yes'. If we are dealing with photons, nuclei and atoms containing an even number of elementary particles then experiments show that model (ii) leads to suitable probabilities. We say, with the hindsight this gives, that the particles are *essentially* indistinguishable. If we are dealing with electrons, protons, neutrons then model (iii) gives suitable probabilities. We say with hindsight that the particles are *essentially* indistinguishable and that each 'state' can only contain a single particle. The first three models are respectively those of Maxwell–Boltzmann statistics, Bose–Einstein statistics and Fermi–Dirac statistics. It was the failure of model (i) to predict good answers that led to its abandonment in certain fields and the invention of the others to replace it in those fields. Model (iv) has no obvious real counterpart.

9. DESIGN OF EXPERIMENTS

Since it is obviously important, or at the very least helpful, to know at the outset suitable values for the probabilities in a model and since only in the case of symmetry *can* we know at the outset suitable probabilities, we take great care when designing real experiments to introduce as nearly perfect symmetry as we can so that actual symmetry can be present in a suitable model.

9.1 Experiment 1. For instance in experiments on the growth of rats we might take fifty pairs of identical twins and from each pair select one to be given treatment A and one to be given treatment B. We would not even trust ourselves to allot the treatments symmetrically, without dice or coins to throw. The reader may feel that if the rats are taken from the cages, in each of which a pair had been brought up together, and the first rat allotted treatment A whether it looked the healthier or more lively or not, then symmetry would be practically achieved, but we do not know that the first rat is not going to be the less agile and so more easily caught one, or even the one that prefers the light near the front of the cage compared to the dark at the back.

Even allotting alternately A, B, A, B. ... might coincide with the rhythm of windows or heaters down the lines of cages and produce some unsuspected asymmetry.

9.2 Experiment 2. In experiments with four varieties of wheat we would plant them in some pattern such as:

$$
\begin{array}{cccc}
A & D & B & C \\
C & B & A & D \\
D & A & C & B \\
B & C & D & A
\end{array}
$$

Patterns of this type are called Latin squares. The object is to get exactly one of each variety in each row and one in each column, so that any linear change of soil or sunlight or humidity in either direction (and hence, by vector properties, in any direction) will have symmetrical effects on each variety. There are many distinct 4×4 Latin squares and the experimenter would choose one of them by a method whose symmetry was itself based on that of coins or dice.

9.3 Experiment 3. In carrying out political or economic surveys symmetry is much harder to achieve; the aim is that every relevant sub-group of the group to be surveyed shall be symmetrically treated, so that once again basic probabilities can be allotted in the model. Sometimes the subgroups are taken to be groups of persons of a single occupation, or persons of a single income range, or of a single town, and this is called stratified sampling, but sometimes the subgroups are individual people.

The literature is full of the errors perpetrated, like telephoning people (this only selects from those who have telephones), stopping them in the street (this only selects from those who are ever in the streets), visiting houses while some people are at work, and so on. One way is to have a complete list of the names of all the people in the town and to choose by coins, dice, packs of cards (or the more sophisticated methods to be mentioned later) those to be interviewed, and then to pursue them relentlessly until they are interviewed.

10. RANDOMNESS

10.1 Importance. A very large body of work has been done on the properties of samples selected by symmetrical methods of one sort or another, but nothing useful can be said about unknown collections of measurements where symmetry has not been preserved. If, on the one hand, you throw a die and I merely choose a number between 1 and 6 at will, the only thing that can be said about their sum is that it lies between 2 and 12, but, on the other hand, we shall see that if we both throw dice then the sum will be 7 more often than any other number in the long-run.

Symmetry in probability problems is thus of very great interest, and the interest is *not* confined to studying games of chance.

It is almost impossible to achieve equally likely outcomes from asking people to choose 'at random' a single-figure number; it is well known that 3 and 7 predominate (and the reasons for these choices would make an interesting psychological study). It is very difficult to make mechanical devices that serve this purpose of symmetrical treatment well; packs of cards cut at the same card far too often, perhaps because a card gets sticky, or less sticky, or bent or straight each time it appears; and the effects begin to snowball. Slips of paper in drums are little better; the mixing is often very poor. Perhaps the most important case of this was the lottery used in 1940 to determine who were to be conscripted into the armed forces of the U.S.A. Roulette wheels are probably the best simple mechanism, but they will need to be expensive ones. Randomness is too important to leave to chance!

10.2 Random number tables. Very carefully made machines have been constructed to give, as closely as possible, symmetry to ten different outcomes; and by coding these outcomes as 0, 1, 2, ..., 9 tables of *random numbers* have been constructed, printed and sold. The whole idea seems paradoxical at first since once a sequence of digits such as 379644137261 has been printed it is hard to say what we mean by calling it 'random'. Nevertheless experimenters using such tables to allot choices know that in the long run each digit occurs with relative frequency nearly $\frac{1}{10}$ and, in contrast with the table consisting of 0123456789 repeated *ad infinitum*, in any *previously determined* subsequence each digit appears with relative frequency nearly $\frac{1}{10}$.

Typical frequencies from such a table have been analysed in Table 2.7.

Table 2.7. Frequencies of digits from Random Number Table

Digit	Relative frequencies in the first N digits			
	$N = 50$	$N = 100$	$N = 200$	$N = 400$
1	0·10	0·09	0·110	0·1025
3	0·00	0·06	0·080	0·1000
5	0·16	0·14	0·100	0·1025
7	0·08	0·07	0·080	0·0925
9	0·08	0·10	0·100	0·1075
Even digits	0·58	0·54	0·530	0·4950

To obtain typical frequencies based on probabilities of, say,

$$\text{pr}(A) = 4/119, \quad \text{pr}(B) = 15/119, \quad \text{pr}(C) = 100/119$$

we could proceed as follows:

(i) Block the random digits off in groups of three at a time and read them as numbers x;

(ii) Discard those with $960 \leqslant x \leqslant 999$;

(iii) Subtract multiples of 120 from any others to bring them to the form $0 \leqslant y \leqslant 119$.

(iv) Discard $y = 0$, allot those with

$$1 \leqslant y \leqslant 4 \text{ to event } A,$$
$$5 \leqslant y \leqslant 19 \text{ to event } B,$$
$$20 \leqslant y \leqslant 119 \text{ to event } C.$$

11. HYPOTHESIS AND EXPERIMENT

Once we assign values to the probabilities of certain events and decide in probability terms how to interpret a few phrases like 'independent events', then other events compounded from the earlier ones can have probabilities *calculated* from the axioms.

For instance we can never know the best value to give to the probabilities that in an idealized trial based on Section 2.1 a plastic drawing pin will land point-down, but *if* we allot the value 0·43 to the probability of this event, then we can calculate (by methods which are explained later in the course) exactly what would be the consistent value of, for instance, pr (2 drawing pins fall point-down when 5 such drawing pins are thrown 'independently' of each other). The required value would be

$$10.(1-0·43)^3.(0·43)^2$$

which is 0·342421857. One prediction we might make from this is that *if* the probability of a single idealized drawing pin landing point-down is 0·43 *then* in 10^{12} real trials each of throwing five drawing pins independently about 342,421,857,000 of the trials will show two drawing pins point-down. Many people would be happy to carry out only 100 such trials, say, and check whether about 34 or so showed 'two down'. Even if, however, 342 trials in 1000 or 342,421,857 in 10^9 showed 'two down' we should still not *know* the best probability; but on the other hand even if the first 1000 trials showed none with 'point-down' we should still not *know* that for a long enough run the best probability was not 0·43.

All we can have is a hypothetical value for the probability; this will lead to predictions which may be verified or not to the degree of accuracy that concerns us (and the theory will later give us a basis for deciding how concerned to be). What more can mathematics do for us in the applied field at all? It can never tell us whether some statement about the real world is true; all the statements of mathematics apart from axioms and definitions begin 'If...'.

The axioms and definitions of Euclid's geometrical system were for long thought to be absolute statements about the real world; we now know, however, that their status in relation to the concrete world is that of an excellent mathematical model for the physics of measurement within the limitations of not-cosmic distances, but that over big enough intervals of space and time they form a less satisfactory model. This in no way detracts from the beauty of the purely mathematical theory of, for instance, irrationals, which has essentially sprung from Pythagoras' theorem. Probability theory too has this dual aspect of rich usefulness on the 'applied' side and austere beauty on the 'pure' side.

3

PROBABILITY—DEVELOPMENT

1. INTRODUCTION

The importance of being able to allot probabilities in an experiment is obvious, and that this leads to the use of symmetry has been explained in the last chapter. Whether the allotment is by symmetry or not is, of course, irrelevant to the discussion of how the probabilities of compound events are obtained from those of the events to which they are related; this is only a matter of the probabilities being consistent with one another because they are permissible. In this chapter many of the exercises will be ones in which the probabilities are not given as if they had been estimated from some real experiment, but are to be allotted by the method of symmetry.

The following is a problem which is simple enough to start from, but which is rich enough to contain several lessons.

'A and B take turns at throwing dice. A's turn consists of throwing one die and trying to get a face which is a multiple of three; B's turn consists of throwing two dice and trying to get a face which is a six. Each wins a prize of a penny if he is successful. Is this game fair between A and B?'

Asking whether it is 'fair' means asking about what happens in the long run. Only a child says 'Not fair' if it loses the first one or two rounds.

It appears at first sight that the game is fair; because although B is trying to do something twice as hard as A on each die, he throws two dice in a turn whereas A only throws one. It was precisely a problem of this sort that set Pascal and Fermat thinking in the seventeenth century and led them to produce the idea of 'probability'. (See Exercise 3 M, note on Question 40.)

In order to fix our ideas we reproduce here the actual results of twelve turns each, in a run of the corresponding real game conducted for this purpose. The reader could easily produce more and it will help him to have a clear idea of the nature of the ideal game if he does so; a certain amount of the early discussion uses this game as a concrete example.

Turn	1	2	3	4	5	6
A's result	one	five	three	five	six	four
B's result	three	four	four	one	three	four
	five	five	five	two	five	five
Turn	7	8	9	10	11	12
A's result	three	four	five	one	five	two
B's result	four	six	four	five	three	one
	five	six	six	six	three	six

We note that A got a prize on turns 3, 5, 7;

B got a prize on turns 8, 9, 10, 12.

The fact that B got only *one* prize for his turn 8 may give the reader a hint as to how an investigation of this problem will proceed.

The data in the table suggest all manner of subsidiary problems, such as 'What is the probability that in throwing two dice a four and a five will be obtained?' or even 'What is the probability that in throwing two dice twelve times a four and a five will be obtained four times or more?' (The answer to the former question is '$\frac{1}{18}$' and to the latter 'about $\frac{3}{800}$'.) In fact the game suggested favours A, despite the first-glance-fairness, and certainly despite the run of the first mere twelve turns each of the real game.

2. ANALYSIS

2.1 Trial, Possibilities, Result. We have already used in the last chapter the words 'trial', 'possibilities', 'result', and we summarize them here with explanations of our usage rather than formal definitions.

In analysing A's turn in the game suggested in Section 1 we proceed as follows:

First analysis

Something happens...A throws a die;

A decides to record a certain type of outcome...the name of the upper-most face.

There are still many *possibilities* of this type for A to find himself recording...{one, two, three, four, five, six}.

The first *result* of A's that actually occurred was...'one'.

In this analysis the first three lines really specify the *trial* which A is carrying out.

What A decides to record depends on his interest in the matter. It might even be whether the die shatters or not; or whether it rolls under the bed or not. In Section 1 he is interested in whether he has a multiple of three or not, so we might analyse A's turn by recording something different.

Second analysis

Something happens...A throws a die;

A decides to record a certain type of outcome...whether a multiple of three occurs or not;

Set of possibilities...{Yes, No};

The first result of A's that actually occurred was...'No'.

A's record of his results in the first twelve turns now reads:

No, No, Yes, No, Yes, No, Yes, No, No, No, No, No.

The second analysis rivets our attention onto A's interest.

45

2.2 Events. When we have described a trial and its set of possibilities we are usually interested in some subset of the possibilities. In this way A's interest in his turn in Section 1 could be stated as follows:

First analysis

Trial:　　　　　　　　　　throwing one die and recording the face;

Set of possibilities:　　{one, two, three, four, five, six};

Subset of possibilities: {three, six}.

Second analysis

Trial:　　　　　　　　　　throwing one die and recording whether a multiple of three occurs or not;

Set of possibilities:　　{Yes, No};

Subset of possibilities: {Yes}.

A subset of possibilities enables us to analyse what we mean by the *occurrence of an event* as follows:

First analysis

If the result of a particular turn belongs to the given subset {three, six}, then the event 'the number on the face is a multiple of three' has occurred in that turn. In A's first turn the event did not occur, in A's third turn it did.

Second analysis

If the result of a particular turn belongs to the given subset {Yes}, then the event 'the number on the face is a multiple of three' has occurred in that turn. In A's first turn the event did not occur, in A's third turn it did.

To summarize: we say that an *event* can be specified by naming a subset of the possibilities. The *event occurs* in a particular turn if the result of that turn is a possibility belonging to that subset; the *event does not occur* if the result is not a possibility belonging to that subset.

2.3 Choice of analysis. We return to an important point here. We have seen that given a certain action (for instance, throwing one die) we have been able to analyse the situation in two ways each with its set of possibilities, and each set of possibilities having a subset which specifies the given event (for instance, the number on the face is a multiple of three). If we are going to be able to give a numerical value to the probability of an event without carrying out a long sequence of trials to suggest a good choice, then we must build up a body of experience to tell us which analysis, if any, suggests appropriate symmetries to take.

3. POSSIBILITY SPACES

In order to investigate the laws of probability for compound events, that is for various combinations of single events, we need a mathematical description which expresses the bare facts of the case with which we are faced, and the easiest description is diagrammatic.

3.1 Diagrams in algebra and geometry. It is possible to develop geometry without any diagrams and to speak purely in terms of such statements as

'*PQR* collinear, *QRS* collinear ⇒ *PRS* collinear'

or '**R** is linearly dependent on **P** and **Q**, and **S** is linearly

dependent on **Q** and **R** ⇒ **S** is linearly dependent on **P** and **R**'.

It would, however, be helpful to have such diagrams as Figure 3.1. The diagrams do not constitute the proofs but they suggest lines of approach and in simple cases allow calculations to be quickly made.

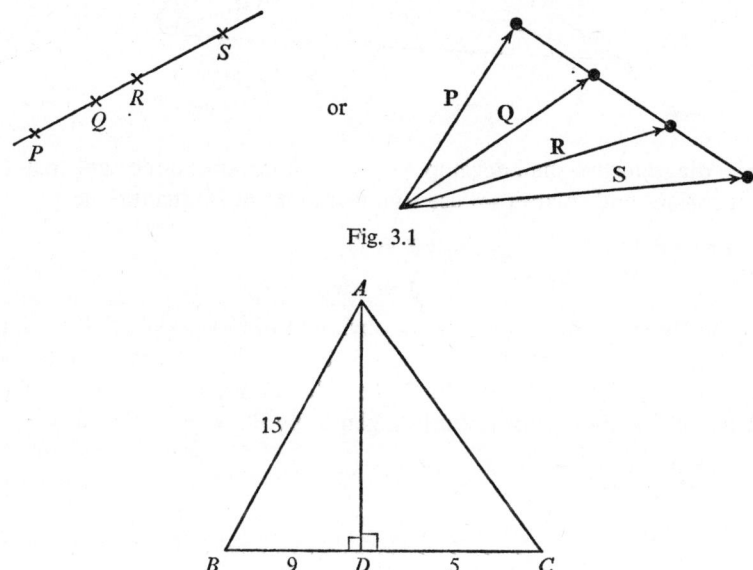

Fig. 3.1

Fig. 3.2

Thus the diagram Figure 3.2, although not drawn to scale, conveys a mass of information and unless it has been most misleadingly drawn would let us calculate fairly quickly that $AC = 13$. It would be possible but tedious to put all the statements in terms of a few accepted theorems (and, intellectually, a grand discipline), but for many purposes it is unnecessary; our geometrical 'intuition', whatever that may mean, carries us along. Sometimes intuition fails, as in the well-known paradox where Figure 3.3, with the data as indicated, apparently leads to

$$LC = MB, \quad AL = AM$$

so that $AC = AB$ and *all* triangles are isosceles.

47

There are correspondingly well known paradoxes of probability too, where contradictory initial assumptions are made unwittingly. As experience develops, however, 'intuition' fades away. Someone whose 'intuitions' are predominantly right is usually someone whose experience is large and who does not recognize individually the early steps, but takes them at a run and calls it 'intuition'.

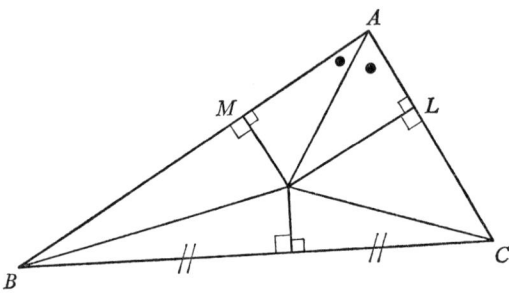

Fig. 3.3

The diagrams we shall develop are no less (or more) important than the number-lines and -planes we use to illustrate such inequalities as

$$x+y \leqslant 2,$$

$$y \geqslant x+1.$$

From these it is not *immediately* obvious to most people that we have $x \leqslant \frac{1}{2}$ however we choose the value of y, but that there is no limitation on the value of y provided we arrange the value of x suitably.

This information is conveyed in a glance by the Figure 3.4.

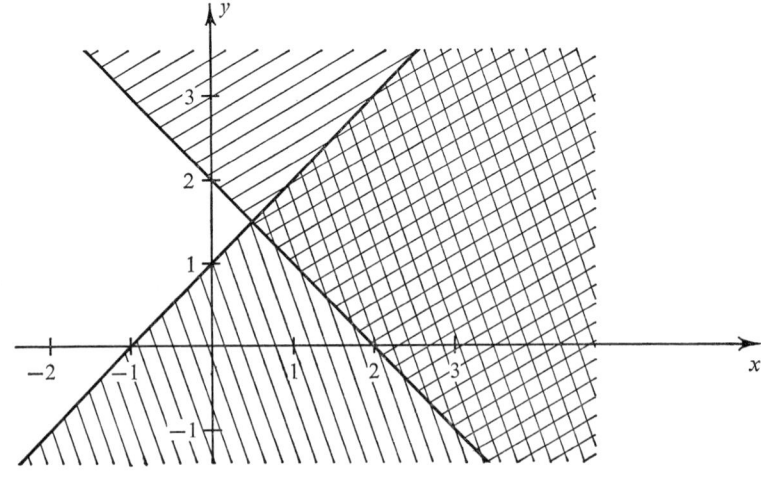

Fig. 3.4

It is our purpose next to explain a form of diagram which will allow us to make correct deductions about probabilities.

3.2 Possibility diagrams. If we take a pack of cards and draw a card we may name the suit obtained as Spades, Hearts, Diamonds, Clubs. We might record these as *S, H, D, C* and mark them on a diagram, as in Figure 3.5; or we might if we wanted use a number code, as in Figure 3.6.

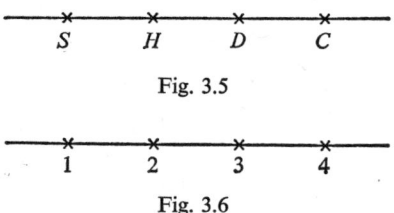

Fig. 3.5

Fig. 3.6

(Those who play bridge will feel this latter arrangement natural as it records the order of precedence in bidding.)

In the first case we have:

Trial: choosing a card and recording the suit.
Set of possibilities: {*S, H, D, C*}.

In the second case:

Trial: choosing a card and recording the suit.
Set of possibilities: {1, 2, 3, 4}.

In these cases we do not have essentially different analyses, but, merely different notations.

We now consider two trials together.

Trial 1: George throws a die and records the face.
Set of possibilities: {1, 2, 3, 4, 5, 6}.

Trial 2: Harry draws a card and records the suit.
Set of possibilities: {*S, H, D, C*}.

We could set up a diagram for the trial consisting of both these actions taking place, as in Figure 3.7.

There are 24 possible outcomes by this analysis and each cross in Figure 3.7 represents one of them; for instance the circled cross represents the outcome when George throws a two and Harry draws a Diamond. We see that an outcome related to a compound event may be mapped onto a single point. Sets related to the events *A*: 'George throws a two'; *B*: 'Harry draws a Diamond' and *C*: 'George throws a two or Harry draws a Diamond' are shown in Figure 3.8. We will use a convention that a script letter will name the set of possibilities that defines (or is defined by) the event with the corresponding capital roman letter as name.

It is important to realize that although compound events often do consist of two events following one another in time (whether by causation or not), sequence is not an essential feature of compound events, or indeed

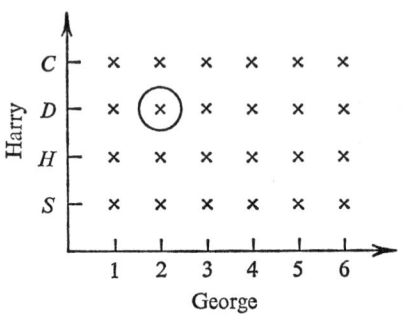

Fig. 3.7

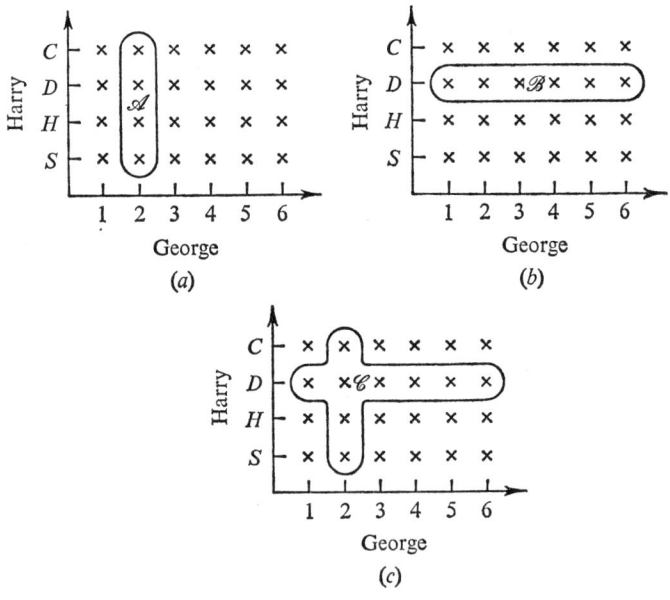

Fig. 3.8

of two-dimensional possibility spaces. We return to the beginning of this section to justify this statement. A significant change of notation there would be to think in terms of major suits (Spades and Hearts) and minor suits (Diamonds and Clubs) and in terms of black suits (Spades and Clubs)

50

and red suits (Hearts and Diamonds). We might then use the following codes:

$$\text{Spades} \to S \to (\text{major, black}) \to (M, B);$$
$$\text{Hearts} \to H \to (\text{major, red}) \quad \to (M, R);$$
$$\text{Diamonds} \to D \to (\text{minor, red}) \quad \to (m, R);$$
$$\text{Clubs} \to C \to (\text{minor, black}) \to (m, B).$$

We now have the means of setting up a two-dimensional diagram if we like (see Figure 3.9). The bottom right cross for instance, is related to the compound event 'We draw a major suit and do not draw a red suit'. Thus we see that we need not consider sequence or even two physical objects to get compound events or a two-dimensional diagram. The compound event we have named occurs, when it does at all, in the single drawing of a single card.

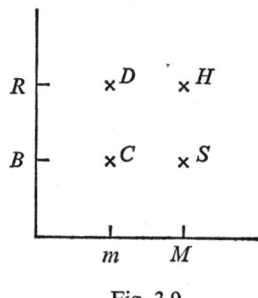

Fig. 3.9

In these diagrams we have used points to represent possible outcomes or possibilities. A set of all such points necessary to represent the possible outcomes of a trial we call a *possibility-space* for the trial. We use the phrase 'possibility-space' even when there is no geometrical representation involved. Most writers use the phrase 'sample-space', but it seems helpful to get away from the overworked word 'sample'; and 'possibility-space' adequately describes what we are dealing with.

We will consider a *third trial* in more detail:

We throw a pair of distinguishable dice, one made of ebony the other of ivory, in the course of some game, say, in which the two events of interest are $A =$ 'total score is 7' and $B =$ 'difference of scores does not exceed 1'.

We might set up a diagram in more than one way.

First analysis: Let $d =$ difference of scores;

$t =$ total of scores.

We will map the possibilities onto $\{(d, t)\}$, but the values of d and t are restricted by the fact that if $x =$ score on ebony die, and $y =$ score on ivory die, then

$$t = x+y, \quad d = |x-y|.$$

so that d, t are even or odd together. This restricts the possibility space to the 21 points shown in Figure 3.10; so that for instance

$$\mathscr{A} = \{(1, 7), (3, 7), (5, 7)\}.$$

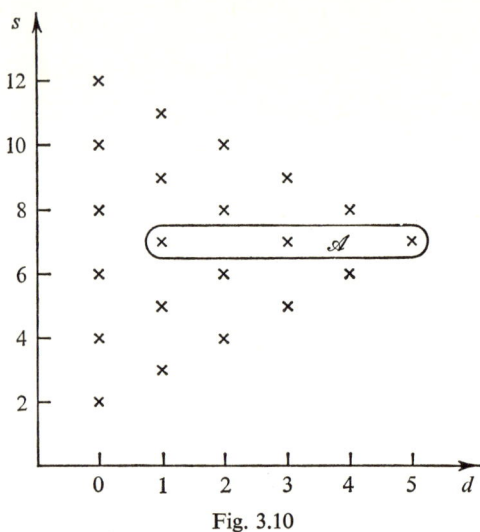

Fig. 3.10

Second analysis: This is in terms of the x, y already discussed. In this case the values range from 1 to 6 for each variable and there are 36 possibilities. In this diagram (Figure 3.11)

$$\mathscr{A} = \{(1, 6), (2, 5), (3, 4), (4, 3), (5, 2), (6, 1)\}.$$

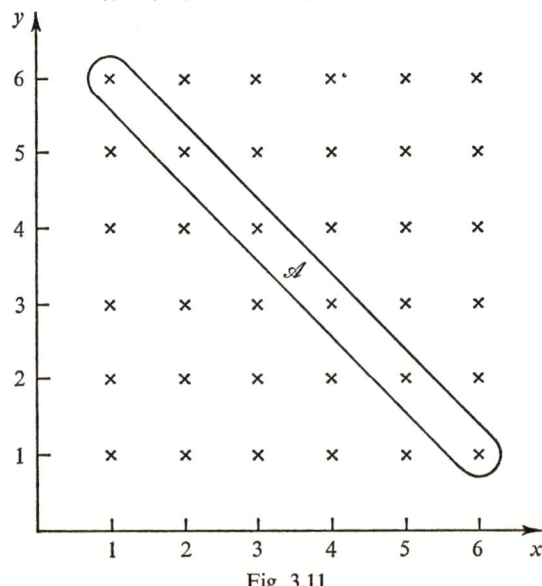

Fig. 3.11

For our next example we go back to the drawing pin experiment of Chapter 2, Section 5.2 where a teacupful of mixed pins was used. Any fall of a drawing pin could be classified as the occurrence of either point-Down (*D*) or point-Up (*U*), and also as the occurrence of either White (*W*), or Coloured (*C*), or Brass (*B*).

We could represent the situation in the following diagram (Figure 3.12) which rather resembles a mixture of Table 2.4 and Figure 2.3, both from Chapter 2.

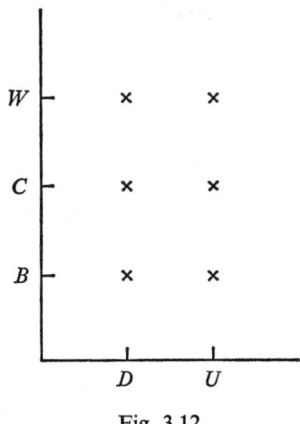

Fig. 3.12

The points here represent the outcomes which lead to the following compound events: (*D* and *W*), (*D* and *C*), (*D* and *B*), (*U* and *W*), (*U* and *C*), (*U* and *D*).

The particular feature of this analysis is that these events are exclusive and exhaustive, and furthermore there is not one of them that can be broken down into a set of exclusive events which exhaust it. They can be thought of as representing an ultimate analysis of this problem—we call them elementary events.

3.3   Summary. We summarize our usage and notation. The diagrams are to be compared with Figure 3.13 where two sets of points $\mathscr{A}$ and $\mathscr{B}$ are illustrated.

(i) The event '*A* does not occur' is called not-*A* and is written *A'*; points corresponding to it fall in the complement of $\mathscr{A}$, written $\mathscr{A}'$, as shown in Figure 3.14.

(ii) The event '*A* occurs and *B* occurs' is called (*A* and *B*) and is written (*A* and *B*); points corresponding to it fall in the set ($\mathscr{A} \cap \mathscr{B}$), as shown in Figure 3.15.

(iii) The event '*A* occurs or *B* occurs or both occur' is called (*A* or *B*) and is written (*A* or *B*); points corresponding to it fall in the set ($\mathscr{A} \cup \mathscr{B}$), as shown in Figure 3.16.

(iv) Notice that if two events *A*, *B* are related in such a way that whenever *A* occurs *B* must occur then the set of points representing *A* falls wholly inside the set representing *B*; a situation which we write as $\mathscr{A} \subseteq \mathscr{B}$.

(v) If *A* and *B* are *exclusive* events then $\mathscr{A}$ and $\mathscr{B}$ have no points in common in the possibility space.

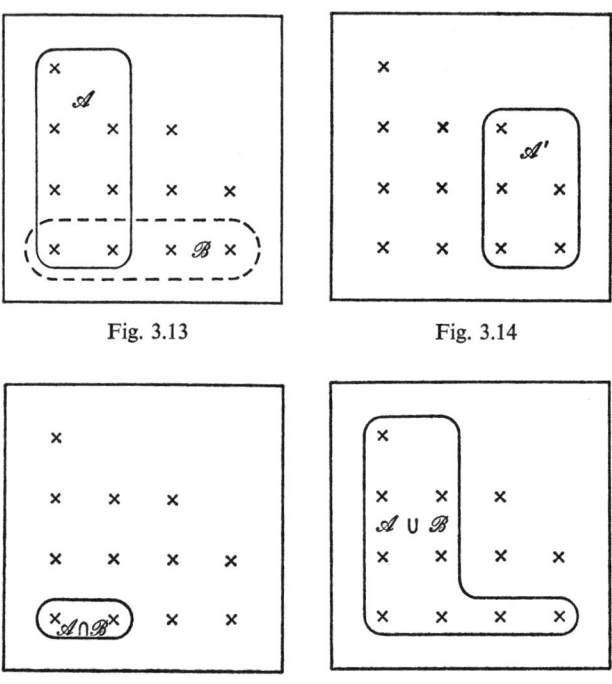

Fig. 3.13   Fig. 3.14

Fig. 3.15   Fig. 3.16

(vi) If *B* and *C* are *exhaustive* then every point of the possibility space is in one or other or both of $\mathscr{B}$ and $\mathscr{C}$.

(vii) If *B* and *C* *exhaust* *A* then every point of $\mathscr{A}$ is in one or other or both of $\mathscr{B}$ and $\mathscr{C}$.

3.4   Events and sets. As we have used the word 'event' here there is a distinction between the event and the set of points representing it. Mathematics abounds in important distinctions, such as between the following pairs:

The set of numbers between 0 and 1;   the inequalities $0 < x < 1$.
The number zero;   the equation $x = 0$.
The null set;   the number zero.

The number zero;	{zero}.
The circle which is the boundary of a disc;	the disc itself.
The point (2, 1);	the vector $\begin{pmatrix} 2 \\ 1 \end{pmatrix}$.

The last of these pairs, however, illustrates another feature of mathematical language—that mathematicians take over words for their own use. In some ways (2, 1) is merely an ordered number pair, but one of the choices of possible sets of rules for handling it makes the structure of its theory indistinguishable from that of a point in two-dimensional Euclidean space. So when these rules of operation apply mathematicians call (2, 1) 'a point'. In the same way it is possible to take the process of abstraction in probability theory one stage further than we do here, and call the sets of points 'events'. The more advanced texts do this and the reader will have no difficulty in following them when he comes to read them. It just seems appropriate at this level to halt the process of abstraction at this half-way stage.

4. ALLOTMENT OF PROBABILITIES

4.1 Finite sets of outcomes. The definition of a trial and the choice of a possibility-space we know from Chapter 2 to be insufficient to *determine* a suitable allotment of probabilities. The following, however, are allotments for the possibility spaces so far discussed which the reader can convince himself are suitable.

For Figures 3.5 and 3.9. The points refer to elementary events; on the grounds of symmetry between suits we allot $\frac{1}{4}$ to each.

For Figures 3.7 and 3.8. The points refer to elementary events and there is double symmetry: first between the six faces of the die and secondly, and quite separately, between the four suits; on these grounds we allot $\frac{1}{24}$ to each elementary event. The reader should check that this gives pr $(A) = \frac{4}{24}$, pr $(B) = \frac{6}{24}$ and pr $(C) = \frac{9}{24}$, and that these are obtainable separately by (i) the symmetry between the six faces, (ii) the symmetry between the four suits and (iii) the theorem pr $(A$ or $B) = $ pr $(A)+$pr $(B)-$pr $(A$ and $B)$.

For Figures 3.10 and 3.11. In Figure 3.10 the points do not refer to elementary events, as Figure 3.11 reveals, and we seem dangerously short of symmetry.

In Figure 3.11 the points represent elementary events and on grounds of symmetry between the dice we allot $\frac{1}{36}$ to each elementary event. Then pr (total is 7) $= \frac{6}{36}$ and pr (scores are within 1 of each other) $= \frac{16}{36}$.

A study of Figure 3.10 and the way some of its points refer to two elementary events, each of which is represented in Figure 3.11, would show that if it were taken together with its reflexion in the line $d = 0$ and rotated

55

through 45° clockwise about (0, 2) it would resemble Figure 3.11. Thus if we persist in using Figure 3.10 it seems we should allot $\frac{1}{36}$ to each of the points of $d = 0$ and $\frac{1}{18}$ to each other point. Axioms I and III are obviously satisfied by this, and since the total of the probabilities allotted is

$$(6 \times \tfrac{1}{36}) + (15 \times \tfrac{1}{18}),$$

we have satisfied Axiom II. The allotment is thus permissible.

For Figure 3.12. The points refer to elementary events but we have no way of allotting probabilities here except from observations of long runs. Such a long run as we have made suggests Figure 3.17; which is equivalent to Figure 3.18.

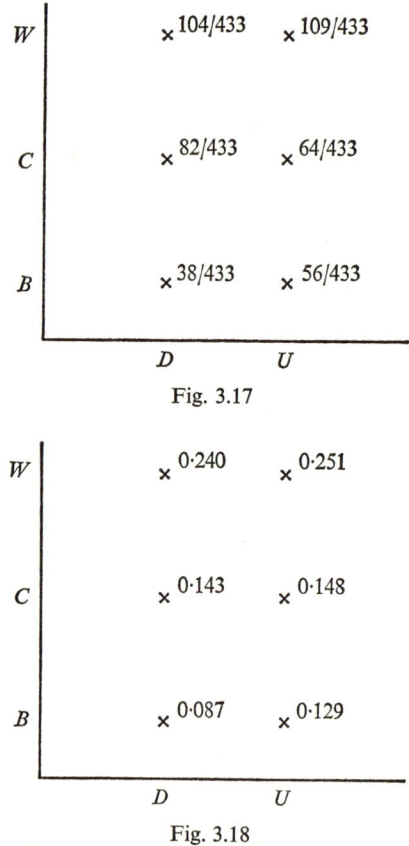

Fig. 3.17

Fig. 3.18

If we wanted to simulate the experiment we could draw up figures with proportional integers, as in Figure 3.19, and allot random numbers (between 001 and 998) to the events as the numbers fall in the ranges shown in Figure 3.20.

Figure 3.19 *resembles* a possibility space with 998 points, each representing one of 998 equally likely outcomes. Making the event (*B* and *D*), for instance, subdivisible in this way is, however, a confusing idea and no more meaning should be read into Figure 3.19 than that it leads to Figure 3.20 which represents a *sampling scheme* for simulating the drawing pin experiment. Such sampling schemes will be used later when we wish to discover the properties of various types of possibility spaces, or to simulate experiments with awkward allotments of probability; see Hammersley and Handscomb: *Monte Carlo Methods*, Methuen, 1964, Chapter 1.

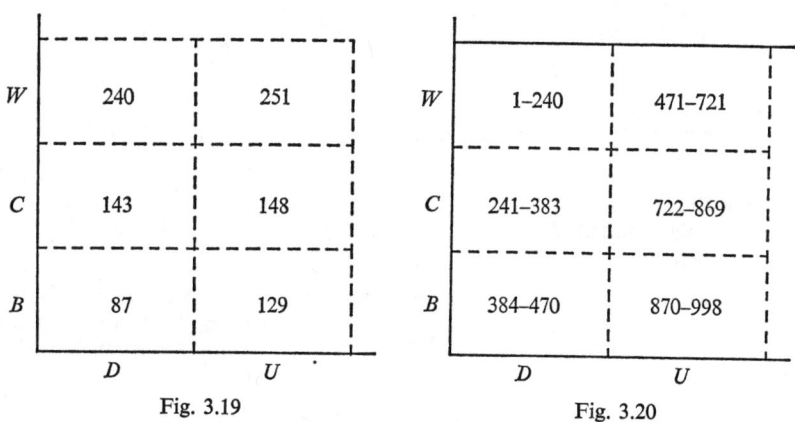

Fig. 3.19   Fig. 3.20

4.2 Infinite set of outcomes. If the possibility space represents an infinite set of outcomes, such as all the real numbers between 0 and 90, for instance, then an allotment of probabilities on grounds of symmetry is fraught with difficulties, since the choice of method of selecting the outcome can have an unexpected effect.

Consider the problem of allotting a probability to the event of getting an angle between 0° and 45° when all acute angles are possible. We may specify a method which treats symmetrically all magnitudes of angle between 0° and 90°, or all magnitudes of sine between 0 and 1, or all magnitudes of cosine between 1 and 0, or all magnitudes of the tangent of the half-angle (a commonly used parameter) between 0 and 1. We could then reasonably allot probabilities $\frac{1}{2}$, 0·7071, $1 - 0·7071$, or 0·4142 to the event. 'The angle is between 0° and 45°', on these four approaches; and the number of other approaches is legion.

The difficulties spring from the fact that an infinite set can be mapped by a 1–1 correspondence onto a proper subset of itself. In this volume, however, we will confine ourselves to possibility spaces with a finite number of points, or else will make the allotment of probabilities unambiguous.

Exercise 3 A

Keep all the working (including the diagrams of the possibility spaces) in any of the first six questions below. It will be needed later.

In each of Questions 1 to 6 draw a diagram of a possibility space; mark the sets $\mathscr{A}$ and $\mathscr{B}$ of points which refer to the given events (i) and (ii) ($\mathscr{A}$ corresponding to the first event). Allot values on grounds of symmetry (that is assuming all coins, dice, cards to be 'fair') to the following:

> pr (A), pr (B), pr (A'), pr (B')
>
> pr $(A$ and $B)$, pr $(A$ and $B')$, pr $(A'$ and $B)$, pr $(A'$ and $B')$
>
> pr $(A$ and $A)$, pr $(A$ and $A')$,
>
> pr $(A$ or $B)$, pr $(A$ or $B')$, pr $(A'$ or $B)$, pr $(A'$ or $B')$
>
> pr $(B$ or $B)$, pr $(B$ or $B')$

1 R. George throws a die and Harry draws a card. The score on the die and the suit of the card are recorded: (i) Harry draws a Club; (ii) George throws a five.

2 R. A pack of cards is taken and a card is drawn. The card is examined and the name of the suit is recorded. (Spades and Hearts are major suits, Spades and Clubs are black suits.) (i) A card of a major suit is drawn; (ii) a card of a red suit is drawn.

3. A pack of cards is taken and one card is drawn, the card is examined and the face-value recorded. (The honours are Ace, King, Queen, Jack, Ten.) (i) An honour is drawn; (ii) a card whose face value is between 5 and 10, inclusive is drawn.

4 R. A pair of dice is thrown as in the example in the text (Section 3.2). (i) Less than four is scored on the ebony die; (ii) scores which differ by two or less occur.

5 T. A domino set consists of twenty-eight dominoes, each having a number from 0 to 6 on each end. All combinations of numbers are present but the ends are indistinguishable and no repetitions of pairs of numbers occur. One domino is examined and the number pair is recorded. (i) A 'double' is drawn; (ii) a number pair with at least one 2 or one 3 is drawn.

Explain the relevance of this question to a problem involving two photons in a system with seven possible energy-states, the behaviour of the photons being governed by Bose–Einstein Statistics (see Chapter 2, Section 8.4).

6. A bag containing three cupro-nickel coins (say 50p, 10p, 5p) and two bronze coins (say 1p, ½p) is available. One coin is taken from the bag into each hand, and the colour in each hand is recorded. (i) A cupro-nickel coin is in the left hand; (ii) a bronze coin is in the right hand.

7. In the game suggested in Section 1:

(i) Why is it not fair between A and B?

(ii) What is the probability that B will throw 'a four and a five' in a single turn?

(iii) What is the probability that B will score a total of 7 in a single turn?

(iv) What is the probability that B will throw a total of 7 in one or other or both of his first two turns? [Hint: set up a new possibility space for two *turns* by B.]

8 F. Explain how the following could be used as a probability model that would simulate reasonably well the drawing pin experiment of Chapter 2, Section 5.2.

A pack of cards *without aces* is available. A single card is drawn and the following events are defined:

> J = (a Spade is drawn);
> K = (a Club is drawn);
> L = (a 2, or 3, ... or 8 of Hearts is drawn);
> M = (a 2, or 3, ... or 8 of Diamonds is drawn);
> N = (a red 9 or 10 is drawn);
> P = (a red face-card is drawn).

[In general, if symmetry is not present in a problem the probabilities, as estimated, of the elementary events can be used to construct an equivalent experiment with N equiprobable outcomes and a probability model can be set up by putting discs into an urn (which as Hammersley and Handscomb remark in *Monte Carlo Methods* is almost certainly a jam-jar), different colours of disc identifying different elementary events. Single discs can then be drawn at random (and replaced after each drawing) to represent carrying out the original experiment. This is called an *urn-model*. In practice a more satisfactory procedure is to use random numbers in the way described in Chapter 2, Section 10.2 and Chapter 3, Section 4.1.]

9. Referring to the original experiment of Chapter 2 (introduced in Section 2.1) as E_1 and to the second experiment of Chapter 2 (introduced in Section 5.2) as E_2, do you consider that it would be suitable to allot equal values to pr (pin is plastic topped and point down: E_i) for $i = 1$ and $i = 2$. If 'brass' were substituted for 'plastic' do you consider that equal values could be suitably allotted?

10. Prove from first principles that if the occurrence of event A implies the occurrence of event B then pr $(A) \leqslant$ pr (B).

4.3   **Summary.** Throughout Exercise 3 A the reader has been using

Axiom I: $0 \leqslant$ pr $(A) \leqslant 1.$

Axiom II: For a set of exclusive exhaustive events $\{A_i\}$ $(i = 1, ..., n)$

$$\sum_{i=1}^{n} \text{pr} (A_i) = 1.$$

Axiom III: For a set of exclusive events $\{A_i\}$ $(i=1, ..., n)$

$$\text{pr} (A_1 \text{ or } A_2 \text{ or } ... A_n) = \sum_{i=1}^{n} \text{pr} (A_i).$$

He has also been frequently employing the counting rules which can be stated as

$$n(\mathscr{A}') = n(\mathscr{E}) - n(\mathscr{A}),$$
$$n(\mathscr{A} \cup \mathscr{B}) = n(\mathscr{A}) + n(\mathscr{B}) - n(\mathscr{A} \cap \mathscr{B}).$$

If we divide these results through by $n(\mathscr{E})$ we get

$$\frac{n(\mathscr{A}')}{n(\mathscr{E})} = 1 - \frac{n(\mathscr{A})}{n(\mathscr{E})},$$

$$\frac{n(\mathscr{A} \cup \mathscr{B})}{n(\mathscr{E})} = \frac{n(\mathscr{A})}{n(\mathscr{E})} + \frac{n(\mathscr{B})}{n(\mathscr{E})} - \frac{n(\mathscr{A} \cap \mathscr{B})}{n(\mathscr{E})}$$

and these match the following general probability theorems

$$\text{pr}\,(A') = 1 - \text{pr}\,(A),$$

$$\text{pr}\,(A \text{ or } B) = \text{pr}\,(A) + \text{pr}\,(B) - \text{pr}\,(A \text{ and } B).$$

It is important to remember, though, that these two theorems are not in any way dependent on diagrams with equally probable outcomes but are deducible from the axioms. It is, however, the resemblance of the behaviour of sets of points in diagrams such as we have drawn to the behaviour of probabilities that makes the diagrams useful in calculating with probabilities.

For a pair of exclusive events A, B Axiom III reduces to

$$\text{pr}\,(A \text{ or } B) = \text{pr}\,(A) + \text{pr}\,(B)$$

which is the form to which also the second theorem reduces (since exclusive events A, B satisfy $\text{pr}\,(A \text{ and } B) = 0$).

In the next sections we will be investigating $\text{pr}\,(A \text{ and } B)$ in general situations; and all possible combinations of pairs of events will then have been covered.

5. CONDITIONAL PROBABILITY

5.1 Restricted possibility space. From Table 2.4, and again using initial letters, consider

$$\frac{\text{fr}\,(B \text{ and } D)}{\text{fr}\,(D)} = \frac{38}{166 + 38} = \frac{38}{204}.$$

In a long run of throws in which we examined the pins that fell point-down we would expect about $\frac{38}{204}$ of them to turn out to be brass ones, so that $\text{fr}\,(B \text{ and } D)/\text{fr}\,(D)$ behaves in a frequency sense in the same general way as $\text{fr}\,(B)/N$ does. We could imagine the drawing pins scattered and a blindfolded man selecting a point-down one at random; he would expect $\text{fr}\,(B \text{ and } D)/\text{fr}\,(D)$ to give rise to probability-like quantities that he could use to predict whether his pin would be a brass one or not. In fact if the information about the pins landing point-up had never been given we would from the very start have regarded $\text{fr}\,(B \text{ and } D)/\text{fr}\,(D)$ as (frequency of brass)/(total frequency); and for all the reader knows, the author was throughout the experiment allowing some of the pins to fall on a bowl of flour which held many of them with their shafts *horizontal*, but did not choose to mention the fact.

The only *essential* difference in the frequency-ratios (that is apart from their different *numerical* values in this case) is the narrower specification of the universal set to which we are relating the outcomes. We are using a frequency ratio not of all brass pins to all pins but of all point-down brass pins to all point-down pins. We are saying: given that the pins concerned are point-down the ratio for brass pins is $\frac{38}{204}$.

5.2 Restriction within a single experiment.

Notice that we cannot carry out a simpler experiment to find this ratio. We have to throw all the pins and *then* identify those point-down and *then* identify the brass ones among them. It is not possible to say 'let us save a stage by only throwing those which are going to land point-down and then identify the brass ones among them'.

The original experiment, on the other hand, defines various other ratios, as for instance fr $(B$ and $D)$/fr (B), whose value is $\frac{38}{94}$. In this case we *could* say 'let us throw the brass pins and identify those that are point-down'. Then we should have: given that the pins concerned are brass, the ratio for point-down pins is $\frac{38}{94}$. The reader may find this a simpler situation to grasp but it is more restricted than it needs to be.

5.3 Conditional probability.

We return now to the ratio $\dfrac{\text{fr }(B \text{ and } D)}{\text{fr }(D)}$ which we thought might give rise to probability-like quantities and note that it can be written as $\dfrac{\text{fr }(B \text{ and } D)/N}{\text{fr }(D)/N}$, which suggests that $\dfrac{\text{pr }(B \text{ and } D)}{\text{pr }(D)}$ is the quantity which might have the looked-for probability qualities; it suggests a probability associated with an ideal event B amid a restricted class of outcomes. Before we call it a probability we must verify that it satisfies the axioms for probabilities. The proofs are easy to construct by anyone who has studied carefully the correspondence between probabilities and relative frequencies.

For the general proofs we shall require pr $(D) \neq 0$ (so that further pr $(D) > 0$, by Axiom I). We will temporarily write $\dfrac{\text{pr }(B \text{ and } D)}{\text{pr }(D)}$ as $\text{pr}_D(B)$.

We have:

For Axiom I:

(a)   pr $(B$ and $D) \geqslant 0$, pr $(D) > 0 \Rightarrow \text{pr }_D (B) \geqslant 0$.

(b)   pr $(D) = \text{pr }(B \text{ and D}) + \text{pr }(B' \text{ and } D)$ by Axiom III

$\Rightarrow$ pr $(D) \geqslant \text{pr }(B \text{ and } D)$

$\Rightarrow \qquad 1 \geqslant \dfrac{\text{pr }(B \text{ and } D)}{\text{pr }(D)}$ since pr $(D) > 0$

$\Rightarrow \qquad 1 \geqslant \text{pr}_D (B).$

61

Combining (*a*) and (*b*) we have

$$0 \leqslant \text{pr}_D(B) \leqslant 1.$$

For Axiom II:

$$\text{pr}(D) = \text{pr}(B \text{ and } D) + \text{pr}(B' \text{ and } D) \quad \text{by Axiom III, as above,}$$

$$\Rightarrow \quad 1 = \frac{\text{pr}(B \text{ and } D)}{\text{pr}(D)} + \frac{\text{pr}(B' \text{ and } D)}{\text{pr}(D)} \quad \text{since pr}(D) \neq 0$$

$$\Rightarrow \quad 1 = \text{pr}_D(B) + \text{pr}_D(B').$$

For Axiom III:

Let $B = (G \text{ or } H)$, where G, H are exclusive, then $(G \text{ and } D), (H \text{ and } D)$ are exclusive also.

We construct pr $(B \text{ and } D)$ as follows:

$$(B \text{ and } D) = (G \text{ or } H) \text{ and } D$$

$$\Rightarrow \quad (B \text{ and } D) = (G \text{ and } D) \text{ or } (H \text{ and } D)$$

$$\Rightarrow \text{pr}(B \text{ and } D) = \text{pr}(G \text{ and } D) + \text{pr}(H \text{ and } D) \quad \text{by Axiom III}$$

$$\Rightarrow \quad \text{pr}_D(B) = \text{pr}_D(G) + \text{pr}_D(H) \quad \text{on division by pr}(D) \ (\neq 0).$$

Thus we may regard $\text{pr}_D(B)$ as a probability concerned with B since it satisfies the three axioms. The numerical value of $\text{pr}_D(B)$ will not necessarily be that of pr (B), but then we have never suggested that the axioms were only capable of being satisfied by one number for each event. It is not an event itself which is described by a probability but the relation of an event to a whole set of events arising from an experiment. Our original notation was designed to make this clear when we wrote prob$(A:E)$ for a probability. We now have the situation envisaged in Chapter 2, Section 5.1 namely that the conditions giving rise to the events are different; provided that the fundamental experiment is unchanged we still need not mention it, but henceforward we will use the standard notation pr $(B|D)$ for what we have written here as $\text{pr}_D(B)$.

We thus define

$$\text{pr}(B|D) = \frac{\text{pr}(B \text{ and } D)}{p(D)} \quad \text{whenever pr}(D) \neq 0.$$

The symbol pr $(B|D)$ is read as 'the conditional probability of B, given D' or even as 'the probability of B given D'.

5.4 Allotment of values: symmetry absent. Deductions from the definition are

$$\text{pr}(B \text{ and } D) = \text{pr}(D) . \text{pr}(B|D)$$

and

$$\text{pr}(D) = \text{pr}(B \text{ and } D) / \text{pr}(B|D),$$

but they do not allow us to calculate pr $(B \text{ and } D)$ without first knowing pr $(B|D)$, which is *defined* in terms of pr $(B \text{ and } D)$!

How can we know the numerical value of the conditional probability in order to substitute it on the right-hand side of those equations (of these last three) in which it appears on that side? Since a conditional probability is a probability in the full sense, there are many permissible allotments of values and the reader should be prepared for the statement that in general we do not know its value and that we often select, on the basis of experiments, values for all three probabilities simultaneously in a way that satisfies the relation between them. We can no more expect to have prior knowledge about how subsets of events are related than we can expect to have it about the events themselves. Thus in the drawing pin problem no one could tell, before the experiment, that the frequencies would be as they appear on the left of Table 3.1; and fr $(D$ and $C)$ is not deducible from a mere knowledge of fr (D) and fr (C).

Table 3.1. (Derived from Table 2.4)

Frequencies	An allotment of probabilities
$N = 433$	—
fr $(D) = 204$	pr $(D) = \frac{204}{433}$
fr $(C) = 213$	pr $(C) = \frac{213}{433}$
fr $(D$ and $C) =$ fr $(C$ and $D) = 104$	pr $(C$ and $D) = \frac{104}{433}$
$\dfrac{\text{fr } (D \text{ and } C)}{\text{fr } (C)} = \dfrac{104}{213}$	pr $(D\|C) = \dfrac{104/433}{213/433} = \dfrac{104}{213}$
$\dfrac{\text{fr } (C \text{ and } D)}{\text{fr } (D)} = \dfrac{104}{204}$	pr $(C\|D) = \dfrac{104/433}{204/433} = \dfrac{104}{204}$

An allotment of values to probabilities by using frequency ratios directly like this automatically ensures that they will be permissible values since the axioms themselves were based on the properties of frequency ratios.

5.5 Allotment of values: symmetry present. In all cases we are greatly helped by the use of diagrams, but in cases where appropriate symmetry is present we can actually allot numerical values from the diagrams (though the allotment still, effectively, involves, in the case of events A, B, a knowledge of pr $(A$ and $B)$).

We look again at the situation partly represented by Figure 3.11, and fully represented by Figure 3.21; we have sets $\mathscr{A}$ and $\mathscr{B}$ and will investigate the relationship of $\mathscr{A} \cap \mathscr{B}$ to $\mathscr{A}$.

Cut, or imagine cut, a piece of paper with a slit in it so that when it is placed over Figure 3.21 all possibility points will be obscured except those in $\mathscr{A}$. We illustrate the result in Figure 3.22. We then have a possibility space whose universal set is $\mathscr{A}$. This possibility space contains a subset $\mathscr{C}$ (which a moment's reflexion shows to be $\mathscr{A} \cap \mathscr{B}$) and the relationship of $\mathscr{A} \cap \mathscr{B}$ to $\mathscr{A}$ is seen to be merely that of a subset of a set to the set itself.

It is, in essence, the relationship of $\mathscr{B}$ to $\mathscr{E}$, as we can see formally by the following scheme:

$$(\mathscr{A} \cap \mathscr{B}) \text{ corresponds to } (\mathscr{A} \cap \mathscr{E})$$

and

$$\mathscr{B} \text{ corresponds to } \mathscr{E}.$$

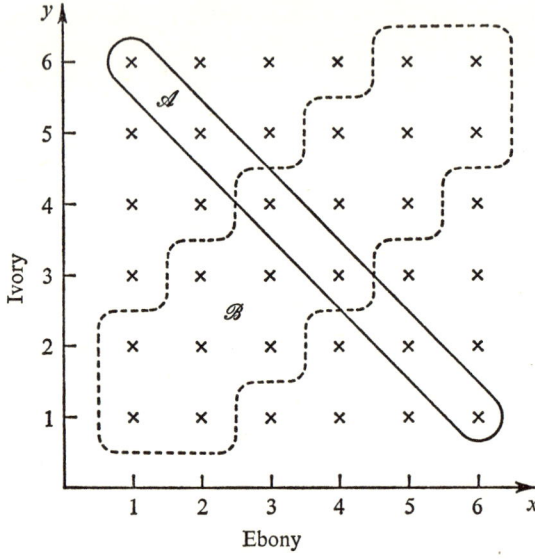

Fig. 3.21

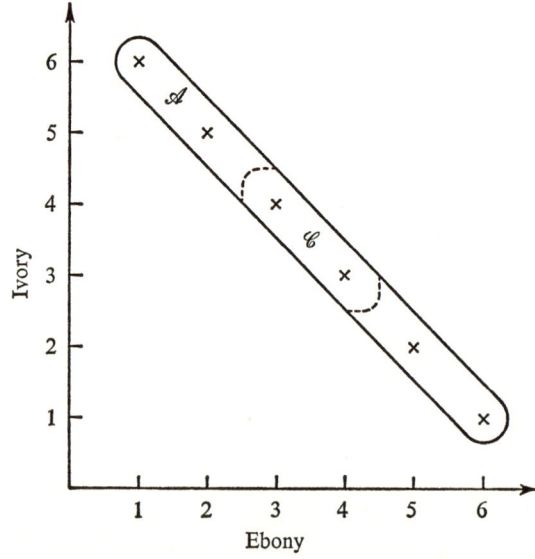

Fig. 3.22

The definition of pr $(B|A)$ is $\dfrac{\text{pr }(B \text{ and } A)}{\text{pr }(A)}$, which gives $\dfrac{n(\mathscr{B} \cap \mathscr{A})/n(\mathscr{E})}{n(\mathscr{A})/n(\mathscr{E})}$;

reducing to $\dfrac{n(\mathscr{B} \cap \mathscr{A})}{n(\mathscr{A})}$, in exactly the way which we would like.

5.6 Absence of order.

Note that despite the word 'given' in the phrase 'probability of B, given A' there is no necessity for an order in time, or causation, and we can perfectly well have pr $(A|B)$ and pr $(B|A)$ defined simultaneously by

$$\frac{\text{pr }(A \text{ and } B)}{\text{pr }(B)} \quad \text{and} \quad \frac{\text{pr }(B \text{ and } A)}{\text{pr }(A)},$$

and in case of symmetry evaluate them simultaneously by:

$$\frac{n(\mathscr{A} \cap \mathscr{B})}{n(\mathscr{B})} \quad \text{and} \quad \frac{n(\mathscr{B} \cap \mathscr{A})}{n(\mathscr{A})}.$$

5.7. Peculiarity of the notation.

Notice that not only is there no set written $\mathscr{B}|\mathscr{A}$ but there is no event $B|A$ either, so that there is nothing to add to the summary of compound events in Section 3.3.

Exercise 3 B

1–6. For each of the events A and B defined in Questions 1–6 of Exercise 3A, allot values to the following:

$$\text{pr }(B|A); \ \text{pr }(A|B); \ \text{pr }(B|A'); \ \text{pr }(A|B')$$
$$\text{pr }(B'|A); \ \text{pr }(A'|B); \ \text{pr }(B'|A'); \ \text{pr }(A'|B').$$

Among these probabilities are various pairs that sum to unity. These exemplify a single general result. State this result in terms of the conditional probabilities of general events A and B.

7. Construct a diagram to illustrate a possibility space for the following: A suit is drawn from a pack of cards (Spade, Diamond, Heart, Club) and a die is thrown (one, two, three, four, five, six). The event A occurs if the name of the suit and the name of the face have a letter in common. The event B occurs if the number of letters in the name of the suit is *any* one of the set of numbers denoted by the names of the faces.

(Use the following mapping:

$$\text{outcome} \to (x, y) \text{ such that}$$

suit is Spade	$\Rightarrow x = 2,$
suit is Diamond	$\Rightarrow x = 1,$
suit is Heart	$\Rightarrow x = 3,$
suit is Club	$\Rightarrow x = 4,$

and the faces are mapped onto the y-values in the obvious way.)

Allot values to the probabilities and proceed as in the first six questions of the exercise.

5.8   Two useful devices. It is simpler to use conditional probabilities
in the form $\quad\quad\quad \mathrm{pr}\,(A \text{ and } B) = \mathrm{pr}\,(A).\mathrm{pr}\,(B|A)$
than in the form $\quad\quad \mathrm{pr}\,(A \text{ and } B) = \mathrm{pr}\,(A|B).\mathrm{pr}\,(B),$
because in the first form we can think of the process as finding the 'fraction'
of the possibilities that give (A and B)—namely $\mathrm{pr}\,(A$ and B)—by first
considering the 'fraction' of the possibilities that give A—namely $\mathrm{pr}\,(A)$—
and then multiplying that fraction by the 'fraction' of the possibilities that,
giving A already, also give B—namely $\mathrm{pr}\,(B|A)$.

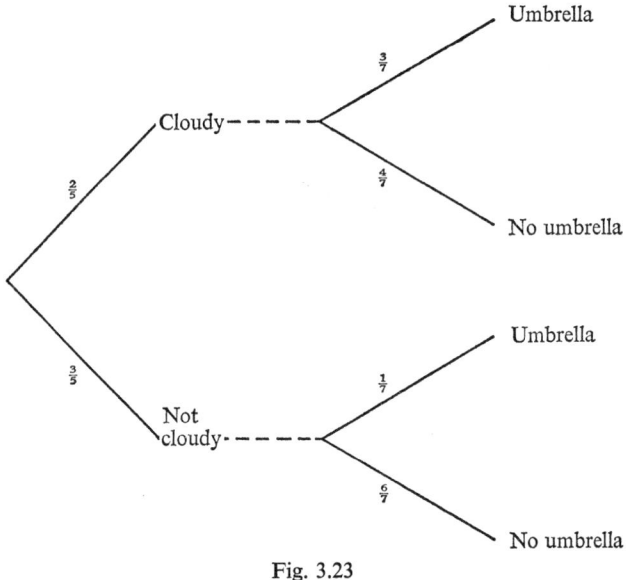

Fig. 3.23

For instance, let us consider the probability that on any particular day
it is cloudy and I shall be carrying an umbrella. Suppose that $\frac{2}{5}$ of *all* days
are cloudy and that on $\frac{3}{7}$ of cloudy days I carry an umbrella; then on $\frac{2}{5} \times \frac{3}{7}$
of *all* days it is cloudy and I am carrying an umbrella. [It might be that on
days when it is *not cloudy* I only carry an umbrella with probability $\frac{1}{7}$, say.
In that case we could say that my carrying of an umbrella and the cloudiness are not independent of one another, but that is an idea we shall
develop later.]

One way of manipulating the probabilities correctly is the tree-diagram,
which is almost self-explanatory. For instance in the situation about
cloudiness and umbrellas we could represent the data as in Figure 3.23.
Writing C for 'it is Cloudy' and U for 'I have my Umbrella', notice that
we have used in the diagram $\mathrm{pr}\,(U'|C) = 1 - \mathrm{pr}\,(U|C)$; this was verified by
the reader in Exercise 3 B and was proved in Section 5.3 in connexion with
Axiom II.

In order to calculate the probability that it-is-*not-cloudy*-and-I-have-*no-umbrella*, we follow the branches which lead from the extreme left through the phrases 'not cloudy', 'no umbrella'. The number beside a branch is the probability of the event at the right-hand end of the branch given the occurrence of *all* the results which precede it when the route is traced from the extreme left. Thus:

$$\text{pr ('not cloudy' and 'no umbrella')} = \tfrac{3}{5} \times \tfrac{6}{7} = \tfrac{18}{35}.$$

The difficulties are:

(i) that the diagram seems to introduce the unnecessary idea of time-sequence and thus of some events preceding others;

(ii) despite the actual lack of a time-sequence, reading the branches from right to left means nothing in terms of probabilities; that is to say $\tfrac{6}{7}$ is *not* pr $(C'|U')$.

For an alternative layout for the information we could consider a diagram with 35 points (we choose 35 to eliminate fractions in the subsequent numbers). In Figure 3.24, C refers to cloudiness and U refers to being with umbrella. We first arrange $\tfrac{2}{5} \times 35$ points to illustrate the probability of it being cloudy and $\tfrac{3}{5} \times 35$ to illustrate the probability of it being not-cloudy. Then within each of these sets we arrange the correct numbers of points to illustrate the probability of being with an umbrella. *Notice that the individual points do not represent possibilities here. The device is purely to make clear the relationship of the fractions involved in calculating probabilities* (see Section 4.1). *The possibility space consists of four points only.*

From this diagram we read off pr $(C'|U') = \tfrac{18}{26}$.

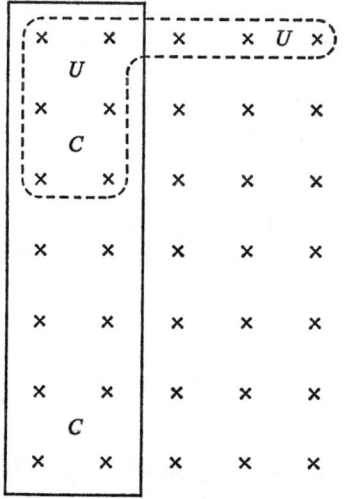

Fig. 3.24

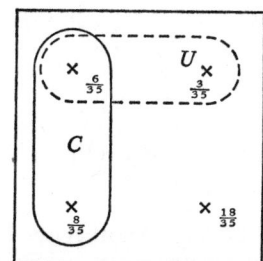

Fig. 3.25

There is no more information to be gleaned from Figure 3.24 than there is from the possibility space, with probabilities allotted which we have given as Figure 3.25.

The reader should contrast this with the situation in Section 3.2 where the second analysis of the third trial (the trial involving two dice) led to a possibility space of 36 points (Figure 3.11), while the first analysis led to one with only 21 (Figure 3.10). In that case both were possibility spaces, with each point referring to an event distinguishably different from that referred to by each other point. There was, naturally, more information to be gleaned from the diagram with more points, (that is the one from the finer analysis), than from the other diagram.

5.9 A reminder. We re-emphasize that the idea of conditional probability is not a new one, since 'probability' only acquires a meaning in the context of *some* initial conditions (though they may often be implicit) which restrict the possibilities to those of interest. The difference is merely that in writing a conditional probability we are imposing explicitly some further conditions, and so are *seen* to be restricting further the set of possibilities in which to seek those that define our event.

5.10 Worked examples.

Example 1. A bag contains 6 red and 4 green discs; 3 are randomly removed in turn without replacements. Find the probability that there is a green one among them.

First solution. The possible sequences of withdrawals are shown on the tree diagram (Figure 3.26). The probabilities are marked along the possible sequences. Notice the simple pattern of the denominators, and the more complicated pattern of the numerators.

For the route that is marked with arrows we have:

pr (first two discs are red and third is green)
= pr (first is red). pr (red then green|first is red)
= pr (first is red). pr (second is red|first is red). pr (third is green|
  first 2 are red)
= $\frac{6}{10} \times \frac{5}{9} \times \frac{4}{8}$.

The sequences ending at the asterisks represent *exclusive* events and the required probability is then obtained by summing the probabilities of these exclusive events.

Second solution. Another set of *exclusive* events leading to the event we seek is the following:

(i) the first is green;

(ii) the first is red and the second is green;

(iii) the first two are red and the third is green.

From these we get probabilities (i) $\frac{4}{10}$, (ii) $\frac{6}{10} \times \frac{4}{9}$, (iii) $\frac{6}{10} \times \frac{5}{9} \times \frac{4}{8}$ (the probabilities of the compound events being obtained as in the first solution). The required probability is then obtained by summing the probabilities of these exclusive events.

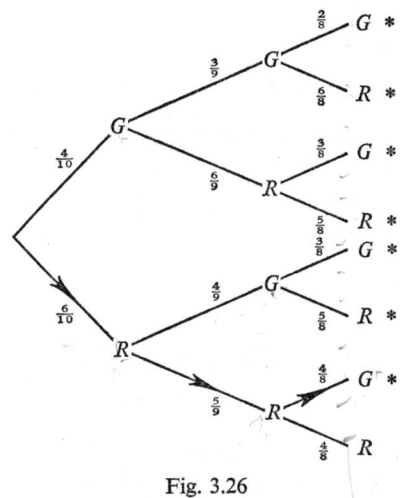

Fig. 3.26

Note on the relationship of the first solution to the second solution. The first of the compound events in the second solution can be obtained from the following sequences, whose occurrences are *exclusive*:

$$GRR, \quad GRG, \quad GGR, \quad GGG$$

We can then verify that:

$$\text{pr}\,(GRR) + \text{pr}\,(GRG) + \text{pr}\,(GGR) + \text{pr}\,(GGG)$$
$$= (\tfrac{4}{10} \times \tfrac{6}{9} \times \tfrac{5}{8}) + (\tfrac{4}{10} \times \tfrac{6}{9} \times \tfrac{3}{8}) + (\tfrac{4}{10} \times \tfrac{3}{9} \times \tfrac{6}{8}) + (\tfrac{4}{10} \times \tfrac{3}{9} \times \tfrac{2}{8})$$
$$= \tfrac{4}{10} \times \tfrac{6}{9} \times (\tfrac{5}{8} + \tfrac{3}{8}) \qquad + \tfrac{4}{10} \times \tfrac{3}{9} \times (\tfrac{6}{8} + \tfrac{2}{8})$$
$$= \tfrac{4}{10} \times \tfrac{6}{9} \qquad\qquad\quad + \tfrac{4}{10} \times \tfrac{3}{9}$$
$$= \tfrac{4}{10} \times (\tfrac{6}{9} + \tfrac{3}{9})$$
$$= \tfrac{4}{10}$$
$$= \text{pr (first is green)}.$$

Third solution. The occurrence of at least one green disc and the occurrence of no green discs are a pair of *exclusive, exhaustive* events.

The occurrence of no green discs arises only from the sequence *RRR*.

$$\text{pr}\,(RRR) = \tfrac{6}{10} \times \tfrac{5}{9} \times \tfrac{4}{8} = \tfrac{1}{6}.$$

So pr (there is a green disc) $= 1 - \tfrac{1}{6} = \tfrac{5}{6}$.

This exemplifies the very important technique of using the probabilities of exclusive, exhaustive events to solve a problem by an indirect approach. It will be much needed in Exercise 3 M.

Exercise 3 C

In the following Questions 1–6:

(i) Draw the tree-diagram with A preceding B, marking all the branch probabilities.

(ii) Calculate pr $(A$ and $B)$, pr $(A$ and $B')$, pr $(A'$ and $B)$, pr $(A'$ and $B')$.

(iii) Draw a possibility space to fit the data.

(iv) Draw a tree-diagram with B preceding A, marking all the branch probabilities.

1. pr $(A') = \frac{2}{5}$; pr $(B|A) = 0$; pr $(B'|A') = 0$.

2 R. pr $(A') = \frac{1}{5}$; pr $(B'|A') = 0$; pr $(B'|A) = \frac{2}{5}$.

3 R. pr $(A) = \frac{1}{6}$; pr $(B'|A) = 1$; pr $(B'|A') = \frac{3}{5}$.

4. pr $(A) = \frac{3}{10}$; pr $(B'|A) = \frac{2}{3}$; pr $(B|A') = \frac{3}{7}$.

5. pr $(A) = \frac{1}{4}$; pr $(B|A) = \frac{1}{3}$; pr $(B'|A') = \frac{3}{4}$.

6 R. pr $(A) = \frac{1}{3}$; pr $(B|A) = \frac{1}{3}$; pr $(B|A') = \frac{1}{3}$.

7. At a party I have a $\frac{3}{4}$ chance of having a cracker to pull but a $\frac{2}{3}$ chance of there not being a paper hat in the cracker even if I get a cracker at all. If there *is* a hat in the cracker I have only a 50 per cent chance of reaching it first. What is the probability that I shall get a hat? that I shan't get a hat?

8 F. There are delivered to a child, at its christening, by its fairy godmother:

2 packets each containing 2 silver coins (1 large and 1 small); 3 packets each containing 3 gold coins (2 large and 1 small). The instructions are that the packets are to be opened simultaneously and from the heap one coin chosen at random. A witch who is present persuades the child to choose an unopened packet at random and from that packet choose a coin at random.

Under each method of selection what are the probabilities that the child receives: (i) a large gold coin; (ii) a small silver coin?

*9. Three nicely balanced wheels are mounted on horizontal axes, and each has its circumference marked off into 32 equal parts.

Wheel A has 7 parts marked 'Score 6' and 25 parts marked 'Score 3'
Wheel B 16   ,,   ,,   'Score 5' ,, 16   ,,   ,,   'Score 2'
Wheel C 25   ,,   ,,   'Score 4' ,, 7   ,,   ,,   'Score 1'

You and your opponent are to choose a different wheel each. Your wheels will then be spun and the point which ends at the bottom will be clearly defined against a fixed pointer. Your score will be determined by the score for the sector in which the lowest point falls.

You have first choice of wheel. Which do you choose?

10 T. A pair of fair coins is first thrown to decide whether next to throw one die or two dice. A number of dice equal to the number of heads will be thrown (and *both* coins will be rethrown if a double tail occurs).

(i) What is the probability of a score of 5 given that two dice were thrown?

(ii) What is the probability that two dice were thrown given that a score of 5 was obtained?

(iii) Explain the interpretation of this second probability in terms of frequencies in repeated real trials.

(iv) What scores (if any) are more likely to have arisen if one die was thrown than if two were thrown?

(v) If the only information you had (apart from the rules of play) was that the score was 5, would you rather bet that one head had occurred or two?

6. INDEPENDENT EVENTS

6.1 Introduction. If 'every time we decide to take tea in the garden the neighbour lights a smelly bonfire' we may be pardoned for thinking that the neighbour is doing it deliberately; but let us analyse more closely what we mean by our complaint.

We mean that if the neighbour lights a bonfire about one week-end in four when we are *not* having tea in the garden then we expect him to be lighting one about one weekend in four when we *are* having tea in the garden—a higher rate would make us suspect a connexion between the events. [In this case of course a little thought shows that the events may well not be independent since both will tend to occur on dry days; though this does not of course imply that they *cause* each other in any way.]

In short: if the events are independent we expect the relative frequency of his lighting a bonfire given that we *are* in the garden to be no different from the relative frequency given that we *are not* in the garden. It would be unrealistic to look for exact equalities in actual frequencies, but we examine the drawing pin data of Table 2.4 in the light of what we have said.

The white and coloured pins were all plastic-topped and among them we expect their landing point-down to be independent of whether they are white or not.

Table 3.2 shows the results for the plastic pins; we have coded the coloured ones 'not-white' and the 'up' ones 'not down'.

Table 3.2

	D (down)	D' (up)	Total
W (white)	104	109	213
W' (coloured)	62	64	126
Total	166	173	339

We note that

$$\frac{\text{fr}(D \text{ and } W)}{\text{fr}(W)} = \frac{104}{213} = 0.488$$

$$\frac{\text{fr}(D \text{ and } W')}{\text{fr}(W')} = \frac{62}{126} = 0.492,$$

so the values are effectively equal as we would expect. When all pins are considered we would expect different rates between brass and plastic pins, as the heavy brass head is likely to affect the dynamics of the fall.

Table 3.3 shows the results for all pins: we have coded the plastic ones 'not-brass'.

Table 3.3

	D (down)	D' (up)	Total
B' (plastic)	166	173	339
B (brass)	38	56	94
Total	204	229	433

and we note

$$\frac{\text{fr}(D \text{ and } B)}{\text{fr}(B)} = \frac{38}{94} = 0.404$$

$$\frac{\text{fr}(D \text{ and } B')}{\text{fr}(B')} = \frac{166}{339} = 0.489.$$

The difference in the ratios seems meaningfully large.

6.2 Definition. The reader will notice that we have come near to making statements about conditional probabilities and indeed the cloudy/umbrella situation of Section 5.8 hinted at such an approach. There is, however, another way of looking at independence, and we describe it in terms of a different experiment.

C throws a 'fair' coin and D throws a 'fair' die; we expect that about half of C's results will be heads, say, and that the presence or absence of a head will not affect the die, so that the die will show a four, say, in one sixth of that subset of occasions when C throws a head. Thus we expect that about $\frac{1}{2} \times \frac{1}{6}$ of *all* throws will result in 'head by C' *and* 'four by D'. Calling the events A and B, respectively, we expect

$$\frac{\text{fr}(A \text{ and } B)}{N} \text{ to be nearly equal to } \frac{\text{fr}(A)}{N} \times \frac{\text{fr}(B)}{N}.$$

Notice that there is no inner arithmetical connexion which ensures *exact equality*—as there is for instance with

$$\text{fr}(A \text{ or } B) = \text{fr}(A) + \text{fr}(B) - \text{fr}(A \text{ and } B).$$

If we express this second approach in probability terms, it seems a simpler one in that it is symmetrical between A and B and does not involve conditional probabilities explicitly.

We will, therefore, frame our definition of independence, for events in an ideal experiment as follows:

Events A and B are (statistically) independent

$$\Leftrightarrow \text{pr}(A \text{ and } B) = \text{pr}(A) . \text{pr}(B).$$

6.3   Relation to introductory approach: A test for independence.

Our first job is to show that this definition is equivalent to the earlier approach, in which we were moving towards conditional probabilities. The equivalence means that the two results below can be used as *tests* for independence of two events in a model.

$$\text{pr } (A \text{ and } B) = \text{pr } (A).\text{pr } (B)$$

$\Leftrightarrow \text{pr } (A \text{ and } B) = [\text{pr } (A \text{ and } B) + \text{pr } (A \text{ and } B')].\text{pr } (B)$ by Axiom III

$\Leftrightarrow \text{pr } (A \text{ and } B) [1 - \text{pr } (B)] = \text{pr } (A \text{ and } B').\text{pr } (B)$

$\Leftrightarrow \text{pr } (A \text{ and } B).\text{pr } (B') = \text{pr } (A \text{ and } B').\text{pr } (B)$ by Axiom II

$\Leftrightarrow \dfrac{\text{pr } (A \text{ and } B)}{\text{pr } (B)} = \dfrac{\text{pr } (A \text{ and } B')}{\text{pr } (B')}$ provided pr (B) is neither 1 nor 0

$\Leftrightarrow \text{pr } (A|B) = \text{pr } (A|B').$

By the symmetry between A and B we can also deduce that for independent events A, B: $\text{pr } (B|A) = \text{pr } (B|A')$.

The usefulness of these results as tests is well brought out by a comparison between the brief calculations that followed Table 3.2 (calculations which effectively used these tests) and the calculations below. Those below go back to the definition to test whether falling point-down could be assumed to be independent of being white, among plastic-topped pins.

From Table 3.2 we have:

$$\frac{\text{fr } (W \text{ and } D)}{N} = \frac{104}{339} = 0.306,$$

$$\frac{\text{fr } (W)}{N} \times \frac{\text{fr } (D)}{N} = \frac{213}{339} \times \frac{166}{339} = 0.628 \times 0.489 = 0.307.$$

The approximate equality indicates that we may take a model in which the events are independent.

The working is considerably longer than before, and may make the reader wonder why, after all, we choose our definition without conditional probabilities.

6.4   Extension of definition.

The reason for choosing the definition as we did is that it is in a form which can be generalized to define the independence of many events in the following way:

Events $A_1, A_2, A_3, ..., A_n$ are independent $\Leftrightarrow$, for *any* subset $A_i, A_j, ..., A_k$ of the events, we have $\text{pr } (A_i \text{ and } A_j \text{ and } ... A_k) = \text{pr } (A_i).\text{pr } (A_j) ... .\text{pr } (A_k)$.

The structure of the various right-hand sides is what is important. They are products of factors, each of which is a function of *only one* event. Furthermore, without the reference to '*any* subset' unlooked-for difficulties occur when more than two events are concerned, as the reader can discover for himself in Exercise 3E, Questions 11, 12, 13.

Exercise 3D

1–7 R. In which of the seven questions of Exercise 3 B was the event A independent of the event B?

8 R. Which of the following pairs of events are independent?

    (i) Trial: Throwing one die.

         Events: A = a throw of six on a die,

                   B = a throw of three on the *same throw*.

   (ii) Trial: Throwing a *pair of dice* simultaneously.

         Events: A = a throw of three on the first die,

                   B = a throw of six on the second die.

 (iii) Trial: Throwing one die twice in succession.

         Events: A = a score of one on the first throw,

                   B = a total score of 7 on the two throws.

      Same trial:

         Events: C = a score of one on the first throw,

                   D = a total score of 8 on the two throws.

      Same trial:

         Events: E = a score of two on the first throw,

                   F = a total score of 8 on the two throws.

9 F. Use the results of the previous questions to guess a connexion between pr $(A|B)$ and pr (A) when the events A, B are independent, and prove it if you can.

10 T. It is important not to confuse *exclusive* events with *independent* events as people are apt to do if they think loosely of 'events that have nothing to do with one another'.

    What can you say about pr $(A$ and $B)$ when (i) A and B are independent; (ii) A and B are exclusive.

    A single die is thrown and the outcomes are recorded:

$$A = \text{a score of 3 or less is obtained.}$$

    Define events C and D such that A and C are independent and A and D are exclusive.

6.5 Another test for independence. One result which it is very useful to have as a test for the independence of two events is derived in the following way, and may well have been discovered by the reader in Exercise 3 D, Question 9.

    For any two events A, B:

$$\text{pr } (B \text{ and } A) = \text{pr } (B).\text{pr } (A|B), \text{ by definition of pr } (A|B),$$

but, for independent events A, B:

$$\text{pr } (B \text{ and } A) = \text{pr } (B).\text{pr } (A), \text{ by definition of independence,}$$

so A, B are independent events $\Leftrightarrow$ pr $(A|B)$ = pr (A).

We can obviously interchange A and B to get:

B, A are independent events $\Leftrightarrow$ pr $(B|A) = $ pr (B).

This is sometimes an easier test to apply than the one in Section 6.3, as the reader can convince himself by looking again at Exercise 3 D.

Either of the forms of the test sounds correct in ordinary usage; for instance the first says that we would expect that in a long run of trials in which independent events A, B occur the proportion, within those trials in which B occurs, of times that A occurs would be the same as the overall proportion of times that A occurs. This agrees with our notions about smelly bonfires, and generally seems a natural result from English usage; but of course it needed proving, since we have certain definitions to obey and we cannot allow our natural impulses unrestricted sway.

6.6 Relationship between the tests. In the case of possibility spaces with equiprobable outcomes the relationship of the test of Section 6.3 to the test of Section 6.5 can easily be illustrated on a diagram.

In the course of the illustration we use the result

$$\frac{a}{b} = \frac{c}{d} \Leftrightarrow \frac{a}{b} = \frac{a \pm c}{b \pm d},$$

which generalizes the well known arithmetical fact 'if I have twice as much money in my left hand as you do in yours and if I have twice as much in my right hand as you do in yours, then although I may have different amounts in my two hands, I have twice as much money all told as you do'; in other words it generalizes the following sorts of statement:

$$\frac{8}{4} = \frac{6}{3} \Rightarrow \frac{8}{4} = \frac{8+6}{4+3},$$

or, more elaborate,

$$\frac{5}{13} = \frac{15}{39} \Leftrightarrow \frac{5}{13} = \frac{5 \pm 15}{13 \pm 39}.$$

If the reader is not convinced of this result already, and even if he is, he should prove it by taking the algebraic statement, setting each of the fractions on the left equal to λ, and reducing the extreme right-hand expression to an expression in terms of b, d, λ. This proves the '$\Rightarrow$' implication. The '$\Leftarrow$' implication is in fact the same result rewritten, since

$$\frac{c}{d} = \frac{a-(a-c)}{b-(b-d)}.$$

The way in which we use this algebra is suggested by Figure 3.27 and the arithmetic below.

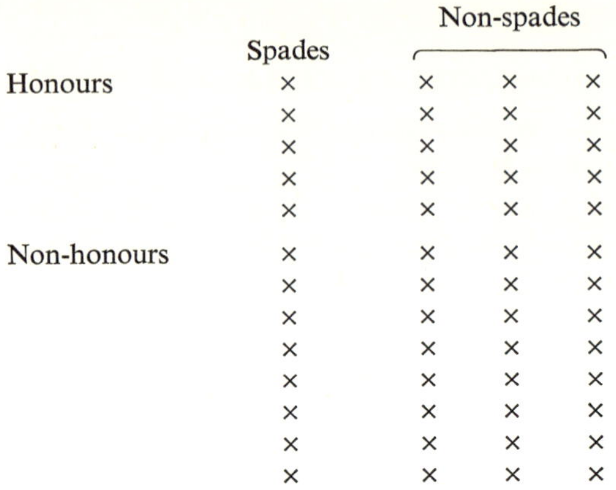

	Spades	Non-spades		
Honours	×	×	×	×
	×	×	×	×
	×	×	×	×
	×	×	×	×
	×	×	×	×
Non-honours	×	×	×	×
	×	×	×	×
	×	×	×	×
	×	×	×	×
	×	×	×	×
	×	×	×	×
	×	×	×	×
	×	×	×	×

Fig. 3.27

$$\frac{5}{13} = \frac{15}{39} \Leftrightarrow \frac{5}{13} = \frac{5+15}{13+39} = \frac{20}{52}.$$

We have just illustrated the fact that if the probability of drawing an honour given that we draw a Spade is equal to the probability of drawing an honour given that we do not draw a Spade, then each is simply the probability of drawing an honour.

The reader probably uses this fact instinctively, being more familiar with the fact that there are five honours in a suit than with the fact there are twenty honours in a pack.

In more general terms the illustration would run:

$$\text{pr}(A|B) = \text{pr}(A|B')$$

$$\Leftrightarrow \frac{n(\mathscr{A} \cap \mathscr{B})}{n(\mathscr{B})} = \frac{n(\mathscr{A} \cap \mathscr{B}')}{n(\mathscr{B}')} \text{ since the possibility-space is of equi-probable points}$$

$$\Leftrightarrow \frac{n(\mathscr{A} \cap \mathscr{B})}{n(\mathscr{B})} = \frac{n(\mathscr{A} \cap \mathscr{B}) + n(\mathscr{A} \cap \mathscr{B}')}{n(\mathscr{B}) + n(\mathscr{B}')}$$

$$\Leftrightarrow \frac{n(\mathscr{A} \cap \mathscr{B})}{n(\mathscr{B})} = \frac{n(\mathscr{A})}{n(\mathscr{E})}$$

$$\Leftrightarrow \text{pr}(A|B) = \text{pr}(A).$$

6.7   Consequence of the definition. When dealing with two general events A, B, without symmetry, from a real experiment, we cannot uniquely allot values to the probabilities in a model and so cannot *prove* from the definition of the experiment that in the model we will find that pr $(A$ and $B)$ satisfies pr $(A$ and $B) = $ pr (A).pr (B). If we believe that the events A, B ought to be taken as independent in the model, then we must allot, to the probabilities, values which satisfy pr $(A$ and $B) = $ pr (A).pr (B). We have

76

only two ways of telling whether we ought to take A, B as independent in our model:

(i) if the real events can be seen to be unrelated: for instance, the hooting of a car outside my window in a given period, and the ringing of my telephone in that period;

(ii) if the approximate equality of the relevant frequency ratios is sufficiently convincing.

In the case of those real events for which the model has symmetry, however, we can allot values to probabilities and can prove that various events in the model must be called statistically independent. This sometimes leads to some unexpected conclusions.

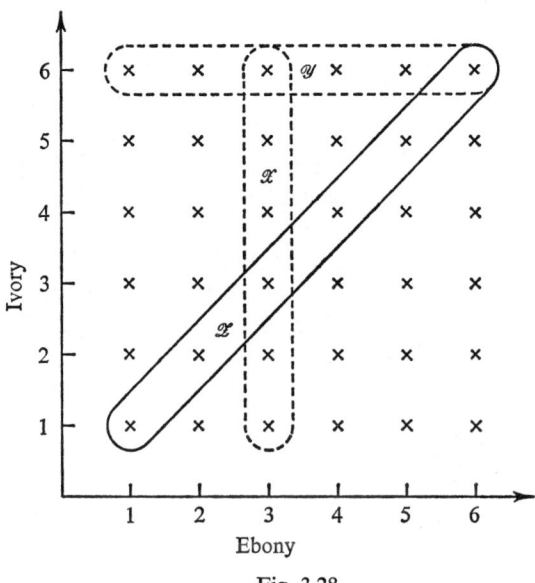

Fig. 3.28

It is clear that if

$\qquad X$ = occurrence of a three on the ebony die of Figure 3.28

and $\qquad Y$ = occurrence of a six on the ivory die of Figure 3.28

then X and Y are independent events in ordinary English usage, since there is no connexion, causal or otherwise, between them. The reader should verify, before proceeding, that X is independent of Y in the statistical sense indicated above.

Just as we had in another problem (Exercise 3B, Question 1) when dealing with conditional probabilities, so we have here a situation where the experiment can be subdivided into two trials, each of which will determine, quite separately from the other, the occurrence or not of one of the events. A player x could throw the ebony die and a player y the ivory one, and they

77

could compare notes afterwards. This is suggested by 'independent' in ordinary usage, but, as we shall see, is an unnecessarily restricted situation to which to confine the idea of statistical independence. The variety of statistically independent event is wider than might be guessed.

It is true that:

Events are independent in ordinary usage

$\Rightarrow$ Events are statistically independent,

but the converse is false, that is to say, statistically independent events may or may not be independent in ordinary usage.

For instance, consider:

Y as above

and $Z =$ occurrence of a double on the pair of dice.

(Z is also illustrated on Figure 3.28.)

Now suppose that a third player, called z, throws the ebony die and that player y again throws the ivory die, and that z hopes for a double (event Z) and y hopes for a six (event Y). Player z needs to know the result of y's throw before he can tell whether or not he (z) has been successful, and so it would *seem* that the success or failure of z is linked with the success or failure of y (in a way in which the success of failure of x was not) and that the events Z and Y are not independent. Whether or not this is a sound argument about the ordinary usage of the word 'independent' is one matter, but it is a false argument about the statistical usage of the word. We can see this by looking more closely: even if z does not see his die until he has heard y exclaim with pleasure or dismay at whether he (y) has succeeded in throwing a six, nevertheless z has no reason to be apprehensive or elated about his own chance of getting a double, whichever noise he hears from y. In symbols we have:

$$\text{pr}\,(Z\,|\,Y) = \frac{n(\mathscr{Z} \cap \mathscr{Y})}{n(\mathscr{Y})} = \frac{1}{6}$$

and $$\text{pr}\,(Z\,|\,Y') = \frac{n(\mathscr{Z} \cap \mathscr{Y}')}{n(\mathscr{Y}')} = \frac{5}{30} = \frac{1}{6}$$

so that by our first test we still have that Z is statistically independent of Y.

We thus see that two events defined *by the same trial* may have the relationship of independence. We do not need to be dealing with separate 'independent' (in the ordinary sense) trials. This situation follows from the fact that conditional probabilities can be defined for events arising from the same trial, as we have seen.

6.8   Independent trials. Naturally there are many occasions when we have the simple situation in which an experiment can be broken down into successive trials, sometimes the results of which do, and sometimes the results of which do not, affect the probabilities of the results of others.

Thus the experiment of finding the voting intentions of people taken effectively simultaneously and at random places over a large constituency can be broken down into several interviews with people $x, y, \ldots, z$; and we will have:

pr (y says 'Conservative'|x says 'Conservative')
= pr (y says 'Conservative'|x says 'Not-Conservative')
= pr (y says 'Conservative'),

and the events are statistically independent.

If a secret ballot is taken at a Conservative Club then again the above probabilities will be equal *to each other* (though *the experiment being different,* they will *not* be equal to what they were in the constituency-experiment). On the other hand if the ballot in the Club was not secret we might find:

pr (y says 'Conservative'|x says 'Conservative')
$\neq$ pr (y says 'Conservative'|x says 'Not-Conservative')

because y might now be emboldened to speak out when he heard x's views, or might feel that he had to close the ranks despite his doubts. In this last case we should not have statistically independent events (nor, for that matter, realistically 'independent' interviews). This is becoming an increasingly serious matter, in considering whether opinion polls perhaps *influence* the voting figures that they intend merely to *measure*.

Again, if a machine is running *steadily* in a factory and items $x, y, \ldots, z$, are examined at random intervals then

pr (y is defective|x is defective) = pr (y is defective|x is not defective)
                                       = pr (y is defective)

and we have statistically independent events.

Another example is provided by a rat running in a T-maze. The rat advances up a corridor and comes to a T-junction. If it goes left it goes to an empty cell, if it goes right it gets an electric shock. Then we shall *not* have independent trials, for:

pr (rat goes right at second attempt | rat goes left at first attempt)
$\neq$ pr (rat goes right at second attempt | rat goes right at first attempt).

When the trials are such that the probability of an event occurring in any one of them is not a function of the result of any other one then we say *the trials are independent*. This is a topic which we explore in Chapter 7.

It is much easier to think of trials that are not independent, or seem likely not to be independent, than to think of independent trials. Indeed, it might be thought that only artificial situations produce independent trials, so that the notion is not a very useful one. That is to misunderstand one of the roles of a mathematical model. If the results of a real experiment do not fit those predicted by a model based on independent trials then we have learnt something about the real experiment: for instance that voting figures 'snowball' and are in a sense affected by the very taking of an opinion-poll; or that the machine is producing defectives at varying rates so that the process is 'out of control'; or that the rats are learning about the maze.

These conclusions could only be convincingly established if we could compare the real results with what *would* have happened if the connexions were absent. A hypothesis, which we set up as a sort of Aunt Sally, that the connexions are absent and that any oddities are due to chance is called a *Null Hypothesis*. The later parts of the course will have much to say about such hypotheses.

7. CONTINGENCY TABLES

In Figure 3.18 we had a possibility space with an allotment of probabilities, that was suggested as suitable by an experiment and that could have been more conveniently displayed in the form shown in Table 3.4 (the bottom right-hand total would be 1·000, except for rounding errors).

Table 3.4

	Down	Up	Total
White	0·240	0·251	0·491
Coloured	0·143	0·148	0·291
Brass	0·087	0·129	0·216
Total	0·470	0·528	0·998

From this table we can read off such consistent allotments as:

$$\text{pr (pin is coloured}|\text{pin is down)} = \frac{0\cdot143}{0\cdot470}$$

or
$$\text{pr (pin is down}|\text{pin is coloured)} = \frac{0\cdot143}{0\cdot291}$$

or
$$\text{pr (pin is coloured)} = \frac{0\cdot291}{0\cdot998}.$$

The last of these, which is read off the totals round the margin, is sometimes called the *marginal probability* of the pin being coloured, to distinguish it from the two conditional probabilities of the pin being coloured. Figure 3.25 might have been conveniently expressed as in Table 3.5.

Table 3.5

	Cloudy	Uncloudy	Total
With umbrella	$\frac{6}{35}$	$\frac{3}{35}$	$\frac{9}{35}$
Without umbrella	$\frac{8}{35}$	$\frac{18}{35}$	$\frac{26}{35}$
Total	$\frac{14}{35}$	$\frac{21}{35}$	$\frac{35}{35}$

Then　　　　　　　　pr (without umbrella | cloudy) $= \frac{8}{14}$.

Figure 3.27 could be given as in Table 3.6. In this, numbers of equally-probable possibilities have been entered, but we still read off

$$\text{pr (not-honour | Spade)} = \tfrac{8}{13}$$

in a similar way.

Table 3.6

	Spade	Not-Spade	Total
Honour	5	15	20
Not-honour	8	24	32
Total	13	39	52

When such a table is being used to record directly information about the results of an experiment it is called a *contingency table*. Tables 3.2 and 3.3 were contingency tables.

The layout is useful whether data, or numbers of *equiprobable* outcomes in various subsets in a possibility space, or probabilities are entered.

Exercise 3 E

1 R. In a certain group of 1000 people 20 per cent are men and 80 per cent are women. Of the men 4 per cent are colour-blind and of the women 1 per cent are colour-blind. The events 'a Male is chosen', 'a Female is chosen', 'a Normally colour-sighted person is chosen', 'a Colour-blind person is chosen', are represented by M, F, N, C.

One person is chosen at random *from this group*. Find

$$\text{pr } (M \text{ and } N), \quad \text{pr } (F \text{ and } C), \quad \text{pr } (M|N), \quad \text{pr } (F|C).$$

Which pairs of events M, F, N, C are independent?

2 R. Twenty-four people went on a picnic; eight got sunburned in the morning, six got stung by mosquitoes in the afternoon. What was the maximum number of people who could have got home unscathed? If eleven got home unscathed, what were the probabilities that

(i) a random sunburned person interviewed at mid-day would be stung by nightfall?

(ii) that a random stung person interviewed at nightfall would turn out to have been sunburned?

(iii) that a random person interviewed at nightfall would turn out to have been both burned and stung?

3 R. Another twenty-four people rashly go on a picnic and again eight get sunburned and six get stung. This time 12 get home unscathed. Show that getting sunburned and getting stung were independent events.

4 F. 343 patients are habitual sufferers from catarrh, and one autumn 96 of them are given a new treatment designed to prevent catarrh in the winter. Of these 96 people seven do not develop catarrh and of the untreated people 229 do develop catarrh. Show that these figures are consistent with the theory that the treatment has no effect.

5. I am going on holiday on 6th August and have left the arrangements rather late. There are possibilities that my train ticket, my passport, my traveller's cheques will fail to arrive on time. The probabilities allotted to these events are based on a large survey of delay times and are $\frac{1}{3}, \frac{1}{4}, \frac{3}{5}$ respectively.

What is the probability that on 6th August
    (i) I shall be without all three things?
    (ii) I shall be without my traveller's cheques but otherwise complete?
    (iii) I shall be without my traveller's cheques whatever else I may or may not have?

How am I most likely to be equipped on 6th August?

6 F. In throwing a die repeatedly what is the chance
    (i) That the first appearance of a 'one' is on the second throw? Is on the third throw?
    (ii) That no 'one' appears in the first three throws?
    (iii) That the first 'one' is on the fifth throw or later?

7 T. (a) Evaluate
    (i) The probability of throwing six heads in a row with an unbiased coin.
    (ii) The probability of throwing seven heads in a row with the coin.
    (iii) The probability of throwing seven heads in a row, given that six of them have already occurred.

(b) What does the man in the street's 'Law of Averages' say about the chance of throwing that seventh head after six have been thrown?

(c) If in fact the coin eventually came down heads 1000 times in succession what would you begin to suspect had happened and what would you estimate to be the probability of 1001 heads in succession?

8 T. Prove from the definitions that
    (i) $\operatorname{pr}(A \text{ and } B|A) = \operatorname{pr}(B|A)$;
    [Notice that there is no ambiguity on the left-side, since $B|A$ is not a symbol for an event.]
    (ii) A and B are independent $\Leftrightarrow$ A and B' are independent;
    (iii) $\operatorname{pr}(A|A) = 1$.

9 T. From the symmetry of the definition, we have that

'A is statistically independent of B

        $\Leftrightarrow B$ is statistically independent of A',

but prove that the relation '...is statistically independent of...' is neither reflexive nor transitive.

10 T. In the following contingency table:

	B	B'	Total
A	x_{11}	x_{12}	y_{10}
A'	x_{21}	x_{22}	y_{20}
Total	y_{01}	y_{02}	N

prove that if

$$\frac{x_{ij}}{N} = \left(\frac{y_{i0}}{N}\right) \cdot \left(\frac{y_{0j}}{N}\right)$$

is true for some ordered pair of values of i, j from the set $\{1, 2\}$ then it is true for the other three pairs and state the results in terms of independence of events.

*11F. A regular tetrahedral die has its faces marked (a), (b), (c) and (abc). When a face marked with a letter a is in contact with the plane on which the tetrahedral die is thrown we say the event A has occurred, and similarly for the other letters. Which of the following statements are true?
  (i) A and B are independent events;
  (ii) A and C are independent events;
  (iii) A and $(B$ and $C)$ are independent events;
  (iv) pr $(B$ and $C) = $ pr (B).pr (C);
  (v) pr $(A$ and B and $C) = $ pr (A).pr (B).pr (C).
(See Exercise 3 M; Question 2.)

*12F. Answer the questions of Question 11 in the following circumstances:

    Trial: Throwing a 'copper' coin and a 'silver' one and recording the results on each.
    Event A: A head occurs on the copper coin.
    Event B: A head occurs on the silver coin.
    Event C: The coins match (both are heads or both are tails).

*13F. Answer the questions of Question 11 in the following circumstances: X, Y, Z are the events defined in Section 6.7 and we take $A \equiv X, B \equiv Y, C \equiv Z$.

8. SUMMARY

In Section 4·3 we remarked that when pr $(A$ and $B)$ had been dealt with then all possible combinations of 2 events had been covered. That is the situation now.

  In general:

$$\text{pr }(A \text{ and } B) = \text{pr }(A).\text{pr }(B|A) = \text{pr }(B).\text{pr }(A|B)$$

by the definition of conditional probability.

  For independent events we have, by the definition of independence,

$$\text{pr }(A \text{ and } B) = \text{pr }(A).\text{pr }(B).$$

Notice that *in both cases* pr $(A$ and $B)$ is obtained by the multiplication of appropriate probabilities, so that the second is a special case of the first, the appropriate probability taking a simple form and the special name of independence then being given.

  Remember, too, that events that are independent in ordinary usage are statistically independent.

Exercise 3 F

1–6 R. For each of Questions 1–6 of Exercise 3 C find pr $(A$ or $B)$, pr $(A$ or $B')$, pr $(A'$ or $B')$ and state which of the following are true.
  (i) A and B are exclusive;
  (ii) A and B are exhaustive;
  (iii) A and B are independent.

7. From the results of Questions 4–6, or similar results, state whether it is possible for a pair of events to be:
  (*a*) exclusive and independent;
  (*b*) exhaustive and independent;
  (*c*) exclusive and exhaustive.
In each case either give an example to show that the double property is possible, or prove the double property to be impossible.

8 F. Prove that
$$\mathscr{B} = (\mathscr{B} \cap \mathscr{A}) \cup (\mathscr{B} \cap \mathscr{A}')$$
but that $(\mathscr{B} \cap \mathscr{A}) \cap (\mathscr{B} \cap \mathscr{A}')$ has no members.
  Hence prove that
$$\mathrm{pr}\,(B) = \mathrm{pr}\,(A).\mathrm{pr}\,(B|A) + \mathrm{pr}\,(A').\mathrm{pr}\,(B|A').$$

9. Prove that $$1 - (1 - a)(1 - b) = a + b - ab.$$

If A and B are independent events rewrite pr $(A) + $ pr $(B) - $ pr $(A).$ pr (B) in two ways and hence prove that
$$\mathrm{pr}\,(A \text{ or } B) = 1 - \mathrm{pr}(A' \text{ and } B').$$

10. By considering $(\mathscr{A} \cup \mathscr{B})$ and $(\mathscr{A}' \cap \mathscr{B}')$ prove that
$$\mathrm{pr}\,(A \text{ or } B) = 1 - \mathrm{pr}\,(A' \text{ and } B')$$
for any two events A, B, independent or not, and hence show that:
$$A \text{ and } B \text{ are exhaustive} \Leftrightarrow A' \text{ and } B' \text{ are exclusive.}$$
Show also that the words 'exhaustive' and 'exclusive' can be interchanged in this statement.
  Draw up typical contingency tables in which a pair of events is (i) exclusive; (ii) exhaustive; (iii) exclusive and exhaustive.

11 R. Complete the following table.

Event A and B are	pr $(A$ and $B) =$	pr $(A$ or $B) =$
General	pr $(A).$pr $(B|A)$	pr $(A) + $pr $(B) - $pr $(A).$pr $(B|A)$
Independent		
Exclusive	0	
Exhaustive		1

★12. On average Lady Bracknell comes to tea once a fortnight and Ernest comes once a week. When Lady Bracknell comes to tea the probability that there will be cucumber sandwiches is $\frac{6}{7}$.
  (i) What is the probability that one or other or both of them comes to tea if their engagement diaries are independent?

(ii) Establish inequalities for the probability that one or other or both of them comes to tea if nothing is known about the relationship between their diaries.

(iii) If Lady **B.** and Ernest operate independently and if the presences of Ernest and cucumber sandwiches are independent and if nevertheless Ernest gets cucumber sandwiches thrice as often as he meets Lady B., find the probability that there will be cucumber sandwiches on a random day. Show that the probability that there will be cucumber sandwiches when Lady B. is absent is $\frac{15}{91}$, and hence verify part of Exercise 3E, Question 9.

13F. The probability that any particular day in July is fine at Puddling Regis is $\frac{3}{4}$, and does not depend on the weather of the previous day.

Draw a tree-diagram and find the number of fine days that is most probable to occur in (i) the first 2 days of July; (ii) the next 2 days of July; (iii) the first 4 days of July; (iv) the first 3 days of July.

14. The probability that the 1st of July is fine at Llandrwnch is $\frac{3}{4}$, but

$$\text{pr (day } (r+1) \text{ is fine} \mid \text{day } r \text{ is wet}) = \tfrac{1}{2} \times \text{pr (day } r \text{ is fine)}$$

and $\text{pr (day } (r+1) \text{ is wet} \mid \text{day } r \text{ is fine}) = \tfrac{1}{2} \times \text{pr (day } r \text{ is wet)}.$

What are the probabilities of the occurrences of various numbers of fine days among the 1st, 2nd, 3rd of July at Llandrwnch?

9. MISCELLANEOUS

9.1   Alternative notations. Sometimes the event $(A$ and $B)$ is written AB to make the rule for independent events look simple. The rule then becomes:
$$\text{pr } (AB) = \text{pr } (A).\text{pr } (B).$$

$(A$ or $B)$ is then written $(A+B)$ and the rule for exclusive events becomes:
$$\text{pr } (A+B) = \text{pr } (A)+\text{pr } (B).$$

We shall not use this notation here but the reader may meet it elsewhere. He should then beware that although he may read '$2+3 = 5$' as 'two and three make five' he must read '$A+B$' as 'A or B' and not as 'A and B'. The addition sign is suggestive of adding to the possible number of outcomes that define the event, and not to the number of conditions to be fulfilled for the compound event to occur.

Another notation that is sometimes used involves writing $(A$ or $B)$ as $(A \vee B)$. The '$\vee$' was originally the initial letter of Latin 'vel' for 'or'. The event $(A$ and $B)$ is then written $(A \wedge B)$, and a pretty parallelism occurs with the notation for the corresponding sets of outcomes in the possibility space: $(\mathscr{A} \cup \mathscr{B})$, $(\mathscr{A} \cap \mathscr{B})$.

Of course if the events are defined as the points of the possibility space it is suitable to write $(A$ and $B)$ as $(A \cap B)$, and to write $(A$ or $B)$ as $(A \cup B)$.

9.2   More than two events: Conditional probabilities. Before leaving the reader to worked examples and exercises it is perhaps worth making explicit the way in which it is useful to compound the events in order to deal with, say, *four events*. The generalization to more and the simplification to three will be obvious. A case of three events occurred in Section 5.9, but in practice the reader will find that taking (the next event) and (all-the-events-so-far) as a pair will be adequate for almost all purposes.

The most useful form of the probability expression for the event (*A* and *B* and *C* and *D*) runs:

$$\text{pr } (A \text{ and } B \text{ and } C \text{ and } D)$$
$$= \text{pr } (A).\text{pr } (B|A).\text{pr } (C|(A \text{ and } B)).\text{pr } (D|(A \text{ and } B \text{ and } C)).$$

The alternative is to write:

$$\text{pr } (A \text{ and } B \text{ and } C \text{ and } D)$$
$$= \text{pr } (A|(B \text{ and } C \text{ and } D)).\text{pr } (B|(C \text{ and } D)).\text{pr } (C|D).\text{pr } (D).$$

The first form leads directly to tree diagrams, as Figure 3.29. The second requires the diagrams to be in form of Figure 3.30, and then to be read: *each branch* from left to right, but *successive branches* from right to left. The risks of confusion are obvious.

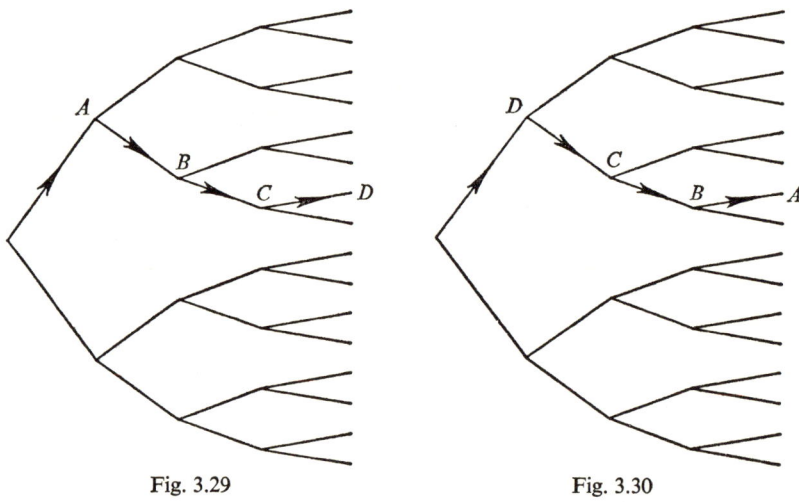

Fig. 3.29   Fig. 3.30

10. SOME WORKED EXAMPLES

Example 2. What is the probability that a person throwing four dice, each marked Ace, King, Queen, Jack, Ten, Nine will have (i) two Aces (and no other repeats); (ii) two of a kind (and no other repeats); (iii) two Aces and two Tens?

Solution. (i) If the Aces occur on the first two dice looked at under some predetermined scheme that has no reference to the *result* of the throw (for

86

instance: the dice may be of different colours, or size or, as a last resort, in different positions on the table) then the probability of such an event is the probability of the occurrence of the sequence

Ace,   Ace,   Non-Ace,   different Non-Ace.

By the method of Example 1 in Section 5.10 this probability is

$$\frac{1}{6} \times \frac{1}{6} \times \frac{5}{6} \times \frac{4}{6} \text{ that is } \frac{20}{6^4}.$$

The Aces, however, might well have been in any of the positions indicated in the table (where X indicates an Ace and N indicates a Non-Ace).

XXNN	NXXN
XNXN	NXNX
XNNX	NNXX

These six arrangements are equally likely since the method of looking at the dice was chosen before the throw and had no connexion with the results, so the required probability is $6 \times \dfrac{20}{6^4}$, that is $\dfrac{5}{54}$.

(ii) If we do not specify that X shall represent Ace then we have any one of six possible choices for it namely, Ace, King, ..., Nine. These are equally likely so the required probability is $6 \times \dfrac{5}{54}$, that is $\dfrac{5}{9}$.

(iii) To return to the scheme for (i), we see that the probability of the occurrence of the sequence $XXNN$ where X stands for Ace, N stands for Ten is

$$\frac{1}{6} \times \frac{1}{6} \times \frac{1}{6} \times \frac{1}{6}.$$

The required probability is thus $6 \times \dfrac{1}{6^4}$, that is $\dfrac{1}{216}$.

(It would be wrong to allow the case X stands for Ten, N stands for Ace as well (which would appear to double the probability) since the resulting sequence Ten, Ten, Ace, Ace, has already been counted (at the bottom of the list) under the first allocation.)

Example 3. In the game mentioned in Section 1 we had: 'A and B take turns at throwing dice...B's turn consists of throwing two dice...'. In the first twelve turns recorded in Section 1 B scored a four and a five four times.

(i) What is the probability that B would score a four and a five sometime among his first twelve turns?

(ii) How many turns should B be granted to give him a $\frac{2}{3}$ chance of scoring a four and a five?

(iii) Find pr (B scores a four and a five exactly once in his first twelve turns).

(iv) Find pr (B scores a four and a five four times or more in his first twelve turns).

Solution. We shall refer to the event 'B scores a four and a five' as 'B succeeds'; thus
$$\text{pr}\,(B \text{ succeeds}) = \tfrac{1}{18}.$$

(i) Consider

    (a) B succeeds at least once in the first twelve throws;

    (b) B fails twelve times consecutively.

These are *exclusive*, *exhaustive* events; and, further, they arise from independent trials so:

$$\text{pr}\,(B \text{ fails twelve times}) = \frac{17}{18} \times \frac{17}{18} \times \ldots \times \frac{17}{18}$$

$$= \left(\frac{17}{18}\right)^{12}.$$

This gives pr (B succeeds sometime in the first twelve throws) $= 1 - \left(\dfrac{17}{18}\right)^{12}$.

$$\log\left(\frac{17}{18}\right) = 1\cdot2304 - 1\cdot2553$$
$$= -0\cdot0249,$$
$$\log\left(\frac{17}{18}\right)^{12} = 12 \times -0\cdot0249$$
$$= -0\cdot2988$$
$$= -1 + 0\cdot7012,$$
$$\left(\frac{17}{18}\right)^{12} = 0\cdot5025,$$
$$1 - \left(\frac{17}{18}\right)^{12} = 0\cdot4975.$$

Thus pr (B succeeds sometime in the first twelve turns) is nearly $\tfrac{1}{2}$.

(ii) We wish to find n so that $1 - \left(\dfrac{17}{18}\right)^{n} = \dfrac{2}{3}$.

We solve
$$\left(\frac{17}{18}\right)^{n} = \frac{1}{3},$$

$$n \log\left(\frac{17}{18}\right) = \log\frac{1}{3},$$

$$n \times (-0\cdot0249) = -0\cdot4771,$$

$$n = \frac{0\cdot4771}{0\cdot0249} \simeq 19\cdot1.$$

With $n = 19$, B has a slightly less than $\dfrac{2}{3}$ chance;

With $n = 20$, we find $\left(\dfrac{17}{18}\right)^{20} \simeq \dfrac{1}{3} \times \dfrac{17}{18} = \dfrac{17}{54}$, B's chance is roughly $\dfrac{37}{54}$.

(iii) To succeed on a previously named throw B must succeed once and fail eleven times. The probability of this is $\dfrac{1}{18}\left(\dfrac{17}{18}\right)^{11}$, whichever throw is named. But we seek not one success on a *named* throw, but one success *at all*. The occurrences of two sequences with their successes falling on different throws are exclusive events. There are twelve such sequences, one for each possible position in which the success might occur, and they exhaust the event we seek, so

$$\text{pr }(B \text{ succeeds exactly once in 12 throws}) = 12 \times \frac{1}{18}\left(\frac{17}{18}\right)^{11}$$

$$= \frac{12}{17}\left(\frac{17}{18}\right)^{12}$$

$$= \frac{12}{17} \times 0\cdot5025$$

$$= 0\cdot354.$$

(iv) To solve this part of the problem we need to be able to compute the number of sequences that include 3 successes; 2 successes; 1 success; 0 successes. Two typical sequences that include 3 successes are:

$$FFFSFSFSFFFF \qquad FFFSFSFFFSFF.$$

(This calculation is left to Exercise 7D, Question 8 in a chapter on Sequences of Events. We merely quote here that there are 220; 66; 12; 1 sequences with 3; 2; 1; 0 successes.) We would again use the technique of considering a set of exclusive exhaustive events to deal with the event we require.

Example 4. A game is played between two players throwing a die alternately. Ann plays first, and wins if she throws a one or a two before Bob throws a one or a two or a three. Bob wins if he throws a one or a two or a three before Ann throws a one or a two. The game stops as soon as either player wins. If neither player wins after three throws each the game is drawn. What are the chances of each player winning, and what are the chances of a draw?

Solution. It will not alter the result of the game if we allow each player three throws (alternately) and then adjudicate on the game afterwards; though this might mean that some later throws were a waste of time.

Thus the sequences

$$3, \quad 5, \quad 1, \quad 2, \quad 1, \quad 3$$
$$3, \quad 5, \quad 3, \quad 2, \quad 1, \quad 3$$
$$3, \quad 5, \quad 3, \quad 4, \quad 6, \quad 6$$

give respectively   Ann wins on her second throw.

Bob wins on his second throw.

The game is drawn.

The object of this manoeuvre is to see that the probabilities of success or failure at each throw that occurs at all are *independent* of the results of previous trials, despite the fact that in the game as originally defined Bob would not have a second throw at all in the sequence starting 3, 5, 1.

We write   'Ann succeeds on a throw' $= A$, pr $(A) = \frac{1}{3}$;

      'Ann fails on a throw' $= X$, pr $(X) = \frac{2}{3}$;

      'Bob succeeds on a throw' $= B$, pr $(B) = \frac{1}{2}$;

      'Bob fails on a throw' $= Y$, pr $(Y) = \frac{1}{2}$.

(*a*) Ann can win in any one of three ways: (i) on her first throw; (ii) on her second throw; (iii) on her third throw. These ways are *exclusive* since a win stops a game.

(i) Ann wins on her first throw is a simple event;

$$\text{pr (Ann wins on her first throw)} = \text{pr }(A) = \tfrac{1}{3}.$$

(ii) Ann wins on her second throw is the compound event consisting of the sequence X, Y, A.

These events are *independent* thus

$$\text{pr (Ann wins on her second throw)} = (\tfrac{2}{3} \times \tfrac{1}{2}) \times \tfrac{1}{3};$$

(iii) Similarly

$$\text{pr (Ann wins on her third throw)} = (\tfrac{2}{3} \times \tfrac{1}{2})^2 \times \tfrac{1}{3}.$$

We have

$$\text{pr (Ann wins)} = \tfrac{1}{3} + (\tfrac{2}{3} \times \tfrac{1}{2}) \times \tfrac{1}{3} + (\tfrac{2}{3} \times \tfrac{1}{2})^2 \times \tfrac{1}{3}$$

$$= \tfrac{1}{3}[1 + k + k^2], \text{ where } k = \tfrac{2}{3} \times \tfrac{1}{2} = \tfrac{1}{3},$$

$$= \frac{1}{3} \cdot \left[\frac{1 - k^3}{1 - k}\right]$$

$$= \frac{1}{3} \cdot \left[\frac{1 - \frac{1}{27}}{1 - \frac{1}{3}}\right]$$

$$= \frac{1}{3} \cdot \frac{26}{27} \cdot \frac{3}{2}$$

$$= \frac{13}{27}.$$

We have chosen this rather long-winded way of setting out the calculations to enable the reader to make the generalization to the case where the game does not stop for a draw until each player has had n throws.

(*b*) Bob wins on his first throw requires the sequence X, B.

Bob wins on his second throw requires the sequence X, Y, X, B.

Bob wins on his third throw requires the sequence X, Y, X, Y, X, B.
Thus pr (Bob wins) $= (\frac{2}{3} \times \frac{1}{2})[1+k+k^2]$, where, again, $k = \frac{1}{3}$

$$= \tfrac{13}{27}, \text{ as for Ann.}$$

(*c*) The events (Ann wins), (Bob wins), (the game is drawn) are exclusive and exhaustive thus:

$$\text{pr (Ann wins)} + \text{pr (Bob wins)} + \text{pr (game is drawn)} = 1$$

and $$\text{pr (game is drawn)} = 1 - 2 \times \tfrac{13}{27}$$

$$= \tfrac{1}{27}.$$

Exercise 3M

Miscellaneous Exercise

1R. Given that the probabilities of the events P, Q, R are p, q, r respectively and that the events are completely independent of one another (that is the occurrence of any combination of two of them does not affect the occurrence of the third (see Question 2)), find the probabilities that (i) P, Q occur and R does not; (ii) Q occurs and P, R do not; (iii) at least one of them occurs; (iv) exactly two of them occur.

2T. If pr $(A$ and $B) =$ pr (A). pr (B) then A and B are statistically independent according to the definition of Section 6.2.

(i) If we have two or more events A_i and if any pair of them are independent we say that the events A_i are *pairwise independent*. Show that the pairwise independence of A_i $(i=1, 2, 3)$ is not a sufficient condition for

$$\text{pr } (A_1 \text{ and } A_2 \text{ and } A_3) = \text{pr } (A_1).\text{pr } (A_2).\text{pr } (A_3).$$

(ii) Show that if

(*a*) A_i $(i=1, 2, 3)$ are pairwise independent,

and (*b*) pr $(A_1$ and A_2 and $A_3) =$ pr (A_1). pr (A_2). pr (A_3),

then pr $(A_1) =$ pr $(A_1 \mid B_{23})$, where B_{23} is any of $(A_2$ and $A_3)$, $(A_2$ and $A_3')$, $(A_2'$ and $A_3)$, $(A_2'$ and $A_3')$.

[By symmetry we can then interchange the suffices and we say: the events A_i $(i=1, 2, 3)$ are *completely independent*.]

3. The fairy godmother delivers to her next godchild (see Exercise 3C, Question 8): s packets each containing x_1 large and y_1 small silver coins; g packets each containing x_2 large and y_2 small gold coins. Her instructions and the witch's are the same as before but she has arranged the packets so that the probability of getting a large gold coin is unaffected by whether her instructions or the witch's are obeyed. Find any condition on the values of s, g, x_1, x_2, y_1, y_2. [See Question 19, on genetics.]

4. If a letter is chosen at random from a 19th Century English text then the probabilities that it is T, H, or A are respectively 0·090, 0·059, 0·078.

If a digram (that is, an ordered pair of letters without a word-ending between them) is chosen at random the probabilities that it is TH, HA, or AT, are respectively 0·034, 0·012, 0·013. If a trigram (that is an ordered triple of letters without a word-ending between them) is chosen at random the probabilities that it is THA, or HAT are respectively 0·004, 0·003.

(i) If T, H occurred independently of one another find the probability that an ordered letter pair would be TH.

(ii) You are reading a text letter by letter and have just observed a letter T not at the end of a word, what is the probability that the next letter will be H? (Assume that the probability of obtaining a letter is independent of whether or not it is the last letter of a word, although this is not in fact so.)

(iii) Suppose TH occurs not at the end of a word, what is the probability that the next letter will be A? What would be the probability if only H had been observed?

(iv) A game is played as follows between two players. Each has a suitable text chosen at random. Player 1 looks for the first occurrence on a certain page of letter H. Player 2 looks for the first occurrence on a certain page of letter A. Player 1 wins if his H is followed by an A. Player 2 wins if his A is followed by a T. If neither or both of these events occur the result is a draw. Suggest suitable simple stakes to make the game fair.

[The probability data are from P. Valerio: *De la Cryptographie*, Paris, 1893, pp. 202, 204, quoted in W. W. Rouse Ball: *Mathematical Recreations and Essays*, Macmillan, 11th Edition, 1947, p. 396.]

5T. A security firm on a regular run carries valuables by placing the valuables in one car and a dummy box in another car. One car is sent by a town route and the other by a country route. The robbers have only the resources to ambush one of the routes. If they ambush the town route and meet the car it will escape with probability 0·8, if they ambush the country route and meet the car it will escape with probability 0·6. The probability that the valuables are on the town route is selected by the firm and is x. The probability that the town route is ambushed is selected by the robbers and is y. Firm and robbers act at random within the framework of these probabilities; otherwise a leakage of information about a fixed sequence of choices of route would help the opposition.

(i) Prove that the probability that the valuables get through is given by

$$z = \frac{4\frac{1}{3} - 3(x - \frac{2}{3})(y - \frac{2}{3})}{5}.$$

(ii) On the (x, y)-plane mark the values of $60z$ for values of (x, y) as follows: $x = 0(\frac{1}{6})1$; $y = 0(\frac{1}{6})1$. Sketch some contours of equal values of $60z$. (A three-dimensional graph displaying z as a function of x and y would be part of a hyperboloid and you will see that it would have a saddle point.)

(iii) Next take x to be fixed and show that if $x < \frac{2}{3}$ then the graph of z against y is a straight line and the minimum value of z that the robbers can attain by varying y is $\frac{3 + 2x}{5}$. Show also that if $x > \frac{2}{3}$ then the minimum value of z as y varies is $\frac{5 - x}{5}$. Let this minimum value of z for given x be $m(x)$. What value of x should the firm select and fix in order to maximize $m(x)$? What is then the probability that the valuables get through?

(iv) What value of y should the robbers select and fix in order to minimize $M(y)$ where $M(y)$ is the maximum value of z attainable by varying x for given y? What is then the probability that the valuables get through? Identify the values in relation to the surface that would be formed by a three-dimensional graph of z against x and y.

92

(v)   Compare the result of the optimum strategies investigated in parts (iii) and (iv) with what would happen if the valuables were always sent by the town route as being more easily protected.

[This is an example of 'The Theory of Games', of which the first extensive account was published by Von Neumann and Morganstern in *Theory of Games and Economic Behaviour*, Princeton 1944. J. D. Williams, *The Compleat Strategyst*, McGraw-Hill, 1954 is an excellent elementary and readable account.]

6. T. Atkins has left the army and now works in an office; he doesn't like carrying an umbrella. There is a $\frac{1}{5}$ chance that any one day will be rainy. On a rainy day Atkins will certainly take his umbrella if his wife has reminded him, but he will only take it one day in three if she hasn't. There is a 50:50 chance that Mrs Atkins will tell her husband to take his umbrella whatever the weather. If it is not rainy and his wife reminds him, Atkins will take the umbrella two times in three to please her, otherwise he doesn't take it.

What is the probability that on a given office day Atkins is with umbrella?

7. In a certain machine the following actions are possible. The top can blow or not; the spiggots can go up or down or remain still; and the cam-level can wiggle or thump. The top blows with probability $\frac{1}{3}$. The spiggots go up with probability $\frac{1}{2}$ if the top blows; they cannot go down if the top blows; otherwise they go up or go down or remain still with equal probabilities. The cam lever wiggles if the spiggots go up, otherwise it wiggles with probability $\frac{1}{6}$.

The machine is robust and works provided the cam-lever is thumping. What is the probability that the machine works?

8. The following questions refer to the machine described in Question 7.

(i)   What is the probability that the top blows if the spiggots go up?

(ii)   What is the probability that the top blows if the spiggots go up and the cam-lever wiggles?

(iii)   What is the probability that the top does not blow if the spiggots go up?

(iv)   If all that you can observe is that the spiggots go up, then at what odds would you be prepared to bet that the top is not blowing?

9F. In a certain multiple choice exam each question has p 'answers' to choose from and *one* of them is correct. There is a probability θ that a particular candidate knows the correct answer to the first question. If he does know the answer he will certainly mark it correctly. If he does not know the answer he will mark one of the choices at random.

If the candidate has marked the correct answer, what is the probability that he really knew the answer?

10T. From the contingency table

	B	B'	
A	x	y	$x+y$
A	z	t	$z+t$
	$x+z$	$y+t$	N

Show that
$$\text{pr}\,(A\,|\,B) = \frac{\text{pr}\,(A \text{ and } B)}{\text{pr}\,(A \text{ and } B) + \text{pr}\,(A' \text{ and } B)}$$

and hence show that

$$pr\,(A\,|\,B) = \frac{pr\,(A).pr\,(B\,|\,A)}{pr\,(A).pr\,(B\,|\,A) + pr\,(A').pr\,(B\,|\,A')}.$$

(This result is known as Bayes' Theorem. If A preceded B, causally or in time, then pr (A), pr (A'), pr $(B\,|\,A)$, pr $(B\,|\,A')$ are sometimes referred to as *a priori* probabilities, and pr $(A\,|\,B)$ is called an *a posteriori* probability.)

11. From Question 6 we shall refer to the events

'it is a rainy day' as R;
'his wife reminds him' as W;
'he takes his umbrella' as U.

Prove that $\quad$ pr $(U\,|\,W) = \frac{11}{15}$; $\quad$ pr $(U\,|\,W') = \frac{1}{15}$;
$\quad\quad$ pr $(U\,|\,R) = \frac{2}{3}$; $\quad\quad$ pr $(U\,|\,R') = \frac{1}{4}$.

Use Bayes' Theorem (see Question 10) to find the *a posteriori* probabilities pr $(W\,|\,U)$ and pr $(R\,|\,U)$. If you knew all the *a priori* probabilities as given in Question 6 and if you further observed that Atkins was carrying his umbrella but you knew nothing else, then at what odds would you bet that Atkins' wife had been at him again about his umbrella? If you knew further that it was not a rainy day how would you bet?

12. In looking back through the records of our village annual croquet competition I see that I won in 1950, but I forget whether the final was against Miss Hoop or Mr Mallet, though they certainly played each other in the semi-final. It was a small club and our chances in any contest never varied; Miss Hoop had a $\frac{2}{3}$ chance of defeating Mr Mallet, and I had a $\frac{1}{4}$ chance of defeating Miss Hoop and an even chance of defeating Mr Mallet. What is the chance that my final was against Miss Hoop?

13F. Prove that

$$\begin{pmatrix} pr\,(B) \\ pr\,(B') \end{pmatrix} = \begin{pmatrix} pr\,(B\,|\,A) & pr\,(B\,|\,A') \\ pr\,(B'\,|\,A) & pr\,(B'\,|\,A') \end{pmatrix} \begin{pmatrix} pr\,(A) \\ pr\,(A') \end{pmatrix}.$$

If the 2×2 matrix is written $\begin{pmatrix} a & b \\ c & d \end{pmatrix}$ prove that its determinant has value

$1 - (b+c)$ and that this lies 'between' -1 and 1. What relationships exist between the events A and B if the value is -1; 0; 1? How many independent elements are there in the given matrix? Write it in a form which uses only two parameters and show that its square is of the same form. [See Question 57.]

14. (i) In a game which is played by throwing two dice what are the probabilities of throwing a total of (a) 5 or 6, (b) at least 10, (c) at most 9?

(ii) What are the probabilities of getting a total of 5 or 6 if three dice are thrown?

(iii) From a consideration of the possible extra scores (due to the third die) to be added to the totals in a diagram of the type of Figure 3.11 (or by any other means) find the value of t which makes $g(t+1) - g(t)$ largest where $g(t) =$ probability of scoring a total of t with three dice.

(iv) Use the result of part (iii) to produce the approximate shape of the graph of $g(t)$ plotted against t.

15. A die is loaded so that the chance of throwing a four is $K/3$ and the chance of throwing a three is $(1-K)/3$. The chances of throwing each of the other faces are $\frac{1}{6}$. The die is thrown twice.

What values of K maximize the chances of throwing a total of (i) 6, (ii) 7, (iii) 8?

16 T. Discs occurring in a collection D of discs have probabilities q, and $1-q$, of being red, and green, respectively. The collection D of discs is so large that the withdrawal of members from it does not affect the probabilities of the occurrences of the two colours.

(i) Two of the discs are chosen at random and pasted together to form a disc of double thickness. Find the probability of its having two red; two green faces; one of each.

(ii) One of these randomly formed double discs is to be spun like a fair coin. Find the probabilities, evaluated before the result of the pasting is known, that the upper face is red; is green.

(iii) Two such randomly formed double discs are spun; their lower halves are discarded and their upper halves pasted together to form a new double disc. Show that the probabilities of having two red; two green faces; one of each on the new double disc are respectively equal to the probabilities of these events on the original double discs, and that pr (a face will be red) $= q$, as in the original collection D.

[This trial provides a mathematical model of inheritance by genes. The single discs represent genes and the double discs represent gene-types or genotypes which can occur when each gene has the possibility of being of one of two sorts, here represented as red or green. The first double disc represents the gene-pair of one parent and the two double discs represent the gene-pairs of both parents. Disregarding mutations, the genotype of an individual is fixed for life. When mating occurs one gene is chosen at random from each parent and this is represented by spinning two double discs and selecting their upper faces. The two genes (one from each parent) go to form the gene-pair of the offspring and this is represented by the formation of the new double disc. The results of this question so far show that with mating which is independent of the genotype of the parents (called random mating) the probabilities of the occurrences of each gene remain fixed from generation to generation and may be calculated as if the genes were *not* paired off, in the same way that the original collection D consisted of discs not paired off.]

(iv) In three collections of men the following frequencies of different genotypes of blood groups were observed:

	MM	MN	NN
American Indian (Pueblo)	83	46	11
From Brooklyn U.S.A.	541	903	405
Australian Aboriginal	3	44	55

(From Boyd, in *Tabulae Biologicae*, **17**, 230, 235 (1939).)

For each collection estimate the probability of the occurrences of the genes, M, N, as being exactly equal to the relative frequency of the occurrence of the given gene in the collection.

(v) For each collection use the model, and the values of the probabilities from (iv), to calculate the expected numbers of the different genotypes, assuming them to be proportional to the probability of the occurrences of the genotypes when random mating occurs. (The degree of agreement between the observed and hypothetically-calculated (called *expected*) frequencies is a matter which is treated later in the course. The agreement here seems close, confirming the value of the model.)

17. If random mating is taking place and an individual is selected at random from the collection, show that $\frac{1}{2}$ is the biggest possible value of pr (2 genes are different); and that this occurs when the two genes are equally probable.

18 T. The two genes in a certain gene-pair are each either A or a. The individuals with genotypes Aa or AA can be distinguished by appearance from those with genotype aa, but not from each other. (The characteristic produced by gene A is said to be *dominant*.)

(i) In a collection K of plants, pr (given gene is a) = $\frac{1}{2}$. What is pr (given gene is A)? Assuming that the collection K had been formed by random mating calculate pr (genotype of a given plant is aa).

(ii) In a collection L of plants of the same species, pr (gene is A) = $\frac{2}{3}$, what are the probabilities of the occurrences of each genotype if the mating that formed L was random?

(iii) In a collection M of plants of the same species, the relative frequency of the individuals of genotypes either Aa or AA can be found by appearance and is 19 per cent. Estimate pr (gene is a) if the mating that formed M was random, and estimate the probabilities of each genotype.

19 T. (See Question 18.)

A botanist takes samples of 12, 144 and 100 plants from the collections K, L, M respectively and, without realizing the different genetic constitutions of the collections, he bulks them.

(i) If one of the 256 plants is chosen at random from the bulked collection calculate the probabilities of the occurrence of each genotype.

If a gene is chosen at random from the randomly chosen plant, calculate pr (gene is a), pr (gene is A). Verify that these probabilities are equal to those which would be obtained by using a model in which all the genes could be detached from their plants and bulked and one gene then chosen at random from the collection of genes.

[Compare the situation with those in Exercise 3 C, Question 8 and Exercise 3 M, Question 3, and notice that it corresponds to the case $x_1 + y_1 = x_2 + y_2 = 2$ in the latter.]

(ii) In a hypothetical collection arising from random mating what value of pr (gene is a) would give rise to the actual value of pr (genotype is aa) that the botanist observes in his bulked collection? With this value of pr (gene is a) calculate the probabilities of the occurrences of each of the other two genotypes in the hypothetical collection. Calculate also the expected proportions of the frequencies of individuals of the two distinguishable appearances in the hypothetical collection.

(iii) With the actual value of pr (gene is a) that exists in the bulked collection and with random mating now being carried out by the botanist in his laboratory what are the probabilities of the occurrence of the three different genotypes among the offspring? What are the expected proportions of frequencies of individuals of the two distinguishable appearances?

20. Repeat Question 19 if the botanist's samples from K, L, M were of sizes 52, 189, 200 respectively.

21 T. The two genes of one of the many gene-pairs of a man can each be either gene B or gene b. A child receives at random one gene from each of its parents. If both parents are of genotype bb then the children are also of genotype bb, (and we are said to have a true-breeding line). If a child receives one gene B in its gene-pair it is resistant to malaria, but if it has two genes B it suffers from sickle-cell-anaemia and there is a probability θ that it will die before reaching the adult-stage, and so will have no children. Thus a true-breeding malaria-resistant line is impossible if $\theta = 1$.

Starting from a population of parents all of whom are of type Bb, and taking $\theta = 1$, calculate successively pr (gene is b) in the adult stage of the children, grand-children, great-grandchildren, assuming that mating among the adult stage is independent of genotype then existing. (This is not *strictly* random mating among the offspring since no offspring of type BB are mated.)

[Note the 'selection pressure' against the disadvantageous gene, which will lead to its ultimate elimination under random mating.]

22 T. There is a probability θ that a person of genotype BB will die of anaemia before having children and a probability ϕ that a person of genotype bb will die of malaria before having children.

(i) Starting from a population of parents all of whom are of type Bb and taking $\theta = \frac{3}{5}$, $\phi = \frac{1}{3}$ find pr (gene is b) in the adult stage of the children and grand-children, assuming the mating among the adult stage is independent of genotype then existing.

(ii) Starting from a generation of parents with pr (gene is b) $= q$, find pr (gene is b) in the next adult stage with general values of θ and ϕ.

(iii) If pr (gene is b) $= q$ and is unchanged in two successive adult stages find the possible values of q in terms of θ and ϕ. We then have a type of equilibrium situation in terms of gene ratios in the collection.

[This is an example of Natural Selection operating to produce a certain gene ratio dependent on the differing survival values of the different genotypes; the existence of two or more forms in equilibrium like this is called *polymorphism*.]

⋆23. In an experiment with fowls we are concerned with two gene-pairs, the genes being C, c in one gene-pair I, i in the other. An offspring receives at random one gene of each gene-pair from each of the parents.

Fowls of genotype cc have no black pigment at all. The presence of gene C ensures black pigment in the fowl. The presence of gene I prevents the black pigment from showing in the wing feathers even if it is present. Fowls of genotype ii show the black pigment in the wing feathers if it is present in the fowl.

(i) Write down the various genotypes of the fowls together with the colour of their wings: for example, Cc Ii—white.

(ii) If both parents are of the same genotype and if all the offspring necessarily have the same wing colour as the parents then we have a true-breeding line. Write down the genotypes that give a true breeding line with black wings; with white wings.

(iii) If both parents are true-breeding white-winged, though possibly of different genotypes to each other, show that all the offspring are white-winged, and that the only not-true-breeding ones are of genotype Cc Ii.

(iv) If both parents have genotype Cc Ii, determine the various possible genotypes of the offspring and their probabilities; what are the expected proportions of the frequencies of offspring showing black feathers; white feathers?

(v) If a rival theory of this phenomenon says that there is only one gene-pair controlling the wing colour the genes being, say, B and b and only the fowls of genotype bb show black wings, then prove that the cross of a fowl of genotype Bb with another of that genotype (the only cross that would have parents of the same genotype and yet would *not* breed true) would give a proportion of offspring with black wing feathers, different to that of part (iv).

(vi) Give some simpler methods of distinguishing between the rival theories.

24. Five red discs and one green are placed in a bag and a random choice of four of them is made (leaving two in the bag). What is the probability that exactly three of the withdrawn discs will be red? That exactly two of them will be red?

25 F. (The results of this Question are used in Chapter 12, on Estimation.)

(i) p red discs and q green are placed in a bag and a random choice of four of them is made. What is the probability that exactly three of the withdrawn discs will be red if

$$(a)\ p = 4, q = 2; \quad (b)\ p = q = 3?$$

(ii) You draw a random sample of four discs from a bag containing six discs each of which is either red or green and the sample consists of three red discs and one green. What numbers of red and green discs in the original bag gave the highest probability of this sample being drawn?

26. Two bad electric bulbs are mixed by you among six good ones, and a friend is going the test them in some random order.

(i) At the start of the process what is the probability of his finding the second bad one (*a*) on the second test; (*b*) on the third test; (*c*) on the eighth test; (*d*) on the seventh test?

(Assume he does not put a bulb back once it is tested.)

(ii) What is the most probable number of bulbs he will have to test if he knows that there are at most two bad ones?

(iii) If he says 'I shall find the second bad bulb at or before the nth bulb' and he wants to have at least a 50 per cent chance of being right, what is the least value of n he can give?

(iv) Would it help if he started at the other end? Well! in the middle then?

27 F. A manufacturer supplies tumblers in boxes of 50. A buyer examines a sample of five tumblers and if *less* than 20 per cent of the sample are cracked he accepts the box. If any one tumbler is cracked he rejects the box.

(i) If there are exactly 20 per cent of the tumblers cracked in one box what is the probability that the buyer will accept it on his test?

(ii) What is the probability that he will reject a box which in fact only contains one cracked tumbler?

(This type of problem is dealt with further in Question 28 and in Exercise 7D, Questions 21, 22, 23.)

28 T. An acquaintance has a coin which he wants to sell to you, as he says it is biased and will help you to win the toss at cricket matches and so on. He cannot for the life of him remember which way the bias goes but no doubt you will be able to find out before the next match (he says). You examine the coin and doubt his claim. You decide to form the *null hypothesis* that it is unbiased and that any fluctuations from an equal number of heads and tails would be due to chance. You furthermore intend to carry out a test to see whether you will need to reject this *null hypothesis*: you will spin the coin five times and will reject your hypothesis (that is, accept his claim) if either 0 heads or 5 heads occur; otherwise you will accept the null hypothesis.

(i) If the true probability of heads is $\frac{1}{2}$, what is the chance that you will reject the null hypothesis? (You will then be said to have made the Type I error by rejecting the null hypothesis when it is true.)

(ii) If the true probability of heads is $\frac{2}{3}$, what is the probability that you will accept the null hypothesis? (You will then be said to have made a Type II error by accepting the null hypothesis when it is false.)

(iii) Plot the graph of

pr (you reject the null hypothesis | probability of heads is θ)

as a function of θ for $0 \leqslant \theta \leqslant 1$.

(This function is called the *power function* of the test.)

(iv) Mark on your graph suitably placed line segments to show the magnitude of (*a*) the probability of the Type II error when $\theta = \frac{2}{3}$; (*b*) the probability of the Type I error.

(v) Sketch on the same diagram the general shape of the graph of the power function of a test which has a smaller probability of the Type I error and, for θ sufficiently different from $\frac{1}{2}$, a smaller probability of a Type II error. What has happened to the probability of a Type II error for θ nearly $\frac{1}{2}$?

(vi) Draw the graph for a test that would have excellent characteristics over the Type I error and most Type II errors.

29. A poker die has faces Ace, King, Queen, Jack, Ten, Nine. Five such dice are considered simultaneously. The possible outcomes are classified into sets, called 'hands', and the following discussion shows how the hands relevant to this problem are defined.

One possible outcome, for instance, is Ace, Nine, Jack, Nine, Nine. This may first be coded as $BACAA$ by preserving only the distinctions between faces and not the actual face names. It is then further coded as $AAABC$ by disregarding order. If by choice of A, B, C an outcome can be put in this form (for instance King, Queen, King, Jack, King $\rightarrow ABACA \rightarrow AAABC$) then we have an outcome of the hand called 'Three of a kind' or 'Threes'.

The other definitions relevant here are indicated by: King, Queen, Queen, Queen, King, $\rightarrow BAAAB \rightarrow AAABB$, called 'Full House'; Nine, Jack, Nine, Nine, Nine $\rightarrow ABAAA \rightarrow AAAAB$, called 'Fours'; Jack, Jack, Jack, Jack, Jack $\rightarrow AAAAA$, called 'Fives'. For the purposes of further codification we will, in this question, call 'Threes' type 1, 'Full House' type 2, 'Fours' type 3 and 'Fives' type 4.

Given type i we throw all the non-A dice. Find the probability, a_{ij}, that the resulting hand will be of type j. Arrange the probabilities in a 4×4 matrix being careful over the definition when $i = 4$. Interpret

$$\sum_{j \geqslant i} a_{ij} \text{ for each value of } i.$$

30. In a deal of 52 cards to four bridge players the cards are dealt clockwise, the player on the dealer's left getting the first card of each round. Find approximately (in the form $1/N$, where N is given to two significant figures) the probability that

(i) the dealer gets the whole of one suit;

(ii) the player on the left of the dealer gets the whole of one suit;

(iii) someone gets the whole of one suit;

(iv) someone gets all the spades;
(v) the dealer gets all the spades;
(vi) everyone gets a whole suit;
(vii) that each person gets a suit previously named by him.

What would be the effects on the answers if the method of dealing were to give 13 cards to a player before giving 13 to the next; the players being taken clockwise from the dealer's left as before?

31. Four letters are written and four envelopes addressed. If the letters are now put at random into the envelopes, find the probability that all the letters will be in the wrong envelopes.

32. There are n objects to be marked. An inexhaustible supply of N different kinds of mark is available and the marks are allotted with probabilities $1/N$, the allotments being completely independent of each other. The probability that the n objects are all allotted *different* marks is $b(n, N)$.

(i) Interpret the case: $n = 20$, $N = 365$, in everyday terms.

(ii) Find $b(20, 365)$. This is quite simply calculated on a slide rule obtaining $b(r, 365)$ from $b(r-1, 365)$.

(iii) Find the least value of n such that $b(n, 365) < \frac{1}{2}$ and interpret this result.
[This 'Birthdays Property' was first noted by H. Davenport.]

★33. With the definition of Question 32 make a table of $b(n, N)$ for $n = 1$ and for $n/N = 0.1$ (0.1) 0.5, $N = 10, 50, 100$. (Use tables for the factorials.)

Assuming the function to be continuous sketch $b(n, N)$ against n/N for the three values of N, and from each of your graphs estimate the value of n/N for $b(n, N) = \frac{1}{2}$. Let this value be $k(N)$. Make a table of $k(N)$ against N and sketch the graph of $k(N)$ against $\log N$. [Use the result $k(365) = 0.063$, obtainable from Question 32.] From your graph estimate $k(1000)$.

★34. (i) With the definitions in the previous question estimate k for $N = 13$. Let n_0 be the integer nearest kN. Shuffle and cut a pack of cards n_0 times and record the face-values (ignoring differences of suit) to find whether a repetition of face-value occurs. Find the proportion of occasions when at least one repetition occurs, and compare with the value predicted by the theory.

(ii) If the above experiment is altered by not continuing any set of n_0 cuttings once a repetition has occurred, but starting a new set of n_0 cuttings on the next cut, then obviously the theory still provides a suitable model. Does it provide a suitable model for the following arrangement?

A large number of cuttings are recorded in sequence and *every* set of n_0 consecutive results within it is examined for existence of a repetition?

35F. A quarter of the population of Muddleton supports the Anarchist party. The Anarchist party agent takes a poll of three people in the town. What is the probability that he finds at least one of them to be Anarchist, so that he over-estimates Anarchist support in the town?

What assumptions have you made about the poll? Could the problem be solved without these assumptions?

36. If I investigate 1000 samples, each of three tennis balls, and find that in two of the samples there are three with mangy fur, and if I assume a constant proportion of defectives and random packing into 'threes', how many of the next 1000 samples of three tennis balls are likely to be perfect?

37. A marksman scores 400 bulls in 500 shots. If his shooting remains as before, what is the probability that he will score exactly four bulls in the next five shots, and is he more likely to exceed this rate than to fall short? Express the three relevant probabilities in decimal form to two significant figures each.

38. Five dice are thrown. What is the probability of getting two or more sixes?

39. A die has the faces marked Ace, King, Queen, Jack, Ten, Nine. What is the probability of throwing exactly 4 aces, with one throw of five such dice? What is the probability of throwing exactly four faces the same with five such dice? At least four the same?

40. Which is more likely to occur:
   (i) A throw of six at least once in 4 throws of a single die?
   (ii) A throw of double-six at least once in 24 throws, each of two dice?
[The Chevalier de Méré noticed a difference in the two frequencies and wrote to Pascal for an explanation. The Chevalier's argument was for equality, and ran
   '(i) A throw of six in a single die occurs once in 6 throws so it will occur 4 times out of 6 when I throw 4 times.
   (ii) A throw of double-six occurs once in 36 throws so it will occur 24 times out of 36 if I throw 24 times.'
Pascal's resulting correspondence with Fermat opened the subject of Probability (1654). 'Un problème relatif aux jeux de hasard, proposé à un austère janséniste par un homme du monde, a été l'origine du calcul des probabilités.' S. D. Poisson, *Recherches sur la probabilité des jugements*, 1837, p. 1]

41. 'The Cholera will have killed by the end of the year about one person in every thousand. Therefore it is a Thousand to one (supposing the Cholera to travel at the same rate) that any person does not die of the Cholera in any one year. This calculation is for the Mass, but if you are prudent, temperate and rich your chance is at least five times as good that you do not die of cholera; in other words 5000 to 1. and if it is 5000 to 1 that you do not die of Cholera in a year, it is not far from two Million to one that you do not die any one day of Cholera.'
[Extract from a letter of Sidney Smith to Lady Grey during the cholera epidemic of 1832.]
Assume that the probability of getting Cholera in a year was 1/5000.
   (i) If Sidney Smith was using a Chevalier de Méré type of argument (see Question 40) to obtain his estimate of 'not far from two Million to one' what would be his estimate to two significant figures?
   (ii) Give an expression for the correct value, without necessarily evaluating it.
   (iii) Use the fact that for small x/n the value of

$$1 - (1 - x)^{1/n}$$

is roughly x/n (with proportional error about $\frac{1}{2}(x/n)$) to show that Sidney Smith's two-significant-figure value is correct to two significant figures.
[Thus the Chevalier de Méré's type of error becomes less important for large numbers of trials with low probability of success.]

42F. What is the probability of throwing a score that is *more* than two different from the mean score (7) in a throw of two dice? What is the probability that in

101

four games, each consisting of throwing two dice, such a score will occur once? At least once?

How many games of two throws each are needed to make the occurrence of such a score '95 per cent certain'?

43 F. A machine is turning out defective items with probability $1/K$. If n items are examined what is the probability of finding (i) no defectives; (ii) one or fewer defectives; (iii) two or more defectives?

If $n = 2K$, use the fact that $\left(1 + \dfrac{x}{y}\right)^y \to e^x$ as $y \to \infty$ to find the approximate values of these probabilities for large K.

⋆44 F. Five coins are tossed. Give the probabilities of each possible number of heads.

45. In 1964 an assessment was being made by the (English) Department of Education and Science (using a large number of schools) of the proportion of people leaving who were going to a University. One school had seven leavers of whom two were going to University but in giving this information the school remarked that their figures were misleading as their usual proportion was something like $\frac{3}{4}$.

What is the probability that a sample as unrepresentative as this would occur in a randomly chosen year, and what action would you take if you were conducting the survey?

46. I operate a gambling system on a simple head/tail bet with a fair coin, by doubling my stake when I lose. Show that when I first win at all I shall be in profit by 1 unit of stake. I then withdraw. Of course I may be ruined by losing consecutively until all my initial capital is gone, but I am prepared to accept a probability of about 1/250 that this will occur. If my initial capital is £1, what should be my unit of stake?

$$\left[\sum_{n=1}^{N} r^n = \frac{1 - r^N}{1 - r} = \frac{r^N - 1}{r - 1}. \right]$$

47. A, B, C have equal claims for an award. They arrange that each will throw a coin, and that the man whose coin falls unlike the other two will win. (That is: 'odd man out wins'). If all three coins fall alike they will throw again. If there is still no decision they will wait until the next day, and repeat the process.

At the start of the procedure what are the probabilities that
  (i) A will win at the first attempt?
  (ii) There will be no winner at the first attempt?
  (iii) B will win at the second attempt?
  (iv) Someone will win at the second attempt?
  (v) A will win on the first day?
  (vi) No decision will be reached on the first day?
  (vii) A decision will be reached by the end of the second day?
  (viii) No decision will ever be reached?

48. Harry and Tom decide to play a game as follows: Each has a coin and they spin alternately. Harry starts and the first to get a head wins. They play indefinitely, stopping only when someone wins.

Show, without evaluating either probability, that

$$\text{pr (Harry wins)} = 2 \times \text{pr (Tom wins)}.$$

It turns out that Harry has an appointment for which he is already late, and he says 'I'll still take first throw, but in fact I'll throw my coin and *leave it lying*; don't you move it, but each time my turn comes simply take my throw to be whatever the coin says when I leave it'. Whom does the game favour now? Interpret the situation.

49. In a modified version of Section 10, Example 4, Ann and Bob select $\operatorname{pr}(A) = a$, $\operatorname{pr}(B) = b$ and they play until one of them wins; but if, after n turns each, neither has yet won then they draw. Show, without evaluating either probability, that

$$\frac{\operatorname{pr}\,(\text{Ann wins})}{\operatorname{pr}\,(\text{Bob wins})} = \frac{a}{(1-a).b}.$$

50. In a modified version of Section 10, Example 4, Ann and Bob select $\operatorname{pr}(A) = a$, $\operatorname{pr}(B) = b$ and they play indefinitely, stopping only when one of them wins. Assume that

$$\lim_{n\to\infty} (1+k+k^2+\ldots+k^n) = \frac{1}{1-k}, \ |k| < 1$$

and show that

$$\operatorname{pr}\,(\text{Ann wins}) = \frac{a}{a+b-ab},$$

$$\operatorname{pr}\,(\text{Bob wins}) = \frac{(1-a)\,b}{a+b-ab},$$

$$\operatorname{pr}\,(\text{the game is drawn}) = 0.$$

Show that $\operatorname{pr}\,(A) = \frac{1}{3}$, $\operatorname{pr}\,(B) = \frac{1}{2}$ still gives a fair game. Interpret the situation in which $a = \frac{1}{2}$ and the game is fair.

*51. (In this question assume the results of Question 49.)

(i) Ann and Bob arrange that, as in Section 10, Example 4, $a/b = 2/3$, but they also have $a+b = 1$; show that the game is no longer fair.

(ii) If $a/b = 1/1$ and $a+b = 1$ show that $\operatorname{pr}\,(\text{Ann wins}) = 2\times\operatorname{pr}\,(\text{Bob wins})$, but show also that if $a/b = 1/2$ and $a+b = 1$ then $\operatorname{pr}(\text{Ann wins}) < \operatorname{pr}(\text{Bob wins})$

(iii) Put in turn

$$a = \frac{u_r}{v_r} = \frac{u_{r-1}+u_{r-2}}{v_{r-1}+v_{r-2}},$$

where $u_0 = 1, \ u_1 = 1; \quad v_0 = 2, \ v_1 = 3$ and retain $a+b = 1$.

Show that

$$\frac{\operatorname{pr}\,(\text{Ann wins})}{\operatorname{pr}\,(\text{Bob wins})} = \frac{2}{1}, \frac{3}{4}, \frac{10}{9}, \frac{24}{25}, \frac{65}{64} \quad \text{for } r = 0, 1, \ldots, 4.$$

For $r = 4$, explain how to use a pack of cards to have a suitable random device for Ann and Bob to get the right odds.

(The numbers in the sequences (u_r) (v_r) are Fibonacci Numbers.)

(iv) Solve the equations

$$\operatorname{pr}\,(\text{Ann wins}) = \operatorname{pr}\,(\text{Bob wins}),$$

$$a+b = 1,$$

to find b for an exactly fair game and compare with $1-(u_4/v_4)$.

52. (In this Question assume the results of Question 50.)

(i) Show that if $a = b$ and both are small enough then $\dfrac{\operatorname{pr}\,(\text{Ann wins})}{\operatorname{pr}\,(\text{Bob wins})} \simeq 1,$

but that if $a = b$ and both are large enough then $\dfrac{\text{pr (Ann wins)}}{\text{pr (Bob wins)}}$ is as large as we please.

(ii) Explain these results and the fact that, for $a = b = \frac{1}{2}$.

$$\text{pr (Ann wins)} > \text{pr (Bob wins)}$$

in general terms, *as for a layman who can understand the original game but not the later mathematics.*

53. In a certain line or queue people join just before the exact end of a minute, or not at all, and at most one person joins at a time; no-one leaves. The queue starts empty; that is to say the first opportunity for people to join is just before the end of the first minute. At each opportunity of joining the probability that someone joins is $\frac{1}{3}$. Find the probability that at the end of the third minute the number of people in the queue is 0; 1; 2; 3; 4. What is the most likely number of people in the queue at the end of the fourth minute?

54 F. A queue is formed as in Question 53 but just before the end of each minute the probability that someone joins is λ (independent of t). The probability that at the end of the tth minute there are n individuals in the queue is $q(n, t)$. Prove that

$$q(n+1, t+1) = q(n, t).\lambda + q(n+1, t).(1-\lambda) \quad \text{for} \quad 0 \leqslant n \leqslant t.$$

Define the initial conditions. Why do we have $n \leqslant t$?

55. In the queue of Question 54 there are the additional facts that people are served without leaving the queue and then leave the queue just before the exact end of a minute or not at all; at most one person leaves at a time; and just before the end of each minute the probability that someone leaves is $\frac{1}{3}$ provided there was someone in the queue during the minute, but independently of an arrival or otherwise at that moment.

(i) Find $f(r, s) = \text{pr (queue increases by } r \,|\, s \text{ people in the queue)}$ for the following pairs

$$(r, s) = (1, 0); \quad (0, 0); \quad (1, 1); \quad (0, 1); \quad (-1, 1).$$

(ii) Find the probability that at the end of the third minute the number of people in the queue is 0; 1; 2; 3; 4. During which minute is it first more likely that there are people in the queue than not?

56 F. A certain system has two states called L and R. At the end of each unit interval of time it may or may not switch instantaneously from one state to the other. It never switches at other times. At any given instant the probability of a switch does not depend on which instant is given but only on the state (such a condition makes the process a *Markov process*, the chain of successive states being called a *Markov chain*).

Project Question on Markov processes. Take a die and operate as follows:
    (i)   States L, R are having the die in the left, right hand respectively.
    (ii) Start with the die in the left hand.
    (iii) Throw the die and determine the transitions as follows:
        (*a*) If in state L and a *one* is thrown, change to state R;
            If in state R and a *one* or *two* is thrown, change to state L.
    (iv) Record the states, the initial one being called the state at instant $t = 0$.
    (v)   Determine the state at $t = 50$.

Repeat with the following transition rules:

(b) L to R: if *one* or *two* is thrown;
　　 R to L: if *one, two, three* or *four* is thrown.

(c) L to R: if *one, two* or *three* is thrown;
　　 R to L: if anything at all is thrown.

(Note that with ingenuity the same series of 50 throws of a die can be used in all cases provided the throws are recorded. The states and changes can be interpreted afterwards by examining the sequence.)

Suggest a physical situation for which method (c) is a suitable mathematical model.

The Markov chains set up suggest various problems such as:

(A) At the start of the experiment what is the probability that at the nth instant the state will be L? will be R? and how are these probabilities affected if the chain starts at state R? Take n to be large.

(B) What are the probabilities of various lengths of run in each state?

(C) What is the mean length of run in each state?

(D) What is the mean number of throws needed to reach each state from the other?

(E) How are the results affected by changing the transition probabilities (note that in cases a, b, c above the ratios of the transition probabilities are equal.)

The reader may find connexions between the problems but most of them are too hard to solve here. Experimental results can be obtained to suggest likely sizes for the answers. Random Number tables are preferable to dice for long runs.

The matter is more fully taken up in the next Question, in Chapter 7 and in Chapter 10.

57F. In a certain Markov process (see Question 56) the probabilities that the system is in states L, R at the nth instant are u_n, v_n respectively. The probabilities of transition from L to R and from R to L, are a, b respectively.

Prove that

(i) $u_1 = (1-a).u_0 + b.v_0$;

(ii) $\begin{pmatrix} u_n \\ v_n \end{pmatrix} = \begin{pmatrix} 1-a & b \\ a & 1-b \end{pmatrix} \begin{pmatrix} u_{n-1} \\ v_{n-1} \end{pmatrix}$;

(iii) $\mathbf{w}_n = (A)^n.\mathbf{w}_0$, where $\mathbf{w}_n$ has been written for $\begin{pmatrix} u_n \\ v_n \end{pmatrix}$ and A has been written for $\begin{pmatrix} 1-a & b \\ a & 1-b \end{pmatrix}$;

(iv) put $a = \frac{2}{4}, b = \frac{1}{4}, \mathbf{w}_0 = \begin{pmatrix} 1 \\ 0 \end{pmatrix}$ and calculate $\mathbf{w}_n$, $(A)^n$, for $n = 1, 2, 3, 4$.

(v) By putting $u_{n-1} = u_n$ and $v_{n-1} = v_n$, determine what must be the limiting values of u_n, v_n as $n \to \infty$, if such limiting values exist at all, in terms of general a, b.

(vi) The genes C and c of a certain gene-pair *mutate* (that is: change) into each other with the probabilities below

Change	Probability of occurrence in unit time
C → c	10^{-6}
c → C	2×10^{-6}

Find pr (gene is C) after a time has passed long enough for equilibrium to be reached, and assuming no selection operates against any of the genotypes.

58*F.* A burglar has a chance q of being arrested on any particular day and the value of q is independent of the day in question.

(i) Find the chance, evaluated at the moment of escape, that he will (*a*) escape arrest for the first 10 days; (*b*) be caught in *more* than 10 days; (*c*) be caught in *more* than 9 days.

(ii) Interpret the value of the difference between the last two probabilities calculated.

(iii) If his chance of being arrested during the first 28 days is $\frac{1}{16}$, find the chance of his being arrested in (*a*) the first week; (*b*) the first t weeks; (*c*) the first day; (*d*) any particular day.

$\left(\text{If you want to use a linear approximation for the value of } \left(\frac{15}{16}\right)^x \text{ near } x = 0\right.$
you can use that the derivative of

$$\left(\frac{15}{16}\right)^x \text{ is } -0{\cdot}0695 \left(\frac{15}{16}\right)^x.\Bigg)$$

(iv) Sketch the graph of pr (he is arrested in first x weeks) against x.

59. A certain type of steel strip contains blemishes dispersed along its length at random in such a way that the chance of a blemish occurring in a particular millimetre of length is independent of the millimetre in question.

$$\text{pr (exactly } one \text{ blemish in a given millimetre)} = \tfrac{2}{1000}$$

and $\quad$ pr (more than one blemish in a particular mm.) $= 0$.

Find $\quad$ (i) pr (exactly one blemish in a given 3 mm.);
(ii) pr (there is exactly one blemish in a given 500 mm.);
(iii) $q(0), q(1), q(2)$, where $q(r) = $ pr (exactly r blemishes in a given metre). (Suggested method: find the ratios $q(0): q(1): q(2)$ and calculate $q(0)$ using $\left(1 - \frac{x}{n}\right)^n \simeq e^{-x}$ for small (x/n).)

(iv) pr (*more* than two blemishes in a given metre).
[See Chapter 8 for a full discussion of this type of problem.]

60. The chance of throwing a six with a weighted dice is $\frac{3}{20}$. Find the chance of throwing exactly one six in N throws. Regarding this as a function of a continuous variable N, sketch its graph, and hence estimate the most likely number of throws in which to score exactly one six. Verify your guess by comparing (without necessarily calculating) the actual probabilities.

$\left(\text{The derivative of } \left(\frac{17}{20}\right)^x \text{ is } -0{\cdot}16 \left(\frac{17}{20}\right)^x.\right)$

61. A particle is buffeted forwards or backwards by unit steps, if at all, at two successive blows. At each blow we have

$$\begin{aligned}
\text{pr (particle is buffeted forward)} &= p \quad (0 \leqslant p \leqslant \tfrac{1}{2}), \\
\text{pr (particle remains fixed)} &= 1 - 2p, \\
\text{pr (particle is buffeted backwards)} &= p.
\end{aligned}$$

(i) What are the possible ordered pairs of successive moves of the particle and what are their probabilities?

106

(ii) Prove that after the two blows

$$f(p) = \text{pr (particle is 0 steps away)} = 1-4p+6p^2,$$
$$g(p) = \text{pr (particle is 1 step away)} = 4p-8p^2,$$
$$h(p) = \text{pr (particle is 2 steps away)} = 2p^2.$$

(iii) Choose p to minimize the probability that the particle is where it started.

(iv) Choose p to maximize the probability that the particle is one step from its start.

(v) Sketch the three probabilities as functions of p on the same axes and give the range of values of p for which one step is the most likely distance of the particle from its start.

62. A puppy is being trained to move towards a whistle and this it does by fits and starts. It either remains fixed or moves forward one step when the whistle is blown. It is rewarded after each move forward so that the probability that it will remain fixed at the next blast is multiplied by k $(0 \leqslant k \leqslant 1)$. The probability of remaining fixed is originally p and only changes after a reward.

(i) Find the probabilities that after three blasts it will be 0; 1; 2; 3 steps forward.

(ii) If $k = 0$ (so that the puppy learns completely after one reward) sketch the graphs of these probabilities as functions of p on the same axes. Show that whatever the value of p the puppy is more likely to be 3 steps forward than 1 or 2; that for $p^3+p > 1$ it is better to bet on 0 steps than 3; and that for $p > \sqrt[3]{\frac{1}{2}}$ or $p < \frac{1}{2}$ it is advantageous to bet on the result at evens.

(iii) If $k = 1$ (so that the puppy is indifferent to reward) sketch the graphs again, and show that the ranges of values of p to give 0, 1, 2, 3, steps most likely are:

$$\tfrac{3}{4} < p \leqslant 1, \quad \tfrac{1}{2} < p < \tfrac{3}{4}, \quad \tfrac{1}{4} < p < \tfrac{1}{2}, \quad 0 \leqslant p < \tfrac{1}{4}.$$

Show also that it is only safe to bet on the number of steps at evens if $p > \sqrt[3]{\frac{1}{2}}$ or $p < 1-\sqrt[3]{\frac{1}{2}}$.

63. A square tray of side 11 in has a rim all round it and is marked off into 16 equal squares by three lines ruled parallel to each side. A coin of diameter 1 in is placed on the tray at random (the rim preventing the centre from being less than $\frac{1}{2}$ in from the edge). If the probability that the centre occupies a point in a given other area of tray is proportional to the area whatever the position or shape of the area, find the probability that the coin-edge does *not* cut one of the ruled lines.

64. By selecting integer-pairs (x, y) satisfying $0 \leqslant x \leqslant 9$ and $0 \leqslant y \leqslant 9$ and estimating the probability that $(x+\frac{1}{2})^2+(y+\frac{1}{2})^2 < 100$ in terms of the area of the first quadrant of the circle $x^2+y^2 = 100$, use a table of random numbers to estimate π.

[When we attempt to solve a determinate numerical problem by setting up a related problem involving probabilities (called a stochastic problem) and using random numbers, we are said to be using a *Monte Carlo method* of solution.]

65. Expose the fallacy in the following and supply a correct argument. 'A point is chosen in a line. Divide the line into n equal parts. The probability that the point is not in the first part is

$$1-\frac{1}{n}.$$

Similarly for the other parts. Hence the probability that the point is not on the line at all is

$$\left(1 - \frac{1}{n}\right)^{n},$$

66 R. D'Alembert once asserted that in throwing two coins the probability of getting a double head was $\frac{1}{3}$. For, he said, the possible outcomes are
   (i) A double head; (ii) a double tail; (iii) one of each.
   Put him right.

*67. A man throws three coins and argues thus:

   pr (all three are the same)
   = pr (two of them are the same and the third is the same as the pair).

Now since there are only two sides to a coin not all the coins are different so at least two of them are the same, so

   pr (two of them are the same) = 1.

Further since the pair which are the same are either both heads or both tails,

   pr (the third is the same as the pair | two of them are the same) = $\frac{1}{2}$.

Hence the laws for compound events give us

   pr (all three are the same)
   = pr (two of them are the same)
      × pr (the third is the same as the pair | two of them are the same)
   = $1 \times \frac{1}{2}$
   = $\frac{1}{2}$.

Expose the fallacy.

68 R. From a letter to the press:
   'There is one reason for large families which [your correspondent] omits to mention; the wish of a couple whose children are of the same sex to produce one (just one) of the opposite sex. This urge has a greater effect now that the equality of the sexes makes girls as worth having as boys. But it is not easy to fulfil: ignoring hereditary tendencies to produce more children of one sex than the other, two-thirds of couples have either two boys or two girls. Only half of such couples going on to a third child can expect it to be of the other sex. . . .'
   Write a reply *suitable for the daily press*, taking up the point of probability theory raised.

108

4

DESCRIPTION OF DATA

1. INTRODUCTION

When we make measurements or count things we do not, as the reader knows well, always find ourselves (as we basically did find ourselves in the last chapter) merely investigating the proportion of occasions when some event or other occurs. Examples of other sorts of measurement were suggested in Chapter 1, and we there mentioned the important art of selecting a few numbers (called statistics) to represent in their various ways certain characteristics of large collections of measurements, collections of so many measurements perhaps that we should otherwise be bewildered by them as trees and unable to see them as a wood.

A statistic may be used in a purely descriptive way to record the essentials of what we have found and perhaps eventually to help us to distinguish, if they are distinguishable, two different collections of measurements; or it may be part of an apparatus of prediction using the theory of probability to suggest the likely behaviour of other collections of similar measurements.

In this chapter we will mainly consider two characteristics of a collection of measurements, and will discuss how we can quantify them by the use of descriptive statistics. To speak loosely we aim to provide the means of answering the questions 'Which are the bigger?' and 'Which are the more varied?': for instance 'Does this fuel or that have the higher calorific value?' or 'Are incomes more varied in Italy than in Britain?'.

Before attempting to condense a collection of measurements we must have a suitable method of recording the collection uncondensed, and this leads us first to investigate the general nature of such collections of numbers as we will meet.

2. POPULATION AND
FREQUENCY FUNCTION

2.1 Populations. In elementary courses we learn how to prepare a frequency table and represent it on a frequency diagram; our purpose in this section is to look at the processes involved in the light of the notions of *sets* and *functions*. The processes are fundamental to the whole theory of statistics, and though it must be admitted that it is not really necessary to analyse our actions quite as carefully as we do here, there is, perhaps, a certain satisfaction in laying bare the rods and cogwheels of a mechanism.

Consider the situation in which £28 is shared among ten people as follows:

Alice £4, Brian £3½, Cynthia £4, David £0, Elizabeth £4, Frank £3½, Kate £2, Victor £3, Jean £1, Geoffrey £3.

The reader may be tempted to say 'The £28 is *distributed* among the ten people', but we have avoided that word here. It is used as a technical term in many different ways in the field of statistics but this natural usage is not one of them. A discussion of the word and the meanings occurs in Chapter 5, Section 5.

We have defined above a set and a function. The set consists of the ten people concerned. The function has this set of ten people for its domain. The image of a person in the domain consists of a number, namely the number of pounds received by the person.

Now we consider more closely the collection of images of this function. It consists of the collection of numbers $(4, 3½, 4, 0, 4, 3½, 2, 3, 1, 3)$.

Several members of the set of people are mapped onto the *same* number of pounds given, so that the collection of images is, because of these repetitions, *not* a set. To help us to see this distinction let us consider a set of, for instance, cups.

Such a set may contain 3 cups, the number of the set is then 3; but we cannot say that the *same* cup pointed at three times comprises a *set* whose number is 3. [The *occasions* on which it was pointed at are, of course, not the same as each other, and *they do* comprise a set whose number is 3.] If we are counting the members of a set of cups we ignore a cup that has been pointed at already. Now, to return to our collection of images: the number 4 is the same number however many times it is pointed at, and in our collection of images it is 'pointed at' three times. We are *not*, however, to ignore it at the second and third pointings as we should have to if we were interested only in the *set* of numbers pointed at. It is the very possibility of repetitions that we study.

A collection of numbers, such as our collection of images, with possible repetitions but no notion of 'order', we will call a *population*. It is one of the basic notions of the branch of mathematics called 'Statistics'. Just as 'Calculus' deals with a certain class of functions, and as 'Geometry' deals with points, lines and planes, so 'Statistics' deals with populations.

2.2 Samples. If we took the set of all households in a city, say Nottingham, and counted the number of rooms that each household had (with suitable definitions of rooms and households) we should again get a population (of numbers of rooms, this time). This population, however, might be thought of as only a part of the larger population which would be obtained by listing the number of rooms to a household throughout the whole U.K. When we particularly want to concentrate on a population as part of a larger one we call it a *sample from the* (*larger*) *population,* and the larger population we call the *parent population.*

To return to our example, the population of rooms-to-a-household in the U.K. is itself merely a sample of the population obtained from households in Western Europe, and so on, and we see that the terms 'sample' and 'parent population' are not absolute. In this case, however, the process ultimately terminates because there is only a finite number of households in the world.

The word 'sample' is widely used to mean merely 'subset', and we will not deprive ourselves of the right to use it in that way too. The context will always make the distinction obvious.

2.3   Some particular populations. We illustrate the generality of the terms and concepts by examining some other ways in which a population might arise.

(i) We take a set of men, say an army intake, and weigh them. The numbers of kilograms in their weights form a population.

(ii) We observe the night sky and consider the set of 'novae' (new stars) that appear between now and, say, the year 1999. The numbers of years *between* the appearances form a population. The numbers of years *from the start* of our experiment *until* a nova appears form another population (obviously mathematically related to the first).

(iii) We take a set, {throws of a pair of dice}, and count 'the score' on each of 500 throws. These numbers form a population.

(iv) We take the set, {throws of a pair of dice until a double six occurs}, and we count the number of throws involved. We go on throwing the dice in this way for fifty occurrences of a double-six. The numbers obtained form a population.

(v) We take the set of all births in the U.K. in 1967 and write down 0 for a female and 1 for a male. The collection of 0's and 1's forms a population. In this case the numbers do not *measure* anything, though they could be said to *count* the males.

(vi) We take a set of 1000 throws each of a 'silver' coin and a 'copper' coin and we record the heads and tails, coding them as follows:

Silver coin	Copper coin	Code
Head	Head	(0, 0)
Head	Tail	(0, 1)
Tail	Head	(1, 0)
Tail	Tail	(1, 1)

The resulting population is a *bivariate population,* that is a population of ordered number-pairs (which provides an extension from our first ideas).

(vii) We take the same army intake and measure their heights as well. The weights and heights lead to a bivariate population.

111

In each case we shall have a population with a finite number of members and, therefore, inevitably, with members separated in value from each other. Such populations we call *discrete*. If a population consists of numbers actually obtained in some experiment, then it obviously contains only a finite number of members and is therefore discrete.

We might, however, think of population (i) as a sample drawn from a parent population of numbers measuring all conceivable army intake weights, whether yet measured or not. We are at liberty to say that, in practice, any measurement that gets made will be made to the nearest whole number of some small unit or other, and that we will select a discrete population as our parent population, but it turns out to be much simpler mathematically to say that the weights could conceivably have any value in an interval. In that case we are considering a different sort of parent population. If the members of a population can have any value in an interval or set of intervals, then we say that the population is *continuous*.

Another type of population arises in considering a suitable parent population for population (iv). If we thought of the numbers of throws that it *might* need for a double six to occur then there is no greatest such number. So although there is a greatest member of population (iv), that population would be best thought of as a sample from a parent population with no greatest member. [The parent population would of course be chosen to be discrete.] Each population of (ii) would best be thought of as a sample from a continuous parent population with no greatest member; although, of course, it is itself finite and has a greatest member.

In later chapters it will be very important to distinguish between on the one hand populations that arise as finite (and thus discrete) collections of numbers as a result of actual measurements, counts or codings, and on the other hand the parent populations that we select for them. We will then reserve '*sample*' for the name of the first of these kinds of population. It has probably occurred to the reader that in probability terms, if the occurrence of a particular number in a sample is thought of as a possible outcome of a trial, then the search for a suitable parent population is the search for a suitable possibility space. In this chapter, however, the relationship between sample and population does not concern us and we will almost entirely use 'population', as being a general term. We shall see that the handling of descriptive statistics is entirely arithmetical and algebraic, and owes nothing to any theory of probability or estimation. Descriptive statistics are the objective bricks and mortar of the subject.

2.4 Frequency functions. The reader should be familiar with the process of drawing up a *frequency table*. For the population of the money-sharing situation of Section 2.1, this is shown by

Number of pounds	Occurrences in population	Frequency
0	/	1
1	/	1
2	/	1
3	//	2
$3\frac{1}{2}$	//	2
4	///	3

If we write this as:

x	$f(x)$
0	1
1	1
2	1
3	2
$3\frac{1}{2}$	2
4	3

we see that we have a function defined by the population of numbers of pounds issued. In words: If x is any number then $f(x)$ is defined to be the number of times that x occurs in the population; (it follows that if x does not occur in the population then $f(x) = 0$). The value of $f(x)$ is in fact the *frequency* of x in the population and f is called the *frequency function*. A frequency function is a perfectly ordinary function such as we are accustomed to, but with two restrictions: the domain is a set of numbers (or, perhaps, n-tuples); the codomain is the set of non-negative integers. In this way a finite population always uniquely defines a function.

What may not be so obvious is that any function f mapping a set of numbers $\{x\}$ into the set of non-negative integers always defines a population. To obtain the population we need only collect together every x, with $f(x)$ repetitions. The relation between 'population' and 'frequency function' is that which Keats said holds between Beauty and Truth! This point is taken up in Section *2.7.

In the steps from the original set of objects or events to the frequency function two mappings are involved:

(i) A mapping which measures (or even arbitrarily gives a numerical code to) the original objects or events and so produces a population of numbers (or, it might be, number pairs or n-tuples).

The same set of objects or events could, of course, be measured for a different characteristic to provide a different population. For instance the men of the army intake might be measured for age. This possibility exercises the statistician's applied mathematical skill.

(ii) A mapping which is generated by the very population itself to provide the frequency function which totally identifies the population and has nothing to do with the original objects or events. This is the hunting ground of mathematical statistics, and is where the statistician exercises his more purely mathematical skill.

A diagram will conveniently summarize all this for our money-sharing data.

In Figure 4.1 and Exercise 4A, the following notation is used:

> S is the set of objects or events
> N = number of members of S
> m is the mapping from S into a set of real numbers x (or number-pairs etc.) It defines the measurements or codings made.

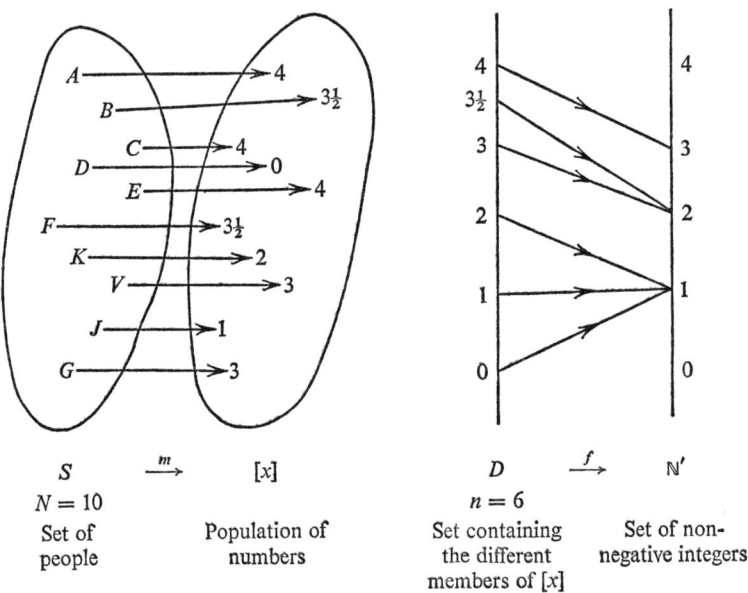

S	$\xrightarrow{m}$	$[x]$	D	$\xrightarrow{f}$	N'
$N = 10$			$n = 6$		
Set of people		Population of numbers	Set containing the different members of $[x]$		Set of non-negative integers

Fig. 4.1

> $[x] = [x_1, x_2, ..., x_n]$ is the population of images of members of S under m. The *size* of $[x]$ is N. If we quote only a typical value from the population, or only the set of different values whose possible repetitions give us the population, we will use *square* surrounding brackets to indicate the fact. These brackets will not be used when, as occasionally happens, the population is written out in full with all its repetitions; thus $[x] = (4, 3\frac{1}{2}, 4, 0, 4, 3\frac{1}{2}, 2, 3, 1, 3)$.
>
> $D = \{x_1, x_2, ..., x_n\}$ is a set containing the different members of $[x]$.
> n = number of members of D.
> f is the frequency function for $[x]$.
> N' is the set of non-negative integers.

2.5   Representations of a frequency function. A frequency function, f, like any other function may be specified in several useful ways:

(i) By means of a formula for $f(x)$: this does not often happen in a practical case and would not be suitable with our money data. It comes into its own in idealized populations, in which frequency ratios are modelled by probabilities.

(ii) By means of a table of values of $f(x)$—called here *the frequency table*. We list a set D of values of x in which we are interested, using integral suffices, i, for reference purposes. D is not unique, but it must contain all the different members of $[x]$.

i	x_i	$f(x_i)$
1	0	1
2	1	1
3	2	1
4	$2\frac{1}{2}$	0
5	3	2
6	$3\frac{1}{2}$	2
7	4	3

We will use consecutive values of i running from 1 up to n, which is thus the number of the set D and is not less than the number of different values of x which actually occur in the population. If $f(x_i) \neq 0$ for $1 \leqslant i \leqslant n$ then n is exactly the number of different values of x which occur in the population. In Figure 4.1 we had $n = 6$, in the above table we have $n = 7$. In general n is *not* the *size* of the population. The *size* of the population is $\sum_{i=1}^{i=n} f(x_i)$ and we have written it as N.

(iii) By means of a diagram—which we will call the *frequency diagram*. It is usual to allow some licence in how the frequency diagram is constructed in order to convey the information easily to the eye.

It seems helpful in the present case to extend the set D by thinking of the unit of issue as £$\frac{1}{2}$, and to display that $f(x) = 0$ for $x = \frac{1}{2}, 1\frac{1}{2}, 2\frac{1}{2}$. We now have $n = 9$. Two presentations follow, the second being more helpful.

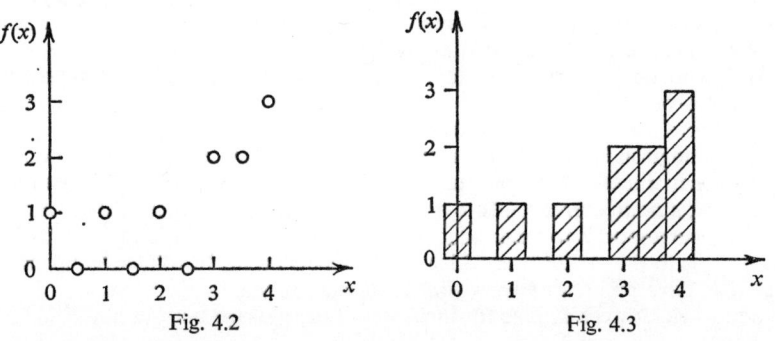

Fig. 4.2   Fig. 4.3

Note that Figure 4.2 is more nearly the graph of $f(x)$. Figure 4.3 is an example of a bar frequency diagram. It is *not* a *histogram*. The distinction is explained in Chapter 5.

Exercise 4 A

1 R. *S* is a set of events or of physical objects.

N is the number of members of *S*.

m is a mapping from *S* into a set of real numbers, or pairs, or triples (etc.) of real numbers. It defines the observations of the experimenter.

[*x*] is the population of images of the members of *S*, under *m*.

D is a set containing the different members of [*x*].

n = the number of members of *D*.

f is the frequency function for the population [*x*].

Identify *S*, *N*, *m*, [*x*], *D*, *n*, *f* for the following sets of data.

   (*a*) In a study of *Paris quadrifolia* (Herb Paris):

Leaves per plant	4	5	6	7
Frequency	10	1	0	1

   (*b*) Rainfall in 1964 (measured in inches):

Jan.	Feb.	Mar.	Apr.	May	June	Jul.	Aug.	Sep.	Oct.	Nov.	Dec.
4·2	3·4	1·8	2·4	2·6	2·4	4·5	4·2	2·4	1·3	3·8	3·6

   (*c*) Throwing of two coins:

   2 heads 3 times;   1 head and 1 tail 7 times;   2 tails 5 times.

2 R. Take the data collected in Exercise 1 A, Question 2 and produce the separate frequency diagrams of the population of guessed ranks, for correct rank = 1, 2, 5, 8 in turn. For each population identify *S*, *N*, *D*, *n*.

3 R. Take a novel and consider two passages each long enough to contain exactly 100 different nouns. Let $f_1(x)$, $f_2(x)$, $f_3(x)$, be the numbers of nouns which occur *x* times in the first, and second and amalgamated passages respectively. Draw up the resulting frequency tables and diagrams and keep them for further use. If different members of a class take different passages, varying (*a*) part of book, (*b*) book, (*c*) author, then interesting populations will be available for later study.

This type of frequency function was first studied by G. U. Yule in a book *The Statistical Study of Literary Vocabulary* (Cambridge University Press, 1944). Yule suggested the following rules for uniformity:

   (i) *Verbal nouns:* in passages such as (*a*) 'the torturing of Peacham', (*b*) 'the propriety of torturing prisoners', treat 'torturing' as a noun in (*a*); but not as a noun in (*b*), since there it governs an object.

   (ii) Do not count words like Lady, Duke when used as titles of address and do not count proper nouns.

   (iii) Ignore the distinction between singular and plural.

   (iv) Distinguish between homonyms such as page (boy) and page (of a book).

   [Frequency functions similar to these occur in the statistics of insurance companies; a noun corresponds to a motorist and the occurrence of the noun in a passage corresponds to an accident to the motorist. The length of the passage corresponds to the period over which the accidents occur. Insurance companies, of course, have premium payers with no accidents while we do not enumerate the nouns which were eligible for inclusion in the passage but did not occur.]

★4. With the notation of Question 1, identify *S*, *N*, *m*, [*x*], *D*, *n*, *f* in Question 3.

*2.6   **Sets and populations.**† We have seen that a completely general method can be laid down once and for all, and that by it any *subsequently* specified population defines a function (a function which in fact contains all the information that there is about the population at all). If, because no repetitions occur in it, the population is in fact indistinguishable from a set, then the function defined from the members of the set by this method is obviously a constant function (having all its values equal to 1). It may be of interest to notice that no amount of ingenuity in the construction of a different general method of defining a function will alter this situation; that is to say: it is impossible to lay down beforehand a general procedure which will enable every set S to define a function from its members into the set of integers unless that function is a constant function. For the members of S have no essential property beyond that of inclusion in S and this property fails to distinguish them from one another. Another way of seeing this is to say that if every set S could define a non-constant function from its members into the set of integers then the images of the members could provide a method of ordering any set S and we know that order is not an essential property of a set.

Consider for instance the set $A = \{a, 2, £, ?\}$. If $x \in A$ we could define

$g_1: x \to x$,
$g_2: x \to \{x\}'$, the complement being taken with respect to A,
$g_3: x \to \{$all subsets of A that contain $x\}$,

so that, for instance:

$g_1(a) = a$,
$g_2(2) = \{a, £, ?\}$,
$g_3(£) = \{\{£\}, \{a, £\}, \{2, £\}, \{£, ?\}, \{a, 2, £\}, \{a, £, ?\}, \{2, £, ?\}, \{a, 2, £, ?\}\}$

and then even the functions

$$f_1: x \to n(\{g_1(x)\}),$$
$$f_2: x \to n(g_2(x)),$$
$$f_3: x \to n(g_3(x)),$$

are all constant functions on the set A, in fact

$$f_1(x) = 1, \quad f_2(x) = 3, \quad f_3(x) = 8, \quad \text{for all } x \in A.$$

The reader can test to what extent he has followed this argument by evaluating $f_1(x), f_2(x), f_3(x)$ for $x \in B$ when B contains N members.

*2.7   **Populations and functions.** We have seen that there is a distinction between a population and a set, but we now see that there is essentially none between a population and its frequency function.

† Starred sections may be omitted at a first reading.

If the population of the money-sharing situation was written successively as follows (the reader is to discover for himself the rules of procedure):

$$(4, 3\tfrac{1}{2}, 4, 0, 4, 3\tfrac{1}{2}, 2, 3, 1, 3),$$

$$(4, 4, 4, 3\tfrac{1}{2}, 3\tfrac{1}{2}, 0, 2, 3, 3, 1),$$

$$(4(3), 3\tfrac{1}{2}(2), 0(1), 2(1), 3(2), 1(1)),$$

$$\{(4, 3), (3\tfrac{1}{2}, 2), (0, 1), (2, 1), (3, 2), (1, 1)\},$$

then we should have a smooth transition from a collection of numbers with possible repetitions to a set of ordered pairs. This last form is a fairly sophisticated but common enough way of defining a function. In this case the function is effectively the frequency function for the given population— only 'effectively' since we have already defined frequency functions as having all numbers for their domain, whereas this function has for its domain only the set of numbers actually occurring in the population.

We have thus shown that it is logically unnecessary for us to indulge in separate names for 'population' and 'frequency-function', since each defines the other uniquely. We shall, however, find that having two words is convenient for distinguishing slightly different aspects of the concept.

3. POSITION AND AVERAGE

3.1 Introduction. We now limit our attention to populations whose members are single numbers, as opposed to number-pairs or even points in n-dimensional space.

The first descriptive statistics we will construct will be ones to measure how big the members of a population are 'on average', or, in other words, to describe what is the position along the x-axis of the bulk of the frequency diagram for $[x]$.

To fix our attention we will consider data derived from the following two sets of objects:

{the first 27 sentences of *The History of Mr. Polly*}

{the first 27 sentences of *The Way of All Flesh*}

We take x words to be the length of a sentence and so we have two populations of values of x. We call these populations *THOMP, TWOAF*. They give rise to two frequency functions which we denote by h and w respectively. We obtain the frequency table, Table 4.1.

3.2 Comparison of positions. Our question is: 'Which set contains the longer sentences?'

A glance shows that the sentences of *TWOAF* are longer than those of *THOMP*. If the frequency diagrams were plotted on the same scale the

118

main bulk of the ordinates for *TWOAF* would be to the right of that for *THOMP*. The position of the bulk of the ordinates is an indication of the average sentence length. We can describe this more precisely in many ways. For instance, by comparing one of the five following statistics:

(i) *The mean of the extreme values of x.*

The extremes of *THOMP* are 3 and 37, with mean 20.

The extremes of *TWOAF* are 4 and 122, with mean 63.

(ii) *The modes.* The modes of a population of values of x are those values of x which occur most frequently (that is, are most '*à la mode*'). Here they are the commonest sentence lengths. They are values of x, *not* frequencies.

Table 4.1

x	$h(x)$	$w(x)$	x	$h(x)$	$w(x)$
3	2	0	23	0	1
4	2	1	24	2	2
5	2	0	25	0	2
6	1	0	26	0	1
7	1	0	27	1	0
10	1	1	28	0	1
11	1	0	29	0	2
12	1	0	31	0	1
14	2	1	32	0	1
15	2	1	35	0	1
16	1	0	37	1	0
17	2	0	39	0	1
18	0	2	40	0	2
19	0	1	41	0	1
20	1	0	46	0	2
21	3	0	73	0	1
22	1	0	122	0	1

The mode of *THOMP* is unique. It is 21. $h(21) = 3$.

The modes of *TWOAF* are 18, 24, 25, 29, 40, 46. For each of them we have $w(x) = 2$.

THOMP is *unimodal*, *TWOAF* is *multimodal*.

(iii) *The median.* To find the median of a population of values of x, arrange the population in ascending order of size of x (that is, starting with the smallest) and roughly speaking the median is then the central value in the list (though Exercise 4B, Question 8 indicates minor snags which can arise).

More precisely: If the population size, N, is odd the median is unique and is the $(\frac{1}{2}(N+1))$th value in the list.

If N is even the median may be indeterminate, being any value b between the $(\frac{1}{2}N)$th and $(\frac{1}{2}N+1)$th values in the list (strictly between if these values are different and are not merely one of the possible repetitions).

119

The median of *THOMP* is 15.

The median of *TWOAF* is 28.

(iv) *The mean of the quartiles.* Roughly speaking the first and third quartiles are the $(\frac{1}{4}(N+1))$th values of x from each end, once the population has been ordered. Together with the median they split the population into four equally sized parts.

The quartiles of *THOMP* are 6 and 21, with mean 13·5.

The quartiles of *TWOAF* are 19 and 40, with mean 29·5.

(v) *The means.* The mean of a population is calculated by adding together all the members of the population *with their repetitions* and dividing by the number of them, that is by the population-size. In symbols: if the population is $(y_1, y_2, y_3, ..., y_N)$ where y_j are not necessarily all different, then the mean is given by

$$m = \frac{1}{N} \sum_{j=1}^{N} y_j.$$

If a frequency table is made out and $\{x_1, x_2, ..., x_n\}$ are the *different* values of y_j, then we have the form

$$m = \frac{1}{N} \sum_{i=1}^{n} x_i f(x_i) = \sum_{i=1}^{n} x_i f(x_i) \bigg/ \sum_{i=1}^{n} f(x_i).$$

The mean is uniquely defined for any finite population. The calculations are excessively tedious unless the short cuts shown in Section 4 are employed, or unless a calculating machine is available.

The mean of *THOMP* is 14·6.

The mean of *TWOAF* is 32·3.

The mean will play such an important role in the later work that a special notation is useful for it. If we have a population of values of x, which we are writing, for convenience, as $[x]$, then it is further convenient to write $\bar{x}$ for the mean of the population.

It is important to understand that *each* of the statistics (i) to (v) gives a measure of the 'average' of a population, and that 'average' is a loose term which here corresponds in geometrical terms to the idea of the 'position' of the main bulk of the ordinates in a bar frequency diagram. There are other measures of 'average' which we do not deal with here and which are appropriate for other types of problems, such as averaging rates of flow of fluid or rates of growth.

In Section 3.3 we show how different types of population produce different relationships between mean, median and mode.

3.3 Some population types. To put on record certain broad types of population and to show a variety of relationships between their statistics of position we consider a few simple examples. The positions of the mode,

120

median and mean are shown by different arrows. It is often said that these three statistics occur in this very order in frequency diagrams, but our examples show that the order is not universal.

(i) *A symmetrical population with a central tendency.*

i	x_i	$f(x_i)$
1	3	1
2	4	2
3	5	3
4	6	2
5	7	1

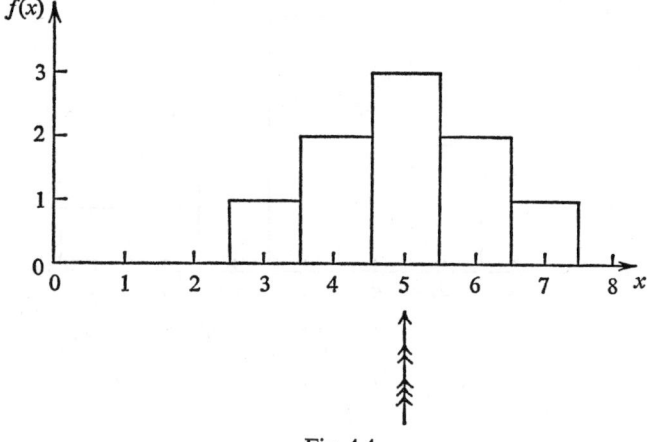

Fig. 4.4

Obviously all the measures coincide on the axis of symmetry. Repeated measurements of the same or similar objects or phenomena often give a frequency diagram broadly resembling this, in being symmetrical and centrally humped. This particularly regular pattern leads us to call the diagram, and by extension the function and even the population, *triangular*.

(ii) *A positively-skewed population.* This is so called because the long tail of the frequency diagram is to the right-hand or positive side. (See over.)

i	x_i	$f(x_i)$	$x_i f(x_i)$
1	3	2	6
2	4	5	20
3	5	4	20
4	6	2	12
5	7	1	7
6	8	1	8
		15	73

$\bar{x} = \text{mean}[x] = \frac{73}{15} = 4 \cdot 9$; $\text{median}[x] = 5$; $\text{mode}[x] = 4$.

121

Later in the course we shall deal with an important type of population (called χ^2; spelt *chi-squared*) which is positively skewed. If Figure 4.5 is reflected in an axis parallel to the $f(x)$-axis, then a negatively-skewed population is represented; the scores in all the completed first innings (whether 'all out' or 'declared') of First-class County Cricket matches form a negatively-skewed population and a cricketer could show why, but a less obscure example would be the number of fingers on a hand.

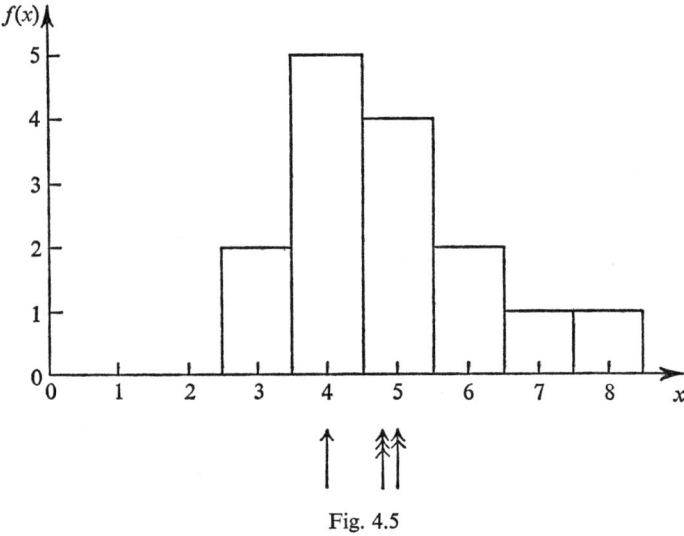

Fig. 4.5

(iii) *A positively-J-shaped population.* This very positively-skewed population is unfortunately named, since it is a *negatively*-J-shaped population that more resembles the letter J. In each case the mode is at one extreme and the value with smallest frequency is at the other extreme.

i	x_i	$f(x_i)$	$x_i f(x_i)$
1	0	5	0
2	1	3	3
3	2	3	6
4	3	2	6
5	4	1	4
6	5	1	5
		15	24

$\bar{x} = \text{mean}[x] = \frac{24}{15} = 1 \cdot 6$; median$[x] = 1$; mode$[x] = 0$.

This is a very commonly occurring type of population. Some examples arise from

    (*a*) the ages in months at death of babies dying in their first year of life;

(b) the numbers of goals scored by each side in football matches;

(c) incomes bracketed in thousands of pounds (the process of bracketing is dealt with in Chapter 5);

(d) the distance-from-the-mean of the members of almost any type of population whose frequency diagram has a single hump.

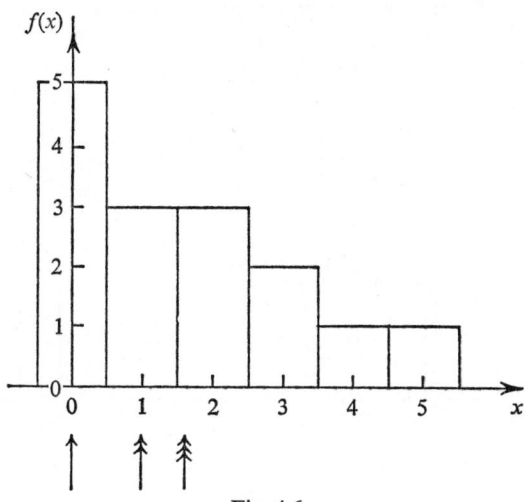

Fig. 4.6

(iv) *An ill-constructed population.* Below is a population that is unlikely to occur in a well-designed experiment.

i	x_i	$f(x_i)$	$x_i f(x_i)$
1	3	1	3
2	4	1	4
3	5	2	10
4	6	2	12
5	7	3	21
6	15	2	30
		11	80

$\bar{x} = \text{mean}[x] = \frac{80}{11} = 7\cdot3$; median$[x] = 6$; mode$[x] = 7$.

Note the disturbing effect which $x_6 = 15$, with $f(x_6) \neq 0$, has on the mean. The occurrence of a population like this suggests that the underlying set contained physical objects or events which should not have been crudely averaged: like the ages of a schoolclass and its teacher.

For example, the figures might refer to sampled hairdressing expenses in a certain workshop of 100 employees where, either by chance or because they were easily interviewed, two of the four girl typists were included in the sample. There are really two populations, one of men's and one of

123

women's expenses. This is an important error in experimental design and may well occur without such an obvious warning in the frequency diagram. A correction could be employed by investigating more of the men's lower expenses to obtain in the sample the same proportion of members of each of the parent populations. Such sampling (which is *not* 'simple random') is one form of what is called *stratified sampling*; differing judgements on stratification partly account for the wide differences in the estimates of various well-conducted political polls.

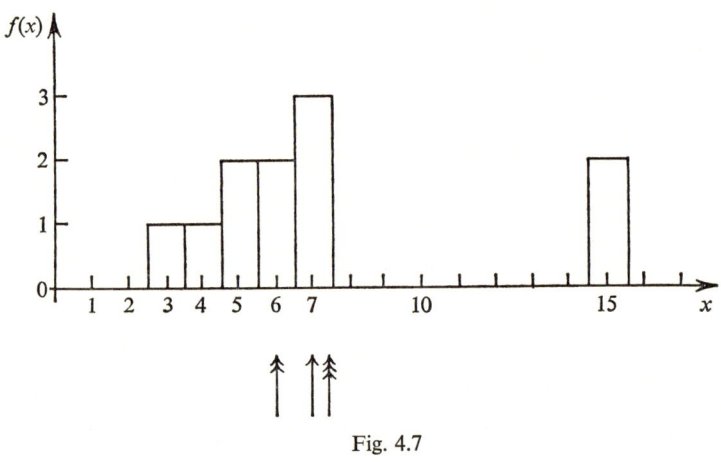

Fig. 4.7

Exercise 4B

1R. Mark the mean, median and mode on the frequency diagrams of each of the following populations.

(i) $x =$ number of yeast-cells in a square of a haemocytometer: (part of 'Student's' data)

x_i	1	3	4	5	6	7	8	9	11
$f(x_i)$	3	4	3	3	6	3	1	1	1

(ii) $x =$ number of petals on a flower of *Ranunculus bulbosus*:

x_i	5	6	7	8	9	10
$f(x_i)$	11	5	2	1	1	1

2R. Construct simple frequency functions to fit each of the following sets of statistics, indicating the functions by drawing the bar frequency diagrams. (*Hint:* It is helpful to fix the median, mode and mean in that order; this reflects the fact that the mean takes account of *all* the data. It is said to be a more *efficient* statistic for measuring position than the others.) You need, in fact, have no more than seven members in each population.

124

		Median	Mode	Mean
(i)	First population	1	-1	0
(ii)	Second population	-1	0	1
(iii)	Third population	0	1	1

3F. Use geometrical transformations on the frequency diagram to produce, from the second population above, populations with the following statistics (describing clearly the transformations you use):

		Median	Mode	Mean
(i)	Fourth population	-2	0	2
(ii)	Fifth population	-2	-1	0
(iii)	Sixth population	-1	$-\frac{1}{2}$	0

4T. The following population is given:

$$1, 1, 0, 2, 0, 3, 0, 4, 8, 6, 5.$$

(a) If one member is to be increased to make the population have mean equal to 3, what must be the increase?

(b) If, instead, two numbers were to be increased to make the population have mean equal to 3 what can you say about the increases?

(c) If, instead of increasing a member, an extra member was to be added to the population to make the mean equal to 3, what would the member have to be?

(d) If, instead of adding one member, two members were to be added to make the mean equal to 3, what could you say about them?

(e) Could you have added a member to the original population to make the mean equal to 2?

5T. (a) A sample of size 20 has mean 3·4 and another two numbers are adjoined to the sample. If each is 4·5, what is the mean of the new sample?

(b) If 'median' is read for 'mean' throughout the data of part (a), what can you say about the new median?

(c) Repeat part (b) with 'mode' in place of 'median'?

6R. Find the mean, median and mode of each of the samples of Exercise 4A, Question 3 and of the sample obtained by combining them. Find also these statistics for the biggest samples available in the class by combining passages from any single book or any single author. [Keep your results.]

7T. (a) Two samples are given whose sizes and means are respectively N_1, N_2 and m_1, m_2. It is decided to combine the samples into a single sample; prove that the mean of the single sample so obtained is $\dfrac{N_1 m_1 + N_2 m_2}{N_1 + N_2}$.

(b) Generalize the result of part (a) to a combination of n samples, expressing your result by using the Σ-notation. Discuss the application of the result to the three analyses of the results of Exercise 1A, Question 1.

(c) Calling the n samples of part (b) the 'elementary samples' show that the mean of the sample obtained by combining all the elementary samples into a single sample is equal to the mean of that population in which the frequency of a number is equal to the sum of the sizes of the elementary samples which have that number as mean.

8 *T*. If median, quartiles, sextiles, etc. are being used for precise purposes like allotting grades and passes in examinations then some care needs to be taken over the details of their exact definitions. For our purposes these are sufficiently exemplified, when the population is discrete, by this question.

The table shows six frequency functions, *p*, *q*, *r*, *s*, *t*, *u*.

i	x_i	$p(x_i)$	$q(x_i)$	$r(x_i)$	$s(x_i)$	$t(x_i)$	$u(x_i)$
1	3	2	1	1	1	1	1
2	4	2	2	2	2	2	3
3	5	1	2	3	2	3	3
4	6	4	4	3	3	2	1

(*a*) For each population complete the following, tabulating your answers clearly.

(i) State the median. (It is called *b* below.)

(ii) Answer 'yes' or 'no' to: 'Are there the same number in the population having $x > b$ as have $x < b$?'

(iii) Answer 'yes' or 'no' to: 'Are there the same number in the population having $x \geqslant b$ as have $x \leqslant b$?'

(*b*) Is it possible to have a population whose size is an even number, whose frequency function *f* is such that *f*(median) is an odd number, and such that the answer to part (ii) is 'yes'? Prove your result.

4. THE MEAN

Amid all the possible measures of position the mean has, in Exercise 4B, shown itself as the easiest to develop theoretically. Since our use of measures of position is intended ultimately to be more than merely descriptive, so that we shall need to know what *might* happen to the value of a statistic and not merely what *has* happened, it is the mean we will investigate further now.

4.1 Working-zero and change of scale. If we wish to find 'the average age' of four girls who give their ages in years as $15\frac{7}{12}$, $16\frac{1}{2}$, $16\frac{1}{3}$, $16\frac{1}{4}$ then we can avoid the calculations if we can look up their 'average age' of a year ago in a record somewhere. All the girls are now a year older and the 'average age' is greater by one year. Failing this, we could count in months from, say, 16 years old and find the excess months to be -5, $+6$, $+4$ and $+3$, giving a total of $+8$ which, shared among the four girls, gives them $+2$ each. The 'average age' is thus 16 years and 2 months or $16\frac{1}{6}$ years.

Both these simple methods rely on the effects on the mean of a translation of the population (by addition or subtraction of some number from them all), and the second method relies also on the effects of a change of scale. We take this matter up in this section to introduce some ideas of change of origin and scale which will be constantly useful throughout the book.

126

Consider the population and calculations given by:

i	x_i	$f(x_i)$	$x_i f(x_i)$
1	12	20	240
2	17	50	850
3	22	40	880
4	27	20	540
5	32	10	320
6	37	10	370
		150	3200

The frequency diagram, with certain additions which are explained below, is shown in Figure 4.8.

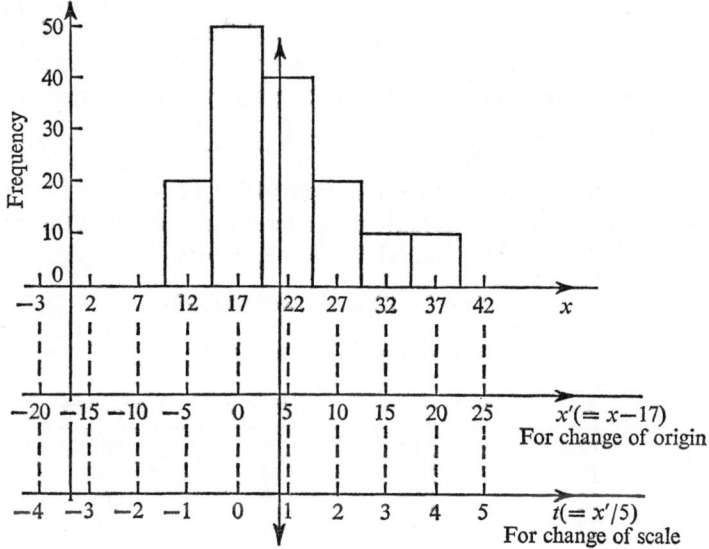

Fig. 4.8

The additions in Figure 4.8 comprise values of x' and of t marked along axes below the axis of x. The numbers x, x' and t are related to one another by the transformations:

$$x' = x-17; \quad t = x'/5,$$

from which we derive:

$$t = (x-17)/5 \quad \text{or} \quad x = 5t+17.$$

The mean of $[x]$ can be seen from the table to be $\frac{3200}{150}$, that is $21\cdot3$. This implies that the arrowed line $x = 21\cdot3$ gives a 'central' line through the diagram and from the way the axes of x and t are laid out in the diagram we can see that $t \simeq 1$ will also represent this central line (in fact we can see

127

from the relevant formula above that $t = (21\cdot3 - 17)/5$ is the accurate value of t giving the central line).

All this suggests that we can reverse the process to find the mean of $[x]$: we first find the mean of $[t]$ (using for each value of t the frequency of the corresponding value of x, as suggested by the diagram) and we then transform the value of t back to a value of x.

In symbols $\bar{t} = \Sigma t_i f(x_i)/N$, which is the usual formula for a mean, and $\bar{x} = 5\bar{t} + 17$, which is deduced from $t = (x - 17)/5$. The values of t are simpler than those of x so we have a desirable short cut. k may be called a *working-zero*; it is sometimes, misleadingly, called an assumed mean, but there is no assumption that it is the mean, the process is merely one of change of origin.

Geometrical experience tells us that the process is perfectly general, and that if we take k to be the number subtracted from all values of x to get values of x' (producing a *translation*), and take c to be the number by which we divide the values of x' (producing a *scale-change*), then we shall have
$$t = (x-k)/c, \quad \bar{x} = c\bar{t} + k.$$

This is the basis of the solution of Exercise 4B, Question 3.

For an instance of the generality of the result, look again at the population earlier in this section, and take $k = -3, c = 5$.

We get
$$t = (x-(-3))/5, \quad \bar{x} = 5\bar{t} + (-3)$$

and these transformations give the following table.

x	$x-(-3) = x'$	$x'/5 = t$	Frequency
12	15	3	20
17	20	4	50
22	25	5	40
27	30	6	20
32	35	7	10
37	40	8	10

Now since a change of frequency scale does not affect the position of the mean, the last two columns effectively give us population (ii) of Section 3.3 (as the reader should now verify) and the corresponding mean we have already obtained as $\frac{73}{15}$.

From that we obtain
$$\bar{x} = 5 \times \tfrac{73}{15} + (-3)$$
$$= 24\cdot3 - 3$$
$$= 21\cdot3, \text{ as before.}$$

The average age of four schoolgirls was obtained in the second method by taking $k = 16, c = \frac{1}{12}$.

Exercise 4 C

1R. Calculate mentally to two decimal places the mean of $[x]$ in the following cases:
  (i) $[x] = (104, 103, 98, 98, 100\frac{1}{2}, 100\frac{1}{2}, 99, 102, 103, 102\frac{1}{2}, 105, 97\frac{1}{2})$.
  (ii) $[x] = (825, 925, 750, 875, 850, 775, 725, 825)$.

2T. Prove algebraically the property stated in Section 4.1, that if all the frequencies in a population are divided by some constant then the mean is unaffected.
  Is the same true of (a) the mode? (b) the mean of the quartiles?

3T. Prove from the definitions of $\bar{x}$ and $\bar{t}$ that if
$$t = (x-k)/c, \quad \text{then} \quad \bar{x} = c\bar{t}+k.$$

4F. Prove directly from properties of sums that
$$\Sigma(t_i+1)f(x_i) = \Sigma t_i f(x_i)+\Sigma f(x_i).$$

5T. (i) Prove algebraically that $\bar{x} = \dfrac{1}{N}\Sigma(x_i-k)f(x_i)+k$ for arbitrary number k
and frequency function f.
  (ii) Deduce that if
$$M_1(k) = \frac{1}{N}\Sigma(x_i-k).f(x_i) \quad \text{then} \quad M_1(k) = 0 \Leftrightarrow k = \bar{x}.$$

$[M_1(k)$ is called the *first moment* of $[x]$ about $x = k$.]

6T. If we are given a horizontal sheet of material of negligible weight, balanced about a knife edge, and a mass y is placed on the sheet at a distance d from the knife-edge then the weight of the mass will exert a leverage or turning effect about the knife-edge. Doubling the mass or its distance from the knife-edge will double the turning effect, whatever may have been the original value of the mass or distance. The formula 'yd units' for the turning effect is the only one apart from change of units which will have this doubling property and it then follows from this formula that we need not confine ourselves to a factor 2 in the proportionality mentioned above.
  (i) Show that $\Sigma x_i f(x_i)$ is the turning effect about the $f(x)$-axis of masses $f(x_i)$ placed at the points $x = x_i$. (We can then think of the masses as giving a mechanical model of the population, or alternatively we may say that the mathematics we are doing provides a mathematical model for parts of both statistics and mechanics.)
  (ii) Explain how the moment function M_1, defined in Question 5(ii), is useful when evaluating the turning effect about $x = k$ of the masses in our mechanical model.
  (iii) What further assumptions do you need to make about masses of rectangular sheets when you say that if the bar frequency diagram were cut out of the sheet metal it would balance about a knife edge laid along $x = \bar{x}$?
  (iv) If b is the median of the population what can you say about a scissor-cut made along the line $x = b$?

4.2 Mean value of $g(x)$. In Section 4.1 we expressed a new quantity t as a function of x in the form $t = (x-k)/c$ and found the mean of $[t]$ by treating the frequency of a value of x as the frequency of the value of t to which it gave rise, that is to say we calculated quantities like $\Sigma t_i f(x_i)/\Sigma f(x_i)$. It is because each value of t arose from and gave rise to a unique value of x that we had the property that frequency of x_i = frequency of corresponding t_i. Notice that we avoid writing $f(x_i) = f(t_i)$, since we are using f as a functional symbol and not as a mere abbreviation of 'frequency'. Thus in Section 4.1 the frequency with which $t = 12$ is zero but the frequency with which $x = 12$ is 20, so $f(12) = 20$, and not $f(12) = 0$.

We should notice that, for some functions g, $f(x_i)$ is not necessarily the frequency of $g(x_i)$ as properly defined; the reason being that more than one value of x might give rise to the same value of $g(x)$. We have seen why the method of working in Section 4.1 gives the correct value for the mean of $[t]$ and an example may help to show that the frequencies combine correctly to give the mean of $[g(x)]$, even for a more complicated function, g, than that which gave t.

Consider the population $(-10, -9, 9, 9, 10, 12)$ with frequency table:

x	$f(x)$
-10	1
-9	1
9	2
10	1
12	1

If $g(x) = x^2$ then the population of values of x^2 is $(100, 81, 81, 81, 100, 144)$, and we have

$$\overline{x^2} = \frac{(81 \times 3) + (100 \times 2) + (144 \times 1)}{6}.$$

Obviously the frequencies have added up as necessary to let us write this as:

$$\overline{x^2} = \frac{((-10)^2 \times 1) + ((-9)^2 \times 1) + (9^2 \times 2) + (10^2 \times 1) + (12^2 \times 1)}{6}.$$

We can see from the table for $f(x)$ that this is of the form

$$\overline{g(x)} = \Sigma g(x_i) . f(x_i)/\Sigma f(x_i)$$

and our method is justified. We note that $\bar{x} = 3 \cdot 5$, $\overline{x^2} = 97\frac{5}{6}$ so that $\overline{(x^2)} \neq (\bar{x})^2$.

In general $\overline{g(x)} \neq g(\bar{x})$; that is to say the operations on the population of taking the mean and making the mapping g are not, in general, commutative.

Exercise 4D

1 R. In Chapter 3 the reader has learnt that the theoretical or idealized frequencies of the various possible scores x when two dice are thrown $36q$ times are given by:

$$f(x) = (6 - |x - 7|)\, q \quad \text{for} \quad 2 \leqslant x \leqslant 12.$$

[This is another triangular population.]
  For such $[x]$ find $\bar{x}$ and the mean value of $(x - \bar{x})^2$.

2 T. Find as general a function g as you can such that $\overline{g(x)} = g(\bar{x})$, whatever $[x]$ may be. Prove algebraically that the function has the property required and illustrate geometrically in the frequency diagram.

3 R. For $n = 6, 9, 12$ in turn find the mean value of $\sin^2\theta$, where $\theta = \dfrac{k}{n}.360°$, and integral values of k from 0 to $n-1$ have equal frequencies.

4 R. A number of light bulbs of equal brilliance are equally spaced along concentric circular rings of radii 1, 2, 3, 4 units. The number of bulbs per unit length of arc is the same for each arc. The illumination at any point due to a bulb at distance x units from the point is k/x^2. Find the mean illumination per bulb at the centre of the circle. What would have to be the radius of a single arc of these bulbs giving the same mean illumination per bulb? Would it depend on the spacing of the bulbs along the arc? On the angle subtended by the arc if it was not a complete circle? On the brilliance of the type of bulb used?

5. MERITS OF EACH STATISTIC OF POSITION

If frequency diagrams were drawn for $h(x)$ and $w(x)$ of Section 3.1 on the same diagram then each of the five statistics of Section 3.2 would provide a method of stating the position along the x-axis of the bulk of each diagram. Any one of them may be called an *average* value of x, the word 'average' not having any precise definition in this context. Each has its merits and demerits, as a descriptive statistic; that is as a statistic which conveniently summarizes a mass of data for comparison with another mass of similar data. These merits and demerits may differ from those that the statistic has when, having been calculated from a sample, it is to be used either to predict the likely properties of the parent population from which it is believed the sample was drawn, or to predict the likely properties of another sample not yet drawn. This problem of estimation was mentioned in Chapter 1 and when we take it up later in the course we shall principally use the mean for describing the position of a population.

  In really complicated experimental situations the median, quartiles and other statistics which only depend on arranging the sample in order (called *order-statistics*) are used, but the full theory is beyond the scope of this course. The methods are called *distribution-free*, which means that they are usable when little is known about the population. (See Exercise 1 A, Question 11 for a situation where the mean is unobtainable.)

We proceed now to a list of pros and cons of the statistics so far mentioned.

Means of extremes.

For: Very quick to find. (No frequency table needed.)

Against: Depends on only two values in the population. Any reliability it has is surpassed by the only slightly more complicated mean of the quartiles (see Section 3.3, population (iv)).

Not suitable for algebraic development, at this elementary level.

Mode.

For: Quick to find (but a frequency table has to be constructed).

If a small business makes only one size of some article, say of clothing, then that size which fits the modal size of person might sell most (see Section 3.3, populations (ii) and (iii)).

Against: In effect uses only a few members of the population and can be very misleading in a small population; the median is more *efficient* (which is a technical term roughly speaking meaning 'wastes less of the data') and is as quick to find.

Not suitable for algebraic development.

Median.

For: Quick to find (but a frequency table has to be constructed).

Usable even when the data have been coarsely grouped (see Exercise 1A, Question 11).

Not much affected by values of x exceptionally large or small; and used commonly, for instance, in quoting an average salary in an industry where the mean would be considerably affected by a few tycoons (see Section 3.3, population (iv)).

Against: Still only partly efficient (but hence its advantage over the mean in Section 3.3, population (iv)).

Not readily suitable for algebraic development at this elementary level (but see Exercise 7D, Question 19).

Mean of the quartiles.

For: As for the median, but preferable since considerably more efficient.

Against: As for the median.

The Mean (that is of the whole population).

For: Takes into account all the members of the population.

Does not require a frequency table to be constructed (this is a useful property when a calculator is being used but is less useful when the calculations are being done by hand since a frequency table reduces the number of separate calculations necessary).

The overwhelming advantage, however, is that its theory can readily

132

be developed algebraically, as has been partly seen in Exercise 4B, Questions 4, 5, 7.

Against: Sensitive to any marked departure from a simple bell-shaped frequency diagram (see Section 3.3, population (iv)), to which, however, many populations conform.

6. SIGNIFICANCE

If two populations are being compared then, whatever their sizes may be, the statistics calculated from them give us condensed information about them. If, however, the populations are being regarded as samples from a couple of parent populations and we intend to use the statistics to estimate some property of the parent populations, then we can place less reliance on estimates derived from smaller samples than on those derived from larger. Some statistics are less sensitive to size in this respect than others and for a reasonably bell-shaped frequency diagram the median, mean of the quartiles and mean, in that order, are increasingly efficient statistics. The efficiency, as it is called, is connected with the proportion of the data in the sample which the statistic in effect uses.

A further treatment of this occurs in Chapter 11 on Samples and Chapter 12 on Estimation. At present the samples or populations are being kept small in order to keep the calculations within reasonable limits, in the hope that meanings of the terms will not be obscured by heavy arithmetic.

7. SPREAD AND VARIABILITY

7.1 Comparison of spreads. Having considered, in Section 3.2, ways of answering the question: 'Which author wrote the longer sentences?' We now ask the question: 'Which author varied his sentence-length more, in the passages under consideration?'

The question of 'how varied?' is next in importance to that of how big; in fact it is virtually useless to know merely the mean of a population without knowing how clustered round that mean are the members of the population. A good example of this is provided by the following pair of experiments.

First: A single die is thrown and the score is recorded. This is done 72 times. The numbers will form a population whose mean is about 3·5 and whose extremes are about 1 and 6.

Secondly: Two dice are thrown as a pair and the total score is recorded. This is done 72 times. When these scores are halved they form a population whose mean is about 3·5 and whose extremes are about 1 and 6.

The reader's study of probability in Chapters 2 and 3 will lead him to expect quite different frequency functions from these experiments, although both populations will be characterized by much the same mean.

Before reading on the reader should carry out these experiments and keep the results, in the form of frequency tables, for later reference.

A second glance at the table of Section 3.1 shows that the sentence-length in *TWOAF* is more varied than the sentence-length in *THOMP*. If the frequency functions were plotted on frequency diagrams the ordinates for *TWOAF* would be more spread out along the x-axis than those for *THOMP*. The spread of the ordinates along the x-axis gives an indication of the variability of the sentence-length. We can describe this more precisely in many ways. We list two in this section and consider more fully two other ways later.

(i) *The Range.* This is the difference of the extremes of the population.

The range of *THOMP* is $37-3$, that is 34.

The range of *TWOAF* is $122-4$, that is 118.

The range is dependent on only two of the values of x and this makes it less and less efficient for samples of larger and larger size; for instance, it is no help in distinguishing between the populations produced by the two experiments above. It is, however, so simple that it can be easily used in industrial quality control and as a result there is a large body of theory about its behaviour in samples of *small* size.

(ii) *The Interquartile Range.* (IQR.) This is the difference of the first and third quartiles of the population.

The IQR of *THOMP* is $21-6$, that is 15,

The IQR of *TWOAF* is $40-19$, that is 21.

Roughly speaking it is the length of the interval that covers the central 50% of the population. The general characteristics of this statistic correspond to those of the median (see Section 5).

Both these statistics are order-statistics (see Section 5) and the reader used them in Exercise 1A, Question 11. They are useful to us as simple guides and we will develop the IQR a little further when considering interval-estimation.

7.2 Deviations from a central value: mean absolute deviation. An altogether more fruitful approach is to consider deviations from some central value and to devise a statistic which takes account of them all. We will develop this approach to compare the two populations

$$[y] = (78, 81, 82, 85, 87, 87, 88),$$

$$[z] = (63, 64, 67, 70, 71, 71, 72, 73, 73, 73, 73),$$

(noting in passing that they have the same range (10) and the same inter-quartile-range (6)).

There are, of course, several central values (or averages) which we might select as origins from which to measure our deviations; that was the message of Section 3; but Exercise 4B, Questions 4, 5, 7 showed us how well the mean yielded to algebra and it is the mean we will use here. Inspection shows that $\bar{y} = 84$, $\bar{z} = 70$.

Subtracting these from the members of the corresponding populations we get:

[[y]-deviations] $= (-6, -3, -2, +1, +3, +3, +4)$,

[[z]-deviations] $= (-7, -6, -3, 0, +1, +1, +2, +3, +3, +3, +3)$

and it is helpful to plot them on a diagram as in Figure 4.9. Looking at the figure it is difficult to judge by eye which of the populations is the more scattered, and different people will reach different conclusions by basing their opinions on different features of the two patterns.

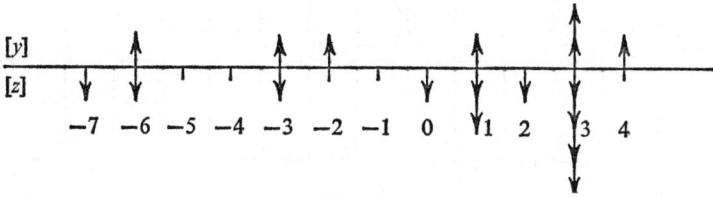

Fig. 4.9

To return to well-defined mathematical processes, we shall obviously get nowhere useful if we merely sum deviations, since those on the left will tend to cancel those on the right whereas, because they are all parts of the overall scatter or variability, we want them to reinforce. For this reason we might drop the negative signs of those on the left, that is to say, we might take the *absolute deviations*. We should get

Sum of absolute deviations

for [y] $= 6+3+2+1+3+3+4 = 22$,

for [z] $= 7+6+3+0+1+1+2+3+3+3+3 = 32$.

We obviously ought to take the *mean* of the absolute deviations if we are to construct a statistic that will make use of all the deviations, and yet will allow sensible comparison of populations of *different sizes*. This gives:

Mean absolute deviation from mean

for [y] $= \frac{22}{7} = 3 \cdot 1$,

for [z] $= \frac{32}{11} = 2 \cdot 9$.

[y] seems to be the more scattered.

So far, so good; but the reader has only to try this process for the population consisting of the numbers a, b, c, *not knowing their order of size*, to see that the algebra of the method is outrageously clumsy. If the mean is m then we cannot tell whether the absolute value of $(a-m)$, written $|a-m|$,

135

has the value $(a-m)$ or the value $(m-a)$. The several possible cases would have to be dealt with separately.

7.3 Deviations from a central value: variance. There is, however, another way of getting all the signs positive, so that the effects of the deviations reinforce whether they are deviations to the left or to the right: we could square all the deviations. Figure 4.10 is the same as Figure 4.9, except for the inclusion of a scale of squared-deviations.

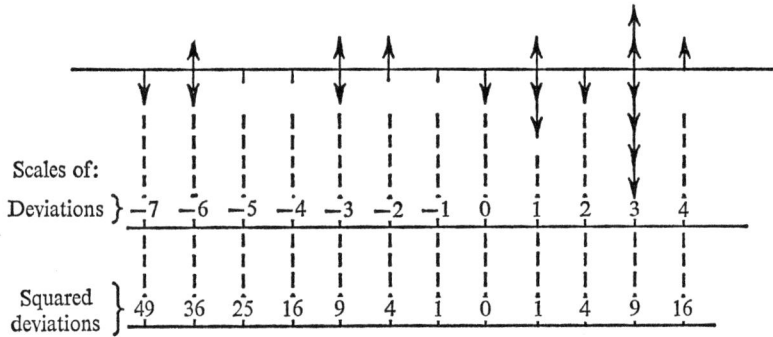

Fig. 4.10

Sum of squared-deviations

for $[y] = 36+9+4+1+9+9+16 = 84,$

for $[z] = 49+36+9+0+1+1+4+9+9+9+9 = 136.$

(The fact that the sum for $[y]$ has the same numerical value as $\bar{y}$ is mere coincidence, as the results for $[z]$ show.)

It follows that

Mean squared-deviations from mean of $[y] = \frac{84}{7} = 12{\cdot}0,$

of $[z] = \frac{136}{11} = 12{\cdot}4.$

It now seems that population $[z]$ is more scattered. This apparently conflicts with the result from the mean absolute deviation, but it is senseless to ask which is correct. Both conflict with the conclusions suggested by both the range and the interquartile range. There is no right or wrong in these matters, there is only more or less suitable. One might as well ask whether an elephant or a giraffe is bigger: it depends on the criterion. Nevertheless hence come the popular gibes 'You can prove anything by statistics' and 'There are lies, damned lies and statistics'. What is certain, of course, is that having chosen, in a particular type of problem, a measure of spread on some grounds or other we must stick to it in that type of problem. It is the statistic defined as the mean of the squares of the deviations from the mean that we will develop. It is called the *variance*. The variance itself and its square root (called the *standard deviation*, often

136

abbreviated to S.D.) will be the precisely defined statistics that we will use to express information about scatter, spread or variability (these being the loose terms which correspond here to the loose terms 'position' and 'average' in Section 3).

Exercise 4E

1R. Find, (i) the variance directly from the definition as the mean of the squared-deviations-from-the-mean, and (ii) the standard deviation, for the following populations:

    (a) (3, 3, 6);
    (b) (2, 2, 8);
    (c) (0, 6, 6);
    (d) (1, 2, 3, 4, 5);
    (e) (2, 4, 6, 8, 10);
    (f) (3, 5, 7, 9);
    (g) (2, 3, 4, 5, 6, 7, 8, 9, 10).

2R. 16 samples, of 100 housewives each, were selected in a certain market research project and the numbers of wives using the brand of detergent concerned were:

    18, 61, 56, 22, 31, 9, 72, 23, 68, 68, 24, 19, 19, 63, 14, 21.

These numbers can be seen to fall into two distinct sub-populations, find the mean and standard deviation of each.

3R. Construct an asymmetrical frequency diagram for a population with mean zero and variance 2.

4R. Without actually calculating any of the three statistics mentioned, construct a frequency function for a population having the same size, mean and variance as the one with the following frequency function:

x_i	$f(x_i)$
0	1
1	6
2	15
3	20
4	15
5	6
6	1

[Note on this frequency function: In a certain trial, A or not-A can occur and the corresponding probabilities are each $\frac{1}{2}$. A certain experiment consists of six such trials and x is the number of occurrences of A in the experiment. 2^6 such experiments are carried out. If the relative frequency of occurrence of a given value of x is found to be exactly equal to the probability of that value of x in the model, then the frequency function is the one given in this question. The reader may have already investigated the case of the experiment consisting of five such trials in Exercise 3M, Question 44, and the general case will be dealt with in Chapter 7. The type of population is called *binomial*.]

5R. (*a*) Find the mean and variance from the frequency tables produced by each of the experiments of Section 7.1.

(*b*) Construct the theoretically ideal frequency tables for these experiments and calculate the mean and variance from these tables.

(*c*) What would have been the likely effects on your four answers if instead of using 72 throws you had used (i) 7200 throws? (ii) 36 throws?

(This question shows the usefulness of the variances in distinguishing between two significantly related populations.)

6T. Calculate the mean and variance of each of the populations given by the following tables:

x_i	$f(x_i)$	$g(x_i)$	$h(x_i)$
-3	0	1	1
-2	3	1	2
-1	6	5	5
0	7	11	6
1	6	5	10
2	3	1	1
3	0	1	0

Draw bar frequency diagrams.

(This question shows that even the mean and variance may not give enough information to distinguish between populations. Other statistics are discussed in Exercise 6M, Questions 16, 17 and 18.)

⋆7T. (*a*) By considering the underlying algebra for Question 1, populations (*e*), (*f*) and (*g*), show that if two populations have equal means and have sizes N_1, N_2 and variances v_1 and v_2, then the population obtained by amalgamating them has variance given by

$$v = (N_1 v_1 + N_2 v_2)/(N_1 + N_2).$$

(*b*) Construct a pair of simple populations with unequal means to find whether or not the above relationship between variances sometimes breaks down.

(*c*) Writing the s.d. corresponding to v_1, v_2 and v as S_1, S_2 and S respectively, write down that relationship between the s.d. that corresponds to the above relationship between the variances.

8T. Calculate the mean absolute deviation from the median of each of the populations [*y*], [*z*] in Section 7.2 and note that the values of this statistic are smaller than the corresponding values calculated in Section 7.2 (where the deviations were taken from the mean).

9F. By considering the underlying algebra for Question 1, populations (*d*) and (*e*), prove that if all the values in a population are multiplied by *c* then the variance is multiplied by c^2, but the s.d. is multiplied by *c*.

10F. Tests are held in two subjects for eleven candidates for a competition, and the raw marks are:

Candidate	Maths.	French
A	56	20
B	51	25
C	29	23
D	29	12
E	28	23
F	26	16
G	22	7
H	21	18
I	21	7
J	19	5
K	16	10

(a) By the use of a *linear* transformation for each population of marks transform them to (i) two populations with the same range, (ii) two populations with the same interquartile range.

(b) Transform the collections of marks to two orders of merit.

(c) Assuming that the orders of merit from each subject are similar enough to justify combining the marks in some way, use each transformed collection to produce a possible final order of merit of the candidates. What would be your reactions if you were candidate D, and K was your great rival?

11 F. Given that the variance of the population $[y] = (78, 81, 82, 85, 87, 87, 88)$ is 12·0 (so that s.d. $= \sqrt{12\cdot0} = 3\cdot46$) *write down* the variance and s.d. for each of the following related populations:

(a) (77, 80, 81, 84, 86, 86, 87);

(b) (0, 3, 4, 7, 9, 9, 10);

(c) (10, 7, 6, 3, 1, 1, 0);

(d) (0, 6, 8, 14, 18, 18, 20);

(e) (−3, 6, 9, 18, 24, 24, 27);

(f) (7·7, 8·0, 8·1, 8·4, 8·6, 8·6, 8·7).

★12 F. State the most general relation you can find between y and a new variable t such that the variance of $[t]$ can be *written down* in terms of the variance of $[y]$. State also the resulting relation between (a) the two variances; (b) the square roots of the two variances, (that is: the two s.d.).

8. VARIANCE

8.1 Effect of translation. The subtraction of an arbitrary constant k, say, from all the values of x to obtain a short cut in the calculations of the mean of $[x]$ must not be confused with the subtraction of some central measure from the values of x to obtain a measure of the spread of $[x]$. When calculating the mean we have to restore the subtracted constant since the whole purpose of the mean is to tell us about the actual sizes of the x-values. In measuring the spread, however, it is the very deviations from the central value and not the actual sizes of the x-values that concern us, so there is no question of returning to the original origin. Any satisfactory measure of spread must be unaffected by a translation.

139

The variance is unaffected; for, if all we have done is to translate the population by the subtraction of k, say, then we shall have left unaltered all the intervals between members, and hence all corresponding deviations including those from the mean; we shall consequently have left unaltered their squares for use in the variance. This idea was used in Exercise 4E, Question 11 parts (a) (b), where $k = 1$ and $k = 78$ respectively.

If therefore we have subtracted k from the members of a population to reduce their sizes as a step towards calculating the mean, we might as well use the resulting smaller values in finding the variance. In Figure 4.8 the variance of $[x']$ is equal to the variance of $[x]$; we will write this symbolically as var $[x']$ = var $[x]$.

8.2 Working-zero. The use of a translation is, however, not always necessary, and of a *mere* translation is not always sufficient, for our case.

If on the one hand we have very simple little populations in which the means are known integers (such populations as $[y]$ and $[z]$ in Section 7.2) then we can always calculate the variance without a translation by finding the deviation from the mean directly. Nothing can short-cut that.

If on the other hand the mean is not an integer, then the arithmetic is very tedious, even after a translation. Suppose, for instance, that we have the population
$$[x] = (78, 81, 82, 86, 87, 87, 88).$$

We subtract 84 from all the members to get the population
$$[x'] = (-6, -3, -2, +2, +3, +3, +4),$$

whose mean is quickly seen to be $+\frac{1}{7}$; giving $84\frac{1}{7}$ for the mean of $[x]$. The x-deviations from $\bar{x}$ are $(-6\frac{1}{7}, -3\frac{1}{7}, -2\frac{1}{7}, +1\frac{6}{7}, +2\frac{6}{7}, +2\frac{6}{7}, +3\frac{6}{7})$ so the squared deviations are all non-integral, and the arithmetic will be heavy. If we hope to simplify matters by considering, instead, the translated population then our hope is vain for the x'-deviations from $\bar{x}'$ are necessarily equal to the x-deviations from $\bar{x}$. Indeed this necessary equality can be used as a check on the arithmetic, by the subtraction of $\frac{1}{7}$ from all the members of $[x']$. The translation has not brought us the advantage that we sought in the calculation of the variance, but it has, of course, made all the numbers smaller. The resulting smallness is more than just *convenient* if we are using only a few significant figures from tables, as we shall see more fully in Section 10.3, Note (ix).

It would be useful to have an algebraic result which would enable us to calculate the mean of the squared deviations from some origin *not* necessarily the mean and then apply a correction in order to obtain the variance, that is the mean of the squared deviations from the mean. In practice, however, when finding first the mean of a population we would nearly always have translated the population to obtain zero as a satisfactory origin to work from. We therefore proceed by studying the mean squared-deviation-from-zero (which is merely another way of saying the mean squared-value) and we do not trouble ourselves at this stage with deviations from a general origin.

We try to find how to relate the mean squared-value to the variance by considering a simple algebraic case first. We take the numbers a, b, c with mean m, which thus satisfies $3m = a+b+c$. Then

$$\text{mean squared value} = (a^2+b^2+c^2)/3,$$

whereas

$$3 \times \text{variance} = (a-m)^2+(b-m)^2+(c-m)^2, \quad \text{[this form avoids writing fractions]}$$

$$= (a^2+b^2+c^2)-2m(a+b+c)+3m^2,$$

$$= (a^2+b^2+c^2)-2m.3m+3m^2,$$

$$= (a^2+b^2+c^2)-3m^2,$$

giving

$$\text{variance} = (a^2+b^2+c^2)/3 - m^2$$

that is (speaking loosely):

$$\text{variance} = (\text{mean of squares}) - (\text{square of mean}).$$

This is a remarkably simple result and is true for any population, as we will prove in Section 8.3.

We can rewrite it symbolically as:

$$\text{var}\,[x] = \overline{(x^2)} - (\bar{x})^2$$

or

$$\text{var}\,[x] = \frac{1}{N}\Sigma x_i^2 f(x_i) - (\bar{x})^2.$$

Example 1. To find the variance of $[x]$ given in Section 8.2.

$\text{var}\,[x] = \text{var}\,[x']$, since we have only translated the population,

but

$$\text{var}\,[x'] = (\text{mean of squares}) - (\text{square of mean})$$

$$= \tfrac{1}{7}(36+9+4+4+9+9+16) - (\tfrac{1}{7})^2$$

$$= \tfrac{87}{7} - \tfrac{1}{49}$$

$$= 12\tfrac{20}{49},$$

thus

$$\text{var}\,[x] = 12\tfrac{20}{49}.$$

When using this new result beginners often make the tactical error of not first translating the original population, in other words of not employing a suitable zero by subtracting a convenient constant from all the values.

141

Sometimes the numbers are all small enough already, but in the case above the beginners would get:

$$\text{var } [x] = (\text{mean of squares}) - (\text{square of mean})$$

$$= \tfrac{1}{7}(6084 + 6561 + 6724 + 7396 + 7569 + 7569 + 7744) - (84\tfrac{1}{7})^2$$

$$= \frac{49{,}647}{7} - (7080\tfrac{1}{49})$$

$$= 12\tfrac{20}{49}, \text{ as before, but with considerable trouble (and difficulty over significant figures; see Section 10.3, Note (ix)).}$$

Exercise 4 F

1 R. By subtracting the given value of k from each of the numbers, find new populations with the same variance (and hence s.d.) as each of the given populations, and then use the method of Example 1 to find the variance and s.d. of each population:

(i) $(1, 2, 3, 4)$; $k = 0$. In this case calculate also the variance *directly*, as the mean of the squared-deviations-from-the-mean. [This and the next two cases are *uniform* populations.]

(ii) $(1, 2, 3, 4, 5, 6, 7, 8)$; $k = 4$.

(iii) $(0, 1, 2, 3, \ldots, (n-2), (n-1))$; $k = -1$.

(iv) $(11, 11, 11, 11, 11, 11, 12, 13, 13, 14, 14, 14, 14)$; $k = 12$.

(v)	x	$f(x)$	; $k = 70$.
	69	3	
	70	7	
	71	6	
	72	4	

(vi)	x	$f(x)$	; $k = 700$.
	690	3	
	700	7	
	710	6	
	720	4	

2 T. The following (fictitious, but typically varied) figures show moisture-content, in arbitrary units, of samples from various positions in four coke wagons at a steel works.

			Position		
Wagon	1	2	3	4	5
1	5	7	4	6	3
2	8	12	7	7	1
3	5	9	19	4	3
4	0	2	0	7	1

(i) Find the mean and variance of the moisture-content for each wagon separately.

(ii) Find the mean of these variances and the variance of these means. [These are called the average variance *within* classes and the variance *between* classes; they become of value in the branches of statistics called 'Analysis of Variance', and 'Design of Experiments'.]

(iii) Verify that the sum of the last two statistics calculated is the variance of the whole population of twenty measurements.

(Not as simple a relation connects the S.D., of course.)

8.3 Proofs of two results.

We have so far indicated two results which help us to obtain more simply the variance and standard deviation of a population. They are

(a) variance = (mean of squares) − (square of mean);

(b) translation of a population by addition of a constant does not affect the standard deviation.

Result (a): Variance = (mean of squares) − (square of mean).

Proof: Let the population have members t_i, with frequencies $f(t_i)$. We have

$$N \times \text{variance} = \Sigma(t_i - \bar{t})^2 f(t_i), \quad \text{[this expresses the definition of the variance symbolically]}$$

$$= \Sigma[t_i^2 - \bar{t}t_i - \bar{t}(t_i - \bar{t})] f(t_i),$$

$$= \Sigma t_i^2 f(t_i) - \Sigma\bar{t} . t_i f(t_i) - \Sigma\bar{t}(t_i - \bar{t}) f(t_i),$$

$$= \Sigma t_i^2 f(t_i) - \bar{t} . \Sigma t_i f(t_i) - \bar{t}\Sigma(t_i - \bar{t}) f(t_i),$$

$$= \Sigma t_i^2 f(t_i) - \bar{t} . \Sigma t_i f(t_i) - \bar{t} \times 0.$$

Hence

$$\text{variance} = \frac{1}{N}\Sigma t_i^2 f(t_i) - \bar{t} . \frac{1}{N}\Sigma t_i f(t_i)$$

$$= \frac{1}{N}\Sigma t_i^2 f(t_i) - (\bar{t})^2.$$

Notice that the treatment of the terms after the first differs slightly from that in the exploratory approach of Section 8.2, which, however, provides a sound and perhaps easier, if less 'insight-ful', proof.

Result (b): Translation of a population by addition of a constant does not affect S.D.

[Comment on the result: It is obvious, by what we mean by the word 'spread', that a translation does not affect the spread. The fact that it does not affect the S.D. (as we show below) is an indication that the S.D. (which we constructed *intuitively* in Section 7.3 as a measure of spread) possesses at least one important property of such a measure.]

Proof: Let the old population have members x_i' with frequencies $f(x_i')$.

Let the new population be given by $x = x' + k$, then $\bar{x} = \bar{x}' + k$, as proved in Exercise 4C, Question 5. We have

$$x_i - \bar{x} = (x_i' + k) - (\bar{x}' + k) = (x_i' - \bar{x}')$$

and also   frequency of x_i = frequency of x_i' = $f(x_i')$.

Hence   new variance = $\dfrac{1}{N}\Sigma(x_i-\bar{x})^2 f(x_i')$

$$= \dfrac{1}{N}\Sigma(x_i'-\bar{x}')^2 f(x_i')$$

$$= \text{old variance.}$$

Thus, new S.D. = old S.D., by square rooting.

Exercise 4 G

1 R. Give a general proof of Result (*a*) of Section 8.3 by a method which treats the terms as they were treated in the simple case in Section 8.2.

2 T. (i) By using Results (*b*) and (*a*) of Section 8.3 show that

$$\text{var}\,[x] = \dfrac{1}{N}\Sigma(x_i-k)^2 f(x_i)-(\bar{x}-k)^2$$

for arbitrary k; and explain in words why this is only another form of Result (*a*).

(ii) Prove that if a population [*x*] is given and the mean of the squared deviations from $x = k$ is calculated, then the result is at least as large as the variance.

(iii) Express the result of part (ii) in terms of sums rather than means, and use the new form to define the variance of a population without explicit reference to the mean and then to define the mean with reference to the variance. [This is an example of a definition employing what is called '*The method of least squares*'. Another example occurs in Section 9.]

3 T. [Continuation from Question 2.]

The form $\text{var}\,[x] = \dfrac{1}{N}\Sigma(x_i-k)^2 f(x_i)-(\bar{x}-k)^2$ with $k \neq 0$

is sometimes useful in small calculations which we may wish to do mostly mentally. Thus in Exercise 4F, Question 2, the four means of the moisture contents were 5, 7, 8, 2, having *their* mean as $5\frac{1}{2}$. If we take $k = 5$ then the correction will be $(5\frac{1}{2}-5)^2$, that is $\frac{1}{4}$. We can then calculate the squares of the deviations of 5, 7, 8, 2 from 5, take their mean and subtract $\frac{1}{4}$ (Remark: the variance comes from the *least* sum of squares, hence '*subtract*'). The resulting succession of calculations runs:

$$0;\quad 4;\quad 0+4 = 4;\quad 9;\quad 4+9 = 13;\quad 9;\quad 13+9 = 22;$$
$$22/4 = 5\frac{1}{2};\quad 5\frac{1}{2}-\frac{1}{4} = 5\frac{1}{4}.$$

To practise this three times, calculate with the minimum of writing the variance of the numbers in each of the first three columns of the table in Exercise 4F, Question 2.

4T. (i) Complete the following table.

x	0	1	2	3
$f(x)$	1	4	3	2
$(x-0)$				
$(x-1)$				
$(x-2)$				
$(x-3)$				

(ii) Use the table to calculate for $k = 0, 1, 2, 3$
    (*a*) the mean of the deviations of x from $x = k$, written $M_1(k)$ below;
    (*b*) the mean of the absolute deviations of x from $x = k$, written $D(k)$ below;
    (*c*) the mean of the squared deviations of x from $x = k$, written $M_2(k)$ below.
[$M_r(k)$ is called the rth moment of [x] about $x = k$; compare Exercise 4C, Question 5.]
(iii) Sketch the graphs of $M_1(k)$, $D(k)$ and $M_2(k)$ against k.
(iv) Use the result of Question 2 part (ii) to approximate to the variance and mean of [x] from the graph of $M_2(k)$.
(v) Check and if necessary correct your graph of $D(k)$ by calculating $D(k)$ for $k = \frac{1}{2}, 1\frac{1}{2}, 2\frac{1}{2}$.
(vi) Give two separate ways in which the graph of $M_1(k)$ could be used to approximate to $\bar{x}$.

*5F. A set of data generalized from Exercise 4F, Question 2 is as follows:

	Position 1	Position 2	.	Position j	.	Position l	mean	variance
Wagon 1	x_{11}	x_{12}	.	.	.	x_{1l}	m_1	v_1
Wagon 2	x_{21}	x_{22}	.	.	.	x_{2l}	m_2	v_2
.	.	.				.	.	.
Wagon i	.	.	.	x_{ij}	.	.	m_i	v_i
.	.					.	.	.
Wagon k	x_{k1}	.	.	.	.	x_{kl}	m_k	v_k

The mean of the whole population of kl members is m.
(i) Show that m is also the mean of the population of values of m_i.
(ii) By writing

$$v_i = \frac{1}{l} \sum_{j=1}^{j=l} (x_{ij} - m)^2 - (m_i - m)^2, \quad (i = 1, 2, \dots, k),$$

and summing for $i = 1, 2, \dots, k$, prove that the variance of all the kl members is equal to the sum of the mean of [v_i] and the variance of [m_i]; that is to say 'is equal to the mean of the variances plus the variance of the means'. [In 'Analysis of Variance', terms closely connected with these are used to estimate what are called 'the variance within classes' and 'the variance between classes'. See Exercise 12B, Question 24.]

*6. The tables show measurements of the yields of nine varieties of wheat grouped in three groups of three varieties on various plots.

	Plot 1	Plot 2	Plot 3
variety 1	2	4	6
variety 2	1	5	9
variety 3	3	6	9

	Plot 4	Plot 5	Plot 6		Plot 7	Plot 8	Plot 9
variety 4	12	14	16	variety 7	3·6	4	4·4
variety 5	21	25	29	variety 8	4·2	5	5·8
variety 6	33	36	39	variety 9	5·4	6	6·6

For *each* table calculate (i) the mean of the variances, (ii) the variance of the means, as defined in Question 5; and consider how these quantities tell us something about the importance to be attached to the measured differences between the varieties.

★7T. The mean of 40 readings of a variable was 6·03 and their S.D. was 0·28. The following 10 readings were then added to the list:

5·40, 6·41, 6·18, 6·30, 6·65, 5·58, 6·38, 6·45, 5·81, 6·00.

Obtain the mean and S.D. of the new population of 50 readings. The following method is suggested:

(i) Call the first 40 readings y_i (not necessarily all different) and consider them to form a population $[y]$, having $\bar{y} = m_1$ and $\mathrm{var}[y] = v_1$. Call the next 10 readings z_j and consider them to form a population $[z]$, having $\bar{z} = m_2$ and $\mathrm{var}[z] = v_2$. Let the whole population of 50 readings have mean M and variance V.

(ii) Show that:
$$V = \{\sum_i (y_i - M)^2 + \sum_j (z_j - M)^2\}/50.$$

(iii) Express $\frac{1}{40}\sum_i (y_i - M)^2$ in terms of m_1, v_1, M and $\frac{1}{10}\sum_j (z_j - M)^2$ in terms of m_2, v_2, M.

(iv) Calculate m_2, v_2, M.

(v) Calculate V.

★8T. [This question generalizes the results of Exercise 4E, Question 7, and of Question 5 of the present exercise. In the first of these the means of the sub-populations were equal and in the second the sizes of the sub-populations were equal.]

Two populations have sizes N_1, N_2, means m_1, m_2 and variances v_1 and v_2. They are amalgamated to form a single population with mean M and variance V.

(i) Prove:
$$V = \frac{N_1 v_1 + N_2 v_2}{N_1 + N_2} + \frac{N_1(m_1 - M)^2 + N_2(m_2 - M)^2}{N_1 + N_2},$$
$$M = \frac{N_1 m_1 + N_2 m_2}{N_1 + N_2}.$$

(ii) Show that the results of part (i) may be expressed as follows:
We define two distinct populations: $[x]$. The first being $x_1 = m_1, x_2 = m_2$, the second being $x_1 = v_1, x_2 = v_2$ and each having $f(x_i) = N_i$ for $i = 1, 2$.
We then have
$$V = \bar{v} + \mathrm{var}[m], \quad M = \bar{m}.$$

(iii) Guess and prove a generalization of part (ii).

★9T. (i) In part (i) of Question 8 write $R_i = N_i/\Sigma N_i$ for $i = 1, 2$ and prove:
$$V = \Sigma(v_i R_i) + R_i R_2(m_1 - m_2)^2, \quad M = \Sigma(m_i R_i).$$

(ii) Make a general statement about Exercise 4E, Question 7, part (b).

146

*10 T. Prove from the definition of the median in Section 3.2, and taking care over the various complications suggested by Exercise 4 C, Question 8, that for any [x] the mean absolute deviation from $x = k$ is not less than the mean absolute deviation from the median, thus confirming the result of Exercise 4 E, Question 8.

[*Hint:* The graph of $D(k)$ in Question 4 will suggest the behaviour of the mean absolute deviation.] Compare the form of this result with that of Question 2 part (ii) of the present exercise.

9. LEAST SQUARES: REGRESSION

[This section and Exercise 4 H develop the idea of 'least squares' mentioned in Exercise 4 G, Question 2. They may be left until later if the reader desires.]

We were reminded in Chapter 1, Section 4, that in the course of experimental work we may well have grounds for expecting some relationship between two variables, but that the readings taken in an experiment will not *exactly* satisfy any simple relationship. For instance if x, y are expected to be linearly related we may typically get a graph like that in Figure 4.11.

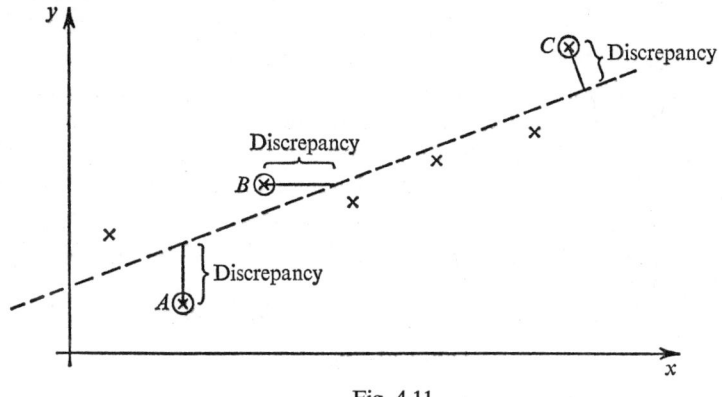

Fig. 4.11

We expect the points to lie on some straight line that can be drawn near them. In practice we may draw by eye 'a good straight line through the points', but we can apply the methods developed in Section 7.3, and their development in Exercise 4 G, Question 2, to set up a criterion for the 'discrepancy' of the points from a proposed line and can, by minimizing this 'discrepancy', select a line which has a claim to be called the 'best fit'.

As in the case of a statistic to measure the similar quantity which is the overall deviation of the members of a population from some single value, there are many different approaches to this discrepancy problem.

In the case of a population we can choose a central point which minimizes the maximum absolute deviation from itself. This would clearly mean choosing the mean of the extremes and we have rejected this 'central measure' for a population on account of its apparent crudity. In the case of fitting a straight line the method has some intuitive appeal but is not very simple to apply. We have to choose a line to minimize the maximum of the absolute discrepancies between the observed values and those 'predicted' by the line of our choice. This might mean choosing a line which, to fit some awkward point, would not pass particularly near any points, and in addition the analysis would be very heavy.

The next suggestion we might make for a population is to choose a central point which minimizes the mean absolute deviation from itself. Exercise 4G, Questions 4 and 10 suggest that this selects the median as the central measure and so we can (perhaps unexpectedly?) obtain the measure by the mere process of counting. In the case of fitting a line however, the method is not so simple to apply, and we look further.

The statistic we will minimize, by choice of line, corresponds to the variance of a population. It is the mean of the squares of the discrepancies between the observed and predicted values. We now have to be precise about what we shall call the discrepancy. Figure 4.11 shows three particular points A, B, C for each of which a different measure of discrepancy is suggested, We will first deal with the type of discrepancy at point A.

Let 'the line of best fit', which we seek, have equation $y = px+q$, where we do not yet know p, q. Let x be the variable we first measure and y be the variable whose value we then try to predict from our line. Let (x_i, y_i) be an *observed* number-pair. The usual names for the variables in this situation are *independent variable* for x, and *dependent variable* for y. The appropriate type of discrepancy is clearly that shown for point A and this discrepancy is

$$\text{(predicted value of } y) - \text{(observed value of } y)$$

that is $(px_i+q)-y_i.$

This we may call the y-error, and the mean squared y-error is

$$\frac{1}{N} \sum_i (px_i+q-y_i)^2.$$

Our problem is to choose p, q to minimize this for the given $[(x_i, y_i)]$.

Sometimes, however, an experiment may be used in such a way that the roles of x and y as independent and dependent variables are interchanged. For instance, we might use a graph relating temperature, x, and electrical resistance, y, of a metal wire in two ways. First, we might predict the resistance, y, when the wire, as part of some mechanism or other, becomes heated to a given temperature, x, and secondly we might use the graph as a thermometer scale to give the temperature, x, say inside a furnace, from the measured resistance, y, of the wire. There is no reason to suppose that

148

the mean square error will be the same as before when this time the values of y are thought of as given and the values of x are thought of as being in error. The typical error in this second situation is shown at B on the graph. The statistic to be minimized by choice of p, q in $y = px+q$ is then

$$\frac{1}{N} \sum_i \left(\frac{y_i-q}{p} - x_i\right)^2.$$

This is more usefully written

$$\frac{1}{N} \sum_i (ry_i+s-x_i)^2,$$

where $r = 1/p$, $s = -q/p$ and the equation of the line is $x = ry+s$. When the first statistic is minimized, $y = px+q$ is called the *regression line of y on x*; when the second statistic is minimized, $x = ry+s$ is called the *regression line of x on y*.

Any attempt to deal with x and y symmetrically by considering the error as measured by the perpendicular distance of (x_i, y_i) from a line, as at point C, is not as meaningful in any given problem as the appropriate one of the two methods already given; and, additionally, leads to very awkward algebra.

The whole topic of the association of two variables, their possible linear correlation and their regression curves is a much larger one than we will deal with here. The purpose of this section is that of throwing a little light on the very powerful method of least squares, by seeing it applied to something other than the variance. The rest of the development we leave to the reader in Exercise 4H.

Exercise 4H

1R. The following values of (x_i, y_i) occur in an experiment

i	x_i	y_i
1	-2	-3
2	-1	-1
3	0	0
4	1	2
5	2	2

and there are theoretical grounds for expecting the relationship to be linear.
  (i) Calculate $\bar{x}$, $\bar{y}$.
  (ii) Take the lines

$$y-\bar{y} = p(x-\bar{x}) \quad \text{or} \quad y = p(x-\bar{x})+\bar{y},$$

and

$$x-\bar{x} = r(y-\bar{y}) \quad \text{or} \quad x = r(y-\bar{y})+\bar{x},$$

and calculate, in terms of p, r, the statistics:

$$\tfrac{1}{5}\Sigma(p(x_i-\bar{x})+\bar{y}-y_i)^2, \quad \tfrac{1}{5}\Sigma(r(y_i-\bar{y})+\bar{x}-x_i)^2.$$

(iii) Using your knowledge of quadratic functions (differentiation is unnecessary) choose p, r to minimize the statistics calculated above.

[Notice that we have in this question arbitrarily limited the class of line which we are considering for the line of best fit to the class of lines through $(\bar{x}, \bar{y})$. Question 4 shows that the line of best fit that we seek is in fact such a line.]

(iv) Draw a graph showing the observed (x_i, y_i) and the two selected lines.

2 R. It may happen that two quantities are only vaguely related and that no precise prediction would be reasonable; the points when plotted occupy quite an extended region of the (x, y)-plane. This sort of situation was indicated in Chapter 1, Section 4. The diagram of plotted points is called a scatter-diagram. Draw the scatter-diagram for the Mathematics marks, x, and the French marks, y, of Exercise 4E, Question 10. Investigate how the lines $x + 2y =$ constant and $2x + y =$ constant are related to the two systems of transforming the marks there employed, and hence to the two choices of 'top scorer'. Could any other system of linear transformations produce some third candidate as top scorer?

[This problem, where neither variable may reasonably be thought of as dependent on the other, is one of *correlation* rather than *regression*.]

3 R. The number, x_i, of millions of pounds spent on road maintenance in Great Britain and the number, y_i, of millions of pounds-weight of tobacco smoked as cigarettes in Great Britain are given in the following table for the years $(1947 + i)$ for $i = 1(1)10$.

i	1	2	3	4	5	6	7	8	9	10
x_i	27	23	27	26	33	36	36	40	42	43
y_i	184	178	182	191	194	199	204	211	216	221

Draw the graph of y_i against x_i. Discuss the use of these figures by a tobacco importer to predict the likely demand for cigarettes in 1958 and 1959, if during 1957 the government had announced that it intended to spend £60,000,000 on road maintenance in 1958, and £30,000,000 in 1959.

*4 T. The regression line of y on x for values (x_i, y_i) is found most easily by changing the origin to the point $(\bar{x}, \bar{y})$; that is by using the transformation:

$$x = \bar{x} + x', \quad y = \bar{y} + y'.$$

The coefficients p, q of the regression line in the form $y' = px' + q$ then result from minimizing $\Sigma(px_i' + q - y_i')^2$, where

$$\Sigma x_i' = \Sigma y_i' = 0.$$

(i) Find the values of p, q that minimize the above statistic and hence show that the line of regression of y on x can be written in the form

$$y - \bar{y} = (v_{xy}/v_{xx}) \cdot (x - \bar{x}),$$

where $\qquad v_{xx} = \Sigma(x_i - \bar{x})^2, \quad v_{xy} = \Sigma(x_i - \bar{x})(y_i - \bar{y}).$

(v_{xy}/N is called the *covariance* of $[(x, y)]$, by analogy with v_{xx}/N, which is the variance of $[x]$.)

(ii) By considering a suitable interchange of x and y and the consequent change of roles as independent and dependent variables, deduce the equation of the regression line of x on y in the form

$$y - \bar{y} = p' \cdot (x - \bar{x}).$$

*5 T. Show that for a population of N number-pairs (x_i, y_i) the relation:

$$\text{covariance } [(x, y)] = \frac{1}{N} \Sigma(x_i - \bar{x})(y_i - \bar{y})$$

can be put in the form

$$\text{cov } [(x, y)] = \frac{1}{N} \Sigma(x_i y_i) - \bar{x}\bar{y},$$

where the abbreviation cov is used.

6 T. (To investigate the behaviour of the regression lines for different extents of linear correlation.)

(i) Find the regression lines of y on x and of x on y for the following four points $(-2h, -h)$, $(-1, 2)$, $(1, -2)$, $(2h, h)$.

(ii) Sketch separately the points and regression lines for $h = 5, 2, 1, \frac{1}{4}, 0$.

(iii) State geometrically how the regression lines behave as the correlation becomes more nearly perfectly linear.

7 T. Find the regression lines for the points given by

$$y = x^2, \quad x = 0, \pm 1, \pm 2$$

and hence show that a wide divergence of the regression lines does *not* mean an absence of relationship. [It only means an absence of *linear* relationship; this problem is one of *curvilinear regression*.]

*8 T. Assuming the results of Question 4 and the generality of the result of Question 6 show that $(v_{xy})/(v_{xx}.v_{yy})^{\frac{1}{2}}$ is a statistic whose value approaches either $+1$ or -1 as the correlation more nearly approaches being perfectly linear. (It is called *the product-moment correlation coefficient between x and y*.)

10. CALCULATION OF STANDARD DEVIATION

10.1 With use of a computer. If we have access to a computer then the mean and variance each have a property which makes them convenient for calculation: we do not need to have a frequency table drawn up in order to calculate them. The mode and median, as measures of position, both need a knowledge of the relative positions of the numbers concerned, as a moment's thought will show; and the interquartile range, as a measure of spread, also needs such a knowledge for the quartiles to be found.

The accompanying flow diagram (Figure 4.12) shows one way of organizing the calculations.

Y denotes a set of stores into which the values of y_j that form the population are read. The y_j are not necessarily distinct, they include possible repetitions.

Z is a working store holding the current value of y_j.

A, B are stores which hold the running totals of Σy_j, Σy_j^2.

151

J counts how many of the data have been used so far.

K holds the (supposed previously known) size of the population.

The loop is concerned wholly with calculating $\sum_1^N y_j$, $\sum_1^N y_j^2$.

The instructions after the loop complete the calculations as follows:

M holds the mean.

U holds the mean of the squares.

C holds the 'correction term': (mean)^2.

V holds the variance.

S holds the standard deviation.

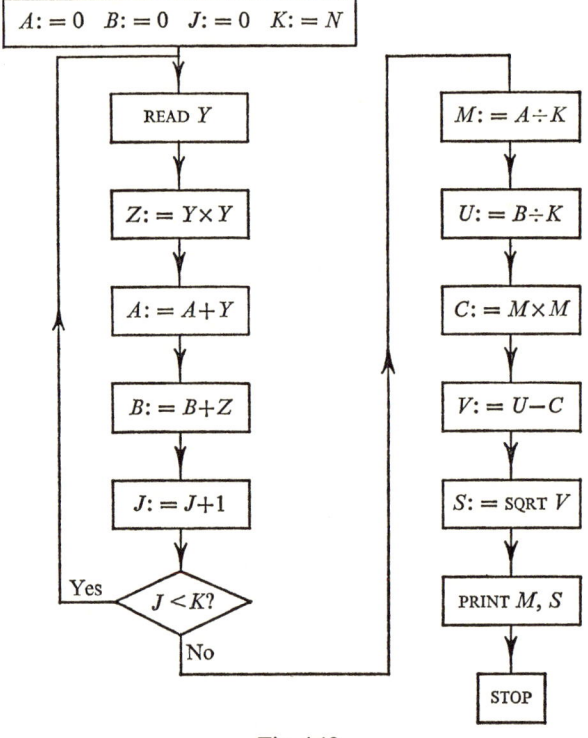

Fig. 4.12

Notice that (i) it is not helpful to translate the population first unless the numbers when squared will be too large for the capacity of the machine.

(ii) If desk calculators are being used then it is helpful to have two of them so that one accumulator can be store A and the other store B. When two such stores are available then each number y_j need only be read once from the data. [Compare the situation with that holding if the mean absolute deviation were to be calculated or if the variance were to be calculated *from its definition*.]

Exercise 4I

1R. (A 'dry-run' to simulate the operation of a computer.)
Construct columns headed as follows:

$$K \quad J \quad Y \quad A \quad Z \quad B \quad M \quad U \quad C \quad V \quad S$$

and fill them systematically according to the flow diagram of Section 10.1 with the following unsorted data:

Division 4 Football Results

Barrow	3	Exeter	0
Carlisle	1	Stockport	0
Chester	0	Accrington	0
Chesterfield	1	York	1
Colchester United	5	Oldham Athletic	1
Gillingham	4	Workington	2
Mansfield	1	Wrexham	2
Southport	0	Hartlepools United	1

[Read the figures row by row and not down a column.]
   What are the mean and S.D. of the number of goals scored by a team?

2. The following data refer to hours of sunshine in June 1962 at a number of Scottish towns. *Use a calculator* to find the mean and S.D.

191	177	203	204	208	183	191	184	140	137	182
135	209	198	193	169	179	214	195	215	222	209
174	233	161	177	187	193	211	187	193	187	163
223	165	136	179	142	141	137	175	165	190	148
161	157	165	175	152	162	149	200	203	151	

★3. Write a flow diagram to calculate the covariance of a bivariate population.
[See Exercise 4H, Question 4.]

10.2 Without a computer: change of scale.

We have noted so far (see Section 8.3) that

   (*a*) variance = (mean of squares) − (square of mean);

   (*b*) translation of a population by addition of a constant does not affect the standard deviation.

   We now consider whether advantage might not be taken of the scales on Figure 4.8 so that we reduce the calculation even more, and use t rather than x'.

   If we have a convenient measure of spread, a change of scale by a factor c ought to change the measure of spread by a factor c; we now show that this convenience is possessed by the standard deviation (though *not* by the variance).

153

We take a numerical case first. Referring to Figure 4.8, we see that since the values of t are obtained from those of x' by multiplication by $\frac{1}{5}$ the corresponding deviations are also obtained by multiplication by $\frac{1}{5}$, and their squares by multiplication by $\frac{1}{25}$.

The variance is obtained from the sum of squares of deviations so

$$\text{var } [t] = \tfrac{1}{25} \times \text{var } [x'].$$

Finally the S.D. is the square root of the variance so, with an obvious notation:

$$\text{stad } [t] = \tfrac{1}{5} \times \text{stad } [x']$$

$$= \tfrac{1}{5} \times \text{stad } [x].$$

The general result is of course much more important than a mere device for shortening calculations. We often find in theoretical work that we have two populations whose values merely differ by a scale factor, the corresponding frequencies remaining unaltered. We may state:

Result (c): Change of scale by a factor c multiplies the S.D. by c (and, incidentally, the variance by c^2).

Proof: Let the old population have members t_i, with frequencies $f(t_i)$. Let the new population be given by $x' = ct$, then $\bar{x}' = c\bar{t}$, from Exercise 4C, Question 3.

We have
$$x_i' - \bar{x}' = ct_i - c\bar{t} = c.(t_i - \bar{t}),$$
$$\text{frequency of } x_i' = \text{frequency of } t_i = f(t_i).$$

Thus
$$\text{New variance} = \frac{1}{N}\Sigma(x_i' - \bar{x}')^2.f(t_i)$$

$$= \frac{1}{N}\Sigma c^2(t_i - \bar{t})^2.f(t_i)$$

$$= c^2 \times \frac{1}{N}\Sigma(t_i - \bar{t})^2.f(t_i)$$

$$= c^2 \times \text{Old variance}.$$

Thus
$$\text{New S.D.} = c \times \text{Old S.D., by square rooting.}$$

We can apply the results successively in the order (b) (c), (a) as follows, to the population already investigated in Section 4.1.

Example 2. Find the mean and S.D. of the population of values of x given (with appended calculations) by the frequency table below.

Solution. We take $x' = x - 17$, $t = x'/5$

so that
$$\begin{cases} \bar{x}' = \bar{x} - 17; \\ \bar{t} = \bar{x}'/5; \end{cases} \quad \begin{cases} \text{stad } [x'] = \text{stad } [x]; \text{ by Result } (b) \\ \text{stad } [t] = \text{stad } [x']/5; \text{ by Result } (c) \end{cases}$$

which give
$$\bar{x} = 5\bar{t} + 17, \quad \text{stad } [x] = 5.\text{stad } [t].$$

154

x_i	$f(x_i)$	x_i'	t_i	$t_i f(x_i)$	$t_i^2 f(x_i)$
12	20	-5	-1	-20	20
17	50	0	0	0	0
22	40	5	1	40	40
27	20	10	2	40	80
32	10	15	3	30	90
37	10	20	4	40	160
	150			$-20+150$ $=130$	390

$$\bar{t} = \tfrac{130}{150}, \quad (\bar{t})^2 = 0.751;$$

$$\bar{x} = (5 \times \tfrac{130}{150}) + 17 = 4.33 + 17 = 21.33;$$

$$\text{var}\,[t] = \tfrac{390}{150} - (\bar{t})^2 = 2.600 - 0.751 = 1.849 \quad \text{(by Result } (a));$$

$$\text{stad}\,[t] = \sqrt{1.849} = 1.360;$$

$$\text{stad}\,[x] = 5 \times 1.360 = 6.80.$$

Result: $\bar{x} = 21.33$, stad $[x] = 6.80$.

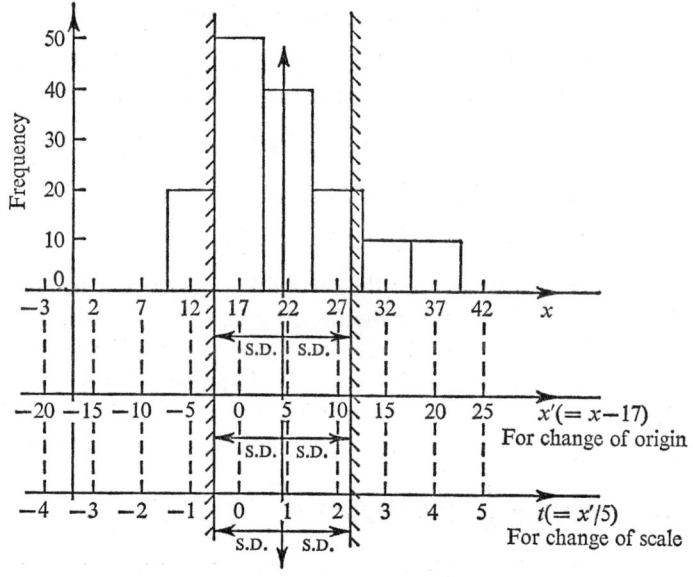

Fig. 4.13

The numerical meanings of the results of this example are best seen by looking at Figure 4.13, which is a reproduction of Figure 4.8 with the addition of an interval showing a range of $1 \times$ S.D. on either side of the mean. This is an interval which can be expressed on the three scales as:

$$0.87 - 1.36 < t < 0.87 + 1.36; \quad 4.3 - 6.8 < x' < 4.3 + 6.8; \quad 21.3 - 6.8 < x < 21.3 + 68.$$

155

Notice that in the calculations it was never necessary finally to *evaluate* $\bar{t}$ since it was only required in a multiple, and squared, and the slide rule calculates *these* directly.

Whichever scale is used, the interval has the same geometric relationship to the frequency diagram, and tells us something about the population itself. This point is taken up in Chapter 6 where the properties of intervals of this form are discussed.

10.3   Without computer: Charlier checks. If we do not have access to calculating machines of any sort then the calculation of the mean and standard deviation can involve quite large tables of figures and it is useful to have a method of checking our accuracy. Such checks were devised by the Norwegian statistician L. V. Charlier and we explain them below by means of a worked example.

The check on the mean simply uses the fact that if 1 is added to all the values then $\Sigma f(x_i)$ is added altogether. Taking $f(x_i)$ to be the frequency for $t = t_i$ in the way indicated in Example 2 we have:

$$\Sigma(t_i+1)f(x_i) = \Sigma t_i f(x_i) + \Sigma f(x_i),$$

and the reader may have proved this by mere algebra in Exercise 4C, Question 4. The check on the variance can take several forms:

$$\Sigma(t_i+1)^2 f(x_i) = \Sigma t_i^2 f(x_i) + 2\Sigma t_i f(x_i) + \Sigma f(x_i),$$

$$\Sigma(t_i+1)^2 f(x_i) = \Sigma t_i^2 f(x_i) + \Sigma(t_i+1)f(x_i) + \Sigma t_i f(x_i),$$

$$\Sigma(t_i+1)t_i f(x_i) = \Sigma t_i^2 f(x_i) + \Sigma t_i f(x_i).$$

The first form is the classic one; the second is rather easier to apply: the author's attention was drawn to the third by C. C. Goldsmith, and it has the advantages not only that it requires fewer terms to be added *in the check* but also that zeros come in the final column both for $t_i = 0$ and for $t_i = -1$. The advantage of either of the first two forms is that the column-sum $\Sigma(t_i+1)^2 f(x_i)$ allows a check *below* the table by recalculating var $[t]$ using

$$\text{var } [t] = \frac{1}{N}\Sigma(t_i+1)^2 f(x_i) - \left(\frac{1}{N}\Sigma(t_i+1)f(x_i)\right)^2.$$

The columns of Example 3 are drawn up for using either of the first two forms of the check; we will, in fact, use the second. The reader will make much better sense of the mass of figures if he follows the notes below. They are arranged, like the *dramatis personae* list in a theatre, in 'order of appearance'.

Example 3. Calculate the mean and standard deviation of the population, [*x*], defined by the first two columns of the table below.

156

Solution. We take $x' = x - 14\frac{1}{2}$, $\ t = x'/4$.

[We omit the column for x' as explained in the notes below.]

x_i	$f(x_i)$	t_i	t_i+1	$t_i f(x_i)$	$(t_i+1)f(x_i)$	$t_i^2 f(x_i)$	$(t_i+1)^2 f(x_i)$
$2\frac{1}{2}$	1	-3	-2	-3	-2	9	4
$6\frac{1}{2}$	8	-2	-1	-16	-8	32	8
$10\frac{1}{2}$	38	-1	0	-38	0	38	0
$14\frac{1}{2}$	81	0	1	0	81	0	81
$18\frac{1}{2}$	79	1	2	79	158	79	316
$22\frac{1}{2}$	32	2	3	64	96	128	288
	239			$-57+143$	$-10+335$	286	697
				$=+86$	$=+325$		

Checks.

$$\Sigma(t_i+1)f(x_i) = 325; \quad \Sigma t_i f(x_i) + \Sigma f(x_i) = 86 + 239 = 325;$$
$$\Sigma(t_i+1)^2 f(x_i) = 697;$$
$$\Sigma t_i^2 f(x_i) + \Sigma(t_i+1)f(x_i) + \Sigma t_i f(x_i) = 286 + 325 + 86 = 697.$$

Calculations.

$$\bar{t} = 86/239; \quad (\bar{t})^2 = 0\cdot129;$$
$$\bar{x} = (4 \times 86/239) + 14\frac{1}{2} = 1\cdot44 + 14\cdot50 = 15\cdot94;$$
$$\text{var}[t] = 286/239 - (\bar{t})^2 = 1\cdot195 - 0\cdot129 = 1\cdot066.$$

Check.

$$\text{var}[t] = 697/239 - (325/239)^2 = 2\cdot92 - 1\cdot85 = 1\cdot07;$$

Calculations.

$$\text{stad}[t] = \sqrt{(1\cdot066)} = 1\cdot033;$$
$$\text{stad}[x] = 4 \times 1\cdot033 = 4\cdot13.$$

Results. $\qquad\qquad \bar{x} = 15\cdot94, \quad \text{stad}(x) = 4\cdot13.$

The reader should have followed the accompanying notes while studying the table and calculations.

(i) Notice the *pattern* in the column headings:

$$\overbrace{t \quad (t+1)} \qquad \overbrace{tf(x) \quad (t+1)f(x)} \qquad \overbrace{t^2 f(x) \quad (t+1)^2 f(x)}$$

(ii) In the body of the table, work down columns and not along rows: this ensures that similar calculations are done together.

(iii) If we see that the values of x_i are at intervals of multiples of some unit we will have taken c to be that unit and will have chosen k to be one of the values of x_i and have written $t_i = 0$ on that level. We then work in the t_i column (up in negative integers and down in positive integers) by counting off the units, thus saving having a column for x_i'.

(iv) It is meaningless to sum any column that does not involve $f(x)$, since it is $f(x)$, and $f(x)$ only, that indicates we are dealing with a population.

(v) The columns $t_i f(x_i)$ and $(t_i+1) f(x_i)$ are written out 'expanded' so that negative and positive values can be separately added.

(vi) It is a good idea to use the check on column $(t_i+1) f(x_i)$ *before* going on to obtain the last two columns, since $t_i f(x_i)$ is needed for $t_i^2 f(x_i)$ and it is as well to have it first checked.

(vii) In evaluating $t_i^2 f(x_i)$ we use $t_i^2 f(x_i) = t_i \times t_i f(x_i)$, so we do not need a column of values of t_i^2.

Similarly for $(t_i+1)^2 f(x_i)$ we use

$$(t_i+1)^2 f(x_i) = (t_i+1) \times (t_i+1) f(x_i).$$

(viii) If a slide rule is being used then both $\bar{t}^2$ and $c \times \bar{t}$, that is to say $(\frac{86}{239})^2$ and $(4 \times \frac{86}{239})$, can be calculated from the same setting of $\frac{86}{239}$, without ever writing down its value.

(ix) The subtraction of the 'correction term' in evaluating the variance can lead to a large loss of significant figures; for instance if $\bar{t} = 15\cdot6$ and var $[t] = 1\cdot12$ then the line for calculating var $[t]$ should read

$$\text{var } [t] = 244\cdot48 - 243\cdot36 = 1\cdot12,$$

but if only three figures were available, from a slide rule, we should have

$$\text{var } [t] = 244 - 243 = 1$$

and so have a 10% error in the variance.

The key to the situation is simply to choose k in a way that ensures that $|\bar{t}|$ is small; it need never be more than 1 or 2 if care is taken over the choice of k. This is another reason, apart from the sheer heaviness of work indicated in Section 8.2, for using a working zero.

(x) Notice the use, in the recalculation of the variance, of the form

$$\text{var } [t] = \frac{1}{N} \Sigma(t_i+1)^2 f(x_i) - \left(\frac{1}{N} \Sigma(t_i+1) f(x_i)\right)^2$$

related to the form proved in Exercise 4G, Question 2, the value of k *in that question* being taken to be -1 *here*.

Exercise 4J

The reader may think it artificial to a degree to be doing questions in which the values of x are numbers like $3\frac{1}{2}$, $11\frac{1}{2}$, $19\frac{1}{2}$, ..., $43\frac{1}{2}$, but by the end of the next chapter, he will see that when readings have been grouped, as they often are in practical cases, it is just such numbers that are apt to arise. There is nothing more to learn about the *organization* of these heavy calculations, so such numbers are included now.

1. (i) If Exercise 4E, Question 11 has not been done already the reader will find he has now been shown the necessary result about scale-change, and should do the question now.

(ii) From Exercise 4F, Question 1 part (iii) find the variance of a set of n numbers spread at equal intervals along a unit length with one member at each end point. Deduce the variance if the numbers remain equally spread but the interval between each extreme number and the end of the unit line is half the interval between pairs of numbers.

2 R. [The answers to this question, and *not* to Question 3, are given.]

Employ a working-zero, change of scale and Charlier's checks to find the mean and standard deviation of each of the following populations, using the layout of Example 3.

(a) x	f(x)	(b) x	f(x)	(c) x	f(x)	(d) x	f(x)
$3\frac{1}{2}$	5	147	2	1·4	100	0	6
$11\frac{1}{2}$	3	153	3	1·8	100	0·1	6
$19\frac{1}{2}$	3	159	3	2·2	400	0·2	4
$27\frac{1}{2}$	2	165	5	2·4	400	0·3	3
$35\frac{1}{2}$	1	171	4	2·6	500	0·6	1
$43\frac{1}{2}$	1	195	1	2·8	600	0·7	1
				3·0	400	0·8	1
				3·4	200	0·9	4
						1·0	4

Find a transformation $y = (x-k)/c$ which converts population (*a*) of this question into population (iii) of Section 3.3. [Population (*d*) of this question is called U-shaped, for obvious reasons; the cloudiness of the sky over Britain on a day, measured in tenths-covered, and the number of representative matches in a particular sport played in by an individual player both have frequency functions of this type. The example here reminds us that the mean is not necessarily one of the numbers actually occurring in the population, and may not even be near the most frequently occurring numbers; hence the popular gibe that the 'average man' or 'man in the street' does not exist.]

3. [The answers to Question 2, and *not* to this question, are given.]

Employ a working-zero, change of scale and Charlier's checks to find the mean and standard deviation of each of the following populations, using the layout of Example 3.

(a) x	f(x)	(b) x	f(x)	(c) x	f(x)	(d) x	f(x)
22·5	2	4	1	0	12	1·3	1
27·5	6	11	1	1	24	1·5	8
32·5	14	18	2	2	36	1·7	23
37·5	16	25	2	3	48	1·9	37
42·5	10	32	3	4	72	2·1	31
47·5	2	88	2	5	60	2·3	16
				6	48	2·5	4
				7	36		
				8	24		
				9	12		

Find a transformation $y = (x-k)/c$ which converts one of these populations into population (iv) of Section 3.3.

5

GROUPED DATA

In Chapter 4 all the populations concerned were ones about which we had full information. There is often much loose thinking about exactly what we are doing when we calculate a mean or a standard deviation, or any other statistic. We may be, of course, and usually are, hoping to be able to say something about the parent population or about another sample that might be drawn from it, but *during the actual calculations we are being sharply, and merely, descriptive*. There are many theories about how best to *use our description* to estimate or to test hypotheses and they will be very important later, but they should not confuse the present clear cut techniques. These techniques of description are equally valid whether we are describing a *sample*, in which we know all the values of x and their frequencies because they have been observed and recorded, or are describing a *parent population*, in which we know what to take for the values of x and their frequencies (or probabilities) because they constitute the hypothesis from which we are starting.

This chapter investigates how to proceed with our descriptions if, as is often the case, we can only pin the values of x down to within certain intervals.

1. CONTINUOUS AND DISCRETE

If we measured the heights, x cm, of 50 ragwort plants (*Senecio jacobaea*) in a field we might get the information given in Table 5.1.

Table 5.1

Range of values of x	Number of plants with x in the given range
20–(25)	2
25–(30)	6
30–(35)	14
35–(40)	16
40–(45)	10
45–(50)	2

[The range 20–(25) means $20 \leqslant x < 25$]

The data have been clearly *grouped*. If we look at the second group (or as

160

it is more usually called in this context: *class*) in greater detail we might find the following values of x:

25·5 (twice),   26,   28·5,   29,   29·5

but we would be wrong if we thought that the data were no longer grouped. '25·5 (twice), 26' merely means that two plants were found with heights in the range 25·25–25·75 and one plant was found in the range 25·75–26·25.

In placing the values of x in these smaller ranges we were supposing, from the different values given, that the measurements were to the nearest $\frac{1}{2}$ cm, but however finely we measured we would ultimately stop at some degree of accuracy, that is to say ranges would be implied. We will use again the notation of Chapter 4, Section 2.4 and say that the function m by which we are mapping the set S of ragwort plants has images which, for different ragwort plants, might lie anywhere in an interval. Thus the set D, which contains these images, has members which might have lain anywhere in an interval. The usual way of expressing this is to say that the *variable x is continuous*, or, as in Chapter 4, Section 2·3, that the parent population is continuous.

In *measuring* heights, lengths, masses, temperatures and other quantities, as opposed to *counting* heads, goals, children and so on, we can safely say we are dealing with a continuous variable, whatever may be our views about atoms and quanta.

Sometimes, however, grouping occurs which is not inevitable but is merely convenient; for instance if we took 15 of the ragwort plants and counted the number, y, of flower-heads on each we might get the following figures:

0 (twice),   1,   3,   5,   9,   14,   15,

16,   20,   22,   24,   29,   35,   46

In order to see, if possible, some pattern here we might arrange the information as in Table 5.2.

Table 5.2

Range of values of y	Number of plants with y in the given range
0– (7)	5
8–(15)	3
16–(23)	3
24–(31)	2
32–(39)	1
40–(47)	1

This view of the frequencies with half-closed eyes, as it were, has allowed us to see a broad pattern in them.

Grouping is certainly not essential for this second type of data, where the values of x are the images of a function m which *counts* and does not *measure*. (Incidentally the frequencies of the values of y are then frequencies

161

of frequencies but this need not worry us; though in Exercise 4A, Question 4 it may have been a source of confusion.)

Where, as here, the only possible values of y are drawn from a discrete set, we say that the *variable y is discrete*, or where suitable, as it is here, the *variable y is integral*. A variable can be discrete without being integral, as for instance when it is the mean score on two dice. As we remarked in Chapter 4, Section 2·3 a sample of actual observations, whether of values of a discrete or of a continuous variable, will always be a finite population and hence a discrete one.

2. APPROXIMATION TO THE MEAN

When we are investigating the mean of a sample and are using grouped data we shall see that the methods of calculation are the same whether the data are inevitably grouped (because of the continuity of the variable) or only grouped for convenience (the variable being discrete). We shall illustrate the theory by using grouped data for *THOMP* and *TWOAF*, where the variable is in fact integral. We have listed in the second column values called the *class-marks*. They are the mid-points of the extreme values which x can take in the classes. Notice that we take the lengths of the classes to be odd numbers, and so we have integral class-marks; (though if the class-intervals are all equal then this is not important as we use a transformed variable).

Table 5.3

		Frequencies from	
Class-interval	Class-mark	*THOMP*	*TWOAF*
$\frac{1}{2}$– $11\frac{1}{2}$	6	10	2
$11\frac{1}{2}$– $22\frac{1}{2}$	17	13	5
$22\frac{1}{2}$– $33\frac{1}{2}$	28	3	11
$33\frac{1}{2}$– $44\frac{1}{2}$	39	1	5
$44\frac{1}{2}$– $55\frac{1}{2}$	50		2
$55\frac{1}{2}$– $66\frac{1}{2}$	61		0
$66\frac{1}{2}$– $77\frac{1}{2}$	72		1
$77\frac{1}{2}$– $88\frac{1}{2}$	83		0
$88\frac{1}{2}$– $99\frac{1}{2}$	94		0
$99\frac{1}{2}$–$110\frac{1}{2}$	105		0
$110\frac{1}{2}$–$121\frac{1}{2}$	116		0
$121\frac{1}{2}$–$132\frac{1}{2}$	127		1

If we were finding the true mean sentence-length in *THOMP*, we would note from Table 4.1 that there are 18 different sentence-lengths in the population and so we would calculate

$$\sum_{1}^{18} x_i h(x_i),$$

the x_i being those for which $h(x_i) \neq 0$.

162

Taking the given grouping by classes, successive classes contain 7, 8, 2, 1 different sentence-lengths, and to display this we could write:

$$\sum_{1}^{18} x_i h(x_i) = \sum_{1}^{7} x_i h(x_i) + \sum_{8}^{15} x_i h(x_i) + \sum_{16}^{17} x_i h(x_i) + \sum_{18}^{18} x_i h(x_i).$$

Now, writing the mean sentence-lengths in the classes as m_1, m_2, m_3, m_4, respectively, we have:

$$\sum_{1}^{7} x_i h(x_i) = 10.m_1 \quad \text{since} \quad \sum_{1}^{7} h(x_i) = 10;$$

and similar relations for

$$\sum_{8}^{15} x_i h(x_i), \quad \sum_{16}^{17} x_i h(x_i) \quad \text{and} \quad \sum_{18}^{18} x_i h(x_i);$$

so that, substituting, we have:

$$\sum_{1}^{18} x_i h(x_i) = 10.m_1 + 13.m_2 + 3.m_3 + 1.m_4$$

and the true mean sentence-length in *THOMP* is

$$\frac{10.m_1 + 13.m_2 + 3.m_3 + 1.m_4}{10 + 13 + 3 + 1}.$$

So far we have only been using the methods of Exercise 4B, Question 7; however, to obtain an approximate value for the mean we do not calculate m_1, m_2, m_3, m_4 but replace them by the class-marks of the corresponding classes. The approximation we are using is that the mean of each class is taken to be the mid-point of that part of the x-axis on which its interval falls. This means that the only assumption we make is that the sentence-lengths are fairly evenly spaced *within each class* and that any errors will tend to cancel. It is thus irrelevant whether the variable is continuous or discrete. For a given population the larger the class-intervals the more likely we are to be in error, for the more scope there is for irregularities within the classes. Also for a given choice of class-intervals the smaller the frequencies within the classes the more likely we are to be in error, for there is less scope for cancelling to occur. For the maximum magnitude of the error see Exercise 5A, Question 7.

Having disposed of the theory of the approximation, we now see in Example 1 that the work can be set out as before.

Example 1. Approximate to the mean of *THOMP*, using the transformation $t = (y-17)/11$, so that $\bar{y} = 11\bar{t} + 17$.

163

Solution.

Class-interval	y_i = class-mark	$f(y_i)$	t_i	$t_i f(y_i)$	$(t_i+1) f(y_i)$
$\frac{1}{2}$–11$\frac{1}{2}$	6	10	−1	−10	0
11$\frac{1}{2}$–22$\frac{1}{2}$	17	13	0	0	13
22$\frac{1}{2}$–33$\frac{1}{2}$	28	3	1	3	6
33$\frac{1}{2}$–44$\frac{1}{2}$	39	1	2	2	3
		27		−10+5 = −5	22

Check. $\qquad \Sigma(t_i+1)f(y_i) = 22;$

$$\Sigma t_i f(y_i) + \Sigma f(y_i) = -5+27 = 22.$$

Calculations. $\quad \bar{t} = -\dfrac{5}{27};$

$$\bar{y} = \left(11 \times -\frac{5}{27}\right) + 17 = -2 \cdot 0 + 17 = 15 \cdot 0.$$

$\bar{y}$ is an approximation to the mean so the mean of *THOMP* $\simeq 15 \cdot 0$. The true mean was given in Chapter 4, Section 3.2 as 14.6.

The theory of the approximation in no way depends upon the class-intervals being of equal length; and our next example uses a grouping, for the data of *TWOAF*, with unequal intervals.

Example 2. Approximate to the mean of *TWOAF*.

Solution. Take
$$t = (y-28)/11; \quad \bar{y} = 11\bar{t}+28.$$

Class-interval	y_i = class-mark	$f(y_i)$	t_i	$t_i f(y_i)$	$(t_i+1) f(y_i)$
$\frac{1}{2}$– 11$\frac{1}{2}$	6	2	−2	−4	−2
11$\frac{1}{2}$– 22$\frac{1}{2}$	17	5	−1	−5	0
22$\frac{1}{2}$– 33$\frac{1}{2}$	28	11	0	0	11
33$\frac{1}{2}$– 44$\frac{1}{2}$	39	5	1	5	10
44$\frac{1}{2}$– 55$\frac{1}{2}$	50	2	2	4	6
55$\frac{1}{2}$–132$\frac{1}{2}$	94	2	6	12	14
		27		−9+21 = +12	−2+41 = +39

Check. $\qquad \Sigma(t_i+1)f(y_i) = 39;$

$$\Sigma t_i f(y_i) + \Sigma f(y_i) = 12+27 = 39.$$

Calculations. $\quad \bar{t} = \dfrac{12}{27};$

$$\bar{y} = \left(11 \times \frac{12}{27}\right) + 28 = 4 \cdot 9 + 28 = 32 \cdot 9.$$

$\bar{y}$ is an approximation to the mean, so the mean of $TWOAF \simeq 32\cdot9$. The true mean was given in Chapter 4, Section 3.2 as 32.3.

A different arrangement of class-intervals will give a different approximation to the true mean, as we see in the next example.

Example 3. Adjust the calculations of Example 2 to find the approximation to the mean based on the grouping of the $TWOAF$ data in Table 5.3.

The last row of the calculation in Example 2 reads:

Class-interval	y_i = class-mark	$f(y_i)$	t_i	$t_i f(y_i)$
$55\frac{1}{2}$–$132\frac{1}{2}$	94	2	6	12

in its place we now have

Class-interval	y_i = class-mark	$f(y_i)$	t_i	$t_i f(y_i)$
$66\frac{1}{2}$– $77\frac{1}{2}$	72	1	4	4
$121\frac{1}{2}$–$132\frac{1}{2}$	127	1	9	9

Thus $\Sigma t_i f(y_i)$ is increased by $(4+9)-12 = 1$ and we have $\bar{t} = \dfrac{12+1}{27}$ instead of $\dfrac{12}{27}$.

$$\bar{y} = (11 \times \tfrac{13}{27}) + 28 = 5\cdot3 + 28 = 33\cdot3.$$

The new approximation to the mean is 33·3.

In the case of grouped data the proper allotment of class-marks may need care, depending on whether the variable is continuous or discrete (see Exercise 5A, Questions 1, 5), but the rest of the arithmetic used in calculating the mean is the same whether the data are grouped or not.

To summarize: in the case of ungrouped data the actual values of the variable are used in the arithmetic and the mean obtained is the actual mean required; in the case of grouped data the class-marks are used and the mean obtained is an approximation to the mean required.

To emphasize the way the methods all converge we rewrite both sets of data of Section 1, using the class-marks, and get:

x_i	$f(x_i)$	and	y_i	$f(y_i)$
22·5	2		$3\frac{1}{2}$	5
27·5	6		$11\frac{1}{2}$	3
32·5	14		$19\frac{1}{2}$	3
37·5	16		$27\frac{1}{2}$	2
42·5	10		$35\frac{1}{2}$	1
47·5	2		$43\frac{1}{2}$	1

The means of these were evaluated in Exercise 4J, Questions 3(a) and Question 2(a) respectively. These questions were, thus, either obtaining the exact means of the populations given, or obtaining approximations to

165

means of other populations, depending what physical situations were being represented by the frequency tables; *and the situations did not have to be known.*

3. HISTOGRAMS

In drawing a diagram to show the grouped data of *TWOAF* in Example 2 we obtain Figure 5.1.

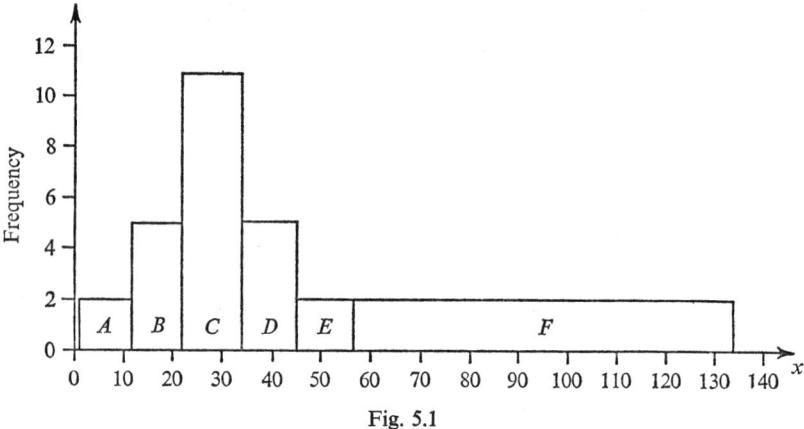

Fig. 5.1

At first sight this seems satisfactory. But the removal of the bar at $x = 55\frac{1}{2}$ would completely alter the interpretation to be placed on the rectangles. The rectangle E indicates that somewhere between $44\frac{1}{2}$ and $55\frac{1}{2}$ there occur two values of x. F has a similar meaning: somewhere between $55\frac{1}{2}$ and $132\frac{1}{2}$ there occur two values of x. Remove the bar, and the combined rectangle (E with F) would indicate that somewhere between $44\frac{1}{2}$ and $132\frac{1}{2}$ there occur two values of x. The indication would be false. We can overcome this difficulty by representing the frequency corresponding to the class by the *area* of the rectangle with class-interval as base, and *not* by the height of the rectangle. The additive properties of the frequencies are then represented by the additive properties of the areas; additive properties which the heights do not possess. In order to keep the area correct the height will be proportionally reduced when the class-interval is increased.

The calculations necessary are:

Class-interval	Length of interval	*TWOAF* frequency	Height of rectangle
$\frac{1}{2}$– $11\frac{1}{2}$	11	2	$\frac{2}{11} = \frac{14}{77} = 0.182$
$11\frac{1}{2}$– $22\frac{1}{2}$	11	5	$\frac{5}{11} = \frac{35}{77} = 0.451$
$22\frac{1}{2}$– $33\frac{1}{2}$	11	11	$1 = \frac{77}{77} = 1.000$
$33\frac{1}{2}$– $44\frac{1}{2}$	11	5	$\frac{5}{11} = \frac{35}{77} = 0.451$
$44\frac{1}{2}$– $55\frac{1}{2}$	11	2	$\frac{2}{11} = \frac{14}{77} = 0.182$
$55\frac{1}{2}$–$132\frac{1}{2}$	77	2	$\frac{2}{77} = \frac{2}{77} = 0.026$

The resulting diagram is Figure 5.2.

The scale up the page has been left in units of $\frac{14}{77}$ to make the relative heights easily comparable; normally it might be expressed in decimal units. We see that in the parts where the class-intervals are of equal lengths the diagram only differs from the frequency diagram in scale up the page.

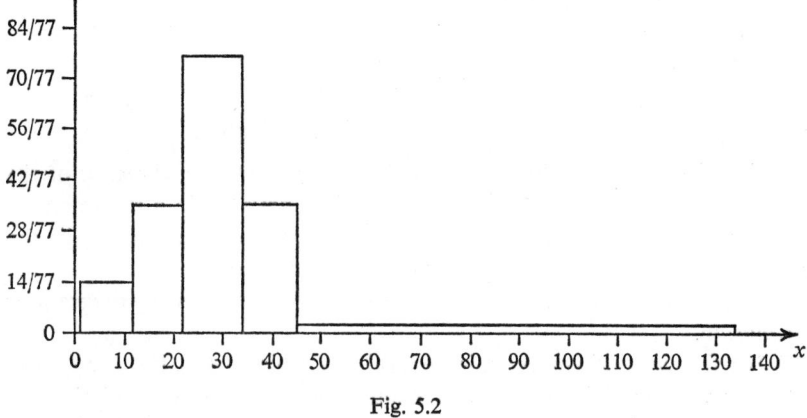

Fig. 5.2

This raises the question as to what it is that the scale up the page measures. The answer to this may be seen in terms of an analogy with housing density along a road. If we wish to make a sensible comparison of the amount by which the road-sides are built up, we will take into account not merely the number of houses passed but the length of the interval of road which contains the houses. We could define a quantity called the housing density, measured by number of houses per unit length of road. Note that this is what is called a *line-density*, it is not an *area-density* (like the density in a census, measured for instance in people per km²), nor yet a *volume-density* (like the 'ordinary' density, of mass per unit volume, measured, for instance, in kg/m³).

Figure 5.3 shows an axis marking every member of the population of sentence-lengths in *TWOAF* with a small bar, doubling the bar when two members occur together.

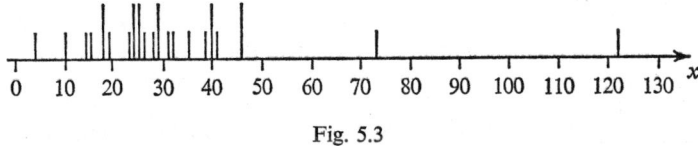

Fig. 5.3

If, instead, the x-scale on Figure 5.3 were taken to measure distance in km travelled along a road from some origin, and each bar represented a house by the road side, then Figure 5.2 would represent the averaged or

167

smoothed out housing density (measured in houses per km) along various intervals of the road. In an exactly comparable way, the quantity which is actually measured by the scale up the page in Figure 5.2 may be called the *frequency-density*.

To obtain the number of houses in any interval we would say:

number of houses in an interval
= (housing density) × (length of interval)

and so here we say

frequency of values of x in an interval
= (frequency-density) × (length of interval).

The right-hand side of this equation measures, as we see, the *area* of the corresponding rectangle in Figure 5.2, and we have again a statement of the fundamental results: (i) We use area to represent frequency; (ii) the 'vertical' scale is of frequency-density.

A diagram with these properties is called a frequency-density diagram or *Histogram*.

A bar frequency diagram is, in general, *not* a histogram, but if *all* class-intervals on the x-axis are of equal lengths then for a given population the histogram and the bar frequency diagram have the same shape, differing only in 'vertical' scale.

By switching our attention from frequency diagrams to histograms we shall be able to deal with a parent population involving a continuous variable whose frequency-density may change continuously. Essentially the method is to consider a smooth curve chosen in such a way that the area between *any* two ordinates represents the idealized relative frequency (that is: probability) with which values of x from the parent population occur between the values of x corresponding to the ordinates; but we do not now go into the details.

Exercise 5 A

1 R. (i) The following function might represent a population of
either (a) numbers of goals scored by 20 hockey clubs,
or (b) lengths (in hr to the nearest hr) of 20 train journeys.

x	1	2	3	4
$f(x)$	7	10	2	1

For (a): Draw the frequency diagram. Calculate the mean and median.
For (b): Assuming a uniform spread within each class, draw a histogram, and approximate to the mean and median.

(ii) Repeat the drawings and calculations for the following frequency function with the same two interpretations.

x	0	1	2	3
$g(x)$	7	10	2	1

(*Warning:* take care over the class-marks in (b).)

168

2. Referring to Exercise 4J, Questions 2, 3 (part (*b*) in each), and taking the values of *x* to be class-marks of grouped data with all the class-intervals equal, discuss in each case whether the ungrouped data are more likely to have referred to a continuous or to an integral variable. State what you think the classes were likely to have been in each case.

3. The following table shows how many high jumpers finally knocked down the bar at various heights. Interpreting carefully the relationship which a height quoted in this table probably has to the *maximum* height able to be jumped by the corresponding jumper, find approximately the mean of the maximum heights of which the jumpers were capable.

height in cm	160·0	162·5	165·0	167·5	170·0	172·5	175·0	177·5
frequency	1	5	13	20	22	17	9	3

4T. The following tables show the same data grouped in different ways. The variable is integral. For each table draw the histogram and approximate to the mean.

(*a*) Class	Frequency	(*b*) Class	Frequency	(*c*) Class	Frequency
0	3	0– 5	18	0– 3	12
1– 3	9	6–20	30	4–10	16
4, 5	6			11–20	20
6–10	10				
11–15	10				
16–20	10				

[In Table (*b*) we see that we have merely amalgamated consecutive classes of equal frequency-density from Table (*a*). This is the simplest alteration we can make to the grouping, and the resulting behaviour of the approximated mean is sometimes described by saying 'the calculation of the mean requires no correction for grouping'. In general, however, the choice of grouping does affect the histogram (and so the approximation), as the results from Table (*c*), from Examples 2, 3 and from Question 5 show.]

5T. [Even when applying a system of classes with all the class-intervals equal in length, we still have two parameters at our disposal, namely (*a*) the length of interval, (*b*) the location of the lowest class limit. Parts (i) and (ii) of this question show the effect of varying (*b*) while keeping (*a*) fixed, whereas part (i) and Example 1 show the effect of varying (*a*) while keeping (*b*) fixed.]

Group the data of *THOMP* into classes of interval length 10, beginning with the first class-interval as (i) $\frac{1}{2}$–10$\frac{1}{2}$, (ii) 1$\frac{1}{2}$–11$\frac{1}{2}$. In each case draw the histogram and approximate to the mean. Compare your results as suggested in the preamble to this question, and with the true mean. (See also Question 8.)

6. (i) The age at death of infants under one year old in U.S.A. in 1917 had the following frequency table.

Age at death in months	Number of infants
0– 1	78,588
1– 2	15,362
2– 3	12,066
3– 6	27,487
6– 9	20,409
9–12	17,112

6–9 months means 6 months 0 days to 8 months 30 days, so to speak. Draw the histogram. The mean is not a very useful statistic here, but would the approximated value obtained from this table underestimate or overestimate the true mean?

(ii) Suggest a general shape of histogram where the means of individual classes do not approximate closely to the class-marks but the overall approximation is good.

7T. Show that the error involved in grouping to approximate to the mean is less than or equal to half the largest class-interval.

8. Construct a simple population that can be grouped in two ways with the following properties.

(a) in each there are the same number of class-intervals with non-zero frequencies.

(b) the class-intervals are all equal in length.

(c) the lowest class limit is less for the second grouping than for the first, but:

(d) the approximated mean based on the second grouping is greater than that based on the first.

9. In the table below are given the class-marks of some measurements of height and of some measurements based on width of muscle in arm, calf, and thigh, for two different types of athlete at the 1964 Olympics. The athletes were 5,000 m specialists and 100 m specialists. The frequencies of each pair of measurements is given. The 5,000 m specialists are quoted first. A dash in the table indicates zero frequency. Thus the extreme right-hand entry in the table indicates that one 5000 m specialist and no 100 m specialists were 186 cm tall and had a muscle width of 24 cm. (This individual was in fact the British athlete Gordon Pirie.)

muscle width in cm	height in cm										
	166	168	170	172	174	176	178	180	182	184	186
20	—	—	1, 0	—	—	—	—	—	—	1, 0	—
21	—	1, 0	—	2, 0	—	1, 0	1, 0	1, 0	1, 0	—	—
22	2, 0	—	—	—	1, 0	1, 0	0, 1	—	—	—	—
23	—	1, 0	1, 0	1, 0	—	—	0, 1	1, 1	—	—	—
24	—	—	—	0, 1	1, 0	—	—	0, 1	—	—	1, 0
25	—	—	0, 1	—	—	0, 1	0, 1	—	—	—	—
26	—	—	—	—	—	—	—	—	—	0, 1	—
27	—	—	—	—	0, 1	0, 1	—	—	—	—	—
28	—	—	—	—	0, 1	—	—	—	—	—	—

(i) Draw a diagram showing height on one axis and muscle width on the other and record the data for each athlete, using different symbols to distinguish the specialities.

(ii) Find approximations to the mean height and muscle width for each type of athlete.

(iii) Mark the position that would be occupied in the diagram by the symbol for an athlete of each type with the approximate mean measurements for his type.

(iv) Consider what the class-intervals are likely to have been, and around each of the symbols for an approximate mean draw a rectangle which would be sure to contain the point for the true mean of the relevant population.

[These are two examples of bivariate populations and the diagrams are often called *scatter-diagrams*.]

[For further reading on measurements of athletes, see *Penguin Science Survey B*, 1965.]

10 *T*. The following (fictitious) grouped data relate to lengths, x min, of breakdowns on a certain mechanism.

Class-interval in mins of lengths of breakdown	Frequency of breakdowns
2·5– 7·5	89
7·5– 20·0	60
20·0– 55·0	43
55·0–150·0	24
150·0–400·0	5

(i) Find the ratio of the biggest value to the smallest value of the frequency density defined by the five classes above.

(ii) Take logarithms (to base 10) of the times in minutes and restate the data. Draw a histogram for the restated data. Give the equation of the line which roughly passes through the tops of the rectangles.

4. APPROXIMATION TO THE VARIANCE

We saw in Section 2 that when data are grouped we can proceed by replacing all the members of a 'class' by the value of the class-mark, and we also noted that the choice of class boundaries affects, in general, the value of our approximation to the mean. In the case of consecutive classes having equal frequency density we saw, however, (in Exercise 5A, Question 4) that the amalgamation of the classes makes no difference to the approximation to the mean. This is sometimes loosely described by saying that in the calculation of the mean *no correction for grouping is necessary*. Exercise 5B, Question 1 shows that even this simple form of grouping of classes of equal frequency density affects the approximation to the variance; a correction for grouping is relevant, and it was devised by Sheppard. It does not, however, always give a better approximation to the variance; the population must satisfy certain conditions. The correction is frequently employed as a matter of routine by persons who do not know when it is likely to help, or even that it may not. This sort of approach to statistics is comparable to preparing the Sunday joint by finding in a dictionary that 'to roast' is 'to cook or heat by exposure to open fire or sun'.

We shall resign ourselves to a degree of uncertainty about our approximation to the variance from a grouped table.

Exercise 5 B

1 *T*. [This question is referred to in Section 4.]

(i) The following data from Exercise 5 A, Question 4 show data grouped in two ways

(*a*) Class	Frequency	(*b*) Class	Frequency
0	3	0– 5	18
1– 3	9	6–20	30
4, 5	6		
6–10	10		
11–15	10		
16–20	10		

By replacing all the values in each class by their class mark obtain an approximation to the variance in each case. Verify that, although the approximation to the mean was unaffected by this amalgamation of neighbouring classes of equal frequency density, the approximation to the variance is affected by amalgamation.

(ii) Give an example of a population of 2 members where the mean is unaffected by your choice of grouping but the variance is affected.

5. APPENDIX ON 'DISTRIBUTION'

This word is used in a variety of ways in statistical literature, and anyone reading on his own is certain to come across it. The following are some of the phrases which could replace it in various contexts:

set of members	frequencies	frequency table
		table
population	frequency function	frequency diagram
	function	diagram
	frequency-density function	frequency-density diagram
	density function	density diagram

It also has several uses in parts of the subject that we have not yet reached.

We have ourselves (in Chapter 4) found it useful to have the jargon 'U-shaped population' when referring to a 'population with a frequency function whose diagram is U-shaped' and so we must be careful to steer clear of pedantry when attacking jargon. Jargon, however, has its dangers in lack of clarity.

It seems best to limit, until later, the use of 'distribution' as follows:

If nine people are measured for height we might find their heights in inches to be: 66, 67, 67, 69, 68, 67, 66, 67, 68. We could classify the people by their heights. For instance we could set aside a particular room for each height-class and could distribute the people into the rooms by their heights. This *distribution* is a process. We should obtain different frequencies of each height. We should be distributing *the people* by their heights. We should not be distributing *the heights,* nor should we be distributing *the frequencies.*

If each person is represented by a stroke then the process of *distribution* could be represented as follows:

66	//
67	////
68	//
69	/

We do not necessarily write

66	66		
67	67	67	67
68	68		
69			

though when we do we are emphasizing the fact that the collection of height-measurements (with their repetitions) is a *population.*

If we write, instead,

66	2
67	4
68	2
69	1

then we are emphasizing the fact that a population gives rise to a *frequency function,* the *table* being a way of saying that under the function $67 \to 4$ and so on.

Exercise 5 C

Select a suitable meaning or meanings for 'distribution' in each of the following invented (but typical) passages:

1. In discrete distributions we can find the median by studying the frequency distribution.

2. Not many statistical distributions are discrete.

3. ...to show the distribution in a diagram....

4. *Distribution of Results in Traffic Experiment.*

5. If we carry out a linear transformation on the original distribution....

6. The diagram shows even more clearly than a table that the distribution is symmetrical.

7. Tables like these are called frequency distributions.

8. If it is the monthly general distribution of rainfall that interests us rather than the weekly....

9. Though these diagrams resemble histograms we will only use the word 'histogram' to refer to frequency distributions.

10. To obtain the frequency distribution we count and tabulate....

11. A histogram of such a distribution is....

12. Frequency distributions can be represented by histograms.

13. Frequency distributions can be regarded as finite samples from infinite populations.

6

POPULATIONS AND STANDARD DEVIATIONS

1. EMPLOYMENT OF STANDARD DEVIATION

1.1 Introduction. We now come to the question 'What does the Standard Deviation (S.D.) really do for us, what does it tell us about a *single* population?' It is clear enough that once we have devised a measure of spread of any sort at all we may use it to *compare* the spreads of two populations, but in the case of the range and the interquartile range we could say something even when dealing with a *single* population; namely that if we placed on the variable axis suitably centred intervals with lengths given by these ranges, then they would cover respectively 100 per cent and 50 per cent of the entire population. There was of course no deduction that an interval of 2 × interquartile range would cover 2 × 50 per cent of the population, but the proportion of the population covered by such an interval would give us interesting information about the lengths of the tails of the frequency diagram. Indeed we could go on to consider other multiples of the interquartile range like $\frac{1}{2}$, $1\frac{1}{2}$ or 3, to paint a picture of the *shape* of the whole frequency diagram.

It is this approach that we now try to adopt with the variance and S.D. The first thing we note is that the variance is useless for this purpose. It has some algebraic properties of a simplicity which the S.D., because of the square-root involved, cannot match, but it is this very absence of the square-root which leaves the variance here floundering. For a moment's consideration shows that if, for instance, we are investigating prices or ages, then the variance will measure the spread in units such as square-pennies or square-years, and that we cannot meaningfully mark these on a diagram with a scale of pennies or years along one axis and a scale of pure numbers for the frequencies (or a scale for frequency densities) along the other. The S.D. raises no such problem; the original units of measurement are restored by the square-root process. This step of square-rooting was taken, apparently quite arbitrarily, in Chapter 4, Section 7.3 when the S.D. was defined almost simultaneously with the fairly naturally occurring variance. Since then the square-root in the S.D. definition has been a help when dealing with change of scale, but the S.D. really comes into its own in the present section.

1.2   Ranges of the form (mean $\pm\lambda$.s.d.). The interquartile range is a measure defined by the very proportion of the population that it covers, so inevitably a suitably placed interval of length $1 \times \text{IQR}$ will cover 50 per cent of the population, whatever the shape of the diagram which gave rise to the IQR. The s.d. is not defined in this way, so we cannot suppose (and it is in fact false) that a definitely placed interval of width, for instance, $1 \times \text{s.d.}$ will always cover a fixed proportion, say $\frac{1}{3}$, of the population, whatever the shape of the diagram that gave rise to the s.d. There is however a general theorem, discovered by Chebyshev or Tchebycheff (and there is a spectrum of spellings between these, so don't despair when searching an index), which tells us that, for instance, strictly inside an interval marked out $2 \times \text{s.d.}$ on either side of the mean we find $\frac{3}{4}$ or more of the population, *whatever may be the shape of the diagram.*

We give an example of a population where the proportion is exactly $\frac{3}{4}$, namely $f(-1) = 1$, $f(0) = 6$, $f(1) = 1$. The reader should verify that s.d. $= 0.5$ so that the relevant range is $-1 < x < 1$, containing exactly $\frac{6}{8}$ of the population.

1.3   Chebyshev's Inequality. Now the s.d. is a function of deviations from the mean, so it is natural to consider ranges formed by laying off a multiple of the s.d. on either side of the mean, that is, ranges given by

$$\bar{x} - \lambda S < x < \bar{x} + \lambda S \quad \text{or} \quad |x - \bar{x}| < \lambda S,$$

where we have written the s.d. as S.

Given a population P and a number λ we can use such a range to divide the population P into two parts:

Subpopulation I, interior to the interval $|x - \bar{x}| < \lambda S$;
Subpopulation E, exterior to the interval $|x - \bar{x}| < \lambda S$ and including also the boundaries of the interval.

Then members x_i of I satisfy $(x_i - \bar{x})^2 < (\lambda S)^2$ and members x_i of E satisfy $(x_i - \bar{x})^2 \geqslant (\lambda S)^2$.

This is shown diagrammatically in Figure 6.1.

The basic idea of the method of proof of the inequality is the replacement of each member of E by a number at the boundary, and the replacement of each member of I by the number at the mean, thus producing a population with a variance lower than the original variance (or equal to it if all members were already at either a boundary or the mean).

Let the number of members of P and of E be N, N_E respectively. In calculating the variance as $\frac{1}{N}\Sigma(x_i - \bar{x})^2 f(x_i)$ part of the sum will derive from the members of E and part from the members of I.

176

Thus we may write

$$NS^2 = N \times \text{variance} = \sum_E (x_i - \bar{x})^2 f(x_i) + \sum_I (x_i - \bar{x})^2 f(x_i).$$

But $\quad \sum_E (x_i - \bar{x})^2 f(x_i) \geqslant \sum_E (\lambda S)^2 f(x_i)$, by the defining property of E and the fact that the square function is here an increasing function,

$$= \lambda^2 S^2 \sum_E f(x_i)$$

$$= \lambda^2 S^2 N_E,$$

whereas $\quad \sum_I (x_i - \bar{x})^2 f(x_i) \geqslant \sum_I 0^2 f(x_i)$, by the non-negativeness of squares,

$$= 0.$$

Thus $\qquad NS^2 \geqslant \lambda^2 S^2 N_E + 0 \quad \text{(using both inequalities)}$

so $\qquad \dfrac{N_E}{N} \leqslant \dfrac{1}{\lambda^2}.$

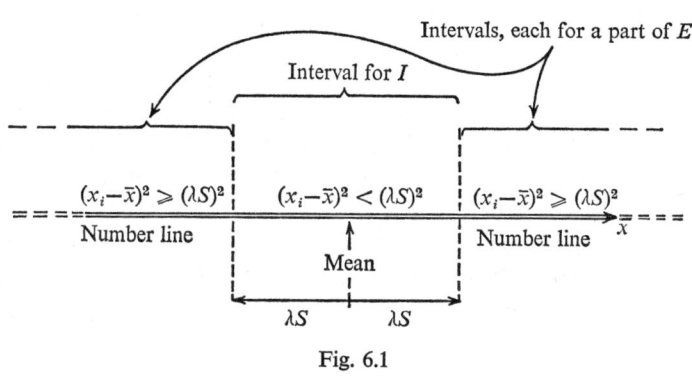

Fig. 6.1

In words: the proportion of the population lying *outside* (or on the boundaries of) an interval extending λS on either side of the mean cannot exceed $1/\lambda^2$; the proportion inside such an interval is therefore at *least* $(1 - 1/\lambda^2)$. See Exercise 6A, Question 2, for a stricter form.

An intuitive understanding of the inequality itself may come from the following argument:

If the reader has found a population for which the inequality is only just satisfied at some value of λ, and if he attempts to alter the population so as to deny the inequality for that value of λ, then he may do one of two things:

First, he may increase the membership of the subpopulation outside the interval, trying to increase the proportion there; at the same time, however, he will be increasing the S.D. so that the interval will be expanding and swallowing members of the population only just outside it, and the status quo may tend to be restored.

Secondly he may increase the membership of the sub-population inside the interval, trying to decrease the S.D. so that the interval will be contracting and excluding members of the population only just inside it; at the same time, however the very proportion of membership of the sub-population inside the interval will be increased by his action and the proportion outside will be reduced; and the status quo may tend to be restored. (See Exercise 6A, Question 1.)

1.4   Standardized population. The changes of scale in Chapter 4, Section 10.2 pose the interesting question as to what scale, if any, is really intrinsic to a population, and we shall see that if we wish to discuss proportions of a given population, the most convenient scale to use for that population is the one which takes its S.D. as unit. We would then choose the mean for origin so that we would be changing the variable from x to t by the formula:

$$t = \frac{x - \bar{x}}{S}.$$

When the scale of the variable is such that the mean is zero and the S.D. is unit, the variable is often said to be *standardized*. This is a deceptive way of referring to the transformation, since the transformation does not so much depend on what is the variable measurer as on what is the frequency function; it might be better to say that the *population* had been standardized. The idea is a very important one indeed in theoretical statistics; t resembles a physicist's non-dimensional quantity.

Exercise 6 A

1 R. Consider the three populations given by the following:

	$f(-1)$	$f(0)$	$f(1)$
1st population	1	5	1
2nd population	1	6	1
3rd population	1	7	1

In each case find the S.D. and the proportion of the population that satisfies $-2 \times$ S.D. $< x < +2 \times$ S.D.

(ii) Construct a population which exactly satisfies Tchebychev's Inequality with $\lambda = 3$; that is to say, has a proportion exactly $\frac{8}{9}$ in the range

$$-3 \times \text{S.D.} < x < +3 \times \text{S.D.}$$

2 T. Prove Tchebycheff's Inequality in the form that the proportion strictly outside the interval $|x - \bar{x}| \leqslant \lambda S$ is strictly less than $1/\lambda^2$. Show this is a tighter form.

3. Show that, with a suitable choice of origin and scales for z, Chebyshev's Inequality can be written:

Relative frequency of values of z having $|z| \geqslant z_0$ is less than z_0^{-2}.

4. Obtain an inequality resembling Chebyshev's but involving an interval $b - \lambda d < x < b + \lambda d$, where b is the median and d the mean absolute deviation of a population. Why can we not deduce immediately an inequality between the mean absolute deviation and the IQR?

5T. Figure 6.2 shows a histogram. A continuous line has been drawn through the tops of the rectangles to suggest the general shape.

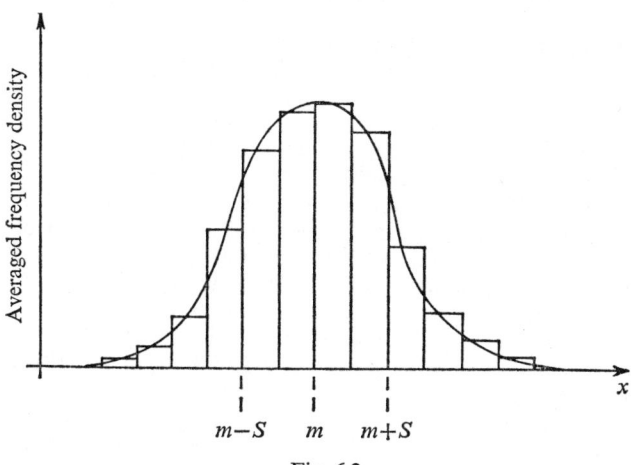

Fig. 6.2

The population size is N, the mean of $[x]$ is m, the s.d. of $[x]$ is S.

(i) Draw four freehand diagrams to show histograms for populations with the statistic-values listed in the table below. Each diagram must contain a copy of Figure 6.2 for comparison; but only the continuous general outlines need be drawn.

Population	Size	Mean	S.D.	Variance
(a)	$N/2$	m	S	S^2
(b)	N	$m/2$	S	S^2
(c)	N	m	$S/2$	$(S^2/4)$
(d)	N	m	$2S$	$4S^2$

(ii) Suggest purely geometrical transformations for obtaining the diagrams for each of the populations (a) to (d) from the original figure.

6. (i) If masses $f(x_i)$ are placed along a line at points given by $x = x_i$, find the moment of inertia of the resulting system about an axis through the centre of mass perpendicular to the line.

(ii) If the radius of gyration about such an axis is K, express the radius of gyration about a parallel axis distant a units away in terms of the parameters of the system.

(iii) Draw up a table showing all the analogies obtained between the properties of various statistics and those of various mechanical quantities.

7 T. In order to compare the variability of two populations of measurements a statistic called the *coefficient of variation* is sometimes computed. The definition is

$$\text{coefficient of variation} = \frac{\text{S.D.}}{\text{mean}} \times 100,$$

the S.D. and mean being measured in the same units. (Thus the coefficient merely expresses the S.D. as a percentage of the mean.)

(i) Discuss critically its use to compare the variability of the following populations of measurements.

    (*a*) The heights of three Lilliputians 5 in, 8 in, 8 in.
        The heights of three Brobdingnagians 60 ft, 75 ft, 75 ft.
    (*b*) Three daily temperatures of 5 °C, 8 °C, 8 °C.
        Three daily temperatures of 28 °C, 31 °C, 31 °C.
    (*c*) Three daily temperatures of -1 °C, 2 °C, 2 °C.
        Three daily temperatures of 0 °C, 3 °C, 3 °C.
    (*d*) Three boiling-point determinations of 5 °C, 8 °C, 8 °C.
        The same three temperatures measured in °Kelvin. [To obtain the temperature in °Kelvin add 273 to the Celsius temperature.]
    (*e*) Three daily temperatures of 68 °F, 95 °F, 95 °F.
        The same temperatures expressed in the Celsius Scale.
    (*f*) The weights of three boys namely 139 lb, 142 lb, 142 lb.
        The heights of the boys namely 5 ft 9 in, 5 ft 10½ in, 5 ft 10½ in.
    (*g*) The lengths of three lines of 5 cm, 8 cm, 8 cm.
        The areas of squares drawn on these lines.

(ii) What transformations applied to a population of numbers will leave the coefficient of variation of the image numbers equal to that of the original numbers?

8. The following table shows heavily approximated values of the mean height and the S.D. of heights (each measured in cm.) for large collections of girls of various ages. [The data refer to the U.K.]

Age in years	Mean height	S.D. of height
1	75	2·5
2	85	2·5
4	105	3·5
6	115	4·5
8	125	5·0
10	135	5·5
12	150	7·5
14	155	6·0
16	160	5·5
18	160	6·0

Draw a graph to show the range of (mean $\pm 2 \times$ S.D.) at each age given, plotted against the age; and a graph of the coefficient of variation against age. (See Question 7.)

2. PRACTICAL CASES AND CHEBYSHEV'S INEQUALITY

The interest of Chebyshev's Inequality is largely theoretical. Its strong point is its generality, but the corresponding weakness is the very high bound it has to place on N_E/N, and the consequent low bound on $1-(N_E/N)$, the proportion inside an interval of (mean $\pm\lambda$ s.d.).

For instance, for almost all the populations we meet in practice, the interval given by (mean $\pm2\times$s.d.) contains a proportion more likely to be about 95 per cent than 75 per cent. Exercise 6B, Question 1 shows a variety of shapes, (J-shaped, bell-shaped, U-shaped and highly irregular) which bear out this observation. We can safely use this $2\times$s.d. range either side of the mean for a rough check on our working, by evaluating the corresponding limits and checking whether about 5 per cent of the population lies outside them. If the proportion outside is too high or the limits very easily include the entire population, then we have probably made an error in the calculations below the tabular stages, that is, below the stages checked by the Charlier checks.

Perhaps the commonest errors are: (*a*) failure to take the square-root of the variance; (*b*) failure to change the scale back to that of the original data.

Example 1. Find the approximate mean and s.d., and the approximate proportion of the population outside the range given by (mean $\pm2\times$s.d.), for a population given by the grouped (and so partial) data below:

Solution. We take

$$t = (x-1\cdot9)/0\cdot2; \quad \bar{x} = (0\cdot2\times\bar{t})+1\cdot9.$$

[We take an opportunity to display the third form of the Charlier checks.]

Class	x_i class-mark	$f(x_i)$	t_i	(t_i+1)	$t_i f(x_i)$	$(t_i+1)f(x_i)$	$t_i^2 f(x_i)$	$(t_i+1)t_i f(x_i)$
1·2–(1·4)	1·3	1	−3	−2	−3	−2	9	6
1·4–(1·6)	1·5	8	−2	−1	−16	−8	32	16
1·6–(1·8)	1·7	23	−1	0	−23	0	23	0
1·8–(2·0)	1·9	37	0	1	0	37	0	0
2·0–(2·2)	2·1	31	1	2	31	62	31	62
2·2–(2·4)	2·3	16	2	3	32	48	64	96
2·4–(2·6)	2·5	4	3	4	12	16	36	48
		120			−42+75	−10+163	195	228
					= +33	= +153		

Charlier checks.

$$\Sigma t_i f(x_i) + \Sigma f(x_i) = 33 + 120 = 153 = \Sigma(t_i + 1) f(x_i),$$
$$\Sigma t_i^2 f(x_i) + \Sigma t_i f(x_i) = 195 + 33 = 228 = \Sigma(t_i + 1) t_i f(x_i).$$

Calculations.

$$\bar{t} = \tfrac{33}{120}, \quad \bar{t^2} = 0\cdot0751,$$
$$\bar{x} = (0\cdot2 \times \tfrac{33}{120}) + 1\cdot9 = 0\cdot055 + 1\cdot9 = 1\cdot955,$$
$$\mathrm{var}\,[t_i] = \tfrac{195}{120} - (\bar{t})^2 = 1\cdot625 - 0\cdot075 = 1\cdot550,$$
$$\mathrm{stad}\,[t_i] = \sqrt{(1\cdot550)} = 1\cdot245,$$
$$S = \mathrm{stad}\,[x_i] = 0\cdot2 \times 1\cdot245 = 0\cdot249,$$
$$\bar{x} - 2S = 1\cdot955 - 0\cdot498 = 1\cdot457,$$
$$\bar{x} + 2S = 1\cdot955 + 0\cdot498 = 2\cdot453.$$

We now assume that within each class the frequency density is constant, so that the parts where the two limits fall may be represented diagrammatically as in Figure 6.3.

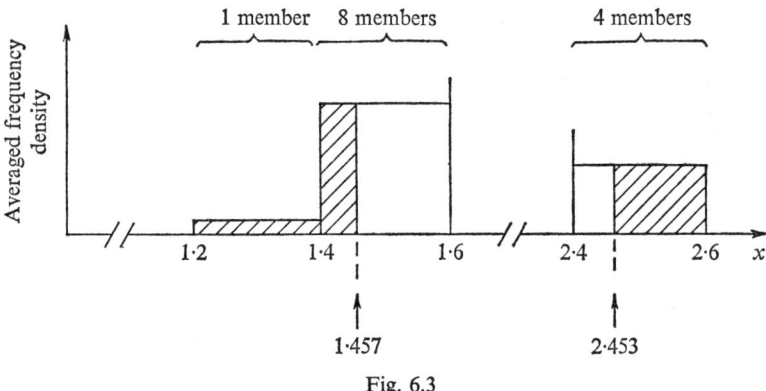

Fig. 6.3

The number of members represented by the total shaded area

$$= 1 + \frac{1\cdot457 - 1\cdot4}{0\cdot2} \times 8 + \frac{2\cdot6 - 2\cdot453}{0\cdot2} \times 4$$

$$= 1 + 2\cdot28 + 2\cdot94$$

$= 6\cdot2$, and this is the approximate proportion of the underlying population P and is 5 per cent of the total membership of 120.

It is very important to realize that if only the proportion outside the limits had been required, and not the actual values of the mean and s.d., then we need never have changed back to the x-variable from the t-variable. This is suggested by the remarks in Chapter 4, Section 10.2 about the scales in Figure 4.13.

The part of the calculations after the checks would then be as follows

$$\bar{t} = \tfrac{33}{120} = 0\cdot275, \quad \overline{t^2} = 0\cdot0751,$$

$$\text{var}\,[t_i] = \tfrac{195}{120} - (\bar{t})^2 = 1\cdot625 - 0\cdot075 = 1\cdot550,$$

$$S_t = \text{stad}\,[t_i] = \sqrt{(1\cdot550)} = 1\cdot245,$$

$$\bar{t} - 2\,.\,S_t = 0\cdot2075 - 2\cdot490 = -2\cdot215,$$

$$\bar{t} + 2\,.\,S_t = 0\cdot275 + 2\cdot490 = 1\cdot765.$$

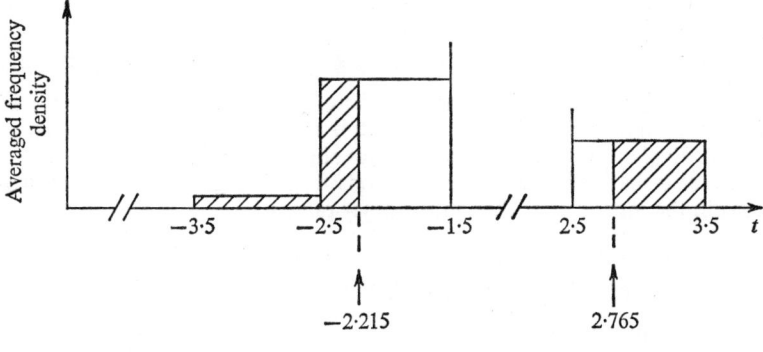

Fig. 6.4

The number of members represented by the total shaded area in Figure 6.4 $= 1 + (-2\cdot215 + 2\cdot5) \times 8 + (3\cdot5 - 2\cdot765) \times 4$

$$= 1 + (0\cdot285 \times 8) + (0\cdot735) \times 4$$

$$= 1 + 2\cdot28 + 2\cdot94, \text{ as before.}$$

Exercise 6 B

1R. (a) For each of the following populations

(i) draw the histogram, assuming that the frequency density is constant in each interval quoted;

(ii) approximate to the mean and S.D., assuming now that each class has all its members at the class-mark (no correction for grouping is needed for the mean, as we saw in Exercise 5A, Question 3, and you are to ignore the correction necessary for the S.D.); *Hint:* once the variance for Population 1 has been obtained, it can be used, with changes of origin, to find quickly the variances of Populations 2, 3;

(iii) mark on the diagram the values of x to the nearest 0·1 corresponding to the approximated values of (mean $\pm \lambda \times$ S.D.), for $\lambda = 1, 2$;

(iv) find the proportion of the *area* of the histogram between these limits; (this approximates to the proportion of the population with values of x between these limits).

183

	Frequencies for			
Intervals of x	Population 1	Population 2	Population 3	Population 4
0·5– 1·5	90	10	—	10
1·5– 2·5	50	10	90	40
2·5– 3·5	40	40	50	0
3·5– 4·5	10	50	40	90
4·5– 5·5	10	90	20	10
5·5– 6·5	—	90	20	50
6·5– 7·5	—	50	40	—
7·5– 8·5	—	40	50	—
8·5– 9·5	—	10	90	—
9·5–10·5	—	10	—	—

(b) The third population is apparently more dispersed than the second, and yet it has a large proportion in the central range of approximately (mean $\pm 2 \times$ s.d.). Explain in general terms why this should be so.

[Notice also that over the four populations there is a greater variety of proportions in the inner ranges than in the outer ranges.]

3. VARIOUS POPULATION TYPES

There are many types of parent population from which we find observations being drawn, and the later parts of the course will be much concerned with them. In this section we mention one in particular and refer to some others in the Exercise 6 M which follows.

Normal Population. There is a particular type of parent population whose diagram is symmetrical and bell-shaped and which is often approximated to in populations of experimental readings. It is called *Normal* or *Gaussian*. The grouped readings in Example 1 were selected from such a population.

(The name Normal is unfortunate since it suggests to the casual reader that other types of population are abnormal, but the use of the capital N should warn the careful reader that 'Normal' is here used as 'Liberal' is in politics, where the opposite of 'Liberal' is not necessarily 'illiberal'.)

Table 6.1 below shows the relevant proportions for this type, and the extreme proportions possible at all in a general population (the extremes being obtained from Chebyshev's Inequality).

Table 6.1

		Proportion included in range	
Range in standardized units	Range on *either* side of mean	Normal population	Any population
$-1 < t < 1$	Within 1 s.d.	68%	Could be none
$-2 < t < 2$	Within 2 s.d.	95%	At least $\frac{3}{4}$
$-3 < t < 3$	Within 3 s.d.	Effectively all	At least $\frac{8}{9}$

Exercise 6M

Miscellaneous Exercise

1F. (i) The following classic data refer to deaths by horse-kick of Prussian cavalrymen in ten corps during 1875–94.

Number of deaths per corps per year x	Frequency with which a corps had this death-rate $f(x)$
0	109
1	65
2	22
3	3
4	1

Calculate the mean and variance. [See also Exercise 8B, Question 2, Exercise 9C, Question 20.]

[f is the frequency function of a nearly *Poisson population*. Such populations can arise in cases like this where it is possible to count how many times during the year a death from horse-kick occurs, but *not* how many times such a death does not occur, as M. J. Moroney has neatly remarked in *Facts from Figures* (Penguin, 1964). Compare the note on binomial populations in Exercise 4E, Question 4, where $6-x$ was the number of times the event did *not* occur in 6 trials.]

(ii) At a certain fire-station in 1964, y calls per day occurred $g(y)$ times, according to the following table.

y	0	1	2	3	4	5	6
$g(y)$	100	65	107	60	30	3	1

Calculate the mean and variance and note how the relationship between them differs from that in part (i).

Discuss a possible reason for the shape of the frequency diagram near $y = 0$.

(iii) Would you anticipate that data from numbers of plants of a given species in, say, a 1 m square would lead to a pattern like that of part (i) or to a pattern like that of part (ii)?

2. We refer to the data collected in Exercise 4A, Question 3. If $f(x)$ be the number of nouns which occur x times in a passage, then the population of values of x defined by taking $f(x)$ as the frequency function will change its characteristics as the length of passage is increased. The mean number of occurrences per noun will increase; its reciprocal, the number of distinct nouns per thousand, (say), occurrences will decrease; the variance will increase; the percentage of nouns used only once will decrease. G. U. Yule has suggested theoretical reasons why the statistic

$$k = \frac{\Sigma x_i^2 f(x_i) - \Sigma x_i f(x_i)}{(\Sigma x_i f(x_i))^2}$$

should be independent of the length of passage selected, but should vary with the degree of variety in the choice of nouns employed by the author. He devised this statistic to find whether there was statistical support for the theory that the author of *De Imitatione Christi* was not Thomas à Kempis but Gerson.

185

(i) Calculate $10^4 k$ for each of the three populations obtained in Exercise 4 A, Question 3.

(ii) If the results from a whole class allow it, compare the value of k for different (*a*) parts of a book, (*b*) books, (*c*) authors.

[Populations of this type occur in insurance problems and resemble amalgamations of several populations of Poisson type with differing means and with the information about frequency of zero omitted.)

3. Examine the first nine verses of The Acts of the Apostles in The Authorized (or King James) Version and in *The New English Bible*. (Exclude all words that, even when not the first word of a sentence, begin with a capital letter. That is, exclude any words *used* as proper names.) Find the mean and s.d. of the lengths of the words measured by number of letters, and compare them for the two populations.

4. Examine the 200 words defined in Question 3 for the Authorized Version, and the first 200 words (not proper names) of any Latin text. Find the mean and s.d. of the number of occurrences per word of the letter 'u' in each text, and compare them for the two populations.

5 *T*. [Improved method of studying sentence-lengths for different authors.] Examine two large collections of sentences for two different authors and construct histograms for the populations of logarithms of sentence-length. [C. B. Williams has shown that for most authors such populations are nearly Normal. A great deal is known about Normal density functions, hence the 'improvement'.]

6. The population P referred to in Example 1 approximates to a Normal population. For this population use the values of $\bar{t}$ and S_t obtained in the Example and find approximately the proportion of the population in the ranges given by $$|t - \bar{t}| < \lambda . S_t \quad \text{for} \quad (a)\ \lambda = 1,\ (b)\ \lambda = \tfrac{1}{2}.$$

Draw the histogram and mark the limits of these intervals.

7 *T*. For the population P referred to in Example 1, find approximately the first and fifth sextiles. Use the results of Table 6.1 to devise a rough method of obtaining the s.d. of a population whose histogram is known to be of the same general shape as that of P, that is to say, suggest a method for finding approximately the s.d. of a population known to be roughly Normal.

8. Do brains shrink with age? The brains of one particular group of persons divided into an old and a young set were distributed into classes by their weights as follows:

Wt range/gm	800–	950–	1050–	1100–	1150–	1200–	1250–	1300–	1350–	1400–
Frequency (old)	3	6	8	19	11	14	14	3	6	1
Frequency (young)	0	2	14	22	45	54	55	23	13	10

Illustrate by histograms and give the ranges of 2 s.d. on either side of the mean for each populations of weights. Comment on the use of these weights, which must have been obtained *post mortem*, to answer the question asked.

9 *F*. From a Normal population of values, x, with variance 8 a set of 100 samples of size 2 was taken. Each sample had its variance, y, calculated, and the population of variances of the samples gave rise to the following grouped frequency table.

186

Interval of values of y	Frequency of y in interval
0·00– 0·05	9
0·05– 0·25	11
0·25– 0·75	13
0·75– 2·75	25
2·75– 5·25	17
5·25–10·25	14
10·25–15·25	6
15·25–25·25	5

(i) Draw the histogram, amalgamating the first three classes; and also draw a separate histogram for these classes, with the vertical scale multiplied by $\frac{1}{4}$ and the horizontal scale multiplied by 10. On the first histogram superimpose a sketch of the graph of $\phi(y) = 100(8\pi y e^{\frac{1}{4}y})^{-\frac{1}{2}}$, choosing $y = \frac{1}{2}, 1, 2, 4, 8, 16$.

[A population of this type is called a chi-squared population. There are, of course, theoretical reasons for the apparently bizarre choice of ϕ.]

(ii) Approximate to the mean of the population of variances of samples (that is to $\bar{y}$); and to the variance of the population of variances of samples (that is to var [y]).

(iii) Approximately what fraction of the samples have their variances less than the variance of the parent population of values of x? Note the implication that the variance of such a sample tends to underestimate the variance of the parent population, and explain why this result might be expected.

[We shall see in Chapter 11 that the mean variance of random samples of size N from a parent population of variance V is $\dfrac{N-1}{N}.V$, whatever may be the frequency function of the parent population.]

10. 288 small cuboids of chert found in archaeological diggings in the Indus Valley had the following weights x gm with frequencies $f(x)$.

x	$f(x)$	x	$f(x)$	x	$f(x)$	x	$f(x)$
1·7	6	3·2	2	6·6	1	13·3	1
1·8	4	3·3	4	6·7	3	13·4	6
1·9	2	3·4	17	6·8	22	13·5	9
2·1	1	3·5	6	6·9	10	13·6	23
2·3	1	3·6	3	7·0	2	13·7	17
		3·8	1	7·3	3	13·8	7
		3·9	1			13·9	7
26·6	1	4·0	2			14·0	5
26·8	5	53·6	1			14·1	1
27·0	11	53·8	1			14·2	1
27·2	31	54·0	6				
27·4	17	54·2	10				
27·6	5	54·4	8	134	1		
27·8	4	54·6	4	135	3		
28·0	2	54·8	1	136	6		
28·2	1	55·0	1	138	1		

Find the mean and mode in each group and comment on the meaning of the results.

[The data are deducible from A. E. Berriman: *Historical Metrology* (Dent, 1953).]

11. The following table shows (fictional, but typical) counts of the number of spots on the hind wing of the meadow-brown butterfly (*Maniola jurtina*):

No. of spots	0	1	2	3	4
No. of butterflies (male)	4	15	61	17	3
No. of butterflies (female)	18	15	38	7	2

Find the mean and standard deviation of each population of numbers of spots.

Using an apparently unrelated characteristic, we split the females into two groups, one of which has spot-counts as follows:

No. of spots	0	1	2
No. of butterflies	16	8	6

Find the mean and standard deviation of the number of spots in each group of females and compare them with the values from the group of males.

12. In 1881, statues of Gudea, who in about 2175 B.C. governed the Sumerian city Lagash, were excavated. One of them, now in the Louvre, shows him holding a scale graduated with the following intervals, measured in mm.

$$17·5, \quad 17·7, \quad 16·5, \quad 16·8, \quad 16·0, \quad 17·7, \quad 16·6, \quad 17·2,$$
$$16·5, \quad 16·5, \quad 16·5, \quad 17·3, \quad 16·2, \quad 17·0, \quad 16·5, \quad 16·5.$$

[These figures are given in A. E. Berriman: *Historical Metrology* (Dent, 1953).]

Find the mean and standard deviation of these lengths. Assuming that the mean length is a Sumerian shusi, find the length of 10 Sumerian cubits in English inches (25·4 mm = 1 in, 30 shusi = 1 cubit). Identify this length in the mediaeval English system of units: yards, poles, chains.

What interval would you cover in English units by using the extremes mean $\pm 2 \times$ S.D. instead of the estimated shusi?

Discuss the possible roles of historical connexion and mere coincidence in your results.

13F. [This question is much discussed in Chapter 7.]

Throw a single die 150 times and record *in sequence* the scores. [This produces a sample from a rectangular or uniform population, so called because 'in the long run' the frequency diagram is either a single rectangle or numerous rectangles of uniform height.] It is important *not* to throw two dice and record their scores as, although this cuts the time taken, some bias may occur over which score is recorded first, and the *sequence* is the basis of this investigation. It is permissible to throw recognizably different dice and fix *before-hand* the order in which their scores will be recorded.

The sequence is to be examined for the occurrences of a particular event, such as 'scoring either a "one" or a "six"'. Each time the event occurs it is mapped onto the number of throws *strictly between* its occurrence and the next occurrence. If the sequence *begins* with the event, we write down 0 for the number of throws as a start. For instance, with this event and the sequence of scores:

$$1, \quad 4, \quad 4, \quad 6, \quad 3, \quad 6, \quad 1, \quad 4, \quad 2, \quad 5, \quad 1, \quad 4, \quad 3, \quad 2, \quad 5$$

we obtain the population of values of x:
$$0, \quad 2, \quad 1, \quad 0, \quad 3.$$

We neglect the parts of the sequence after the final occurrence. Consider quite separately the events: (i) scoring either a 'one' or a 'six'; (ii) scoring a 'six'. Find the mode, median, mean and variance for each of the two resulting populations.

In which of the two cases should a gambler bet at evens that the next length of run would be the integer nearest to the mean?

[The populations of lengths of run are called *negative binomial* for reasons explained in Chapter 7.]

14F. From the *sequence* of scores obtained by the method of Question 13, obtain a population as follows: Start from *each score in turn* and count the number of throws *strictly before* the next occurrence of 'scoring either a "one" or a "six"'.

Thus the example sequence in Question 9 gives the population

$$0, \quad 2, \quad 1, \quad 0, \quad 1, \quad 0, \quad 0, \quad 3, \quad 2, \quad 1, \quad 0.$$

Compare the mode, median, mean and variance with those obtained in part (i) of Question 13.

15F. [This question is much discussed in Chapter 7.]

From the sequence of scores obtained by the method of Question 13, obtain a population as follows: Divide the sequence into 30 consecutive blocks of 5 scores; count the occurrences in each block of the event 'scoring either a "one" or a "six"'. Find the mean and variance of the number of occurrences of this event per block.

[The population of number of occurrences is a binomial one as described in Exercise 4E, Question 4.]

16T. For a population $[x]$ with frequency function f and size N we define the rth *moment of the population* about the origin by

$$m'_r = \frac{1}{N}\Sigma x_i^r f(x_i),$$

and the rth *moment about the mean*, $\bar{x}$, by

$$m_r = \frac{1}{N}\Sigma(x_i - \bar{x})^r f(x_i).$$

Prove, by multiplying out the appropriate powers of $(x_i - \bar{x})$, that

$$m_0 = m'_0 = 1,$$
$$m'_1 = \bar{x},$$
$$m'_2 = S^2 + \bar{x}^2,$$
$$m_3 = m'_3 - 3m'_2 m'_1 + 2(m'_1)^3,$$
$$m_4 = m'_4 - 4m'_3 m'_1 + 6m'_2(m'_1)^2 - 3(m'_1)^4.$$

[$\alpha_3 \equiv m_3/S^3$ is often taken as a measure of *skewness*. The *kurtosis* is defined as $\alpha_4 \equiv m_4/S^4$; it indicates the lengths and heights of the tails of the frequency diagram. Since for large samples from a Normal parent population we have $\alpha_4 \simeq 3$, the quantity $(\alpha_4 - 3)$ is sometimes called the *excess kurtosis*.]

17 T. Use the statistics of skewness and kurtosis defined in Question 16 to distinguish between the populations of Exercise 4E, Question 6, namely those for f, g, h below. Calculate also the 'skewness' for the population given by k.

x_i	$f(x_i)$	$g(x_i)$	$h(x_i)$	$k(x_i)$
-3	0	1	1	2
-2	3	1	2	2
-1	6	5	5	0
0	7	11	6	5
1	6	5	10	6
2	3	1	1	0
3	0	1	0	0
4	0	0	0	1

18 T. (Due to P. J. Phair.) Show that the following populations agree in their first 8 moments (as defined in Question 16).

x_i	-4	-3	-2	-1	0	1	2	3	4
$f(x_i)$	2	3	30	4	91	4	30	3	2
$g(x_i)$	1	11	2	60	21	60	2	11	1

19. Devise a Charlier-type check for the columns involved in calculating the skewness.

7

SEQUENCES OF EVENTS—
BERNOUILLI MODEL

1. INTRODUCTION

The reader may have felt that there has been a certain divergence of mathematical content between the end of Chapter 3, where he met such problems as finding the probability that Atkins was with umbrella, and the end of Chapter 6, where he dealt with such things as the mean number of fire-engines called out per day. He may however see a linking thread if he realizes that:

(i) in the Atkins problem there were just two possible events on any day (Atkins was with or without umbrella) and the problem was taken as far as deciding the probabilities of these events from those of other events, the model being outlined by the given patterns of behaviour. The only point of knowing the probabilities would presumably be to estimate such things as in how many weeks of a 50-week year we could expect Atkins to have his umbrella on 5, 4, 3, 2, 1 or 0 days.

Whereas (ii) in the fire-engines problem, the actual frequencies were already presented and any probability model would have worked backwards from them.

If we analyse the problems more closely we could classify them as follows. In each case we make a sequence of trials (not very arduous trials; all we do is wait until the end of the day) at a rate of one per day and for the engine problem record the number of departures there were, while for the Atkins problem we record whether Atkins had his umbrella or not.

2. ATTRIBUTES

In collecting observational data

(i) we sometimes make a sequence of trials whose outcomes are *quantities*; such as (to change the example) 10 in, 9 in, 12 in, 11 in, for the lengths of four fish in a catch;

(ii) we sometimes record quantities only vaguely by, for instance, calling those fish under 10 in long 'rejects' or 'failures', those over 10 in long 'successes' and those 10 in long 'indeterminates'. We then have a list of *qualities*: indeterminate, failure, success, success for the four fish listed above. We note in passing that these qualities can in fact still be arranged in a meaningful order, say of increasing size; but in extreme cases:

(iii) we go further and make trials whose outcomes are *mere* qualities, for which no sensible order suggests itself. For instance the same four fish might be recorded as 'brown trout, grayling, rainbow trout, brown trout'. In the study of statistics qualities such as these last are often called *attributes*; carrying out a trial whose outcome is described in terms of such attributes is called *sampling attributes*. We must remember that in ordinary usage any property that a fish may possess is an attribute of that fish and we could say that a fish that was 10·3 in long had the attribute of 'being 10·3 in long'. 'Attribute' in its technical sense, however, is really short for 'non-quantitative attribute'.

It is clear that many of the data of real life are recorded in terms of attributes and the statistical side of the process of sampling attributes is a correspondingly important one. This chapter is concerned with some aspects of it.

In the abstract study of probability that we have already undertaken the idea of (if not the word) attributes has already been well-established, as the discussion of the Atkins problems in this chapter has shown. We began the study of probability with a slightly different vocabulary because we analysed 'events', but we could have used a different form of words and recorded the event 'Harry draws a Club' as 'Harry's card has the attribute of being a Club'. English is particularly rich in synonyms and periphrases, and it is part of our job as mathematicians to choose as few different words as we conveniently can to help us to discern real, and not merely apparent, similarities and distinctions.

We therefore have no difficulty about the probability aspects of sampling attributes; M. Jourdain had spoken prose all his life and we have been sampling attributes long enough. It is for the statistics of the situation that we need a new weapon.

For a start we remind ourselves that we can make frequency tables of attributes, such as:

Attribute	Frequency	Attribute	Frequency
brown trout	2	failure	1
grayling	1	indeterminate	1
rainbow trout	1	success	2
pike	0		
others	0		

In fact it is the ability to make such frequency tables that led us in Chapter 2, Section 5 to introduce the notion of probability to describe the regularity such tables often display.

We cannot, however, calculate with attributes so that in no sense is there a mean attribute in either table. Even in the second table, although the fish had a mean length of $10\frac{1}{2}$ in, we do not say that the catch was equivalent in quality to four successful fish. The statistics of the situation are handled less directly.

3. DICHOTOMY

We will initially limit our attention to sequences of trials whose outcomes can be specified in terms of one or other of *just two* attributes; for example: the fish is pleasant to eat or not; the plant has whorled leaves or not; Atkins has his umbrella or not. Such a trial is sometimes called a *dichotomy*.

From the mathematical point of view the nature of the dichotomy under investigation is unimportant. We may therefore represent the only two possible outcomes abstractly by symbols such as A, B; so that the complete set of possibilities of one trial is $\{A, B\}$. These are not, of course, necessarily equally likely outcomes. Sometimes A is described as 'success' and B as 'failure' but this must not be taken to imply that we are making any value judgements about the situation. What we are doing is setting up a mathematical model applicable to any dichotomy.

We wish to study the problem of what happens when we repeatedly carry out a dichotomous trial.

Exercise 7 A

1. Name a game in which the scores on the faces of dice are regarded as attributes and one in which they are regarded as numbers.

2 R. Suggest four sets of attributes relevant to discussing the composition of a group of people; and also four sets of measurements.

3 R. We define the set S to be the first 200 words of The Acts of the Apostles in *The New English Bible*. In which of the following can we describe the process in terms of repeatedly making a dichotomous trial? We make a list to show:
    (i)   the numbers of letters per word of the words of the set S;
    (ii) the number of words of length two letters in the set S;
    (iii) the numbers of letters 'e' in each word of S;
    (iv) the number of words containing exactly two letters 'e' in S.

4. In a certain Victorian chess set the only safe distinction between King, Queen, bishop and pawn is the number of waists the pieces have. Kings have 7, Queens 6, bishops 4, and pawns 2. In a certain experiment 5 men are drawn out of a box holding the whole of the set; records are made of
    (i)   how many bishops are obtained;
    (ii) the total number of waists obtained;
    (iii) how many men in the sample have no waists.
  In which of these cases can we describe the experiment in terms of a sequence of identical dichotomous trials?

4. ORDER OF TRIALS

Sometimes when we make repeated trials the order of the trials is important: For instance, if we are examining the level of activity of cosmic radiation to discover whether it is changing, we might have a continuously recording instrument whose needle would kick every time a particle entered the collecting chamber. The trials would amount to the passages of consecutive intervals of, say, 1 second each and success would be recorded when a particle entered during the interval and failure otherwise. If we wanted the set of points in the possibility-space representing success to consist of a *single* point (representing only exactly one entry), we should have to make the interval so short that the probability of multiple entries in the interval was negligible; in that case there might be mechanical difficulties over the time taken for the machine to reset itself and the application of a simple model to the output of that machine might not fit the data well. We will however develop this simple model (named after Poisson) in Chapter 8; its applications range from the arrival of patients at a hospital, or telephone calls at an exchange, or requests for an item from a stock of spare parts, through traffic-flow, to births and deaths in a community.

In gambling situations the order of trials is important, because we obviously need to know more than the average number of successes in a long run of plays; a few runs of failures near the beginning might exhaust our capital before we ever reached the later results, which would in retrospect be seen to have swamped the initial ones (if only we had still been in the game). A study of even traditionally 'fair' games shows that although for many simultaneous players the total gains and losses at a given moment tend to balance, nevertheless the total gain or loss for any individual tends to remain unbalanced for quite unexpectedly long runs of play.

From a *mathematical* point of view gambling is a very widespread activity including the whole of the business and insurance worlds! The feature in common is that decisions are made and money risked on *incomplete* knowledge of the future. To describe these activities as Gambling Situations is picturesque and should not lead the reader to feel that probability theory is of no *practical* use.

Sometimes, on the other hand, when we make repeated trials the order of the trials is unimportant, since they are intended to be effectively simultaneous. For instance if we are collecting voting intentions in a constituency or checking the number of defective items in a box, then a trial consists of examining a person or item and merely counting the successes, and thus losing the pattern by which they were interspersed among the failures.

In inspection procedures in a factory it is possible to choose which

method we will use. We might simply collect, say, 100 items for inspection
and be satisfied if 5 or fewer were rejects. Alternatively we might check
consecutive items as they are produced or delivered until we had 5 rejects,
and then see how long it had taken us to amass them. If it had only taken
about 50 trials, then we should seem to be in a situation where a worse-
than-5 per cent failure-rate was operating, and we could stop sampling
earlier and so save time and money. This is *optional* stopping; individual
gamblers employ optional stopping: either they retire when they have
made what they want, or they are ruined (though *this* may seem to be
stretching the meaning of 'optional').

In terms of the factory procedures the first is also called *direct sampling*
and the second *sequential sampling*; in general terms the second procedure
is sometimes called *inverse sampling*.

5. TWO CONCRETE SITUATIONS

5.1   The data. We next need some concrete data about sequences of
trials, for our analysis. A desk simulation of a simple type can be carried
out with random numbers or dice. We will record the trials in order, so
that we do not throw away any information; the order can always be
destroyed by a later rearrangement, but if we only record orderless trials
(for instance placing all the successes first) we cannot restore the information about the order. This is less obvious than it sounds: if we throw a coin,
known to be fair, 10 times, and get 4 heads and 6 tails, we cannot even say
that *THTHTTHTTH*, where the heads are 'nicely scattered', is a more
likely order than *TTTHHHHTTT*; so we cannot begin to reconstruct a
'likely order' from the orderless data *once the trials are complete*. There is
no useful sense in which we can conceive of indefinite repetitions of experiments with 4 heads and 6 tails arranged in random order which does not
make these two orders equally likely.

In Exercise 6M, Question 13 the following sequence of trials was
proposed: Throw a die and record 'success' if a one or six was thrown and
'failure' otherwise.

The author selected 150 as the number of trials, *before carrying them
out*, and Table 7.1 is the record, with *A* written for success and *B* for
failure.

Table 7.1

BBB   A   BB   AA   BB   A   BB   A   B   A   BBB   AA
BBBB   AA   BBB   AA   B   AA   BBBBBB   A   BB   A
BBBBBB   A   B   AA   B   AAA   BBB   AA   BBBBB   A
BBBB   A   BBBB   AAA   B   A   BB   A   BBB   A   B   AA   B
A   BBBB   AA   BBBBB   AA   BB   A   B   AA   BBBBBBB
AA   B   A   BBBB   A   BBB   A   B   A   B   AAA   BB   A
BBBBBB

For further study a record of 250 trials of the Markov process defined in Exercise 3M, Question 56 is given below. In this sequence of trials there are two 'states', each occurring as the result of a trial; the 'states' of a Markov process are the 'events' of ordinary probability language. Being in states R, L can be called events A, B and we then have the following probability rules:

pr $(R \to R)$ = pr $(A|A)$ = $\frac{2}{3}$, due to *not* throwing a *one* or *two*.

pr $(R \to L)$ = pr $(B|A)$ = $\frac{1}{3}$, due to throwing a *one* or *two*.

pr $(L \to R)$ = pr $(A|B)$ = $\frac{1}{6}$, due to throwing a *one*.

pr $(L \to L)$ = pr $(B|B)$ = $\frac{5}{6}$, due to not throwing a *one*.

The record is given in Table 7.2.

Table 7.2

BBBBBBBB AAAAAAAA BBBBBBBB A BBBBB A
BBBBBBBBB A BBBBBBBBBB AAAAA BBBBBBBB A
BB AAAAAAAAAA BBBB AA BBBBBBBBBBBBB AA
BBBBBBBBB A BBBBBBBBBBBB A BBBBBBBBB AAAA
BBB AAAAAA BBBBB A B AAAA B AAAAA B AAA
BBBBBBBBBB AA BBBBB AAAAAAAAAAAA BB
AAA BBBB AA B AAAA BBB AA BBBBBBBBB A
BB AA BBBBBBBBB AA BBBB A B

It is obvious that the patterns of Table 7.1 and Table 7.2 are quite different. That is the subject matter of this investigation.

5.2 Some questions. For the purposes of simple description imagine these as the records of the success (A) or failure (B) of a player in a game. The player has a 'turn' until he gets a success. The record continues with what is his *next* turn and so on. Thus in Table 7.1 the player has 3 failures before his first success and in Table 7.2 he has 8. In Table 7.1 there are 5 failures altogether before the second success and in Table 7.2 there are 8.

The records suggest such questions as: 'In what proportions of turns would we *not* have to wait through any failures for our success?' and 'How many failures can we expect to wait through before getting our second success?' In these cases we are thinking of a situation where we repeat the trials until something definite happens and we have optional stopping.

In many real situations, however, we repeat trials an arbitrary, but fixed, number of times and then we ask questions such as: 'In what proportions of times can we expect to get 3 successes in 5 trials?' and 'In what proportion of times can we expect to get 3 switches in 5 trials? (as for instance in $A\underline{B}A\underline{A}\underline{B}$ where the switches from A to B, or vice-versa are marked)'.

We have really been asking questions about a few *statistics* of the sequence and we are beginning to get a numerical knife-blade in.

5.3 Towards the answers. Let us make frequency tables of the occurrences of values of the four statistics suggested so far:

Let U denote the number of failures in a turn, that is the number of failures (perhaps none) in a sequence which starts immediately after a success and terminates at the next success. U is called the waiting-time to the first success.

Let V denote the number of failures in a sequence which starts immediately after a success and terminates at the *second* success thereafter. V is called the waiting-time to the second success; (notice that it does *not* count the intervening success). We have chosen the sequences as non-overlapping in our tables.

Let R denote the number of successes in 5 consecutive trials; (we have here divided each sequence into non-overlapping subsequences of 5 trials).

Let S denote the number of switches in 5 consecutive trials; (we have again used non-overlapping subsequences).

At the moment we will call the first experiment 'simple' (giving it its proper technical name later) and the second 'Markov'. We obtain Tables 7.3–6, which should now be carefully examined.

Table 7.3

u	For simple experiment		For Markov experiment	
	$\mathrm{fr}(U=u)$	relative $\mathrm{fr}(U=u)$	$\mathrm{fr}(U=u)$	relative $\mathrm{fr}(U=u)$
0	18	0·35	61	0·693
1	11	0·21	4	0·045
2	7	0·13	3	0·034
3	5	0·10	2	0·023
4	6	0·11	3	0·034
5	2	0·04	2	0·023
6	2	0·04	1	0·011
7	1	0·02		
8			3	0·034
9			3	0·034
10			3	0·034
11			1	0·011
12			1	0·011
13			1	0·011
	52	1·00	88	0·998

The questions asked above can be roughly answered by using the appropriate relative frequencies in the tables (though the runs are not really long enough for any confidence). This corresponds to the fire-engine problem. It would be interesting, however, to adopt the Atkins approach and obtain not merely the relative frequencies of the various values of the statistics in these particular real experiments of 150 and 250 trials, but

197

Table 7.4

	For simple experiment		For Markov experiment	
v	$\mathrm{fr}(V = v)$	rel. $\mathrm{fr}(V = v)$	$\mathrm{fr}(V = v)$	rel. $\mathrm{fr}(V = v)$
0	1	0·04	22	0·50
1	6	0·23	4	0·09
2	3	0·12	1	0·02
3	4	0·15	2	0·05
4	3	0·12	3	0·07
5	5	0·19	1	0·02
6			1	0·02
7	2	0·08		
8	1	0·04	1	0·02
9	1	0·04	1	0·02
10			2	0·05
11			2	0·05
13			2	0·05
19			1	0·02
22			1	0·02
	26	1·01	44	1·00

Table 7.5

	For simple experiment		For Markov experiment	
r	$\mathrm{fr}(R = r)$	rel. $\mathrm{fr}(R = r)$	$\mathrm{fr}(R = r)$	rel. $\mathrm{fr}(R = r)$
0	4	0·13	14	0·28
1	8	0·26	12	0·24
2	11	0·37	10	0·20
3	6	0·20	4	0·08
4	1	0·03	6	0·12
5			4	0·08
	30	0·99	50	1·00

Table 7.6

	For simple experiment		For Markov experiment	
s	$\mathrm{fr}(S = s)$	rel. $\mathrm{fr}(S = s)$	$\mathrm{fr}(S = s)$	rel. $\mathrm{fr}(S = s)$
0	4	0·13	18	0·36
1	8	0·27	21	0·42
2	10	0·33	10	0·20
3	8	0·27	1	0·02
	30	1·00	50	1·00

their probabilities from a consideration of the elementary trials whose repetitions generated the sequences.

6. RANDOM VARIABLES

6.1   The notion. When we are concentrating on obtaining the *probability* of getting a certain value or range of values of a statistic, the statistic is sometimes called a *random variable*. In fairly abstract terms we may say that if a number can take certain values, and which value it takes is determined by the result of an experiment or process whose outcome contains an element of chance, then the number is called a *random variable*.

Thus R denotes the random variable: 'number of successes in 5 trials'; S denotes the random variable: 'number of switches in 5 trials'. We are usually more interested in the predictive possibilities of a random variable than in its mere descriptive use (as a statistic), such as we dealt with in Chapters 3 and 4.

6.2   Generality. As was remarked in Chapter 4, Section 2.3 and hinted at as early as Chapter 3, Section 3.2 a random variable may not measure or count anything at all; it may only supply an arbitrary numerical code. As an apparent instance of this consider a situation where a single dichotomous trial is being carried out and the probability of success is a and of failure is $1-a$. If a random variable takes the value 1 when the trial is successful and 0 when the trial is unsuccessful then, if we denote the random variable by X,
$$\operatorname{pr}(X=0)=1-a, \quad \operatorname{pr}(X=1)=a.$$
(These may be expressed concisely as
$$\operatorname{pr}(X=x)=(1-a)^{1-x}a^{x} \quad (x=0,1),$$
but whether quite such a concise statement is only a party trick or whether it is useful, in the sense that we can do anything with it, is another matter.)

6.3   Notation. Notice our use (at the end of the last section) of 'capital' X and 'lower case' x. We shall often need to use two symbols in this way, and they may as well be alphabetically related. For instance to say 'Let X denote the number of heads when 2 fair coins are thrown, then
$$\operatorname{pr}(X=x)=\frac{1}{2^{|x-1|+1}} \quad \text{for} \quad x=0,1,2.'$$
is less clumsy than to say: 'Let x denote the number of heads when 2 fair coins are thrown, then
$$\operatorname{pr}(x=u)=\frac{1}{2^{|u-1|+1}} \quad \text{for} \quad u=0,1,2.'$$
We need a symbol for the variable and a dummy symbol for use in formulae;

we cannot manage with only one of them. It is false to say 'Let x denote the number of heads when 2 fair coins are thrown, then

$$\text{pr } (x = x) = \frac{1}{2^{|x-1|+1}} \quad \text{for} \quad x = 0, 1, 2.'$$

because pr $(x = x)$ can only have the value 1; x is always equal to x. If the reader will distinguish between, on the one hand, the quantity to which the random variable refers, and, on the other, the values which the random variable from time to time takes, then all will be well.

6.4 As carriers of their probability functions. From a purely mathematical point of view the nature of the experiment which gives rise to the random variable is immaterial, and the entire specification of a random variable X consists of the function p, say, such that

$$p(x) = \text{pr } (X = x).$$

Note that here we are not using p as an abbreviation for 'probability', but as a symbol for an ordinary function (with certain limitations on its values) whose domain is a set of real numbers, and that x is seen in its true colours as a dummy variable. If we had two random variables X, Y in our problem we might write $p(t) = \text{pr } (X=t)$, $q(t) = \text{pr } (Y=t)$, though a more consistent notation might involve suffixing the functional symbol as:

$$p_X(t) = \text{pr } (X=t), \quad p_Y(t) = \text{pr } (Y=t).$$

In this way the random variable is sometimes said to be merely a 'carrier' of its probability function.

The situation is very comparable to that in Chapter 4 where we had a finite collection of values of x, say, $x_1, x_2, ..., x_n$ with possible repetitions and from a mathematical point of view we knew all that there was to know when we knew the frequency function f such that $f(x_i) =$ frequency with which $x = x_i$. The situations look even more comparable if we use the notation X for the variable in that case and x for the typical value, so that $f(x) =$ frequency with which $X = x$. The advantage of suffixes in that chapter was that it made the notation for summations a little more precise while we were explaining the ideas involved.

6.5 Alternative approach. Some writers say that a random variable is a function from the set of outcomes of a trial into a set of numbers. The reader may find this helpful. It is equivalent to saying in mountaineering circles that the variable 'height' is a function from the set of mountains into the set of positive real numbers. If we denote this function by f we can write (it being understood that the functional relation involves feet, say):

$$f(\text{Snowdon}) = 3560, \quad f(\text{Ben Nevis}) = 4406,$$

and we almost find ourselves reading 'height of Snowdon is 3560'. In this mountaineering situation the language of the present text would, however, be:

Let H denote the height of a mountain in feet, then

for Snowdon: $H = 3560$,   for Ben Nevis: $H = 4406$.

H would be called a variable, though not a random one; and 3560, 4406 *values* which it takes.

6.6   Summary. In the way in which we will use the term 'random variable', a random variable takes values which are numbers like the values taken by any other statistic. A random variable is merely a variable about which we know something more than we know about some other variables: namely the probabilities with which it takes its possible values.

It may seem paradoxical that we usually know more about a variable when it is a random variable (that is to say whose value is chosen by chance) than otherwise; but what can you say about an integral variable whose value I shall choose between 1 and 6, except to say that if v denotes the value then $1 \leqslant v \leqslant 6$? About S, the random variable which is your score on throwing a fair die, however, I know that if s is the value, not only do I have $1 \leqslant s \leqslant 6$ but also $\mathrm{pr}\,(S = s) = \frac{1}{6}$ for $s = 1, 2, 3, 4, 5, 6$.

A random variable is sometimes called a *variate*.

6.7   'Continuous', 'Discrete', 'Integral'. There is nothing inherent in the idea of a random variable which prevents it from being able to take any value in an interval or set of intervals of the number line; and, for instance, when we measure the length of life of a radioactive nucleus or the height of a ragwort plant we do behave as if the values could be anywhere in an interval. We ought to call the random variable a continuous-valued random variable but we usually shorten this to a *continuous random variable*. (The allotment of probabilities then requires study.) For now, however, we will restrict our attention to random variables whose values form a discrete, though not necessarily finite, set. We call them *discrete* (-valued) *random variables*. In point of fact they will often have values confined to the set of integers, and are then called *integral*(-valued) *random variables*.

7. TWO CONCRETE SITUATIONS (CONTINUED)

7.1   Some comparisons. A study of the two sequences that we have chosen for our raw-material reminds us of one or two differences, but first a similarity.

In the simple one, the value of $fr\,(A)/N$ is $\frac{52}{150}$, which is 0·35; and in the Markov chain the value of $fr\,(A)/N$ is $\frac{88}{250}$, which is also about 0·35. We expect this equality (if we trust our model), because on the one hand the method of generation of the simple sequence has pr $(A) = \frac{1}{3}$ and so in the long run will lead to about $\frac{1}{3}$ of the results being A; while on the other hand we could prove from Exercise 3 M, Question 57 part (v) that the rather complicated Markov situation leads us to allot a probability of $\frac{1}{3}$ to getting A in a randomly selected trial, if we are *told nothing of the result which precedes our randomly selected one.* Therein lies the important distinction.

If we calculate, for each pair of consecutive results in a chain the fractions:

$$\frac{\text{fr (result is } AA)}{\text{fr (result is } AA) + \text{fr (result is } AB)}$$

and

$$\frac{\text{fr (result is } BA)}{\text{fr (result is } BA) + \text{fr (result is } BB)}$$

we ought to find numbers approximating to:

pr (second member is A|first member is A)

and   pr (second member is A|first member is B).

For the simple chain we have

$$\frac{18}{52} \quad \text{and} \quad \frac{36}{99}, \quad \text{that is } 0.35 \quad \text{and} \quad 0.36;$$

while for the Markov chain we have

$$\frac{61}{88} \quad \text{and} \quad \frac{26}{161}, \quad \text{that is } 0.69 \quad \text{and} \quad 0.16.$$

The corresponding probabilities calculated from the model are:

for the simple chain   0·33   and   0·33;

for the Markov chain   0·67   and   0·17.

So that the probability model provides results which agree well with the observations.

The different behaviours of the two chains remind us that for us to be able to predict the frequencies of compound events it is not enough *merely* that the long-run relative frequencies of successes should be stable; if the trials do not produce results 'independently of each other' then the short-run structure of the chain is altered.

Exercise 7 B

1 R. Obtain the numerical values for the ratios mentioned in Section 7.1 from your own records of Exercise 3 M, Question 56.

7.2 Probability functions for waiting times. We can quite easily obtain the probabilities of various waiting times to the first success in the simple chain as follows.

Let U denote such a waiting time. To get $U = u$ we have u consecutive failures followed by 1 success. Now the events *in* this chain are independent of one another so the probability we seek is $(\frac{2}{3})^u.(\frac{1}{3})$. The successive values of this for $u = 0, 1, 2, \ldots$ are 0·33, 0·22, 0·15, 0·10, 0·07, 0·04, ... and we see from Table 7.3 that the agreement with the relative frequencies of values of U for the simple chain are excellent. What is intriguing is that the probability of a waiting time of, say, 3 is 0·10 *whether the previous event was A* (so that we are now at the beginning of a turn) *or not*. This means that even if we have been waiting already through a succession of half a dozen occurrences of B, we still have a probability of 0·10 that we will have to wait for 3 more B's before an A turns up. This, of course, is merely to say again that the events are independent; the chain is sometimes said to have no memory.

The waiting-time to the second success is hardly more complicated; if the waiting time is given by $V = v$, then we must have v failures and 1 success, arranged among the $(v+1)$ events that precede the second success. This can be done in $(v+1)$ ways, one for each possible position of the success. The probability of any particular sequence of these $(v+1)$ events is $(\frac{2}{3})^v (\frac{1}{3})$ and there are $(v+1)$ possible different orders; finally the sequence terminates with a second success, so the probability we seek is

$$\text{pr} (V = v) = (v+1) (\tfrac{2}{3})^v (\tfrac{1}{3})^2;$$

the first few values are 0·11, 0·15, 0·15, 0·13, 0·11, 0·09, 0·065, ...; again agreement is reasonable considering the fewness of the trials (see Table 7.4).

If the first and second values of the column of relative frequencies are paired and summed, and the third and fourth are paired and summed, then the general fit is improved, but notice that this involves a change of standpoint. We are then asking whether the experiment provides a reasonable realization of the probability model rather than whether the model accurately describes so small an experiment.

The waiting times to later successes than these introduce complications which we will postpone until we have looked briefly at the Markov chain.

For the Markov chain things are, as we would expect, more complicated. The waiting time is defined as starting from a success and this is now important because of the lack of independence. The possible sequences are as shown in Figure 7.1 (for the first three branchings only). We have:

$\text{pr} (U = 0) = \frac{2}{3} \qquad = 0 \cdot 67$

$\text{pr} (U = 1) = \frac{1}{3} \times \frac{1}{6} \qquad = 0 \cdot 06$

$\text{pr} (U = 2) = \frac{1}{3} \times \frac{5}{6} \times \frac{1}{6} = 0 \cdot 05$

and for $u \geqslant 1$

$\text{pr} (U = u) = \frac{1}{3} \times (\frac{5}{6})^{u-1} \times \frac{1}{6}.$

$\text{pr} (V = 0) = \frac{2}{3} \times \frac{2}{3} \qquad\qquad = 0 \cdot 44$

$\text{pr} (V = 1) = (\frac{2}{3} \times \frac{1}{3} \times \frac{1}{6}) \times 2 = 0 \cdot 07$

$\text{pr} (V = 2) = (\frac{2}{3} \times \frac{1}{3} \times \frac{5}{6} \times \frac{1}{6}) \times 2$

$\qquad\qquad + (\frac{1}{3} \times \frac{1}{6} \times \frac{1}{3} \times \frac{1}{6}) = 0 \cdot 06$

Once more the agreements are reasonable, considering the shortness of the run.

We delay further study of the *sequence* of results that form the Markov chain, until Chapter 10.

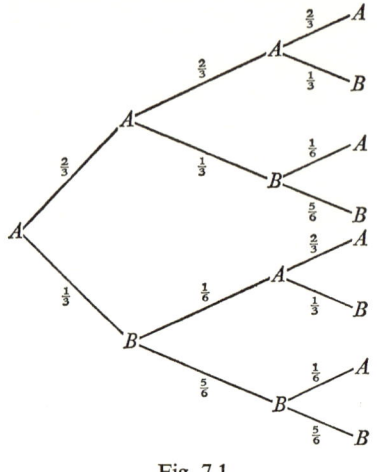

Fig. 7.1

8. PRESENCE OR ABSENCE
OF INDEPENDENCE

We turn our attention back to the simple sequence and see how its features of *constant probabilities and independence of events* makes it different from other sorts of sequences.

8.1 Markov chain. In the Markov chain of our example the probabilities for each of the two possible outcomes of each trials are functions of the result of the previous trial, but not of the results of trials previous to that. Generalizations of Markov chains allow more than two, even an infinity of, different possible outcomes of each trial but the probabilities are to be functions of the results of no *trial except at most the previous one.* (They may in very general and scarcely studied circumstances also be functions of 'time'.) Processes giving rise to Markov chains are called Markov processes and they provide models for situations in: diffusion of gases, Brownian movement of small particles suspended in fluid, genetics, epidemics, and so on.

8.2 Sequences by sampling without replacement—small parent population. Chapter 3, Section 5.9, Example 1 has a bag with 6 red and

204

4 green discs. Results of trials consisting of drawing a disc can be represented by the tree-diagram, Figure 7.2.

This displays that when the sampling is from a finite set and the items are not replaced between draws the probabilities alter in a way that is non-Markovian and time-dependent. Before leaving this example the reader should notice something that he may not have expected. It is that there are equal probabilities for the possible runs of three results which lead to a particular *constitution* of sample of size 3, (the *order* of drawing being irrelevant to the *constitution*). For instance the constitution 2 green, 1 red can be reached by the following *ordered* samples: *GGR*, *GRG*, *RGG* and the probabilities are $\frac{4}{10} \times \frac{3}{9} \times \frac{6}{8}$, $\frac{4}{10} \times \frac{6}{9} \times \frac{3}{8}$, $\frac{6}{10} \times \frac{4}{9} \times \frac{3}{8}$ respectively.

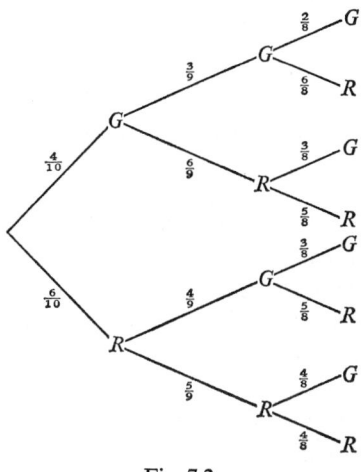

Fig. 7.2

Incidentally, it follows that the probability of drawing exactly 2 greens in 3 draws without replacement is $\frac{3}{10}$, but we shall see later that, with a *constant* probability at *each draw* equal to that which existed at the beginning of the draws, the probability of such a result is 0·288; so that the type of constancy that the drawing method of this section displays is *not* sufficient for the theory which we will be basing on chains as simple as ours.

8.3 Sequences with probabilities functions of time only. We consider once again the bag containing red and green discs, but this time when we draw a disc and examine its colour we then remove a disc of opposite colour from the bag and replace *neither* in the bag, before the next draw. We will have probabilities which are *not* functions of previous results. For instance, if we start with 6 red and 4 green discs then for the third draw there will be 4 red and 2 green discs and in general for the ith random draw we will have

$$\text{pr (disc is red)} = \frac{6-(i-1)}{10-2(i-1)} \quad (1 \leqslant i \leqslant 5),$$

$$\text{pr (disc is green)} = \frac{4-(i-1)}{10-2(i-1)} \quad (1 \leqslant i \leqslant 5),$$

whatever the results of previous draws. The example given is rather artificial (the reader might like to devise a situation it models, involving something other than couples going onto a dance floor), but in real life the probabilities can be functions of time without being functions of previous results, and in this example i effectively measures time. The consecutive trials produce independent events but are not yet simple enough for our present purposes.

8.4 Sequences by sampling without replacement—large parent population. If the bag contained 600 red and 400 green discs then the probabilities concerned in the first three draws would satisfy:

$$598/998 < \text{pr (disc is red)} < 600/998$$
or $\qquad 398/998 < \text{pr (disc is green)} < 400/998,$

and we would almost have constant probabilities. It might well be that a model using constant probabilities would be adequate. This is indeed a very common situation: opinion polls and inspecting items from a large batch are obvious examples.

The use of random number tables could come under this heading, since, although the mechanism which generated them may have had equal probabilities for each digit at each draw, the numbers once printed merely form a large population. If we knew the total number of occurrences of each digit in a particular table we could predict the last exactly, by the time we had worked through the table to there. The Rand Table of random digits contains, however, a million digits!

8.5 Sequences from sampling with replacement. Few natural situations involve *strictly* sampling with replacement, but the constant probabilities which it would produce are what we want. If we wish to say that our model requires constant probabilities of the sort we have been discussing (and not for instance the accidentally equal probabilities of a different sort that arose in the sampling without replacement from a small population as mentioned above), then we simply say 'the sampling is to be with replacement', and add under our breath if we wish 'or the population to be so large that it is effectively so'.

Throwing dice or coins is equivalent to sampling with replacement since, although the faces do not get withdrawn at all, neither are they 'used up' by being merely looked at. That they do not get 'used up' cannot of course be proved by argument and many losses at casinos are attributable to a disbelief that 'reds' and 'blacks' on a roulette-wheel have this property too. The final paragraph of Edgar Allan Poe's *The Mystery of Marie Roget* is relevant here!

It may seem that if we limit our attention to sequences generated with the constant probabilities which we have demanded, then we shall be restricted to investigating games of chance (and well conducted ones at that), but, of course, a carefully designed experiment is scarcely the 'natural situation' with which this section began, and, as we have emphasized before, enormous trouble is taken to *design* experiments so that the very powerful theory which follows from constant probabilities can be employed.

9. BERNOUILLI TRIALS

9.1 Definition. Sequences of trials in which we have independence of trials and a constant probability of success, as in Table 7.1, are often called Bernouillian after their first investigator Jacob Bernouilli (1654–1705) whose *Ars Conjectandi* was published posthumously in 1713. The Bernouilli family produced distinguished mathematicians for some generations in the same way that the della Robbia family produced sculptors and the Bach family produced musicians.

9.2 Typical problems. We now pursue an extension of the problem about the random variables U, V introduced in Section 5.3. We specifically ask what is the probability, in the Bernouilli sequence, that the waiting time to the fourth success is 2, so that the fourth success will occur on the sixth trial; and in solving it shall find that much else falls into our lap.

How *can* the fourth success fall on the sixth trial? It can only do so if 3 successes occur in the first 5 trials *and* another success occurs at the sixth. If we can find the probability of the first event of this pair then, with our independence of trials, we need only multiply it by the (fixed) probability of success at any trial to get the answer we seek. But notice that the probability of three successes in 5 trials was one of the quantities required in the investigation of the random variable R, the number of successes in 5 trials; so our routes are converging.

Problems about S, the number of switches in 5 trials, turn out to be harder.

The probability of a switch from A to B is

$$\text{pr}\,(A \text{ occurs}) \times \text{pr}\,(B \text{ occurs}|A \text{ occured}),$$

but the probability of a switch from B to A is

$$\text{pr}\,(B \text{ occurs}) \times \text{pr}\,(A \text{ occurs}|B \text{ occured}).$$

With the independence of the basic trials these are both equal to

$$\text{pr } (A \text{ occurs}) \times \text{pr } (B \text{ occurs}),$$

that is $\frac{2}{9}$.

Thus if a pair of consecutive trials is picked at random the probability that a switch occurs between them is $\frac{2}{9} + \frac{2}{9}$, that is $\frac{4}{9}$. On the other hand if the first trial of a pair is *known* to be A, then

$$\text{pr (switch occurs)} = \text{pr } (B \text{ occurs}) = \tfrac{2}{3},$$

while if the first trial of a pair is known to be B, then

$$\text{pr (switch occurs)} = \text{pr } (A \text{ occurs}) = \tfrac{1}{3}.$$

Thus the secondary systems of trials consisting of overlapping *pairs* of the original trials, and for which the events are 'switches' or 'non-switches', may not form a sequence of independent trials. They do in fact constitute a Markov sequence of trials as we can see in the following way. The possible sequences of 3 original trials, with their corresponding probabilities, are:

AAA	$1/27$	BAA	$2/27$
AAB	$2/27$	BAB	$4/27$
ABA	$2/27$	BBA	$4/27$
ABB	$4/27$	BBB	$8/27$

and from this table it can be seen that:

pr (switch occurs between 2nd pair | switch occurs between 1st pair)

$= \text{pr } (ABA \text{ or } BAB)/\text{pr } (ABA \text{ or } ABB \text{ or } BAA \text{ or } BAB)$

$= (2+4)/(2+4+2+4)$

$= \frac{6}{12}$,

while

pr (switch occurs between 2nd pair | switch does not occur between 1st pair)

$= \text{pr } (AAB \text{ or } BBA)/\text{pr } (AAA \text{ or } AAB \text{ or } BBA \text{ or } BBB)$

$= (2+4)/(1+2+4+8)$

$= \frac{6}{15}$.

Furthermore the probabilities are clearly not functions of the results of any earlier pairs of original trials, so the Markov condition is satisfied.

9.3 The case of three successes in 5 Bernouilli trials. In what follows we will fix the number of trials and call a succession of 5 trials 'the experiment'.

In the initial stages of finding the probabilities we are only concerned with the number of appropriate arrangements of the symbols A, B, because we first need to find the full possibility space of the outcomes of the experiment before ever allotting probabilities to the elements of the space.

We have already seen that for an experiment of 2 trials the most useful possibility space consists of 4 points {*BB, BA, AB, AA*}. Notice that we are mapping each outcome of the experiment-consisting-of-two-trials onto a *single point* in the possibility space.

Similarly for an experiment of three successive trial we can use as we suggested in Section 9.2 a possibility space:

{*BBB, BBA, BAB, ABB, BAA, ABA, AAB, AAA*}.

It is obviously unimportant what symbols we use to denote the two outcomes of the single elementary trial whose repetitions constitute the experiment. We might, for instance, display the situation for an experiment of 5 trials by drawing a row of 5 cells, one for each trial as in Figure 7.3.

1st trial   2nd trial   3rd trial   4th trial   5th trial

☐   ☐   ☐   ☐   ☐

Fig. 7.3

A particular sequence of results of the 5 trials in the experiment could then be shown by inserting a tick, or check, for each success. The failures are then determined and we could fill their cells with crosses at our leisure.

In this notation

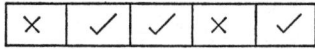

corresponds to   *B A A B A.*

The number of ways in which the symbols *AAABB* can be rearranged is then seen to be number of ways in which 3 cells can be selected from 5 to receive a tick.

The first tick can be put in any one of 5 cells, then the second in any one of 4, and finally the third in any one of 3. This suggests $5 \times 4 \times 3$ ways—namely ticks in cells:

1	2	3	2	1	3	3	1	2	4	1	2	5	1	2
1	2	4	2	1	4	3	1	3	4	1	3	5	1	3
1	2	5	2	1	5	3	1	5	4	1	5	5	1	4
1	3	2	2	3	1	3	2	1	4	2	1	5	2	1
1	3	4	2	3	4	3	2	4	4	2	3	5	2	3
1	3	5	2	3	5	3	2	5	4	2	5	5	2	4
1	4	2	2	4	1	3	4	1	4	3	1	5	3	1
1	4	3	2	4	3	3	4	2	4	3	2	5	3	2
1	4	5	2	4	5	3	4	5	4	3	5	5	3	4
1	5	2	2	5	1	3	5	1	4	5	1	5	4	1
1	5	3	2	5	3	3	5	2	4	5	2	5	4	2
1	5	4	2	5	4	3	5	4	4	5	3	5	4	3

We notice that this table is divided into 5 columns each with 4 blocks containing 3 arrangements.

Of course, within this pattern each solution to our problem occurs many times. To see this, notice the underlined arrangements, all of which give the same *selection* of cells. The number of times a selection is thus represented is the number of ways of permuting the ticks among the same 3 cells, and is obtained (by a process similar to that just explained) as $3 \times 2 \times 1$.

In this way we finally see that the required number of selections is $\dfrac{5 \times 4 \times 3}{1 \times 2 \times 3}$, as elementary algebra books show. This reveals that the number of points in the required subset of the possibility space is $\dfrac{5 \times 4 \times 3}{1 \times 2 \times 3}$, that is 10.

In order to complete our investigation of the probability of getting three successes in 5 trials we need only note that, *with constant probability of success*, each of the 10 outcomes which give us three successes has equal probability.

Let $\quad$ pr (outcome is success) $\equiv$ pr $(A) \equiv a,$ $\quad$ say

then $\quad$ pr (outcome is failure) $\equiv$ pr $(B) = 1 - a \equiv b,$ $\quad$ say.

Each outcome with three successes now has probability $b^2 a^3$ of occurring —for instance the outcome represented earlier (that is $BAABA$) has probability $b.a.a.b.a$ by the product law for compound events.

It follows, from the addition law for exclusive events, that the probability of getting some one of the 3-success outcomes is $10b^2 a^3$.

For the sequence of trials listed in Table 7.1 we have, on theoretical grounds, $\quad a \equiv$ pr (success) $= \frac{2}{6} = \frac{1}{3},$ $\quad b \equiv 1 - a = \frac{2}{3}$

so that

$$\text{pr (3 successes in 5 trials)} = 10 \times (\tfrac{2}{3})^2 \times (\tfrac{1}{3})^3 = 0.165.$$

Notice that the relative frequency of this event in the run of 30 experiments of 5 trials each was found, in Table 7.5, to be 0.20, so the agreement is good (one fewer occurrences would have given a relative frequency of 0.167).

9.4 Other numbers of successes in five trials.

We next turn to the problem of finding pr $(R = r)$ for $r = 0, 1, 2, 4, 5$ as well, and we look for a systematic approach.

A helpful step is to return to the idea already mentioned, in which a random variable denoted by X is given the value 1 when a success occurs and the value 0 when a failure occurs. The result already indicated, $BAABA$, gives rise by this system to the sequence of values of X:01101.

One useful feature is that this sequence of digits 0, 1 can be read as a single binary number, so that there is a 1–1 correspondence between {possible outcomes of 5 trials} and {5-figure binary numbers}, and this correspondence maps each outcome directly onto a single point of a possibility space of numbers.

Furthermore the 5-figure binary numbers range from 0 to $2^5 - 1$, so that there are seen to be 32 possible outcomes of an experiment of 5 dichotomous trials, and we can list them without fear of omissions.

Another point, which becomes very helpful later, is that the value of the random variable R (namely number of successes) whose probability function we seek is the sum of the 5 values, (each 0 or 1), which are given to the random variable X which codes the result of an individual trial and whose probability function was mentioned in Section 6.2. We thus see that we are well on the way to a completely numerical description of this problem in the sampling of attributes, and that as the reader may have suspected in Section 6.2, X was only partly an arbitrary code but contained also the seeds of a possible counting system.

Exercise 7C

1R. List systematically all the possible arrangements of
   (a) 3 zeros and 1 one;
   (b) 2 zeros and 2 ones;
   (c) 1 zero and 3 ones.
How many possible arrangements of zeros and ones are there if exactly 4 places are used?

2R. List systematically by two different systems, each defined, all the arrangements possible with 3 digits either zero or one, and count them.

3. Define 2^5 without reference directly or indirectly to addition or multiplication.

4. Write out all the ternary numbers (that is all the numbers in the scale of 3) of exactly 2 digits (including those beginning with or consisting of a sequence of zeros). Define 3^2 without reference to addition or multiplication. Define also 2^3 similarly. Consider the result of defining 10^3 like this.

5T. A subset of a given set can be defined by examining each member of the set in turn and admitting it or not. Define 2^4 in terms of a set and its subsets. Check your definition by constructing a particular set, and its subsets.

6. In how many ways can a group of people be selected for interview in a firm with 270 employees. Using the approximation $2^{10} \simeq 10^3$ compare this with Eddington's estimate of the number of protons in the universe (10^{79}).

7. Give an example of a problem in sampling attributes which would use ternary numbers (as in Question 4) for its possibility space.

8. How many 4-digit numbers can be formed with the 10 digits 0, 1, 2, ..., 9 if
   (a) repetitions are allowed?
   (b) repetitions are not allowed?
   (c) the last digit must be zero and repetitions are not allowed.

★9. (i) In how many ways can 7 people be seated at a round table if
(*a*) they can sit anywhere? (*b*) two particular people must not sit next to each other?
(ii) In how many ways can 7 beads be strung on a necklace under similar conditions?

9.5 Possibility space for five trials. Using a binary coding we have a table (Table 7.7) of outcomes of five dichotomous trials. The table is ordered by increasing size of binary number. A table ordered by values of R is also given (Table 7.8).

Table 7.7

Outcome	Value of R	Outcome	Value of R	Outcome	Value of R	Outcome	Value of R
00000	0	01000	1	10000	1	11000	2
00001	1	01001	2	10001	2	11001	3
00010	1	01010	2	10010	2	11010	3
00011	2	01011	3	10011	3	11011	4
00100	1	01100	2	10100	2	11100	3
00101	2	01101	3	10101	3	11101	4
00110	2	01110	3	10110	3	11110	4
00111	3	01111	4	10111	4	11111	5

Table 7.8

Value of R	Outcomes
0	00000
1	00001, 00010, 00100, 01000, 10000
2	00011, 00101, 00110, 01001, 01010
	01100, 10001, 10010, 10100, 11000
3	00111, 01011, 01101, 01110, 10011
	10101, 10110, 11001, 11010, 11100
4	01111, 10111, 11011, 11101, 11110
5	11111

This may all be summarized by Table 7.9.

Table 7.9

r	Number of outcomes in which $R = r$
0	1
1	5
2	10
3	10
4	5
5	1

The numbers 1, 5, 10, 10, 5, 1 which have just arisen have no connexion with probabilities allotted to success and failure *in a single trial*; they are characteristic of the classification by numbers of successes of a general sequence of five dichotomous trials, whether independent or not.

10. BINOMIAL THEOREM

10.1 Introduction. The reader may have recognized the numbers 1, 5, 10, 10, 5, 1 as one of the rows of 'Pascal's Triangle' and thus as the coefficients of powers of ξ in the expansion of $(1+\xi)^5$. That they are in fact these numbers is no mere coincidence, as we will now demonstrate.

The coefficients of powers of ξ in $(1+\xi)^5$ are the same as the numerical coefficients in the expansion of $(b+a\xi)^5$ in powers of ξ. How do these coefficients arise?

Consider

$$(b+a\xi)^5 = (b+a\xi)(b+a\xi)(b+a\xi)(b+a\xi)(b+a\xi).$$

The expansion is obtained by adding all the possible products got by choosing *one* term from each bracket. Thus to get the term in ξ^3 we choose terms as underlined below

$$(\underline{b}+a\xi)(b+\underline{a\xi})(b+\underline{a\xi})(\underline{b}+a\xi)(b+\underline{a\xi})$$

or $\quad (\underline{b}+a\xi)(b+\underline{a\xi})(b+\underline{a\xi})(b+\underline{a\xi})(\underline{b}+a\xi)$

and so on.

There are many possible patterns of underlining giving terms in ξ^3, precisely as many in fact as there are ways of placing 3 ticks in 5 cells; the patterns of underlining shown above correspond to

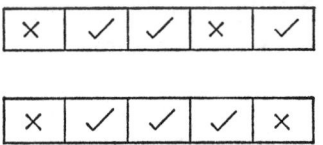

or 01101, 01110.

The structures of the several problems are identical. This means that we shall get a term $b^2 a^3 \xi^3$ occurring in the ten ways listed in Section 9.5, so that the coefficient of ξ^3 in the expansion of $(b+a\xi)^5$ is $10b^2 a^3$. This is the probability of the occurrence of three successes in 5 trials, when at *each trial* the probability of success is a and of failure b.

The argument is obviously quite general as to number of trials in the experiment and as to number of successes. It follows that if the random variable R denotes the number of successes in N trials then

$$\sum_{r=0}^{N} \text{pr}(R=r).\xi^r = (b+a\xi)^N.$$

213

If we knew the values of the numerical coefficients in the expansion of $(b+a\xi)^N$ or, what comes to the same thing, knew the values of the numerical coefficients in the expansion of $(1+\xi)^N$, then this particular problem in sampling attributes would be completely solved.

The statement of how to expand the powers of the binomial term $(1+\xi)$ is known as the *Binomial Theorem*.

10.2 Some standard results. A knowledge of the Binomial Theorem as a standard result of algebra will be assumed and if the reader is not familiar with it he should examine the case of integral exponents in an algebra book.

The following results are important for our purposes.

(*a*) $_NC_r$ is often used as a notation for the number of ways in which *r* items can be marked or selected from *N* initially identical items. It is easy to show that

$$_NC_r = \frac{N(N-1) \dots (N-r+1)}{r!},$$

the numerator accounting for the placing of the marks on a particular *r* item in some order, and the denominator accounting for the fact that once the items are safely marked the order in which they were marked becomes irrelevant.

We, in effect, showed in the previous section that $_NC_r$ is both the coefficient of ξ^r in the expansion of $(1+\xi)^N$ and also the necessary multiplier for the term $b^{N-r}a^r$ in the probability problem.

(*b*) We will use the notation $\binom{N}{r}$ for $_NC_r$ and will extend the range of possible values of *N* to the negative integers, though *r* will always be a positive integer.

It is convenient for counting purposes (ensuring that we write down the correct number of factors in the numerator) to write the factorial term in the denominator as shown; and we define

$$\binom{N}{r} = \frac{N(N-1)(N-2) \dots (N-r+1)}{1.2.3 \dots r}$$

for any integral *N* and for positive integral *r*. Then for any integral *N*

$$(1+\xi)^N = \sum_{r=0}^{N} \binom{N}{r} \xi^r, \quad \text{where} \quad \binom{N}{0} = 1.$$

(*c*) If *N* is a positive integer we can write

$$\binom{N}{r} = \frac{N!}{(N-r)! \, r!}.$$

214

This form is useful:

(i) For numerical purposes with large N, r if tables of factorials or their logarithms are available.

[An interesting and useful fact, whose proof is beyond the scope of this course, is that, for values of n outside our table of factorials, we have available an approximation known as Stirling's Formula:

$$n! \Big/ (2\pi n)^{\frac{1}{2}} \left(\frac{n}{e}\right)^n \to 1 \quad \text{as} \quad n \to \infty$$

or, equivalently,

$$\log_{10}(n!) \sim 0{\cdot}39909 + (n + \tfrac{1}{2}) \log_{10} n - 0{\cdot}4342945 n.]$$

(ii) To remind us that if $0 \leqslant r \leqslant N$, then $\binom{N}{r} = \binom{N}{N-r}$.

(iii) To remind us that for any three positive integers u, v, w with $u = v + w$, we have that $\dfrac{u!}{v!w!}$ is a binomial coefficient in two ways; for example

$$\frac{(N+1)!}{(N-k)!\,(k+1)!} = \binom{N+1}{k+1} = \binom{N+1}{N-k}$$

and we may even notice, for example,

$$\frac{(2N+2)!}{(N+k+1)!\,(N-k)!} = (2N+2) . \binom{2N+1}{N+k+1}.$$

(d) $\dbinom{N}{r} = \dfrac{N(N-1)\dots(N-(r-1)+1)}{1.2\dots(r-1)} . \dfrac{(N-r+1)}{r} = \left(\dfrac{N}{r-1}\right) . \dfrac{N-r+1}{r}.$

This is a useful relation for generating each probability from its predecessor for fixed N.

Thus

$$\text{pr}\,(R = r) = \binom{N}{r} b^{N-r} a^r$$

$$= \binom{N}{r-1} \frac{N-r+1}{r} . b^{N-r+1} a^{r-1} . \frac{a}{b}$$

$$= \text{pr}\,(R = r-1) . \left(\frac{N-r+1}{r}\right) . \left(\frac{a}{b}\right).$$

(e) If N is a negative integer, say $N = -k$ for $k > 0$, then

$$\binom{N}{r} = \frac{(-k)(-k-1)(-k-2)\dots(-k-r+1)}{1.2.3\dots r}$$

$$= (-1)^r \frac{k(k+1)\dots(k+r-1)}{1.2.\dots r}.$$

Coefficients of this form with the terms of the denominator increasing are often met and not recognized as binomial.

Notice further that

$$\binom{-k}{r} = (-1)^r . \frac{(k-1)!}{(k-1)!} \frac{k(k+1) \dots (k+r-1)}{1.2 \dots r}$$

$$= (-1)^r . \frac{(k+r-1)!}{(k-1)! \, r!}$$

$$= (-1)^r . \binom{k+r-1}{r} \quad \text{by result } (c).$$

This is often more useful in the form

$$\binom{N}{r} = (-1)^r \binom{-(N-r+1)}{r} \quad \text{for} \quad r \geqslant 0.$$

All the coefficients are positive in the expansion of $(1-\xi)^{-k}$ for $k > 0$.

(f) The Pascal Triangle has the properties indicated in Figure 7.4.
15, for instance, is a binomial coefficient in many ways: $\binom{6}{2}$, $\binom{6}{4}$ are
obvious; but $\binom{-5}{2}$ and $\binom{-3}{4}$ are deducible as $\binom{-(6-2+1)}{2}$ and
$\binom{-(6-4+1)}{4}$, yet are often overlooked.

Pascal's Triangle is obtainable from the relationship

$$\binom{N+1}{r} = \binom{N}{r-1} + \binom{N}{r},$$

which is obtained by comparing the coefficients of ξ^r on the two sides of
$$(1+\xi)^{N+1} = (1+\xi)^N (1+\xi).$$

If we multiply through by $b^{N-r+1}a^r$ we may rearrange the given result as

$$\binom{N+1}{r} b^{N-r+1}a^r = a . \binom{N}{r-1} b^{N-r+1}a^{r-1} + b . \binom{N}{r} b^{N-r}a^r,$$

then, writing $p_N(r)$ for the probability of getting r successes in N trials
we have: $p_{N+1}(r) = a . p_N(r-1) + b . p_N(r),$
which may be read as:

The probability of getting r successes in $(N+1)$ trials equals the sum of
(the probability of getting $(r-1)$ successes in the first N trials, followed by
one more success) and (the probability of getting r successes in the first
N trials, followed by a failure).

It would be possible to start from this independently provable result
about exclusive exhaustive events each of which is the occurrence of a pair
of independent events, and to build up the earlier results by induction.

The form involving $p_i(j)$ is called a *difference relation*, and the reader has
used one already in Exercise 3M, Question 54 for the study of queues and
Exercise 3M, Question 57 for the study of Markov chains.

216

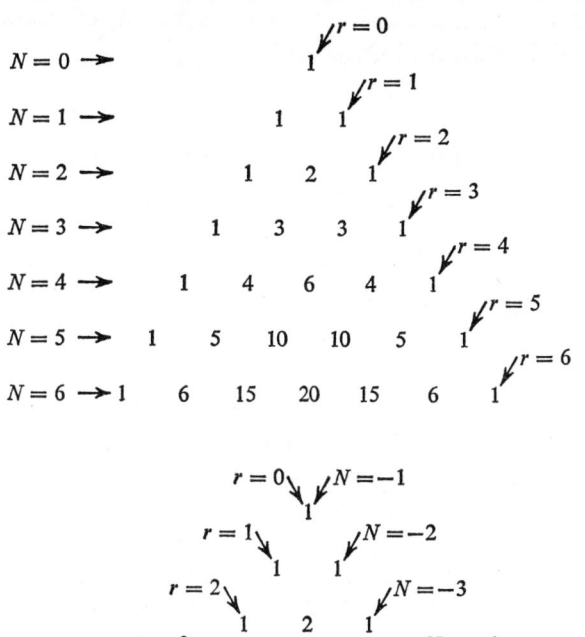

Fig. 7.4

11. EXAMPLES

Example 1. Five coins are spun. Give the probabilities of each number of heads. [This is Exercise 3M, Question 44.]

Solution (to be contrasted with that of Example 5).

Let R denote the number of heads. The following table is self explanatory.

r	$\mathrm{pr}(R = r)$
0	$1 \times (\frac{1}{2})^5 \times (\frac{1}{2})^0 = \frac{1}{32} = 0.031$
1	$5 \times (\frac{1}{2})^4 \times (\frac{1}{2})^1 = \frac{5}{32} = 0.156$
2	$10 \times (\frac{1}{2})^3 \times (\frac{1}{2})^2 = \frac{10}{32} = 0.312$
3	$10 \times (\frac{1}{2})^2 \times (\frac{1}{2})^3 = \frac{10}{32} = 0.312$
4	$5 \times (\frac{1}{2})^1 \times (\frac{1}{2})^4 = \frac{5}{32} = 0.156$
5	$1 \times (\frac{1}{2})^0 \times (\frac{1}{2})^5 = \frac{1}{32} = 0.031$

check: total $= 0.998$

Example 2. Five dice are thrown. Give the probability of getting 2 or more sixes. [This is Exercise 3 M, Question 38.]

Solution. We find the probability of getting no sixes and the probability of getting 1 six. $a \equiv$ pr (six is thrown on a single die) $= \frac{1}{6}$.

Then

$$\text{pr (0 sixes in 5 throws)} = 1.(\tfrac{5}{6})^5.(\tfrac{1}{6})^0,$$

$$\text{pr (1 six in 5 throws)} = 5.(\tfrac{5}{6})^4.(\tfrac{1}{6})^1$$

so that

$$\text{pr (0 or 1 sixes in 5 throws)} = (\tfrac{5}{6})^5 + 5(\tfrac{5}{6})^4.(\tfrac{1}{6})$$

$$= 2(\tfrac{5}{6})^5$$

$$= 0.80$$

The probability we require is $1 - 0.80$, that is 0.20.

Notice the fact, which surprises many people, that with 5 throws we are exactly as likely to get 0 sixes as 1 six.

Example 3. 4 per cent of males are colour-blind. Find the probability that in a random selection of 5 males, 3 are colour-blind. A certain randomly selected household contains 5 males of whom 3 are colour-blind; comment on this situation.

Solution.

$$\text{pr (3 males of 5 are colour-blind)} = \binom{5}{3}\left(\frac{96}{100}\right)^2\left(\frac{4}{100}\right)^3$$

$$= 10(0.922)\,(64 \times 10^{-6})$$

$$= 0.00059$$

$$= 1/1700.$$

Comment. The situation described will happen in about 1 household of 5 males out of every 1700 such households. Either the investigator has lighted upon such a coincidence or the males were not randomly selected. Being in one household they may well be related; if they are, then the result would suggest investigating whether colour-blindness is genetically controlled. In fact, it is.

Example 4. If a machine produces an average of 1 defective item in 5 scattered randomly is it more likely that

  A: a sample of 10 items will show no defectives;

or that

  B: a sample of 20 items will show at most one?

218

Solution.

$$\text{pr}(A) \equiv \text{pr}(0 \text{ defectives in } 10 \text{ items}) = (\tfrac{4}{5})^{10},$$
$$\text{pr}(B) \equiv (0 \text{ or } 1 \text{ defectives in } 20 \text{ items}) = (\tfrac{4}{5})^{20} + 20(\tfrac{4}{5})^{19}.(\tfrac{1}{5}),$$

$$\text{pr}(B)/\text{pr}(A) = (\tfrac{4}{5})^{9}(\tfrac{4}{5} + \tfrac{20}{5})$$
$$= (\tfrac{4}{5})^{10}.6$$
$$= 6/(\tfrac{5}{4})^{10}$$
$$= 6/9 \cdot 3.$$

The first event is more likely.

[Notice the use of $(\tfrac{5}{4})^{10}$ rather than $(\tfrac{4}{5})^{10}$, because the first can be read more simply off most slide rules.]

Example 5. Find the probability of each possible number of successes in a run of 5 trials with $a = \text{pr}$ (success at each trial) $= \tfrac{1}{3}$, $b = 1 - \tfrac{1}{3}$ and compare with the relative frequencies in Table 7.5.

Solution (to be contrasted with that of Example 1).

The method uses the ratio of each probability to the preceding one. We use $\dfrac{a}{b} = \dfrac{1}{2}$ and pr (0 successes) $= \left(\dfrac{2}{3}\right)^{5} = \dfrac{1}{(1\cdot5)^{5}} = 0\cdot132.$

Number of successes	0		1		2	
Probability ratio {		$\frac{5}{1}\cdot\frac{a}{b}$ $\frac{5}{2}$		$\frac{4}{2}\cdot\frac{a}{b}$ 1		$\frac{3}{3}\cdot\frac{a}{b}$ $\frac{1}{2}$
Probability	0·132		0·329		0·329	
Relative freq.						
Simple:	0·13		0·26		0·37	
Markov:	0·28		0·24		0·20	

Number of successes	3		4		5
Probability ratio {		$\frac{2}{4}\cdot\frac{a}{b}$ $\frac{1}{4}$		$\frac{1}{5}\cdot\frac{a}{b}$ $\frac{1}{10}$	
Probability	0·164		0·041		0·004
Relative freq.					
Simple:	0·20		0·03		0·000
Markov:	0·08		0·12		0·08

We notice that for the simple experiment of Table 7.5 the Bernouilli model predicts all the relative frequencies (but one) correctly to one place of decimals but that for the Markov experiment it is an unsatisfactory model.

[When working these successions of values with a slide-rule, it is not worth doing the simple ones mentally (except as a check), since the slide rule will have to be re-set if a step is left out.]

219

Example 6. Samples of size 6 are taken at random from a large consignment in which 30 per cent of the items are defective. Find the most probable number of defectives. Find also the probabilities of obtaining fewer than this number; exactly this number. If 100 samples of size 6 were taken in how many would you expect to find more than the most probable number?

Solution. Compare the solution of Example 5.

$$a \equiv \text{pr (a defective in a trial)} = \frac{3}{10}; \quad b = 1 - \frac{3}{10} = \frac{7}{10}; \quad \left(\frac{a}{b}\right) = \frac{3}{7},$$

$$\text{pr (0 successes)} = \left(\frac{7}{10}\right)^6 = \frac{1}{8 \cdot 50} = 0 \cdot 118.$$

Number of successes	0		1		2		3
Probability ⎰ ratio ⎱		$\frac{6}{1} \cdot \frac{3}{7}$ $\frac{18}{7}$		$\frac{5}{2} \cdot \frac{3}{7}$ $\frac{15}{14}$		$\frac{4}{3} \cdot \frac{3}{7}$ $\frac{4}{7}$	
Probability	0·118		0·302		0·324		(<0·324)

Since $\frac{15}{14} > 1$ and $\frac{4}{7} < 1$ we have that 2 is the most probable number of success;

$$\text{pr (fewer than 2 defectives)} = 0 \cdot 420,$$
$$\text{pr (2 defectives)} \qquad = 0 \cdot 324,$$
$$\text{pr (more than 2 defectives)} = 1 - (0 \cdot 420 + 0 \cdot 324) = 0 \cdot 256.$$

In 100 samples of size 6 we would expect about 26 to show more than 2 defectives.

Example 7. What is the probability of obtaining exactly 50 heads in a throw of 100 coins and then repeating this event 2 times in the next 5 throws?

Solution. We split the problem into two Bernouilli processes.

First process. The trial is throwing one coin; event A is throwing a head; $a = \frac{1}{2}$, $N = 100$, $r = 50$.

$$\text{pr (50 heads in 100 throws)} = \frac{100!}{50! \, 50!} \times \left(\frac{1}{2}\right)^{100}.$$

From tables of factorials

$$\log (100!) = 157 \cdot 9700, \quad \log (50!) = 64 \cdot 4831.$$

We then calculate pr (50 successes in 100 throws) = 0·0796.

Second process. The trial is throwing 100 coins at a time.

Event A is throwing exactly 50 heads, $a = 0 \cdot 0796$, $N = 5$, $r = 2$,

$$\text{pr (2 successes in next 5 trials)} = \binom{5}{2} (0 \cdot 9204)^3 (0 \cdot 0796)^2$$
$$= 10 \, (0 \cdot 780) \, (0 \cdot 00634)$$
$$= 0 \cdot 049.$$

The required probability is thus $0 \cdot 0796 \times 0 \cdot 049$, which is $0 \cdot 039$.

Note that this is *not* the probability of obtaining 50 heads 3 times in the first 6 trials, since we have in effect specified that the first success should occur on the first trial. The later trials are independent of this first trial, of course.

Exercise 7D

1R. (i) Use Pascal's Triangle to write down the first five terms of the expansions of the following in powers of ξ.

$$(1+\xi)^6; \quad (1+\xi)^7; \quad (1-\xi)^{-3}; \quad (1-\xi)^{-4},$$

$$(\tfrac{2}{3}+\tfrac{1}{3}\xi)^6; \quad (\tfrac{2}{3}+\tfrac{1}{3}\xi)^7; \quad (\tfrac{2}{3}\xi)^3(1-\tfrac{1}{3}\xi)^{-3}; \quad (\tfrac{2}{3}\xi)^4(1-\tfrac{1}{3}\xi)^{-4}.$$

(ii) List all the ways in which 28 is expressible as a binomial coefficient.

(iii) Factorize 1001 and show that $\binom{14}{4}, \binom{14}{5}, \binom{14}{6}$ are in arithmetic progression.

(iv) Prove that if p is a prime then $\binom{p}{r}$ is a multiple of p for $0 < r < p$. Is the converse true?

(v) Evaluate $\binom{100}{4}$: (a) directly as $\dfrac{100}{1} \cdot \dfrac{99}{2} \cdot \dfrac{98}{3} \cdot \dfrac{97}{4}$ and (b) as $\left(\dfrac{100!}{4!96!}\right)$ using the table of logs of factorials.

(vi) $\binom{22}{9} = 497{,}420$, find $\binom{22}{10}, \binom{22}{11}, \binom{23}{13}$.

(vii) Given that $_{100}C_{20} \simeq 5 \cdot 36 \times 10^{20}$, find, with slide rule only, $_{100}C_{21}$ and $_{101}C_{20}$.

2R. Calculate for $r = 0, 1, 2, 3, 4$, in turn, $\binom{N}{r} b^{N-r} a^r$ when

(i) $N = 12$, $b = \tfrac{5}{6}$, $a = \tfrac{1}{6}$, given $(\tfrac{5}{6})^{12} = 0 \cdot 1122$;

(ii) $N = 15$, $b = \tfrac{19}{20}$, $a = \tfrac{1}{20}$, given $(\tfrac{19}{20})^{15} = 0 \cdot 5834$;

(iii) $N = 1000$, $b = 0 \cdot 999$, $a = 0 \cdot 001$, given $(0 \cdot 999)^{1000} = 0 \cdot 3677$.

Interpret the answers as probabilities and with the same vertical scales draw bar diagrams to illustrate them.

3. In N trials with probability of success θ let the number of successes be R. Make tables to three decimal places for pr $(R = r)$ against r, for $r = 0, 1, 2, ..., N$ in the following eight cases: $\theta = \tfrac{1}{2}, \tfrac{1}{10}$; $N = 2, 3, 5, 10$. Sketch the graphs of pr $(R = r)$ against r in the form of bar diagrams.

4. (i) Give a simple relation between the coefficients of ξ^r and ξ^{r+1} in the expansion of $(b+a\xi)^N$ in ascending powers of ξ.

(ii) What terms have the greatest coefficients in the expansion in powers of ξ of $(\tfrac{2}{3}+\tfrac{1}{3}\xi)^N$ when $N = 6, 7, 8, 9$?

(iii) What are the most likely numbers of multiples of 3 scored when 6, 7, 8, 9 dice are thrown together?

5T. (i) If 10 coins are thrown together calculate the probabilities of getting $0, 1, 2, ..., 10$ heads (or use results of Question 3).

(ii) If you wanted to state an inclusive range of values in the form $5 \pm d$ for the number of heads in a throw, what are the minimum values of d that would give a probability of at least about $\tfrac{3}{4}$ that you would be right; only about at most a 2 per cent chance of being wrong.

(iii) If you wanted to say that there will be at least N heads, what are the maximum values of N for you to have about at least 95 per cent chance of being right; only about at most a 0·01 chance of being wrong?

(iv) If 10 cards are of two colours chosen at random with equal probabilities, what are the probabilities of a split by colours of 7:3, of 6:4, of 5:5? What is the most likely split?

6. (i) Four dice are thrown and the number of sixes per throw is recorded. Calculate to the nearest integers the likely frequencies of throws with 0, 1, 2, 3, 4 sixes if 25 throws of the four dice are made. What is the probability of obtaining a triple (not necessarily of sixes) or better in a throw of four dice?

(ii) Throw four dice 25 times and compare your frequencies with those calculated from the probabilities. Count also the triples (and 'better' throws) whether of sixes or not.

7R. (i) The Devil offers me the choice of course A or course B to escape some dreadful doom.

A: To cut a pack of cards 5 times in an effort to get exactly 2 picture cards (Jack, Queen, King).

B: To cut a pack of cards 3 times in an effort to get exactly 2 honours (Ten, Jack, Queen, King, Ace).

Should I take course A or course B?

(ii) The Devil applies a squeeze to the system by altering the conditions to getting 3 pictures or 3 honours respectively. Should I choose differently?

8R. (i) In a certain lecture room the probability that a bulb will have become out of order during the vacation is $\frac{1}{4}$. The room has 6 lights, and is unusable if there are fewer than 4 lights working. What is the probability that the room will be unusable on the first day of term?

(ii) In Chapter 3, (Section 10, Example 3, part (iv)) we have players A and B playing under rules of which the following is relevant here 'B's turn consists of throwing two dice'. Use the fact that $(17/18)^{12}$ is about 0·50479 to show that the probability that B scores 'a four and a five' four times or more in his first 12 throws, is about 0·001.

[The problem was suggested by the fact that in Chapter 3, Section 1, the author records just this unusual event.]

9. Two Bernouilli trials are carried out: First trial has $\theta = \frac{1}{2}$, second trial has $\theta = \frac{3}{4}$;

(i) Which is the more likely? That the first trial will yield three or more successes in 4 trials or that the second will yield two or fewer successes in 4 trials. Comment.

(ii) Which is the more likely? That the first will yield six or more successes in 8 trials or that the second will yield four or fewer successes in 8 trials. Comment further.

10R. In an inspection process, repeated random samples are taken from a production line, on which 2 per cent of the items are defective. What proportion of the samples, each of 100 items, will contain at least one defective item? What is the smallest sample size for which the chance is 50 per cent or better of getting not more than one defective item?

11. On average 1 per cent of the articles produced by a certain process are defective, defectives occurring randomly throughout the production. The articles are made up into batches of 144. What percentage of the batches may be expected to include at most one defective?

12. A factory finds that on average 20 per cent of the bolts produced by a given machine will not satisfy a certain requirement. If 10 bolts are selected at random from the day's production of this machine find the probability that: (*a*) exactly two will be defective, (*b*) two or more will be defective, (*c*) five or more will be defective.

13. In Atkins' office there are 10 people all of whom bring umbrellas to the office quite independently of one another and with the same probability as Atkins, that is $\frac{2}{5}$ (see Exercise 3M, Question 6). What are the chances on any one day that there will be 10; 9 or more; 8 or more; 7 or more umbrellas in the office? Atkins' firm is very economical and wishes to install the smallest umbrella stand consistent with its strong beliefs on tidiness. It is only prepared to accept an umbrella or umbrellas out of the stand about once a month (they work a five-day week). What size of umbrella stand does it supply?

14. In an experiment to decide the preferences of woodlice for dry or damp conditions the animals are placed individually in a cage with two compartments (called a choice-chamber). In order to calibrate the instrument, 100 runs are carried out with the two compartments as identically treated as possible so that a Bernouilli trial is the appropriate model. The woodlouse goes north 62 times and south 38 times.

One of the compartments is now made damp and the figures for 12 runs are: north: 9 times and south 3 times.

What is the probability of this or a higher number of north visits occurring in 12 trials of the original supposedly-Bernouilli sequence?

15. The first 51 digits of the decimal form of π are as follows:

$$3\cdot14159\,26535\,89793\,23846\,26433\,83279\,50288\,41971\,69399\,37510.$$

What is the probability that as few or fewer zeros would occur in a random sequence of 51 digits? ($\log_{10} 9 = 0\cdot9542425$.)

16 T. In the crossing of two hybrids we observe a certain biological characteristic, say a curly tail, and Mendelian theory states that three offspring will have this particular dominant characteristic to each one that does not.

(i) From the mating of 2 individuals 9 offspring are observed and all are dominant-looking. What is the probability that this occurs at random from hybrid parents?

(ii) As a general rule we have doubts about the validity of a hypothesis if, by assuming the truth of the hypothesis, we can show that the observed event or worse only had a probability of 5 per cent or less. How many offspring all dominant-looking would we need to observe before being doubtful to the specified extent that both parents were hybrid?

(iii) If we set the rejection level for the hypothesis at a probability of 1 per cent, how many offspring all dominant-looking would we need to observe?

17 T. In a certain type of habitat the proportion of butterflies with wing spots is $\frac{3}{8}$.

(i) In a certain sample of 6 butterflies only 1 has wing spots. What is the probability that butterflies from the given type of habitat would give this or a smaller number of spots on a sample of size 6?

(ii) If a sample of 12 butterflies contained only 2 with wing spots would this lead you to reject at the 5 per cent level the hypothesis that they were from the given type of habitat? [See Question 16.]

18. The author bought some blackberries, and on getting them home found them infested with small maggots. The first 58 blackberries yielded 22 maggots and the author anticipated about 8 in the remaining 20 blackberries. Outline the mathematical model he had instinctively formed and discuss its likely validity. In the event 3 maggots were found in the last 20 blackberries; what light does this throw on the validity of the model; explain why the model may have turned out to fit the facts reasonably well (or not, whichever view you take.)

19 F. The median mark in a certain examination which 300 candidates sat was $55\frac{1}{2}$ per cent.

(i) The eight pupils who sat in the front row of a randomly chosen class had marks: 58, 43, 58, 56, 57, 52, 60, 59 per cent. By judging only on whether or not their marks exceeded the median, determine the probability that in a fully random selection of eight pupils from the candidates the results would be at least as successful as these. Comment.

[This question displays what is called *The Sign Test*; the sign concerned being the numerical sign of (mark − median).]

(ii) In the same examination the pupils in a randomly chosen row from a randomly chosen class had marks 61, 56, 51, 57, 42, 65, 58, 60, 54, 58 per cent. What is the probability of a result at least as *extreme* as this in a fully random selection of ten pupils? Comment.

[In part (i) a 1-*tail test* was applicable but in part (ii) a 2-*tail test* was required.]

20 T. (i) On a certain south-facing hillside the proportion of metre squares which when examined contained a plant of tormentil was 0·4. A botanist suspects that the north-facing hillside is less dense with tormentil plants so he carries out a survey and in 20 metre-squares he observes 4 with tormentil. On the assumption that the south-facing type of vegetation *does* in fact exist on the north-facing hillside and that the tormentil is otherwise randomly distributed over the ground what is the probability of a deficiency as large or larger than that observed? Discuss the botanist's suspicion in the light of this survey.

(ii) Another botanist maintains that the density is unchanged and he calculates the probability of getting a discrepancy of this amount or more. Discuss his conclusions, and compare them with those from part (i).

21 T. A man you meet at the seaside has a 'self-righting dinghy' for sale, which, he says, is heavily biased towards floating top up (that is to say: he maintains that $\theta \equiv$ pr (top-up) $> \frac{1}{2}$). You set up the null hypothesis that *at best* the dinghy floats either way up indifferently and may even favour floating bottom up. You devise a test of a similar sort to that in Exercise 3 M, Question 28: you will throw the dinghy in the water N times and reject the null hypothesis (that is: accept the dinghy) if it lands top up r or more times; and to keep matters simple you consider the cases (*a*) $N = 5$, $r = 5$; (*b*) $N = 5$, $r = 4$; (*c*) $N = 6$, $r = 5$. You next plot the power function of each of these tests on the same diagram.

Investigate the following possible situations.

(i) You need a rescue dinghy of this sort and time is running out. Even if its performance is not above average in righting itself, it would be something to have and if you went without it you might be haunted by the thought that perhaps it was all right.

Which type of error do you wish to avoid? Which test *of the three* do you propose to the salesman? What is the snag about your test when $\theta < \frac{1}{2}$?

(ii) There are plenty of other dinghy salesmen crying their wares and you would look a fool if you bought this one and it didn't work.

Which type of error do you wish to avoid? Which test of the three do you propose to the salesman? What is the snag when $\theta > \frac{1}{2}$?

(iii) Draw the graph for a hypothetical test, (*d*), that would be strong on accepting the null hypothesis for $\theta < \frac{1}{2}$ and strong on rejecting it for $\theta > \frac{1}{2}$. On the same diagram draw the graph for (*e*) the 'test' which simply consists of saying 'I will reject the null-hypothesis out of hand' (that is: 'I will accept the salesman's claim'); and for (*f*) the 'test' which simply says 'I will accept the null-hypothesis out of hand' (that is: 'I will reject the salesman claim without argument'). To which type of error is each of (*e*), (*f*) liable?

[The test of Exercise 3 M, Question 28 is called 2-*tailed*; the first four tests of this question are 1-*tailed*.]

22 T. A test for a certain second-hand article is designed to find out whether more than about 20 per cent are defective. Ten articles are picked at random from a consignment and if 2 or more defectives are found the consignment is rejected on the grounds of this single sample. Such a test is known as a *single sampling scheme*.

(i) Calculate pr (batch is accepted | proportion of defects in batch is really θ).

(ii) This probability expressed as a function of θ is called the *operating characteristic function* of the inspection scheme. Plot its graph against θ.

(iii) What is the relation between the operating characteristic function of the inspection scheme and the power function of the test which the scheme employs? (See Question 23 and also Exercise 3 M, Question 28.) Take care over your definition of the null hypothesis in the test.

23 T. The inspection scheme for the articles in Question 22 is modified as follows:

A sample of 10 articles is inspected, and the following scheme applied:

If (*a*) there are no defectives, the consignment is accepted.
   (*b*) there are 2 or more defectives, the consignment is rejected.
   (*c*) there is 1 defective, a further sample of 10 is inspected and if there are any defectives in that sample then the whole consignment is rejected.

Find the operating characteristic function for this scheme.

[This is called a *double-sampling scheme*; it is possible to generalize the process to multiple sample schemes and if the samples are each of size 1 and the sampling is to be continued until a decision is reached we are said to have a *sequential sampling scheme*. The idea of sequential sampling and the theory behind it were first produced by Wald (1902–50) in 1943. Wald had had no experience of modern statistical theory until 1938, but between 1940 and 1950 he revolutionized the subject.]

24. At least 3 members of a team of 5 housewives have to pass the entrance test if their team is to be allowed to compete in a certain television contest. To pass the test a member has to answer one or more of 3 questions and each member has a probability $\frac{1}{3}$ of answering each of her questions. What is the probability that the team will be allowed to compete?

25. (i) Without writing down the values of any probabilities, show that the probability of the single event of getting exactly 50 heads in each of two throws of 100 coins at a time is less than the probability of the single event of getting 100 heads in a throw of 200 coins at a time.

(ii) What is the chance of throwing 100 coins to land 50 heads and 50 tails?

(iii) What is the chance of throwing 98 coins to land 49 of each?

(iv) Given that the first 98 coins have landed 49 of each, what is the probability that the final result will be 50 of each?

(v) Why do we not have the answer to part (ii) to be $\frac{1}{4}$ that of part (iii)?

26. Two people each throw a coin n times. What is the probability that both have the same number of heads?

Hint: consider a certain coefficient in $(1+\xi)^n (1+\xi)^n$ obtained by two separate methods.

27. An aircraft has 4 engines, 2 on each wing. The probability of any one engine failing on a transatlantic flight is 0·1, and the event of any one engine failing is independent of the behaviour of the other engines. What is the probability of the crew getting wet if

(*a*) the plane will fly on any two engines?

(*b*) the plane requires at least one engine operating on each side?

[These probabilities are highly unrealistic!]

28. A red ball is chosen with probability θ and a white ball with probability $1-\theta$. After a ball has been chosen, it is placed in Box 1 with probability θ' and in Box 2 with probability $1-\theta'$. Show that if this experiment is repeated n times, then the probability that there will be exactly m red balls in Box 1 is

$$\binom{n}{m} (\theta\theta')^m (1-\theta\theta')^{n-m} \quad (m = 0, 1, ..., n).$$

29 F. We have seen in Exercise 3 M, Question 16 that in a collection of plants in which pr (gene is a) $= \theta$, the probabilities of genotypes aa, aA, AA are respectively θ^2, $2\theta(1-\theta)$, $(1-\theta)^2$.

Suppose that the characteristic determined by gene a is recessive so that plants of genotypes aA and AA are indistinguishable from each other but are distinguishable from those of genotype aa. Let us say that aa gives blue flowers and the others white flowers.

(i) In experiment X, a random plant with white flowers is crossed with a plant with blue flowers; let ϕ be the probability of the resulting flowers being blue.

In experiment Y, random plants with white flowers are crossed with each other; show that the probability of getting blue flowers as a result is ϕ^2.

(ii) N different plants are crossed in each experiment; show that the probability of getting r plants with blue flowers in experiment X and s plants with blue flowers in experiment Y is

$$K . \phi^{r+2s}(1-\phi)^{N-r} (1-\phi^2)^{N-s},$$

where K is independent of ϕ.

(iii) By considering the logarithm of the above probability find an equation for the value of ϕ which maximizes it for fixed N, r, s.

(iv) Determine this maximizing value, and the corresponding value of θ, for $N = 72$, $r = 12$, $s = 6$; and, with the maximizing value and $N = 72$, find what values of r and s you would have expected to get in the experiments.

[This question is an example of the *Method of Maximum Likelihood* for estimating parameters. The reader will meet the method in greater detail in Chapter 12.]

30 F. In a certain survey connected with controlling animal populations, seals are being examined for two characteristics: whether they have a particularly valuable fur marking and whether they are suffering from oil-pollution. It is assumed that the proportions of each are fixed over a reasonable length of time and a reasonable area, and furthermore that the characteristics occur inpendently of one another.

Of 10 animals observed 3 are unpolluted valuable, and this seems to make it worth handling the furs commercially. A second survey almost immediately afterwards at the same place shows

	Unpolluted	Polluted
Valuable	2	3
Not valuable	8	22

What is the probability of getting this or a smaller number of unpolluted valuable specimens if the conditions of the hypotheses hold?

*31. If N Bernouilli trials with probability θ of success are carried out, investigate the most probable number of successes for various values of N and θ.

12. WAITING TIMES

12.1 Introduction. In Section 9.2 we contemplated various problems which suggested themselves when a sequence of Bernouilli trials was under consideration. We saw that the problem of how long we should have to wait for, say, the third success required us to investigate first how many successes would occur in a fixed interval. The rest of Section 9 was concerned with a fixed number of trials during which various numbers of successes might occur. That is called direct sampling and led us to a binomial probability function. It is important to distinguish that problem from the related problem in which a fixed number of successes is being sought and they may require various numbers of trials to occur. That is called inverse sampling. It is to the investigation of waiting times in a succession of Bernouilli trials that we now return briefly.

12.2 The probability function. We begin with a simple numerical case. For instance:

pr (waiting time to the 3rd success = 4)

= pr (3rd success is on the 7th throw)

= pr ((2 successes in first 6 throws) and (success on 7th throw))

$$= \binom{6}{2} (1-\theta)^4 \, \theta^2 . \theta.$$

$\binom{6}{2}$, the numerical coefficient in the above expression, is picked out in the Pascal Triangle as follows:

both from: 1   and from: 1

$\quad\quad\quad$ 1   1 $r = 2$ $\quad\quad\quad\quad$ 1   1 $N = -3$

$\quad\quad\quad$ 1   2   1 $\quad\quad\quad\quad\quad$ 1   2   1

$\quad\quad\quad$ 1   3   3   1 $\quad\quad$ $r = 4$ 1   3   3   1

$\quad\quad$ 1   4   6   4   1 $\quad\quad\quad$ 1   4   6   4   1

$\quad\quad$ 1   5   10   10   ..   .. $\quad$ 1   5   10   10   ..   ..

$N = 6 \rightarrow$ 1   6   **15**   .. $\quad\quad\quad\quad$ 1   6   **15**   ..

from which we see that $\binom{6}{2} = \binom{-3}{4}$, so that $\binom{6}{2}$ is also the coefficient of ξ^4 in the expansion of $(1-\xi)^{-3}$.

By a similar argument

$$\text{pr (waiting time to the 3rd success} = 5)$$

$$= \text{pr (3rd success is on the 8th throw)}$$

$$= \binom{7}{2} (1-\theta)^5 \, \theta^2 . \theta$$

and $\binom{7}{2} = \binom{-3}{5}$ = the coefficient of ξ^5 in the expansion of $(1-\xi)^{-3}$.

The argument is quite general as to the length of the waiting-time and the coefficient required in pr (waiting time to the 3rd success $= w$) is the coefficient of ξ^w in $(1-\xi)^{-3}$.

Thus

$$\text{pr (waiting time to 3rd success} = w) = \binom{-3}{w} (1-\theta)^w . \theta^3.$$

The argument is also general as to which success is under consideration and we have:

$$\text{pr (waiting time to the } r\text{th success} = w)$$

$$= \binom{-r}{w} (1-\theta)^w . \theta^r$$

$$= \text{coefficient of } \xi^w \text{ in } (1-(1-\theta)\,\xi)^{-r} . \theta^r$$

$$= \text{coefficient of } \xi^w \text{ in } \left(\frac{\theta}{1-(1-\theta)\xi}\right)^r.$$

The probability function concerned is called an *inverse* (or *negative*) *binomial probability function*.

When $r = 1$ it is sometimes called a *geometric probability function*.

228

Exercise 7E

1T. The following are two sequences each of 30 sets of 5 items listed as success (A) or failure (B). Estimate the probability (which is the same in each sequence) that an item chosen at random from a whole sequence will be a success. By examining the frequency tables for the values of R, U, V defined the Section 5.3, and comparing them with their theoretical values, decide whether or not you think that the items occur in random order in each sequence. Since 'independence' is under question, do not treat the first A as if it was immediately preceded by a (hypothetical) A.

First sequence

ABABB, ABBBB, ABABA, BBABB, ABABB, ABABA, ABBBB, ABABB,
ABBBB, ABBBA, BBBBB, BBABA, BBABA, BBBBA, ABABB, ABBBB,
BBBBB, ABABA, BBBBA, BBBBB, AAAAB, ABABB, ABABA, ABBBB,
BABBB, ABABA, BABBB, ABBBB, BBABA, ABBBA.

Second sequence

AABAA, AAABB, BBBBB, BBBBA, AABBB, BBBBA, AABBA, ABABB,
BBBBB, BBBAB, BBAAB, BBBAA, AABBB, BBBAB, BBBAB, BABAA,
AABBB, BBBAA, ABABB, BBBAB, BBBBB, BBABB, BABAB, BBABB,
BBABB, BABBA, ABABA, ABAAB, BBBAB, BBBBA.

2. (i) Consider sequences of 6 throws of a fair coin and write down the patterns of all such sequences which contain somewhere a run of exactly k heads for $k = 3, 4, 5, 6$ (indicating by crosses those results which do not affect whether or not we have a run of exactly k heads; thus $THHHT\times$ gives a run of exactly 3 heads).

(ii) Verify that for $N = 6$ and $k = 3, 4, 5$, we have:

pr (the longest run is of exactly k heads) $= (N-k+3)/2^{k+2}$.

(iii) Prove that this formula is true for all N and $\frac{1}{2}(N-1) \leqslant k < N$.

(iv) Why do difficulties occur for $k < \frac{1}{2}(N-1)$?

3T. If two large enough samples are drawn at random from a *single* indefinitely large population of values and the members are arranged in order of increasing size, then at each stage the next member to be written down has a probability of $\frac{1}{2}$ of being from each sample.

(i) Show that the probability that k is the length of the run, at start, of members from one of the samples before the first occurrence of a member from the other is $1 \times (\frac{1}{2})^{k-1} \times (\frac{1}{2})$.

(ii) Show that the probability that L (the sum of the lengths of the runs at the two ends of the full ordered list of the samples) is l is

$$\sum_{k=1}^{l-1} (\tfrac{1}{2})^k . (\tfrac{1}{2})^{l-k} \quad (l \geqslant 2).$$

(iii) Find the probability that L is 8 or more, 9 or more and the probability that L is 11 or more, 12 or more and thus prove the properties which form the basis for the very useful test due to Quenouille, quoted in Exericse 1 A, Question 8.

13. MULTINOMIAL PROBABILITY FUNCTION

Suppose the results of a trial can be classified into three types, say A, B, C, and that, for instance, 7 trials are carried out, then we may ask 'What is the probability of getting 2 results of type A, 3 of type B and 2 of type C?' The answer follows from the binomial methods, for if we first classify the results as A or not-A then we have: the number of ways of getting 2 of type A and 5 of type not-A is $\dfrac{7!}{2!\,5!}$. But the ways in which the 5 not-A results occur contain $\dfrac{5!}{3!\,2!}$ ways in which 3 of the type B and 2 of type C occur. It follows that the total number of ways of getting the stated frequencies is

$$\frac{7!}{2!\,5!}\times\frac{5!}{3!\,2!} \quad \text{which is} \quad \frac{7!}{2!\,3!\,2!}.$$

The result can be easily generalized to: the numbers of ways of getting $r_1, r_2, r_3, ..., r_k$ result of types $A_1, A_2, ..., A_k$ in $(r_1+r_2+.+r_k)$ trials is

$$\frac{(r_1+r_2+...+r_k)!}{r_1!\,r_2!\,...\,r_k!}.$$

If the probabilities of types $A_1, A_2, ..., A_k$ are $\theta_1, \theta_2, ..., \theta_k$, and the results of successive trials are independent of each other, then the probability of getting the specified frequencies is

$$\frac{(r_1+r_2+r_3+...+r_k)!}{r_1!\,r_2!\,...\,r_k!}.\theta_1^{r_1}.\theta_2^{r_2}.\theta_3^{r_3} ... \theta_k^{r_k}.$$

We do not pursue this topic beyond a couple of Examples and a very brief exercise.

Example 8. How many ways are there of arranging the letters of the word Mississippi?

Solution. The frequencies are given by the table:

letter	M	I	S	P
frequency	1	4	4	2

The required number is $\dfrac{(1+4+4+2)!}{1!\,4!\,4!\,2!} = 34{,}650$.

Example 9. Items fall into four classes with probabilities $\frac{9}{27}, \frac{6}{27}, \frac{4}{27}, \frac{8}{27}$ respectively. What is the probability that 50 such items should show frequencies 8, 20, 8, 14?

230

Solution. The probability is

$$p = \frac{50!}{8!\,20!\,8!\,14!} \cdot \frac{9^8 \cdot 6^{20} \cdot 4^8 \cdot 8^{14}}{(27)^{50}}$$

$$\simeq 1 \cdot 1 \times 10^{-5}.$$

Exercise 7F

1R. If the river from which the fish of Section 2 were caught was believed to contain brown trout, grayling, rainbow trout, pike and others in large numbers and in the proportions $10:20:4:1:25$, what is the probability of the catch described in that section arising from a random sample? Comment.

2R. If twelve dice are thrown what is probability of getting each face exactly twice?

*3T. What is the applicability of Example 9 to Exercise 7E, Question 1.

*4. Apply the method of Question 3 to the rest of Exercise 7E, Question 1.

8

SEQUENCES OF EVENTS—
POISSON MODEL

1. FAMILY OF BINOMIAL
PROBABILITY FUNCTIONS

The reader has now done a certain amount of work on the probabilities of getting various values of R (the number of successes) in N trials when the probability of success in a given trial is fixed, say a. We have seen that

$$\text{pr}\,(R = r) = \binom{N}{r} (1-a)^{N-r}\, a^r,$$

the right-hand side being an expression containing r, N and a. It is of interest to see how, for a given combination of values of N and a, the whole set of values of probabilities behaves: ten typical graphs are shown in Figure 8.1, and the reader would do well to contemplate them and absorb the pattern they exhibit as a changes with fixed N, and as N changes with fixed a. To specify the pattern the values of both N and a have to be given. We say the binomial probabilities have two (independent) parameters. It further turns out that no two sets of probabilities are identical.

There are, however, two very important approximations available, one of which we will deal with next, and the other of which is the Normal approximation.

2. POISSON APPROXIMATION

2.1 Introduction. If the Bernouilli trial model for a given situation has the value of a selected by the long run estimate of

(number of successes)/(number of trials)

—and this is at the foundation of our ideas of probability—then it is clear that in another number of trials, say N, we will anticipate getting about the proportional number, Na, of successes. (These concepts will be more rigorously defined and dealt with later.) This means that if we have 15 defectives in 240 trials of an item we will estimate a as $\frac{15}{240}$ or $\frac{1}{16}$, and though it is obvious that this does not ensure that we shall get exactly one defective in every alternate sample of 8 items we could ask what is the probability of getting 0, 1, 2, ... defectives in 8 trials. We would expect an 'average' of about $\frac{1}{2}$ in 8 trials.

In problems like this about sampling for defectives, we often have situations where a is small, but N is large enough to give a reasonably-sized probability of finding defectives in the sample.

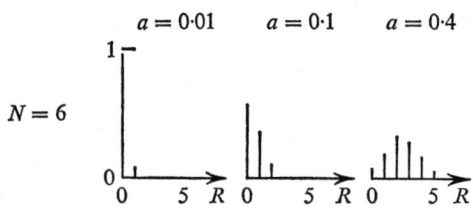

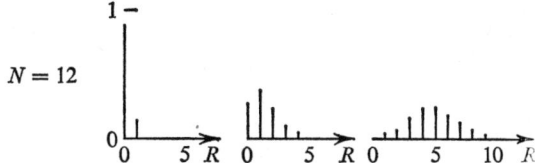

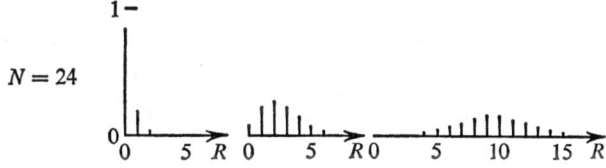

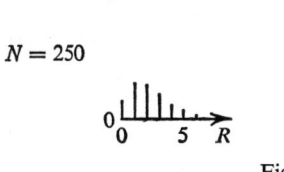

Fig. 8.1

Let us write $Na = \mu$, which is thus an 'average' number of defectives in a sample and keep μ constant while varying N and a. For instance we started with $a = \frac{1}{16}$, $N = 8$, but we might take $a = \frac{1}{32}$, $N = 16$ or even $a = \frac{1}{200}$, $N = 100$ and still get the same value, $\frac{1}{2}$, for μ.

2.2   The approximations. We examine the probabilities. Let R denote the number of successes (that is, findings of defectives).

$$\mathrm{pr}\,(R = 0) = (1-a)^N$$

$$= \left(1-\frac{\mu}{N}\right)^N$$

$$\to e^{-\mu}, \qquad \text{as } N \to \infty, \text{ with } \mu \text{ fixed.}$$

Now, by the usual relation,

$$\mathrm{pr}\,(R = 1) = \mathrm{pr}\,(R = 0).\frac{N}{1}.\frac{a}{b}$$

$$= \mathrm{pr}\,(R = 0).\frac{\mu}{(1-\mu/N)}$$

$$\to e^{-\mu}.\mu \qquad \text{as } N \to \infty, \text{ with } \mu \text{ fixed.}$$

In general

$$\frac{\mathrm{pr}\,(R = r)}{\mathrm{pr}\,(R = r-1)} = \frac{N-r+1}{r}.\frac{a}{1-a}$$

$$= \left(\frac{1-\dfrac{r-1}{N}}{r}\right).\frac{\mu}{\left(1-\dfrac{N}{\mu}\right)}$$

$$\to \frac{1.\mu}{r.1} \qquad \text{as } N \to \infty, \text{ with } \mu \text{ fixed.}$$

The successive probabilities seem to approach the terms:

$$e^{-\mu}; \quad \mu e^{-\mu}; \quad \frac{\mu^2}{2!}e^{-\mu}; \quad \frac{\mu^3}{3!}e^{-\mu}; \quad \ldots; \quad \frac{\mu^r}{r!}e^{-\mu}; \quad \ldots$$

These are known as the *Poisson* approximations to binomial probabilities.

They are nearly the probabilities associated with 0; 1; 2; 3; ...; r; ... defectives in a sample of N items when the average number of defectives in samples is μ. *The remarkable fact is that they involve only μ.* The size of the sample is to be large but the actual size, N, and the rarity of defectives, measured in terms of a, have only entered in the form Na. In Figure 8.1 the shapes for $(a, N) = (0.4, 6)$, $(0.1, 24)$, $(0.01, 250)$ are all very alike; they correspond to $\mu = Na = 2.4$, 2.4, 2.5 respectively. So although no two patterns are in fact identical, there are sets of very similar patterns. We check whether, with all the approximations, we still have a unit sum:

$$\text{Sum of terms} = e^{-\mu}\left(1+\frac{\mu^2}{2!}+\frac{\mu^3}{3!}+\ldots+\frac{\mu^r}{r!}+\ldots\right) = e^{-\mu}.e^{\mu} = 1.$$

Those that are in error one way cancel with those that are in error the other way. Of course if the sample is of size 20, say, our approximations

234

give non-zero values to the probabilities of finding 21, 22, ... defectives!
The fact, however, that the approximations sum to unity shows that they
could be used as a set of probabilities in their own right, and we will have
more to say about that in Section 3.

The approximation is remarkably good even for quite small values of N,
as Tables 8.1–8.3 show.

Table 8.1

$\mu = 5$

pr $(R = r)$

r	Approxi-mation (3 dec. pl.)	$N = 1000$ $a = \frac{1}{200}$	$N = 25$ $a = \frac{1}{5}$	$N = 10$ $a = \frac{1}{2}$	Approxi-mation (2 dec. pl.)
0	0·007	0·007	0·00	0·00	0·01
1	0·034	0·036	0·02	0·01	0·03
2	0·084	0·085	0·07	0·04	0·08
3	0·140	0·141	0·14	0·12	0·14
4	0·175	0·177	0·19	0·20	0·18
5	0·175	0·177	0·20	0·25	0·18
6	0·146	0·147	0·16	0·20	0·15
7	0·104	0·105	0·11	0·12	0·10
8	0·065	0·065	0·06	0·04	0·06
9	0·035	0·035	0·03	0·01	0·04
10	0·017	0·018	0·01	0·00	0·02
Totals	0·982	0·993	0·99	0·99	0·99

In Tables 8.2 and 8.3 the upper of each pair of figures is pr $(R = r)$ and
the lower is pr $(R \leqslant r)$.

Table 8.2

$\mu = 0.6$

r	0	1	2	3	4
Approximations (3 dec. pl.)	0·549	0·329	0·099	0·020	0·003
	0·549	0·878	0·977	0·997	1·000
$N = 12, a = \frac{1}{20}$	0·540	0·341	0·099	0·017	0·002
	0·540	0·882	0·980	0·998	1·000
$N = 6, a = \frac{1}{10}$	0·53	0·35	0·10	0·02	0·00
	0·53	0·89	0·98	1·00	1·00
$N = 3, a = \frac{1}{5}$	0·51	0·38	0·10	0·01	0·00
	0·51	0·90	0·99	1·00	1·00
Approximations (2 dec. pl.)	0·55	0·33	0·10	0·02	0·00
	0·55	0·88	0·98	1·00	1·00

Table 8.3

$\mu = 0.1$

r	0	1	2
Approximations	0·905	0·090	0·005
	0·905	0·995	1·000
$N = 10, a = \frac{1}{100}$	0·904	0·091	0·004
	0·904	0·996	1·000
$N = 2, a = \frac{1}{20}$	0·902	0·095	0·002
	0·902	0·998	1·000
$N = 1, a = \frac{1}{10}$	0·900	0·100	0·000
	0·900	1·000	1·000

Exercise 8 A

1 R. (i) Calculate approximately the probabilities of 0, 1, 2, ... successes in 8 trials with $a = \frac{1}{16}$.

(ii) Calculate approximately the probabilities of 0, 1, 2, ... successes in 100 trials with $a = \frac{1}{50}$.

2 R. Compare the exact (binomial) probabilities with their Poisson approximations for the events of scoring 0, 1, 2, double-sixes in 36 throws of a pair of fair dice. Recalculate Exercise 7 D, Questions 2, 3 using the approximations.

3 T. Draw two axes on graph paper to a scale of 10 cm to a unit, and mark on them values of a across the page

and $1/N$ up the page, for $N = 1, 2, 3, 4, 5, 6, 8, 10, 20, 40$.

You now have a two-dimensional parameter space for a and $1/N$.

Mark the lines $aN = \mu$ for $\mu = 0.1, 0.6, 1, 2, 5$. Mark points to represent each of the probability functions of Tables 8.1, 8.2, 8.3. Make conjectures as to roughly what is the largest value of a, and smallest value of N that would give a reasonable Poisson approximation to the binomial probabilities with $\mu = 2$. Check your guess by evaluating the relevant probabilities.

4 R. In a binomial sequence of trials, $p_N(r)$ is the probability of r successes in N trials.

(i) Explain why

$$p_{200}(3) = p_{100}(0)\, p_{100}(3) + p_{100}(1)\, p_{100}(2) + p_{100}(2)\, p_{100}(1) + p_{100}(3)\, p_{100}(0).$$

(ii) Evaluate the right-hand side of this in terms of the approximations obtained, and simplify the result algebraically.

(iii) Interpret the simplified result as a probability in its own right.

3. POISSON PROCESSES

3.1 The binomial background. We now have a set of probabilities of various numbers of successes but to what *trial* do they refer? In a sense, the individual trial (specified by the probability of success, a) and the number of times it is repeated, N, have vanished away leaving the wraith-like μ behind. And this is a not unhelpful flippancy.

Suppose we take $a = \frac{1}{50}$ and $N = 100$, then we could concoct some data with a table of random numbers, read consecutively in pairs, by picking out occurrences of, say, 33 and 77 (as being easily spotted) and recording the gaps (after each success up to and including the next success) measured in numbers of trials, X. These gap-measures are not quite the waiting times of the Bernouilli sequence theory.

A typical record of the lengths of these gaps runs:

$X = $ 86, 75, 19, 37, 23, 71, 106, 39, 36, 78, 23, 100, 1, 31, 56, 51, 9, 15, 53, 14, 16, 108, 110, 19, 73, 41, 44, 84, 62, 35, 31, 40, 26, 24, 89, 13, 39, 10, 42, 85.

Suppose we now presented these to an investigator as time-gaps: perhaps in milliseconds between arrivals of particles in a Geiger counter; or in minutes between *arrivals* of telephone calls (whatever their ultimate lengths) at a remote switchboard; or in hours between arrivals of appendicitis cases at a hospital; or again as space-gaps: perhaps in feet between flaws in a continuously produced steel pipe; and so on. The investigator is faced with the figures; he might analyse them in various ways. Let us suppose they are presented as lengths of gaps in minutes between telephone calls. He might choose a standard time-unit of, say, 40 minutes, as looking comparable with these lengths of gap, and start measuring off from the start in successive periods of his unit of 40 minutes.

In each 40-minute period he would count the number of calls arriving. Thus the first two 40-minute periods give 0, 0 and the third gives 1 and so on; the work is tedious and may be done by marking points along a scale at the appropriate intervals and then stepping off the 40-minute intervals. If the readings came from a continuously recording instrument like a drum barometer which gave a kick at each call then the spacing out along the scale would already be done. The investigator's table reads as Table 8.4.

Table 8.4

(R represents number of calls in a 40-min period)

r	frequency of $R = r$
0	18
1	22
2	9

From this table he calculates:

mean number of calls per 40-minute period

$$= \bar{r} = \frac{(0 \times 18) + (1 \times 22) + (2 \times 9)}{49}$$

$$= 40/49 = 0{\cdot}817.$$

Now *we* know that the data were concocted from a binomial process in which the mean number of successes in 100 trials was 2; we are not surprised that the mean number of successes in 40 trials is nearly $(\frac{40}{100}) \times 2$, even though the investigator does not use the 'trial' terminology.

If the investigator had taken 80-minute periods his data would have read as in Table 8.5.

Table 8.5

(*S* represents the number of calls in an 80-min period)

s	frequency of $S = s$
0	4
1	7
2	8
3	4
4	1

(Notice that the last call is omitted as the end of its 80-minute period extends into a time where the investigator does not know about arrivals.)

From this the investigator calculates:

$$\text{mean-number of calls in 80-minute period} = \bar{s} = \tfrac{39}{24} = 1 \cdot 62.$$

The *trial*, in fact, may be the passage of a unit time interval of arbitrary fixed length and the recording of the number of occurrences in it; the successive trials occur as the intervals come along. Double the length of the interval, and the mean number of occurrences in it is doubled. There is of course nothing sacred about *doubling* the interval, in fact we began this problem with an investigator unwittingly multiplying by $\frac{2}{5}$ the interval upon which we had based our approximating binomial process of averaging two successes in 100 trials.

3.2 Poisson processes. Let us then idealize our approximating binomial process and suppose that some arbitrary unit of a continuous variable like time (or distance along a line, but we will use 'time' as a general term for the moment) is involved; and that we have a process which retains the independence of the occurrences and produces a mean of λ occurrences in a unit of time. In our model we will furthermore assume that the occurrences are instantaneous. We may think of the real life occurrences as flashes (like lightning strokes or Geiger counter scintillations), so short that they effectively never overlap from one time interval into another. Let us go on subdividing our time intervals until *they* are so short that the probability of two flashes in one interval is effectively zero. We see that the two requirements compete with each other since the shorter the flash the more possible to have two in any given interval, while the shorter the interval the more possible for a flash to overlap and be recorded in two of them.

In real life our requirements cannot be attained; but we are building a mathematical model, and will continue. If our model fits facts then it is justified (and no other justification for a *model* exists either; though as entertaining mathematics it would be basing its plea on other grounds altogether; and many pieces of entertaining mathematics have, much later, turned out as useful models, though that is not *their* justification!).

3.3 Derivation of probabilities. We will denote the passage of time by the increasing value of a variable T and consider the period $0 \leqslant T \leqslant t$, for various numbers t. We write $p(r, t)$ for the probability that there are r flashes in the period $0 \leqslant T \leqslant t$.

Now $p(r, t+\epsilon)$ for $\epsilon > 0$ is the probability of the event 'r flashes occur in a period $0 \leqslant T \leqslant t+\epsilon$' and by considering the breakdown of this event into exclusive, exhaustive events we have, rather generally:

For $r \geqslant 1$:

$$p(r, t+\epsilon) = p(r, t) \times \mathrm{pr} \text{ (no flashes in an interval of length } \epsilon |$$
$$r \text{ flashes up to } T = t)$$

$$+ p(r-1, t) \times \mathrm{pr} \text{ (1 flash in an interval of length } \epsilon |$$
$$(r-1) \text{ flashes up to } T = t)$$

$$+ \sum_{s=0}^{r-2} p(s, t) \times \mathrm{pr} \text{ (}(r-s) \text{ flashes in interval of}$$
$$\text{length } \epsilon | \text{suitable number of}$$
$$\text{flashes up to } T = t).$$

This equation will be referred to as Equation I in this section. Now, for our problem, there are a number of simplifications which we intend to make to limit the generality of the discussion.

(1) The flashes occur independently of other flashes: we can drop the conditional nature of the probabilities.

(2) The flashes effectively occur singly: we can express this by saying pr (multiple flashes in an interval of length ϵ) $= 0(\epsilon^2)$ when the interval ϵ is small.

(3) The flashes occur uniformly (*not* regularly): there is a number λ so that the probability of an occcurrence in interval ϵ is $\lambda\epsilon + 0(\epsilon^2)$, however small ϵ may be, and wherever the interval ϵ may start in the time-scale.

From these we deduce that

pr (no flash in an interval of length $\epsilon | r$ flashes up to $T = t$)

$= \mathrm{pr}$ (no flash in an interval of length ϵ)

$= 1 - \mathrm{pr}$ (1 or more flashes in that interval)

$= 1 - (\lambda\epsilon + 0(\epsilon^2))$.

Now since the term involving multiple flashes in Equation I is a finite sum of terms each of which is $0(\epsilon^2)$, it is itself $0(\epsilon^2)$ and we have

$$p(r, t+\epsilon) = p(r, t)(1-\lambda\epsilon-0(\epsilon^2))$$
$$+p(r-1, t).(\lambda\epsilon+0(\epsilon^2))$$
$$+0(\epsilon^2)$$

giving $\quad p(r, t+\epsilon) = p(r, t)+\epsilon.\lambda[p(r-1, t)-p(r, t)]+0(\epsilon^2)$

from which it follows that, ϵ being arbitrarily small,

$$\frac{d}{dt}p(r, t) = \lambda.[p(r-1, t)-p(r, t)].$$

This is the resulting equation for all $r \geqslant 1$. We call it Equation II.

For $r = 0$, we seek the probability of no flashes in the interval $0 \leqslant T \leqslant t$. This satisfies Equation I modified as follows:

$$p(0, t+\epsilon) = p(0, t)(1-\lambda\epsilon+0(\epsilon^2))+0+0$$

giving $\quad\quad p(0, t+\epsilon) = p(0, t)-\epsilon.\lambda p(0, t)+0(\epsilon^2)$

from which it follows that, ϵ being arbitrarily small,

$$\frac{d}{dt}p(0, t) = -\lambda.p(0, t).$$

This has solution $\quad\quad p(0, t) = p(0, 0).e^{-\lambda t}.$

We take $p(0, 0)$ as 1, meaning that we certainly start with no flashes, and obtain $\quad\quad\quad\quad\quad p(0, t) = e^{-\lambda t}.$

Returning now to Equation II, we put it into a form which will enable us to obtain successively $p(1, t), p(2, t), p(3, t), \ldots$ that is the probabilities of 1, 2, 3, ... flashes occurring in the interval $0 \leqslant T \leqslant t$.

We have

$$\frac{d}{dt}p(r, t)+\lambda p(r, t) = \lambda p(r-1, t),$$

giving

$$\frac{d}{dt}(e^{\lambda t}p(r, t)) = \lambda e^{\lambda t}p(r-1, t)$$

and finally, for $r \geqslant 1$,

$$p(r, t) = e^{-\lambda t}\int_0^t \lambda e^{\lambda x}.p(r-1, x).dx.$$

The choice of lower limit is dictated by wanting $p(r, 0) = 0$ for $r \geqslant 1$, since this expresses the fact that when $t = 0$ the probability of any number of flashes having occurred is to be zero (the time interval then being $0 \leqslant T \leqslant 0$).

The integration then gives

$$p(1, t) = e^{-\lambda t}\int_0^t \lambda e^{\lambda x}e^{-\lambda x}.dx = \lambda te^{-\lambda t},$$

$$p(2, t) = e^{-\lambda t}\int_0^t \lambda e^{\lambda x}(\lambda x e^{-\lambda x}).dx = \frac{\lambda^2 t^2}{2}e^{-\lambda t}$$

and in general $$p(r, t) = (\lambda t)^r \, e^{-\lambda t}/r!$$

This formula is valid for $r \geqslant 0$ and not merely $r \geqslant 1$, by the customary definition of 0! as 1.

Now λ can be considered to be the mean *rate of occurrences* of flashes, and then λt is the mean *number* of flashes in an interval $0 \leqslant T \leqslant t$. We expected in Section 3.1 to be able to get different means of numbers of occurrences in an interval, in proportion to the length of the interval, and we see that we have exactly that property. Let us take a particular value t_0 of t so that $\lambda t_0 = \mu$, then we have chosen an interval giving mean number of flashes as μ. The probabilities of various numbers of flashes in an interval of this *fixed length* are
$$e^{-\mu}, \quad \mu e^{-\mu}, \quad \mu^2 e^{-\mu}/2!, \quad \dots, \quad \mu^r e^{-\mu}/r! \dots$$
for numbers of flashes $= 0, 1, 2, \dots, r, \dots$, exactly as we were led to expect from the approximating binomial approach.

3.4 Summary. The Poisson probabilities thus provide a model for effectively instantaneous events that occur
(i) independently of each other, and
(ii) singly along a continuous axis (whether spatial or of time), and
(iii) uniformly, in the sense that the mean number in an interval is proportional to the length of the interval, whatever its length, and does not depend on the position of the interval along the 'time-axis'.

Such events are called *random events*, and any process which produces them is called a *Poisson process*.

It is meaningless to ask whether the Poisson probabilities are *really* an approximation to the binomial probabilities with small a. They are often indeed a very good approximation, but have we not *really* in this section reversed the ideas and considered a sequence of binomial trials with small a as the starting point for an investigation into random events occurring in continuous time? The Poisson process stands on its own feet as a model for a very important class of physical situations.

Exercise 8 B

1 R. Assuming that large quantities of appropriate data have been collected, discuss the application of the theory of Poisson processes to estimating the probabilities:
(*a*) that in a given year there will be no national election in a given country;
(*b*) that on a given page of a given book there will be 2 or more misprints;
(*c*) that on a given page of a given book there will be exactly one sentence whose full stop is at the end of a line;
(*d*) of various numbers of arrivals in a given minute at one of the paying lines or queues in a supermarket between 10.30 a.m. and 11.30 a.m.;

(e) of various numbers of cars passing in a given minute at a point at the south end of motorway M1 between 5 p.m. and 6 p.m.: (i) travelling South, (ii) travelling North.

(f) of various numbers of *calls* to a fire station in an hour of daylight in February;

(g) of various numbers of *fire-engines called out* in the same hour of daylight in February.

2R. The following data already quoted in Exercise 6M, Question 1 refer to deaths by horse-kick of Prussian cavalrymen in 10 corps during the years 1875–94.

Number of deaths per corps per year	0	1	2	3	4
Frequency	109	65	22	3	1

Would you expect them to be well represented by a Poisson model? Can they in fact be well so represented?

3R. A flying-bomb was not a large object and its arrival at any given spot was a rare event, but over a number of months in the Second World War and observing 576 squares of $\frac{1}{4}$ km each way in South London many flying-bomb hits took place; even today few people realize how many.

The following table shows the number of squares in which 0, 1, 2, 3, 4, 5 or more hits took place.

Number of hits	0	1	2	3	4	5 or more
Frequency	229	211	93	35	7	1

The average number of hits is 0·9323 per square. Take $e^{-0\cdot9323} = 0\cdot3936$ and, by a judicious use of a slide-rule, calculate the numbers obtained from a Poisson model for the process. Compare the calculated and observed frequencies.

4. The disintegration of a radioactive nucleus occurs rarely, but 7·5 seconds may be a very long time in atomic matters, and a great many atoms are usually involved. 3·870 is the mean number per period of 7·5 sec from the following table of observations by Rutherford, Chadwick and Ellis, made in 1920.

Number of disintegrations	0	1	2	3	4	5	6	7	8	9	10 or more
Frequency	57	203	383	525	532	408	273	139	45	27	16

Take $e^{-3\cdot870} = 0\cdot020858$ and calculate the frequencies to be expected from a Poisson model. Compare them with the observed frequencies.

5. Two machine shops, A and B, in a factory are known to be dangerous workplaces, with accidents occurring randomly at a rate of 2 accidents in 3 months in A and 3 accidents in 4 months in B. A safety officer is appointed on 1 January to shop B. In fact he has no effect on the accident rate, but what is the probability nevertheless that he will be able to report 'No accidents so far' on 1 March? As a result of his report to this effect he is transferred to shop A, where he still has no effect. What is the probability that there will be no accidents in shop A in the next month and 1 or more accidents in shop B?

6 T. (i) It is observed over a long period that the currant buns of a certain baker contain, on average, 2 currants per bun, which was fair enough; but after the bakery has changed hands you find that the first 3 buns you buy contain no currants at all. What are your views on whether or not the bakery's standards are slipping?

(ii) You complain to the baker and he protests that all is well. Your next bun, however, also has no currants. When you return to the baker he points out correctly that the probability of a bun having no currants under the old dispensation was 0·1353 or thereabouts, and that events of probability about ⅛ are nothing to be alarmed at. How do you continue?

(iii) I am critically scanning some tables of (purportedly) Random Numbers when I come across the sequence 65 04 65 65. Comment.

(iv) Compare your comments in the previous part with your answer to the baker in part (ii).

7. It is observed over a long period that in 1 week in 20, one or more flying saucers are reported over Loamshire. What is the most probable number of flying saucers to be reported over Loamshire in a year? What are the probabilities that there will be less than this number, exactly this number, more than this number in a year? What assumptions do you make about the reporting of flying saucers?

8. It is known that if the average number of bacteria per litre in suspension in a reservoir is $1,000k$, then the number of bacteria in a 1 cc specimen of the water has a Poisson probability function with parameter k (that is to say: no clumping appears to occur).

(i) Find the probability that of 5 independent specimens, each of size 1 cc, exactly 1 will contain no bacteria.

(ii) If the only record of a certain survey of the water is that of 5 specimens, of 1 cc each, exactly 1 contained no bacteria, find the value of k which makes this probability as large as possible, and show that the probability of the observed event is then about 0·41.

(iii) Show also that the average number of bacteria expected in a specimen of 5 cc is then about 8.

9. In a certain tropical town the climate does not vary all year and the chance that the tallest skyscraper is struck by lightning on any particular day is independent of the day in question. The elevator operator notices that in 1967 only 134 days passed without the building being struck by lightning. On how many days would you expect to find that the building had been struck more than twice; and how often do you estimate that it was struck during the year?

10. (i) In the table, each cell of which represents a 10 metre-square, we show the numbers of plants of a certain species found in the 10 metre-squares. If the plants were randomly scattered with the same mean density, what would be the probability of getting at least as many blank squares? Comment.

0	4	3	0	1
3	4	0	4	4

(ii) Compare with the table

0	4	3	0	0
3	4	0	4	5

11 T. (i) Discuss whether or not the processes producing the accompanying patterns of occupied squares may reasonably be taken to be random or not.

(ii) After analysing the two patterns, attempt to produce out of your head, and without the assistance of any randomizing devices, a random pattern of dots in an area 12 cm × 15 cm. Test your pattern as you did the two given patterns.

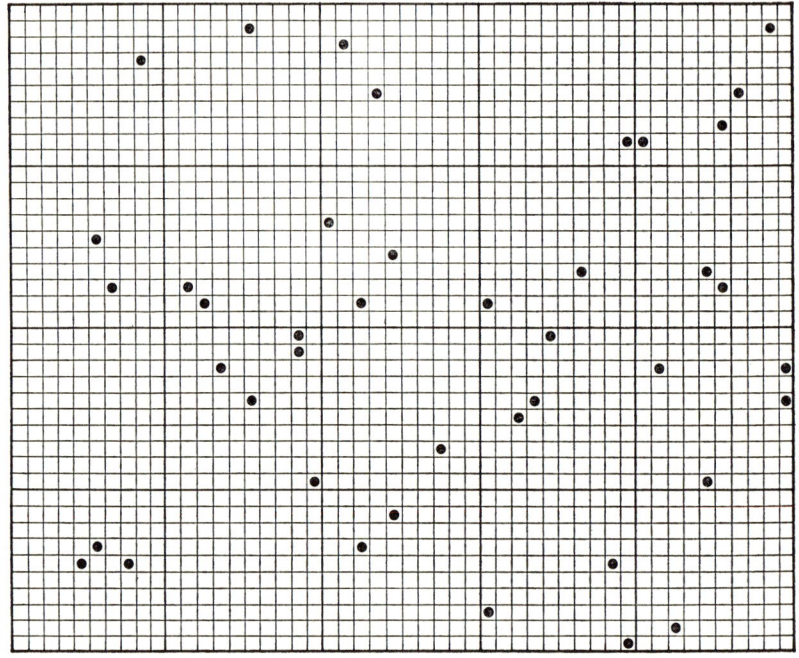

Fig. 8.2. Pattern (a).

12 T. In botanical survey work *frequency* of a species is defined as the probability of finding a plant of the species within a randomly selected square of a given area. It obviously depends on the area of the standard square and on the *density* of the members of the species, in terms of the mean number of plants of the species in a standard square.

The early workers thought that frequency and density were linearly related in a randomly occupied area; show that they are only approximately so.

13. In a certain Poisson process the mean number of flashes in a unit interval is, let us say, $\mu = 0 \cdot 693148$ giving $e^{-\mu} = 0 \cdot 50000$ and pr (there are no flashes in a unit interval) $= \frac{1}{2}$.

(i) Take the passage of a unit interval as a *binomial* trial with

$$\text{pr (success)} = \text{pr (} no \text{ flash)} = \tfrac{1}{2},$$

and find, using only the Binomial Theorem, the probability that in 5 consecutive intervals there will be no flashes.

(ii) Recalculate the probability of part (i), using the theory of Poisson processes, but calculate also, from this theory, the probability that in 5 intervals there will be exactly 1 flash.

(iii) Considering that if there is exactly 1 flash in 5 intervals then 4 of the intervals must contain none, why can you not obtain the second probability of part (ii) using the Binomial Theorem?

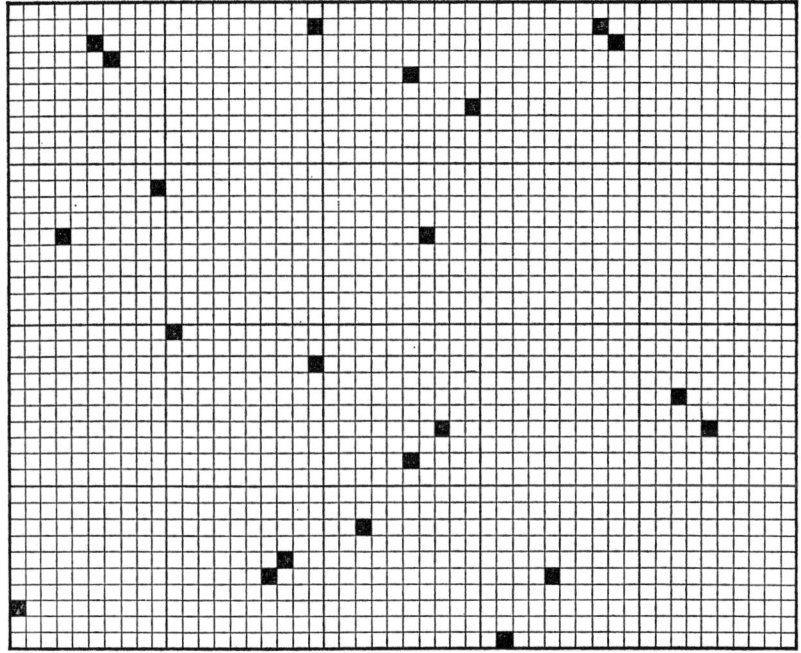

Fig. 8.3. Pattern (*b*).

14. Explain why in a Poisson process we have (with the notation of the chapter) that, for arbitrarily small ϵ and any $n > 0$,

$$\text{pr (no flashes in } 0 \leqslant T \leqslant n\epsilon) = \{1 - [\lambda\epsilon + 0(\epsilon^2)]\}^n.$$

Write $\epsilon = t/n$ and obtain $p(0, t)$ by a limiting process.

15. (i) If Poisson processes A, B produce random events with means μ_1, μ_2 per unit time respectively prove that if they both operate together and independently of each other they are equivalent to a Poisson process with mean $(\mu_1 + \mu_2)$.

(*Hint:* Consider $\sum\limits_{i=0}^{r} p_A(i, t).p_B(r-i, t)$ in the manner of Exercise 8A, Question 4.)

(ii) Consider the relation of this result to the simulation in the text of a Poisson process with mean of $\frac{1}{50}$ telephone call per minute by combining two processes each yielding $\frac{1}{100}$ telephone call per minute, namely looking for 33 and 77 in a table of random numbers read as pairs.

(iii) Discuss the feasibility of obtaining a Poisson process with mean 1 telephone call per minute by extending the search to the occurrences of any of the 100 digit pairs 00 to 99.

(iv) How *would* you simulate a Poisson process with mean representing 1 telephone call per minute?

*16. (i) Sketch on the same diagram the graphs of $e^{-\lambda t}, (\lambda t) e^{-\lambda t}, (\lambda t)^2 e^{-\lambda t}/2!, \ldots$ taking intervals of $1/\lambda$ along the t-axis.

(ii) Prove that each curve (except the first) has its maximum at the only point where it crosses its predecessor.

(iii) Interpret in probability terms the fact that, in each interval given by $k/\lambda \leqslant t \leqslant (k+1)/\lambda$ for $k = 0, 1, 2, \ldots$, the value of $(\lambda t)^k e^{-\lambda t}/k!$ exceeds that of $(\lambda t)^r e^{-\lambda t}/r!$ for every $r \neq k$.

(iv) Interpret in probability terms the set of values of $(\lambda t)^r e^{-\lambda t}/r!$ for two fixed values of t say $t = \mu/\lambda$ and $t = k\mu/\lambda$, and thus confirm that if a Poisson process has mean λ per unit interval of time it will generate Poisson probabilities with means in the ratio $1:k$ if the counts are to be made after times in the ratio $1:k$; that is to say, if the 'trials' are periods with lengths in the ratio $1:k$.

17F. With the notation of this chapter, $e^{-\lambda t}$ is the probability that in the interval $0 \leqslant T \leqslant t$ there is no occurrence of a given random event.

(i) Show that $\Phi(t) = 1 - e^{-\lambda t}$ is the probability that, in a sequence of these random events, at least the *first* event occurs in the interval $0 \leqslant T < t$.

(ii) Interpret $\Phi(t_0 + x) - \Phi(t_0)$, and express it in the form

$$\int_{t_0}^{t_0+x} \phi(t).dt.$$

[ϕ is called the *probability density function* for the *continuous* random variable X which is the length of interval from $t = t_0$ to the next event.]

(iii) If the first event of a sequence of random events occurs when $T = t_1$, use the results of parts (i) and (ii) to show that

$$\text{pr}(t_0 \leqslant t_1 < t_0 + x \mid t_0 \leqslant t_1) = \text{pr}(0 \leqslant t_1 < x).$$

Interpret this result in terms of the 'uniformity' of the Poisson process.

(iv) The uniformity of the Poisson process has led to ϕ being independent of t_0. The same function ϕ thus also determines the probabilities of various lengths of interval from *one event* to the next. Many people are at first surprised by this identity, (but see the corresponding remarks about the random variable U in Chapter 7, Section 7.2).

To examine the property:

(*a*) Draw the cumulative relative frequencies of values of X up to $X = 0, 20, 40, 60, \ldots, 100, 200$, for the data of Section 3.1.

(*b*) Take $\lambda = \frac{1}{50}$, as in the data of Section 3.1, and on the same diagram, plot $\Phi(x) = 1 - e^{-x/50}$ for $x = 0, 20, 40, 60, \ldots, 100, 200$.

18 T. (i) Starting from the relationships of Section 3.4, explain how we can show that $\mathrm{pr}\,(t_0 \leqslant t_1 < t_0 + h \mid t_0 \leqslant t_1) = \mathrm{pr}\,(0 \leqslant t_1 < h)$.

(ii) Show that writing $\mathrm{pr}\,(0 \leqslant t_1 < t) = \Phi(t)$ we have:

$$\frac{\Phi(t+h) - \Phi(t)}{h} = -\frac{\Phi(h)}{h}(\Phi(t) - 1) \quad \text{for all } t.$$

Hence show that, if, for each t, $\dfrac{\Phi(t+h) - \Phi(t)}{h}$ tends to a limit as $h \to 0$, we have

$$\Phi'(t) = -\lambda(\Phi(t) - 1), \quad \text{where} \quad \lambda = \lim_{h \to 0}\left(\frac{\Phi(h)}{h}\right).$$

Solve this D.E. for $\Phi(t)$ with the initial condition $\Phi(0) = 0$.

19. Comment on the following quotation from a newspaper article on precautions against earthquakes and relate your comments to possible models and the results of Question 17.

'In fact, the train of enquiries which I started as a result of this incident shows that a great deal can be done—if it is done in time. Japanese scientists have estimated that an earthquake of the same force that hit Tokyo in 1923, causing the loss of 60,000 lives, could be expected every 70 years or so.

On that calculation the next such earthquake might occur in about 1990, but the scientists do not claim any such precision for their forecast, and the earthquake could come long after that date—or much sooner. Whatever the merits of the 70-year cycle theory, said the Tokyo *Yomiuri* last week, in an article on the forty-fourth anniversary of the 1923 earthquake, there was no mistaking the fact that the earth's crust in this region was extremely unstable.'

9

RANDOM VARIABLE—DISCRETE

1. EXPECTED VALUE

1.1 Introduction. One immediately obvious difference between the two sequences tested in Chapter 7, Tables 7.1 and 7.2, was the number of runs, *whether of A or of B*, in the Markov chain that were long compared with those in the Bernouilli chain. (This is not necessarily a feature of Markov chains, we merely point it out in this case.) We could investigate this difference by considering the number of *switches* that occurred, because the longer the run the fewer the switches, and we already have some values of statistics of the switches collected in Table 7.6. We might ask what is the average number of switches in an experiment of five Bernouilli trials with pr (success) = $\frac{1}{3}$, as there.

In Table 7.6, reproduced in Table 9.1, we have a table of values of S, the number of switches in such an experiment, and the frequencies of these values in thirty repetitions of the experiment.

We have written fr $(S = s) = f(s)$ and $f(s)/N = q(s)$, the relative frequency with which S takes the value s. We have also added a column for $p(s) = $ pr $(S = s)$, the evaluation of this column being deferred until Section 1.2.

Table 9.1

s	$f(s)$	$q(s)$	$p(s)$
0	4	0·13	0·14
1	8	0·27	0·25
2	10	0·33	0·37
3	8	0·27	0·20
4	0	0·00	0·05
	30	1·00	1·01

From this table we can investigate the mean number of switches in 30 actual repetitions of the experiment. We have

$$\bar{s} = \frac{(0 \times 4) + (1 \times 8) + (2 \times 10) + (3 \times 8) + (4 \times 0)}{30}$$

$$= \left(0 \times \frac{4}{30}\right) + \left(1 \times \frac{8}{30}\right) + \left(2 \times \frac{10}{30}\right) + \left(3 \times \frac{8}{30}\right) + \left(4 \times \frac{0}{30}\right)$$

$$= \sum_{\text{all } s} s.q(s).$$

248

It would be interesting to consider the value of $\bar{s}$ that would occur in an ideal situation where the probabilities and relative frequencies exactly matched, and this suggests the consideration of $\sum\limits_{\text{all }s} s.\text{pr}\,(S = s)$, which resembles the previous sum except that $\text{pr}\,(S = s)$ has replaced $q(s)$. The suggested quantity is one which will nearly match the mean of the sample of values of S, that is $\bar{s}$.

We now break off to show the calculations for the values of $p(s)$.

1.2　Calculation of probabilities for numbers of switches. Table 7.7 is here adjusted so that beside each possible arrangement of zeros and ones is written the value of S and the probability of the arrangement:

Outcome	Value of S	Prob.	Outcome	Value of S	Prob.	Outcome	Value of S	Prob.	Outcome	Value of S	Prob.
00000	0	b^5	01000	2	ab^4	10000	1	ab^4	11000	1	a^2b^3
00001	1	ab^4	01001	3	a^2b^3	10001	2	a^2b^3	11001	2	a^3b^2
00010	2	ab^4	01010	4	a^2b^3	10010	3	a^2b^3	11010	3	a^3b^2
00011	1	a^2b^3	01011	3	a^3b^2	10011	2	a^3b^2	11011	2	a^4b
00100	2	ab^4	01100	2	a^2b^3	10100	3	a^2b^3	11100	1	a^3b^2
00101	3	a^2b^3	01101	3	a^3b^2	10101	4	a^3b^2	11101	2	a^4b
00110	2	a^2b^3	01110	2	a^3b^2	10110	3	a^3b^2	11110	1	a^4b
00111	1	a^3b^2	01111	1	a^4b	10111	2	a^4b	11111	0	a^5

It is worth noting from this table that, in contrast with the table for values of R (Table 7.7), a given value of S is not always associated with a sequence having the same probability. The probability function is clearly *not* binomial.

The following table shows the general values of the required probabilities and their numerical values when $a = \frac{1}{3}$ and $b = \frac{2}{3}$, as here.

s	$\text{pr}\,(S = s)$		Numerical value
0	$b^5 + a^5 =$	$b^5 + a^5$	$33/3^5 = 0{\cdot}136$
1	$2(b^4a + b^3a^2 + b^2a^3 + ba^4) =$	$2ba(b + a)\,(b^2 + a^2)$	$60/3^5 = 0{\cdot}247$
2	$3(b^4a + b^3a^2 + b^2a^3 + ba^4) =$	$3ba(b + a)\,(b^2 + a^2)$	$90/3^5 = 0{\cdot}370$
3	$4(b^3a^2 + b^2a^3) =$	$4b^2a^2(b + a)$	$48/3^5 = 0{\cdot}198$
4	$b^3a^2 + b^2a^3 =$	$b^2a^2(b + a)$	$12/3^5 = 0{\cdot}050$
			$\overline{1{\cdot}001}$

1.3　Definition of expected value. The completion of the calculation for $\bar{s}$ gives

$$\bar{s} = (8 + 20 + 24)/30 = 1{\cdot}73$$

while the calculation for $\Sigma s.\text{pr}\,(S = s) = (60 + 180 + 144 + 48)/3^5 = 1{\cdot}78$. This second number is the theoretical quantity of interest, while $\bar{s}$ is reasonably close to it. In a certain sense $\Sigma s . \mathrm{pr}\,(S = s)$ is the 'expected' number of switches, that is the 'expected value' of S in an experiment.

In general, if X is a random variable then the value of the expression $\sum\limits_{\text{all } x} x . \mathrm{pr}\,(X = x)$ is called the *expected value of X* or the *expectation of X*. It is clearly a function of the whole set of all the different possible values of X and their associated probabilities, that is of the parent population of values of X. We will write such a parent population as $[x]$ and use the symbol E for the function.

We ought very fully to write:

$$E_X([x]) = \sum_{\text{all } x} x . \mathrm{pr}\,(X = x)$$

but we will usually abbreviate the left-hand side to $E[x]$, when the random variable, X, to which the values x refer is obvious; or to $E(X)$ if we prefer. As in other circumstances we will often omit the reference to the range of summation. The reader must therefore be prepared to meet:

$$E[x] = \Sigma x . \mathrm{pr}\,(X = x) \quad \text{or} \quad E(X) = \Sigma x . \mathrm{pr}\,(X = x).$$

Example 1. Find the expected value of the number of successes in an experiment of 5 Bernouilli trials such as those of Chapter 7, Section 5 and compare it with the mean of the given sample of 30 values of R, from Table 7.5.

Solution. Let R denote the number of successes;

$$q(r) = \text{relative frequency of } R = r,$$
$$p(r) = \mathrm{pr}\,(R = r),$$

the values of which we have from Table 7.5 and Chapter 7, Example 5, respectively.

r	Approximate $p(r)$	$rp(r)$	$q(r)$	$rq(r)$
0	0·13	0·00	0·13	0·00
1	0·33	0·33	0·26	0·26
2	0·33	0·66	0·37	0·74
3	0·16	0·48	0·20	0·60
4	0·04	0·16	0·03	0·12
5	0·00	0·00	0·00	0·00
		1·63		1·72

Thus $\qquad\qquad E[r] \simeq 1 \cdot 63 \quad \text{and} \quad \bar{r} = 1 \cdot 72.$

Comment on the solution. If we think about the nature of 'probability' we will remember that the meaning to be attached to $\mathrm{pr}\,(A) = \frac{1}{3}$ is that in a long run of N trials there will be nearly $N/3$ successes and that in 5 trials there may be reasonably nearly $\frac{5}{3}$ successes; sometimes there will be more, sometimes less, but if a run of 30 experiments, each of 5 trials, were carried out then the mean value of the number of successes in each experiment might be nearly $\frac{5}{3}$; the theoretically ideal value of this mean would be $\frac{5}{3}$. The question arises whether the ideal approached from this basic stand-point agrees with $E[r]$.

Our calculations give approximately $1\cdot63$ for $E[r]$ but an accurate evaluation, avoiding the approximations in decimalizing, gives

$$E[r] = \left(0 \times \frac{2^5}{3^5}\right) + \left(1 \times 5 \times \frac{2^4}{3^5}\right) + \left(2 \times 10 \times \frac{2^3}{3^5}\right) + \left(3 \times 10 \times \frac{2^2}{3^5}\right)$$
$$+ \left(4 \times 5 \times \frac{2}{3^5}\right) + \left(5 \times \frac{1}{3^5}\right)$$
$$= (0 + 80 + 160 + 120 + 40 + 5)/3^5$$
$$= 405/3^5$$
$$= 5/3.$$

This confirms our view that $E[r]$ is a useful quantity.

1.4 Warning. We must beware that it is the *mean* value of X and not necessarily the *modal* value that approaches $E[x]$, so that when used technically the word 'expected' does not carry all its everyday associations.

Furthermore, even the modal value might not be 'expected' by ordinary usage. For it is simple to show that in a throw of 10,000 coins the expected value of the number of heads (speaking technically) is 5,000. This means that in many experiments each consisting of 10,000 throws of a coin the mean value of the number of heads will be nearly 5,000; and yet we should be surprised, and rightly, if any particular one of the experiments specified beforehand resulted in 5,000 heads, and we might even suspect cheating. In fact the probability of getting *exactly* 5,000 heads in a throw of 10,000 coins is

$$\frac{10,000!}{(5,000!)^2} \times \frac{1}{2^{10,000}}$$

and this is approximately $\dfrac{1}{\sqrt{(5,000.\pi)}}$ as the reader can easily verify by using the approximation, quoted in Chapter 7, Section 10.2, that, for large N, $N! \sim \left(\dfrac{N}{e}\right)^N \sqrt{(2\pi N)}$. So the required probability is about $\frac{1}{125}$, that is roughly the probability of winning a toss seven times in a row.

This point is emphasized by our next example.

Example 2. Calculate the mean value of U in the sample of 52 values of U in Table 7.3 and compare it with $E(U)$.

Solution.

Write relative frequency of $U = u$ as $q(u)$. We have:

u	$q(u)$	$uq(u)$
0	0·35	0
1	0·21	0·21
2	0·13	0·26
3	0·10	0·30
4	0·11	0·44
5	0·04	0·20
6	0·04	0·24
7	0·02	0·14
		1·79

Thus $\bar{u} = 1\cdot79$.

Because of the non-zero probabilities attached to even very large values of U the calculation of $E(U)$ needs a more subtle approach than direct summation. We have:

$$E[u] = \Sigma u \cdot p(u)$$
$$= 0(\tfrac{1}{3}) + 1(\tfrac{2}{3})\,(\tfrac{1}{3}) + 2(\tfrac{2}{3})^2 \cdot \tfrac{1}{3} + 3(\tfrac{2}{3})^3\,\tfrac{1}{3} + \ldots + r(\tfrac{2}{3})^r\,\tfrac{1}{3} + \ldots$$
$$= (\tfrac{2}{3})\,(\tfrac{1}{3})\,\{1 + 2b + 3b^2 + \ldots + rb^{r-1}\ldots\} \quad \text{where } b = \tfrac{2}{3}$$
$$= (\tfrac{2}{3})\,(\tfrac{1}{3})\,\{1 - b\}^{-2} \quad \text{by the Binomial Theorem}$$
$$= \tfrac{2}{3} \cdot \tfrac{1}{3} \cdot (\tfrac{1}{3})^{-2}$$
$$= 2.$$

The values of $E[u]$ and $\bar{u}$ are in reasonable agreement, but the most probable single value of U and the modal value of U in the actual sample are both 0. Although these values 0 may suggest that in ordinary usage 0 is the *expected* value of U, that is *not* the correct technical usage.

1.5   Origin of the term. A game of chance may be thought of as merely a method of defining a random variable by the score, and historically this was where the notion of expectation arose. The amount you could 'expect' to win in a long evening's play was the quantity of interest. This amount was usually obtained by addition of scores or winnings, and addition of scores is directly linked to mean score; thus expected value got attached to mean and not modal score.

Exercise 9 A

1R. Six dice are rolled. If a five or six turns up, you are paid 1 penny. If no five or six turns up, you pay 10 pennies. What is your expected gain?

2R. (i) In a certain game two dice are thrown and I receive a prize of $|7-x|$ units, when x is the score. What is the expected value of my gain?

What constant sum must I pay as an entry fee before each turn if the expected profit or overall gain (that is: prize minus fee) is to be zero?

[A game in which the expected gain is negative is said to be *unfavourable* and one in which the expected gain is positive is said to be *favourable*.]

(ii) Carry out 5 runs, each of 10 turns, and find the mean profit per game for each run.

3 R. [In classical probability theory the case when the expected gain was zero was called '*fair*'.]

(i) A player throws three coins and is paid 1 unit for the first head, then 2 for the second and finally 3 for the third (so that his total payment for two heads is $1 + 2$). What would be a forfeit for 'no heads' if the game is to be 'fair' without an entrance fee?

(ii) Carry out 5 runs, each of 10 plays, of the game and find the mean score per game in *each* of your runs. (Remember: this game can be easily simulated with a table of random numbers.)

4. In a game (similar to one played at St Petersburg) between one player and the Bank the player throws one coin. If a head first appears on the nth throw the player receives 2^{n-1} units of prize. If no head appears after ten throws, the game is declared a draw and the player can retire or start again.

(i) What is the expected winning in a game?

(ii) To make a 'fair' game what should the player pay the Bank as an entry fee, if the rules for a draw are that the entry fee is *refunded*?

5. It is said that at St Petersburg there was no limit to the number of throws for which the game described in Question 4 continued.

(i) What should the player have paid as an entry fee?

(ii) If the Bank will pay only 10,000 units to any one player because of shortage of funds, what is the probability that a player will 'break the Bank'?

(iii) If the game is played 256 times per night, what is the probability that the Bank will not be broken during the course of the night? (Use the Binomial theorem to get an approximate answer.)

(iv) What is the probability that the Bank *will* be broken in a season of 64 nights? (Use the exponential limit.)

6 T. A millionaire arranges in his will for a lottery to take place to dispose of his fortune.

100,000 tickets each costing £20 are for sale. No-one may buy more than one; syndicates holding more than one ticket are forbidden. All the ticket numbers whether corresponding to sold tickets or not are placed in the draw. One number is drawn and the holder of that ticket receives £2 million (so that his gain is £2 million minus £20.)

The proceeds of sales of tickets go to charity; and if no-one holds the ticket for the drawn number, then the fortune goes to charity too.

(i) Calculate the expected value of the gain for a holder of a ticket.

(ii) Is the game (classically) 'fair'?

(iii) Does the game seem one upon which one might spend £20 that one had saved up?

(iv) Answer parts (i) (ii) (iii) again if the millionaire was giving away £20 million like this.

[This question shows a possible inadequacy of the classical concept of a 'fair' game.]

*7. In the problem of three wheels in Exercise 3C, Question 9 each wheel had its circumference divided into 32 parts.

Wheel A had   7 parts scoring 6 and 25 parts scoring 3
    „ B „ 16 „   „ 5 „ 16 „   „   2
    „ C „ 25 „   „ 4 „ 7 „   „   1

Calculate the expected score for each wheel. Does the expected score provide a suitable criterion for choosing a wheel if the game is going to consist of:
    (i) a single spin of his wheel by each player,
or (ii) repeated spins with victories totalled,
or (iii) repeated spins with scores totalled?

8. In the game of crown and anchor the player has three dice each marked Spades, Hearts, Diamonds, Clubs, Crown, Anchor. The player nominates one of these, say Anchor, and places a stake. The three dice are then thrown.

    If 0 Anchors occur player gives stake to Bank.
    If 1   „   „   player keeps stake and gets 1 × stake from bank.
    If 2   „   „   „   „   „   „   „ 2 × „   „   „
    If 3   „   „   „   „   „   „   „ 3 × „   „   „

Find the expected value of what he gets from the Bank.

9. In a certain hand of Bridge I know that my opponents hold 6 trumps between them at the start (while each player holds 13 cards). I also know which trumps they are, and which non-trumps they hold.
    (i) Prove that the probability that one or other of them holds 4 trumps is

$$2\binom{13}{2}\binom{13}{4}\bigg/\binom{26}{6}.$$

    (ii) What is the expected value of the larger number of trumps held by either of them, that is the expected number of rounds that will have to be led to clear both their hands of trumps?

10. In the British Government's Premium Bond prospectuses for Series A and Series B it is implied that prizes of value £x are issued monthly with the following frequencies for each £240,000,000 worth of Bonds issued.

i	x_i	$f_B(x_i)$	$f_A(x_i)$
1	5,000	9	0
2	1,000	90	80
3	500	90	160
4	250	180	320
5	100	270	800
6	50	1,800	1,600
7	25	22,320	16,000

Assume that for lesser amounts of Bonds issued the scales are strictly proportional (this is not quite accurate) and that every Bond issued is immediately eligible for a prize (nor is this). If a man holds one Bond, evaluate separately for Series A and Series B:
    (i) the probability of winning a prize of £250.

(ii) the probability of winning a prize at all.

(iii) the expected value of a prize.

Investigate the position of a man who holds the maximum of 1250 bonds. What is his expected income in a year?

11. One bag contains 2 Red and 1 Green discs, a second bag contains 4 Red and 2 Green discs. One disc is drawn at random from each bag. Each drawn disc is then placed in the other bag. Find the expected number of each colour of disc in the first bag. Find also the expected number of each colour in the first bag if the disc from the second bag was not drawn until the disc from the first bag had had been placed in the second bag. (See a further discussion in Exercise 9B, Question 8.)

12. A circular plane region of radius R ($= nd$) is being searched for an object (of negligible size) which is 'equally likely to be anywhere in the region' (so that by definition, the probability that it is in any particular sub-region is proportional to the area of the sub-region). The following regions (given in polar co-ordinates) are searched in turn for $k = 1, 2, ..., n$.

$$\begin{cases} (k-1)\,d \leqslant r \leqslant kd, \\ 0 \leqslant \theta \leqslant 2\pi. \end{cases}$$

Counting the region in which the object is eventually found, what is the expected number of regions to be searched before the object is found?

13 T. [This problem shows the application of the results of this chapter to the design of an experiment.]

If a large number, N, of people are going to be tested to detect the presence of a certain factor in their blood then various courses of action are possible, of which two are given below:

I: Each person's blood specimen is separately analysed. This requires N analyses.

II: The blood specimens from k (> 1) people can be mixed and then analysed together. If this analysis shows *absence* of factor then *no further* analysis is done for those k people. If the analysis shows *presence* of factor then k *further* analyses are used to test the k people separately in order to obtain an individual reaction from each affected person. The next group of k people is then treated in the same way, until all the people are tested.

(i) Simulate II as follows: Read random digits in groups of four; look for the presence of 0; record the number of analyses necessary to detect the zeros in each group of digits. For example:

5416 3940 9857 0205 6515 7323 5151 7506 3813 5168

leads to the numbers of analyses as follows:

    1   5   1   5   1   1   1   5   1   1

The effect of doubling the size of the groups can be quickly read off this list of numbers of analyses by noting that a pair 1, 1 leads to only 1 analysis but pairs 5, 1 and 1, 5 and 5, 5 lead to nine tests being required (the pairs are, of course, not to overlap). Thus for these figures and a group size of 8 we would need 9, 9, 1, 9, 1 analyses.

The total numbers of analyses are thus

By method I; number = 40

By method II, with $k = 4$; number $= 22$ instead of 40, a saving of 45 per cent

By method II, with $k = 8$; number $= 29$ instead of 40, a saving of 27 per cent.

(ii) Consider a situation where N/k is large, so that the remainder when people are divided in groups of size k is of negligible size compared with N. Suppose that the probability that the factor is present in any one person is θ, and that people's reactions to the test are independent of each other.

Show that, neglecting the residue when groups of size k (> 1) are made up, the expected number of analyses required is given by $q = N\left(1 + \dfrac{1}{k} - (1-\theta)^k\right)$, when method II is adopted. Compare the values obtained in your simulation.

(iii) Show that, *if θ is small enough*, then q is minimized by a value of k somewhere near $\theta^{-\frac{1}{2}}$, and hence find the expected percentage saving in number of analyses when $\theta = 0.01$ and method II rather than method I is adopted.

(iv) What is the best value of k if $\theta = 0.2$ and what is then the percentage saving?

(v) Discuss the case $\theta = 0.3$, and obtain as low an upper bound for θ as you can if a saving is to result from method II. Explain as to a layman why such an upper bound is to be expected.

[This problem was much studied in World War II. It is of course possible in only three weighings to detect a known-to-be-single fake coin mixed among 12, without knowing before whether the fake is over- or under-weight. It has been suggested that this 12-coin problem was planted by German intelligence agents to distract Allied mathematicians during the war!]

14 T. A certain travel agent making train reservations between two cities (one north of the other) finds that the following are the only words needed and that their probabilities are estimated as given.

one	0.20	sleeper	0.06
tourist	0.18	luxury	0.06
seat	0.18	two	0.04
northbound	0.12	facing	0.03
southbound	0.12	back	0.01

The diagram, Fig. 9.1, shows the code system devised by Huffman, as it would operate in this case.

The explanation is as follows: The initial stage of the diagram is the list of words and their probabilities.

Two lines are drawn at a time from left to right and the probability written at their meet is the sum of the probabilities from which they started. The probabilities from which they started are then of no further relevance. The two lines next to be drawn are always decided upon as linking the two smallest remaining probabilities of relevance.

Thus after 0.01 and 0.03 have been joined and 0.04 written at their meet, we next join *this* 0.04 to that beside the word 'two' to get 0.08. The two numbers 0.06 are next joined, and so on. The coding is done by a tree diagram system reading from the number 1.00 towards the required word and writing 1 when an upper branch is taken and 0 when a lower one.

In this case the code runs

one	11	sleeper	1011
tourist	011	luxury	1010
seat	010	two	1001
northbound	001	facing	10001
southbound	000	back	10000

Thus to transmit: 'one tourist northbound sleeper', we send 110110011011.

(i) Verify that any properly coded message can only be decoded in one way; as for instance:

1001101010001010001   (sent without gaps)

or   0101 0101 0001 0011 001   (sent with *arbitrary* gaps for easier copying)

(ii) Calculate the expected number of symbols per word of message.

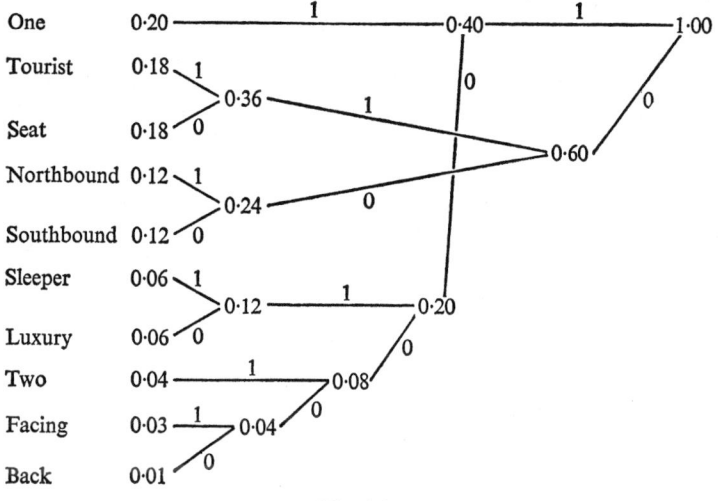

Fig. 9.1

(iii) It has been shown that for transmitting messages with words of different probabilities and retaining the property of part (i) the Huffman code is in a certain sense the most efficient. The number of binary symbols per word is always within 1 unit of the *entropy of the source*, which is defined as $\sum_i -p_i \log_2(p_i)$,

where p_i is the probability of the ith word in the vocabulary and the summation is over all words of the vocabulary. The entropy provides a theoretical limit however complicated the coding system.

Calculate the entropy of this source and verify the closeness of the Huffman expected length.

[For information on this and other matters in Communication Theory see J. R. Pierce *Symbols, Signals and Noise* (Hutchinson, 1962).]

257

2. SAMPLES AND PARENT POPULATIONS

2.1 General. In the theory of finite populations outlined in Chapter 4 we saw that a population of values of a variable gave rise uniquely to a frequency function and that, conversely, given any suitable function (the values had to be non-negative integers) we could call it a frequency function and construct a population to match it. We shall see that we now have a situation analogous to this converse. In the study of a random variable we start with a probability function defined. The domain of the function is a set of numbers containing the values of the random variable and suitable functions are ones with non-negative images $p(x_i)$ satisfying $\Sigma p(x_i) = 1$, (the summation extending over all possible values of the random variable). It seems reasonable to use the word population to refer also to the compound notion of the collection of possible values, x_i, of the random variable together with the probability function. To make a distinction, and one is often needed, we will refer to a population of values obtained from an experiment or survey as a *sample*, and will refer to the idealized or hypothetical population defined by the probability function as the *parent population* (from which the sample was drawn).

Many theoretical books start from the idea of a random variable and regard the observations in an experiment as being a collection of values actually taken by this variable; we have started from the observations and their frequencies, and built up the idea of a random variable and its probability function as a model for the observations.

2.2 Notation. We have up to now rather reserved the use of the letter m for a mean and we now specialize its use slightly further and introduce also the symbol μ.

Suppose we are given a certain experiment which defines a random variable X, then $\bar{x}$ or m is a statistic calculated from a particular sample of values of X, each value occurring with its own frequency in that particular sample; different samples of values of X will define in themselves different frequencies of values of X and will have different means.

Underlying all these samples, however, is the parent population of values of X from which they are all hypothetically drawn; this parent population has a probability function which, we believe, to a certain extent expresses the general pattern of the frequencies in the samples, and hence of the various values obtained for m. If our belief is not justified then the construction of the theory of probability becomes only a pretty mathematical toy.

If, for this parent population, we calculate the values of the various statistics we have been using (mean, median, IQR, variance and so on),

258

but using probabilities in place of relative frequencies, then we get quantities analogous to them, to which, for convenience, we allot the Greek letters (so far as they exist) corresponding to the Roman letters used for the statistics of the sample.

In this way we write:

$$\bar{x} = \Sigma x_i . \left(\frac{\text{fr } (X = x_i)}{N}\right) = m \quad \text{and} \quad E[x] = \Sigma x . \text{pr } (X = x) = \mu.$$

2.3   Parameters. It often happens that the probability function defining the parent population of values of a random variable can be represented as a member of some family of functions, and that one or two of these idealized statistics are enough to identify which member of the family we have. For example, suppose that we are collecting numbered coupons from packets of tea hoping to build up a set and win a prize, and that the coupon numbers are equally likely to be any member of a set of consecutive integers, then we can invent a parent population where all the members are integers having equal probabilities, so that they are all as likely as each other in a sample. In that case, if we specify the extremes (the least and greatest integers possible), then we can completely define the probability function giving the parent population. If the least and greatest integers used on the coupons are α and β then there are $(\beta - \alpha + 1)$ integers involved and the probability function is

$$\text{pr } (X = x) = \begin{cases} 1/(\beta - \alpha + 1) & \alpha \leqslant x \leqslant \beta, \ x \text{ integral,} \\ 0 & \text{all other } x. \end{cases}$$

Alternatively we could specify γ, the midpoint of the interval of possible values of X, and λ, the length of the range of values of X. In that case there are $\lambda + 1$ integers involved and the probability function is

$$\text{pr } (X = x) = \begin{cases} 1/(\lambda + 1) & \gamma - \lambda/2 \leqslant x \leqslant \gamma + \lambda/2, \ x \text{ integral,} \\ 0 & \text{all other } x. \end{cases}$$

The numbers α, β, γ, λ are called parameters of the parent population since a knowledge of a suitable selection of them completely determines the probability function and thus the parent population.

In this case we can obviously vary only two of them independently, so that we need just two of them to specify the population and our population can be represented by a point in a two-dimensional parameter space. The binomial probability function $\text{pr } (R = r) = \binom{N}{r} (1 - \theta)^{N-r} \theta^r$ needs two parameters (for instance N, θ); and the Poisson probability function $\text{pr } (R = r) = e^{-\mu} . \mu^r / r!$ needs only one parameter (for instance μ).

It is clear also that for any sample we get in our coupon-collecting project we could state the extreme occurrences in it, the centre of the interval between the extremes and the range. To help us to keep in mind the relationships between all these quantities (*statistics* of the sample) we could denote them by the Roman letters *a*, *b*, *g*, *l*, to correspond to the letters for the *parameters* of the parent population, to which we have already allotted the Greek letters α, β, γ, λ.

We notice that to:

$$\alpha + \beta = 2\gamma \quad \text{and} \quad \beta - \alpha = \lambda,$$

there would correspond:

$$a + b = 2g \quad \text{and} \quad b - a = l,$$

but we also notice that such simple correspondences would not always occur. For instance defining $m = \bar{x}$ and $\mu = E[x]$, we would have $\mu = \gamma$ but not necessarily $m = g$.

2.4 Estimation. One other word of warning seems necessary here. In the problem of 'estimation' in statistics, we are trying to devise defensible procedures for guessing suitable values to give to parameters of parent populations, on the evidence of the values of various statistics of a sample. It is *not* always most satisfactory to allot to a parameter the value calculated for the corresponding statistic of a sample. A consideration of this problem must wait until Chapters 11 and 12.

2.5 Variance. Just as in Chapter 4 we defined the variance of what we now call a sample of values, x_i, of a random variable X, by:

$$\text{var } [x_i] = \sum_i (x_i - m)^2 \cdot \frac{\text{fr } (X = x_i)}{N}, \text{ the summation being over the}$$
$$\text{observed values of } X;$$

so we now define, for a random variable X with values x:

$$\text{var } [x] = E[(x - \mu)^2] = \Sigma(x - \mu)^2 \cdot \text{pr } (X = x), \text{ the summation being}$$
$$\text{over all values of } X.$$

We call these quantities S^2 and σ^2, respectively. It is a curious omission of the standard notations that, whereas we distinguish between $\bar{x}$ (calculated from a sample) and $E[x]$ (calculated from a parent population), we refer to S^2 and σ^2 by the same operational symbol: var [ ]. On the whole, however, we will use this last symbol more for the σ^2 usage than the S^2 one; confusion is less likely than it would seem at first.

2.6 Function of a random variable. It is helpful now to have the notion of a function of a random variable. For instance, if when a random variable X takes the value x another variable Y takes the value x^2, then Y is a random variable; and it is convenient to write $Y = X^2$ to show the relationship between the variables. In general, if, whenever X takes the value x, a variable Y takes the value $g(x)$, then we write $Y = g(X)$. The relationship between the probability functions of X and $g(X)$ and between their parameters is one of great interest and importance.

2.7 Expected value in general. We define the expected value of a function of a random variable $g(X)$ by

$$E(g(X)) \equiv E[g(x)] \equiv \Sigma g(x) . \text{pr} (X = x)$$

the summation being over all values x. Our definition corresponds to that of the mean of a population of observed values $g(x)$. There are the same minor details to clear up in noticing that, because g may not satisfy $g(x_i) = g(x_2) \Rightarrow x_1 = x_2$, we may not necessarily have

$$\text{pr} (X = x_1) = \text{pr} (g(x) = g(x_i)).$$

The definition, however, leads to the quantity we want, in the same way as $\overline{g(x)}$ was shown to do in Chapter 4, Section 4.2.

We see that

$$\text{var} (X) = E((X - \mu)^2) \quad \text{and} \quad \text{var} [x] = E[(x - \mu)^2]$$

are cases of general expected values where we have

$$g: x \to (x - \mu)^2.$$

Exercise 9 B

1 T. (i) If c is a constant, prove $E[cx] = cE[x]$.
(ii) If a is a constant, prove $E[x + a] = E[x] + a$.
(iii) What is the most general function, g, that you can find to satisfy $E[g(x)] = g(E[x])$?
(iv) Prove $E[g(x) + f(x)] = E[g(x)] + E[f(x)]$
but that in general $E[g(x) . f(x)] \neq E[g(x)] . E[f(x)]$.

2 T. If $E[x] = \mu$, $E[(x - \mu)^2] = \sigma^2$ and k is a constant. Prove
(i) $E[x - \mu] = 0$.
(ii) $\sigma^2 = E[x^2] - (E[x])^2$.
(iii) $\sigma^2 = E[x(x - 1)] - \mu(\mu - 1)$.
(iv) $E[(x - k)^2] = \sigma^2 + (\mu - k)^2$.
Show how to obtain each of these results by special choice of a, b, c in $E[ax^2 + bx + c] = a\sigma^2 + (a\mu^2 + b\mu + c)$, and prove this result from the results of Question 1 together with part (ii) of the present question.

3 T. (i) Prove $\text{var} [cx] = c^2 \text{var} [x]$; $\text{var} [x + a] = \text{var} [x]$.
(ii) If $E[x] = \mu$ and $\text{var} [x] = \sigma^2$ and t is defined by $t = (x - \mu)/\sigma$, prove that $E[t] = 0$ and $\text{var} [t] = 1$.
[A random variable T taking the values t related in this way to the values and parameters of a random variable X is said to be the *standardized random variable* corresponding to X. Compare the remarks in Chapter 6, Section 1.4.]

4 R. With the rules of the game in Exercise 9 A, Question 1,
    (i) find the variance of the player's *winnings* ignoring his entrance fee;
    (ii) if the entrance fee is 2 units, find the variance of his *profits*;
    (iii) if the entrance fee is such as makes the game fair, find the variance of his *profits*.

5 R. In Exercise 9 A, Question 3, find the variance of the player's winnings if the forfeit for scoring no sixes was 18 units. Discuss briefly the relevance of this to the observed mean scores in each of the 5 runs played by your class.

6. (i) If the bank in a gambling den wants a *steady* income from a game, how does it arrange the parameters of the possibility space of the scores?
    (ii) If the bank merely wants a long term profit while making the game superficially attractive to the player, what does it do to the parameters?

7. The discrete variate R has possible values $0, 1, 2, ..., 10$, with probabilities given by

$$\text{pr}\,(R = r) = \begin{cases} k(r+1) & r = 0, 1, ..., 5, \\ k(11-r) & r = 6, 7, ..., 10, \end{cases}$$

where k is a constant.
    (i) Determine the expectation, μ, and standard deviation, σ, of R.
    (ii) Find the probability that, given two independent random observations of values of R, they both lie in the range $\mu \pm \sigma$.

8. From Exercise 9 A, Question 11 find the variance of the number of red discs in the first bag under each method of transfer given there.

9. The rth moment of a random variable X is defined by $\mu'_r = E[x^r]$ and the rth moment about the mean, μ, by

$$\mu_r = E[(x-\mu)^r].$$

Thus
$$\mu_2 = \sigma^2, \ \mu'_1 = \mu.$$

Prove (i) $\mu_0 = \mu'_0 = 1$.
    (ii) $\mu_1 = 0; \ \mu'_1 = \mu$.
    (iii) $\mu'_2 = \sigma^2 + \mu^2$.
    (iv) $\mu_3 = \mu'_3 - 3\mu'_2\mu'_1 + 2(\mu'_1)^3$.
    (v) $\mu_4 = \mu'_4 - 4\mu'_3\mu'_1 + 6\mu'_2(\mu'_1)^2 - 3(\mu'_1)^4$.
[The *skewness* is defined by μ_3/σ^3 and the *kurtosis* by μ_4/σ^4.]

10. Using, as convenient, either the definitions themselves or the results in Question 9, find the skewness and kurtosis of the random variables defined by the number of successes in a sequence of N Bernouilli trials with $\theta = \frac{1}{3}$, for $N = 2$ and $N = 3$.

*11 T. In a certain game devised by Feller we have a situation resembling the St Petersburg game in that for each turn the player has several 'throws'. He either wins at a throw or continues to another throw. In this game, however, the probabilities are $\log 2$ of winning at the first throw, and $1/2^k k(k+1)$ of winning at the $(k+1)$th throw for $k \geqslant 1$. The prize for winning at the first throw is zero and for winning at the $(k+1)$th throw with $k \geqslant 1$ is 2^k units.
    (i) Verify that the probability that the player eventually wins is unity.

    (*Hint:* Consider $2\int_0^{\frac{1}{2}} -\log(1-t) \, . \, dt$ both directly and by integrating the Taylor Series term by term.)

262

(ii) Calculate the player's expected gain.

[The interesting feature of this game is that if the player pays the entrance fee (calculated from the result of part (ii)) to make the game classically 'fair', the probability nevertheless approaches unity that in n plays of the game he will have sustained a loss greater than $(1-\epsilon)\,n/\log_2 n$ for any $\epsilon > 0$, and any n.

For instance take $n = 128$, then there is a probability nearly unity that the player's loss will exceed $(1-\epsilon)\frac{128}{7}$ for any $\epsilon > 0$, and in particular will exceed $\frac{7}{8}.\frac{128}{7}$, say. This is essentially because the *variance of the loss is not finite*; the proofs of these results are beyond the scope of this course.]

3. PROBABILITY GENERATING FUNCTIONS

3.1　Recapitulation. We have met so far three examples of a situation where the probabilities of getting various values of a random variable can be simply expressed as coefficients in a power series.

Thus (i) for R, the number of successes in a fixed number N of Bernouilli trials, with probability θ of success at each trial, we have a function G_R such that

$$G_R(\xi) \equiv \sum_r \mathrm{pr}\,(R = r).\xi^r = \{(1-\theta)+\theta\xi\}^N;$$

(ii) for W, the waiting time to the kth success in an unlimited sequence of Bernouilli trials we have a function G_W such that

$$G_W(\xi) \equiv \sum_w \mathrm{pr}\,(W = w).\xi^w = \theta^k\{1-(1-\theta)\,\xi\}^{-k}$$

and (iii) for X, the number of random events in an interval whose length is such that the expected number of events in intervals of that length is μ, we have a function G_X such that

$$G_X(\xi) \equiv \sum_x \mathrm{pr}\,(X = x).\xi^x = e^{-\mu}.e^{\mu\xi} = e^{-\mu(1-\xi)}.$$

3.2　Definition. These three functions are only the tip of a very large iceberg; the method of expressing and indeed obtaining the probabilities of various values of a random variable as the coefficients in the expansion of a suitable function is a very powerful one. The function having the power series expansion is called the *probability generating function* (p.g.f). or *probability generator* for the random variable.

We may generally, and rather formally, state the following: If R is a random variable defined by any process which gives it integral (not necessarily positive) values r, and if ξ is an arbitrary variable, then whatever the values of pr $(R = r)$ the series $\sum_r \mathrm{pr}\,(R = r).\xi^r$ converges for $\xi = 1$ (since $\sum \mathrm{pr}\,(R = r) = 1$) and hence certainly converges for $|\xi| < 1$, and thus defines a function G_R by

$$\sum_r \mathrm{pr}\,(R = r).\xi^r = G_R(\xi) \quad \text{for} \quad |\xi| \leqslant 1.$$

Starting, as we might, from the function G_R, it is said to *generate* the probabilities.

3.3 Expected values. Let us look at some properties of such functions. Let R be an integral random variable and let pr $(R = r) = p(r)$.
To find $E(R)$ we need

$$0 . p(0) + 1 . p(1) + 2 . p(2) + ...,$$

which is

$$\sum_r r . p(r),$$

(provided the series is convergent).

By inserting, as is common in this sort of work on series, suitable powers of an arbitrary variable ξ we can make this series look like the result of a differentiation by considering:

$$0 . p(0) + 1 p(1) . \xi^0 + 2p(2) . \xi^1 + 3p(3) . \xi^2 + ...$$

$$= \frac{d}{d\xi} \{p(0) . \xi^0 + p(1) . \xi^1 + p(2) . \xi^2 + p(3) . \xi^3 + ...\}$$

$$= \frac{d}{d\xi} G(\xi), \text{ giving the usual definition of a generating function } G,$$

$$= G'(\xi).$$

It now remains to put $\xi = 1$ in this series to get

$$E(R) = G'(1).$$

We apply this to the three cases of G so far mentioned:
(i) When R denotes the number of successes in N Bernouilli trials for fixed N,

$$G_R(\xi) = \{(1-\theta) + \theta\xi\}^N,$$

$$G_R'(\xi) = N\theta\{(1-\theta) + \theta\xi\}^{N-1},$$

$$E(R) = G_R'(1) = N\theta.$$

This is in keeping with our interpretation of θ as the probability of success in a single trial, and of $N\theta$ as the 'average' number of success in a (pre-determined) number N of trials.

The function is called a *binomial p.g.f.*
(ii) When W denotes the waiting time to the kth success in an unlimited sequence of Bernouilli trials.

$$G_W(\xi) = \theta^k\{1 - (1-\theta)\xi\}^{-k},$$

$$G_W'(\xi) = -k\theta^k\{1 - (1-\theta)\xi\}^{-k-1} . -(1-\theta),$$

$$E(W) = G_W'(1) = -k . \theta^k . \theta^{-k-1} . -(1-\theta)$$

$$= k(1-\theta)/\theta.$$

264

We note some consequences of this. Suppose the event consists of throwing a 'two' with a fair die (then $\theta = \frac{1}{6}$), and that we are awaiting the first success (so that $k = 1$). Our result shows that $E(W) = 5$. The interpretation of this is that the *mean* number of 'non-twos' before the first 'two' is 5—in other words 'on average' (in a certain sense) a 'two' comes up about every 6th throw. What if we wait to the 4th success? Then $k = 4$ and the expected waiting time is quadrupled by the formula. We average 20 failures (with 3 successes) before our 4th success; 'on average' the 4th success occurs on every 24th throw. It is important, however, to remember that pr (1st success occurs on 6th throw) = $(\frac{5}{6})^5 . \frac{1}{6} = 0.067$ and that this is considerably less than pr (1st success occurs on 1st throw), which is $\frac{1}{6}$, or about $2\frac{1}{2}$ times as large. The expected value is *not* the modal value, which is 0.

If we look at this property of linearity in k which $E(W)$ possesses and remember that to get, say, 4 successes we have to wait until 1 success occurs and then wait for 3 more (or until 2 occur and then wait for 2 more, and so on), we see that the expected waiting times are additive. This is a property of random variables generally; namely that if a new random variable is constructed by making its *value* the sum of the values of 2 other random variables then the *expected values* of the 3 random variables are also additive. Now that the property is noticed in this case we see that it holds in case (i) where the number of successes in say 6 trials is the sum of the number in the first 2 trials and the number in the next 4, and so for various other combinations. The expected values have the same additive property because of the linearity in N of the expression $N\theta$.

We will return to all this in Section 4, but meanwhile we note that the function G_W is called a *negative binomial* or *inverse binomial p.g.f.*

(iii) When X denotes the number of events arising in a unit interval from a Poisson process with parameter μ:

$$G_X(\xi) = e^{-\mu(1-\xi)},$$
$$G_X'(\xi) = \mu e^{-\mu(1-\xi)},$$
$$E(X) = G_X'(1) = \mu.$$

This confirms the interpretation we have all along placed on μ. The function is a *Poisson p.g.f.*

3.4 Variances. We recall that with the usual notation $\sigma^2 = E[r^2] - \mu^2$. If we are to use the differentiating techniques of Section 3.3 then a term like $p(r).\xi^r$ will give rise to $r(r-1).p(r).\xi^{r-2}$ upon being differentiated twice.

This suggests that we rewrite the formula for σ^2 in the shape:

$$\sigma^2 = E[r(r-1)] + E[r] - \mu^2;$$

a shape already examined in Exercise 9B, Question 2(iii). We proceed as follows:

If G is the p.g.f. for an integral random variable R and $\text{pr}(R = r) = p(r)$ then

$$G(\xi) = p(0) + p(1).\xi + p(2).\xi^2 + p(3).\xi^3 + \ldots + p(r).\xi^r + \ldots,$$

$$G'(\xi) = p(1) + 2.p(2).\xi + 3.p(3).\xi^2 + \ldots + r.p(r).\xi^{r-1} + \ldots,$$

$$G''(\xi) = 1.0.p(1) + 2.1.p(2) + 3.2.p(3).\xi + \ldots + r(r-1).p(r).\xi^{r-2} + \ldots,$$

$$G''(1) = 1.0.p(1) + 2.1.p(2) + 3.2.p(3) + \ldots + r(r-1).p(r) + \ldots$$

$$= E[r(r-1)].$$

Combining this with
$$\sigma^2 = E[r(r-1)] + \mu - \mu^2,$$

we have $\quad \text{var}[r] = G''(1) + \mu - \mu^2 \quad$ where $\mu = G'(1)$.

Proceeding further in our investigation of the random variable R which is the number of successes in a fixed number N of Bernouilli trials with probability θ of success at each trial, we have (from Section 3.3)

$$G'_R(\xi) = N\theta\{(1-\theta) + \theta\xi\}^{N-1}$$

so that $\quad G''_R(\xi) = N(N-1)\,\theta^2\{(1-\theta) + \theta\xi\}^{N-2}$

and $\quad G''_R(1) = N(N-1)\,\theta^2;$

we obtain $\quad \text{var}(R) = N(N-1)\,\theta^2 + N\theta - (N\theta)^2$

$$= N\theta(1-\theta),$$

thus achieving the two famous results on Bernouilli trials: that $E(R) = N\theta$ and var $(R) = N\theta(1-\theta)$. [See Exercise 9C, Question 17, on Poisson processes.]

Exercise 9C

1R. (i) Write out in full the terms of $\sum_{r=0}^{5} r.\text{pr}(R = r)$ where R is the number of successes in five Bernouilli trials with probability of success p and of failure q at each trial. Using the common factor $5p$, simplify the expression and obtain $E(R)$.

(ii) Use the identity $r.\binom{N}{r} = N.\binom{N-1}{r-1}$ to find in a similar way, the value of $E(R)$ for general N, without the use of generating functions.

2R. In Table 7.5 two samples of values of R are recorded. Calculate the variance of R in each one, and compare each with the variance calculated from the Bernouilli model which would fit the values of N and $N\theta$. (Be careful over the meanings to be attached to N and $N\theta$; $N = 5$, $N\theta = \frac{5}{3}$.)

Discuss the agreement of each observed variance with the variance from the model.

266

[We see in Chapter 11 that when the variance of a sample is calculated it tends to be *smaller* than the variance of the parent population, though the effect is negligible for samples of size about 20 or more (and the samples of Table 7.5 are of sizes 30 and 50). As we also see in Chapter 11, this behaviour contrasts with the behaviour of the mean of a sample, which may be expected to be 'on average' equal to the mean of the parent population.]

3. Thirty sets of 4 random digits D were examined and a success was recorded if the digit D satisfied $0 \leqslant D \leqslant k$ for integral k; the results were

number of successes in a set of 4 digits	frequency of sets with this number of successes
0	5
1	14
2	10
3	0
4	1

What value of k fits the data best? Calculate the expected frequencies with this value of k. Calculate also the 'expected' and observed variances of the number of successes.

4. Sixty samples of size 4 are taken after an International Match and the following records made:

x	number of samples with x Welshmen
0	20
1	22
2	8
3	6
4	4

Calculate the mean number of Welshmen per sample in the experiment. Assume that a Bernouilli model holds to give $E[x]$ = observed mean, and calculate the theoretical variance of $[x]$.

Calculate the observed variance of $[x]$ and the theoretical values of the frequencies. Discuss the results. (See Question 21.)

5. 40 sets of N random digits, D, were examined and a success was recorded if D was zero. The following table resulted:

number of zeros in a set	frequency of sets with this number
0	26
1	12
2	2

What do you think was the value of N? Calculate the expected frequencies with this value of N; and the 'expected' and observed variances of the number of zeros in a set.

6. (i) In a certain experiment in sampling of attributes the mean and variance of the number of successes R, are, m, and S^2. On the assumption that the frequencies arise from a Bernouilli process what is your estimate of $1-\theta$, and hence of θ and of N? (Ignore the small effects mentioned in the note on Question 2.)

(ii) Apply the results of part (i) to the following data:

r	frequency of $R = r$
0	32
1	82
2	88
3	62
4	26
5	8
6	2

(iii) Calculate the theoretical values of the frequencies from your assumed model. Comment on the results.

7. Apply the results of Question 6, part (i) to the following problem: Sets of N random digits, D, were examined and a success was recorded if D satisfied $0 \leqslant D \leqslant k$, with the following results:

number of successes in a set	frequency of sets with that number
0	4
1	8
2	7
3	4
4	1

(i) What do you think were N and k?
(ii) Calculate the frequencies corresponding to your model.

8. Apply the results of Question 6 part (i) to the following data and calculate the theoretical frequencies.

r	fr $(R = r)$
0	66
1	29
2	5

9. (i) Adapt the methods of Chapter 6, Section 1.3 to prove Chebyshev's Inequality: that for a random variable X with expectation μ and variance σ^2 we have:

$$\text{pr}\,(|X-\mu| > \lambda\sigma) < 1/\lambda^2 \quad \text{for any } \lambda.$$

(ii) Deduce that, for a sequence of N Bernouilli trials with probability of success θ, the number of successes, R, satisfies

$$\text{pr}\,(|R-N\theta| > kN\theta) < \left(\frac{1-\theta}{\theta}\right) \cdot \frac{1}{N} \cdot \frac{1}{k^2}.$$

10. Show that for an exactly Binomial population of numbers of successes in N trials the largest possible value of the variance is $N/4$

11. Show that for a fixed probability of success at any trial the coefficient of variation of an exactly binomial population is proportional to $N^{-\frac{1}{2}}$. (See Exercise 6A, Question 7.)

12. Find the p.g.f., G, for the score on a single throw of a die. Differentiate $G(\xi)$ with respect to ξ; put $\xi = 1 - h$; let $h \to 0$ in an attempt to find $G'(1)$. Contemplate.

13. (i) Prove that if $H(\xi) = \xi^k G(\xi)$, where G is a p.g.f.

then
$$H'(1) = G'(1) + k$$

and
$$H''(1) + H'(1) - \{H'(1)\}^2 = G''(1) + G'(1) - \{G'(1)\}^2.$$

(ii) Interpret.

(iii) Use the results of parts (i) and (ii) to explain why in an inverse Binomial problem it is more convenient to consider the waiting time to the kth success (as defined so far) than to consider the total number of trials to the kth success.

14T. Prove that
$$(\xi G'(\xi))' = \Sigma r^2 p(r) . \xi^{r-1}$$

and use this result to obtain $\sigma^2 \equiv \mathrm{var}\,[r]$ in terms of G.

15T. If R is an integral random variable taking values r, and ξ is an arbitrary (non-random) variable, interpret $E[\xi^r]$.

16T. Let R be the number of successes in N Bernouilli trials, with probability θ of success at each trial. Write $N\theta = \mu$.

Starting from
$$G_R(\xi) \equiv \{(1-\theta) + \theta\xi\}^N$$
$$= \{1 - \theta(1-\xi)\}^N,$$

substitute for θ and find the limiting form of $G_R(\xi)$ as $N \to \infty$; thus verifying the earlier results about the Poisson Generating Function.

17T. The Poisson probability function for the number, R, of random events occurring in a unit interval is $\mathrm{pr}\,(R = r) = e^{-\mu} . \mu^r / r!$, where μ is a parameter determining the particular member of the family of functions.

(i) By evaluating the series $\sum_0^\infty r . \mathrm{pr}\,(R = r)$ and $\sum_0^\infty r^2 . \mathrm{pr}\,(R = r)$ directly, identify μ as $E(R)$ and prove that $\mathrm{var}\,(R) = E(R)$.

(ii) By using the Poisson p.g.f. $G(\xi) = e^{-\mu(1-\xi)}$ and the methods of Sections 3.3 and 3.4 obtain the two results of part (i).

18. In Chapter 8, Tables 8.4 and 8.5 we have the following data from a Poisson process.

Table 8.4

(R represents the number of calls in a 40-minute period.)

r	frequency of $R = r$
0	18
1	22
2	9

Table 8.5

(S represents the number of calls in an 80-minute period.)

s	frequency of $S = s$
0	4
1	7
2	8
3	4
4	1

By calculating the mean and variance from each table see whether these both show the characteristic of a Poisson probability function that the variance increases in scale with the mean. [Note the unreliability of the variance.]

19. In Exercise 8 B, Question 4 a table of frequencies of various numbers of radioactive decays in periods of 7·5 sec is given. These have already been checked as Nearly Poissonian in that Question. Find the proportion of this large sample of numbers of disintegrations per 7·5 sec that lie within the interval $m \pm 2S$.

20. (i) In Exercise 8 B, Question 2 the well known 'Horse-kick data' are given. Check the equality of mean and variance.

(ii) In a binomial population we have $\mu = N\theta$ and $\sigma^2 = N\theta(1-\theta)$. The Poisson approximation is obtained by taking N to be large and θ to be small. Show that for fixed μ we would anticipate an approximate equality of σ^2 and μ.

(iii) By using the small discrepancy between m and S^2 in the Horse-kick data suggest values of θ and N which would give parameter values equal to the observed values of m and S^2.

21 T. In botanical surveys it is often of interest to decide whether plants occur at random over the area or not. There are two common forms of departure from randomness, that is from good agreement with a Poisson model.

I   Clumping of plants occurs, leading to too many test squares (quadrats) with zero frequency and too many with high frequencies. Such a distribution of plants is called *contagious*. The frequency diagram is 'overdispersed', the plants themselves are physically underdispersed.

II Plants are distributed over the ground too *regularly*, like trees in an orchard. This leads to a deficiency of both high and low frequencies in quadrats. The frequency diagram is 'underdispersed', the plants themselves are overdispersed.

(i) Does occupancy of quadrats in an ordinary binomial pattern exhibit contagion or regularity?
Does occupancy of quadrats with a uniform frequency diagram exhibit contagion or regularity?

(ii) Various tests have been proposed to detect departures from randomness (quite apart from the general test called the 'chi-squared test' which is used to test departure from *any* given frequency pattern).

Suppose that $f(r)$ is the frequency with which the number, R, of plants in a quadrat takes the value r, then we define:

$$B = \mathrm{var}\,[r]/\bar{r} \quad \text{and} \quad \phi = 2f(0)\,f(2)/(f(1))^2.$$

Evaluate these statistics in each of the following idealized cases:

f is proportional to a probability function by which μk plants are spread over k squares and the probability function is

(a) Binomial giving a maximum of, say, N plants in a quadrat.

(b) Uniform from $R = 0$ upwards (so that there are $k/(2\mu+1)$ quadrats with each number of plants from 0 to 2μ).

(c) Poissonian with parameter μ.

[That particular descriptive statistics are 'invented' rather than 'discovered' was one of the main themes underlying Chapter 4 where various ways of describing the position and spread of a population were first discussed. This side of descriptive statistics is well brought out by P. Greig-Smith in *Quantitative Plant Ecology* (2nd Edition, Butterworth, 1964) where no less than ten statistics are quoted, each designed by its inventor to deal with this problem in discrimination between contagious, random and regular distributions of plants.]

(iii) How would you use the statistics B, ϕ to detect (a) contagion, (b) regularity? Which is the more convenient to evaluate from a real survey? What effect, if any, does the size of quadrat have?

(iv) Examine the following data quoted in P. Greig-Smith: *Quantitative Plant Ecology* (2nd Edition, Butterworth, 1964). The first frequency function refers to plants of *Carex flacca* in limestone grassland, the quadrat being 10×10 cm. The second function is artificial (and is slightly adapted from the above source).

(R = No. of shoots per quadrat)

r	0	1	2	3	4	5	6	7	8	9	10	>10
fr $(R = r)$ for *Carex flacca*	134	34	12	8	8	0	1	1	1	0	1	0
fr $(R = r)$ (artificial)	20	75	0	0	0	5	0	0	0	0	0	0

(v) Apply the ideas of this question to the survey of Welshmen in Question 4 and decide whether you think a Bernouilli model suitable.

[We see that the values of B and ϕ would be much more useful to us if we had some sort of scale to tell us how significant was a discrepancy from their expected values. This important point will be taken up later, though for other statistics than these.]

22 T. (i) If W is a random variable with an inverse binomial p.g.f. find var (W), using the notation of Section 3.3.

(ii) Calculate var $[u]$, var $[v]$ from Tables 7.3 and 7.4 for the 'simple experiment' and compare them with the values expected from the Bernouilli model.

23 R. (a) In Chapter 3, Section 1, B throws two dice and records the score. In carrying out a run of this game for illustrative purposes the author obtained 'a four and a five' 4 times in the first 12 turns. In fact the occurrences were on the 2nd, 3rd, 6th and 7th turns.

Before any of the throws took place, what were the expected values of these four numbers?

(b) If the records of a certain finite sequence of Bernouilli trials are shortened by discarding all the results after the last success, find the expectation and variance of the number of results discarded.

24 T. In a certain sequence of Bernouilli trials the probability of success at each trial is $1/k$, where k is an integer.

(i) Find the expected value of the waiting time to the first success. Let it be called N.

(ii) The value of N is calculated by an experimenter beforehand and he then carries out $N+1$ of the Bernouilli trials. Does a binomial function provide an appropriate probability model for investigating the number of successes in the $N+1$ trials?

(iii) What is the probability that the experimenter will find no successes in his $N+1$ trials, and what is the expectation and variance of the number of successes in the $N+1$ trials?

(iv) What is the expectation and variance of the number of successes in a sequence of Bernouilli trials which stops at the first success? Are these values obtainable from a binomial probability function?

25. Let L denote the length of run started by the first result of a sequence of Bernouilli trials (whether the first trial results in a success or a failure).

(i) Find $\mathrm{pr}\,(L = 1)$.

(ii) Show that $G_L(\xi) = \left\{\dfrac{1-\theta}{1-\theta\xi}+\dfrac{\theta}{1-(1-\theta)\,\xi}\right\}-1.$

(iii) Prove that $E(L)$ is a weighted mean of the expected number of trials to the first *success* and the expected number of trials to the first *failure*; the weights being respectively the probability of *failure* and of *success*. Explain the meaning of this result in words.

(iv) Find var (L).

⋆26. Three children, A, B, C, in a car are looking at inn signs as they go along. A proportion $1-\phi$ of the signs contain references to Royalty and such references effectively occur independently of each other along the route. They play a game of collecting such references.

The signs are allotted to the children in rotation; to A, then to B, then to C. The children continue to collect until *the end* of that round in which the last of them to collect a Royalty sign collects his or her first such sign.

(i) Prove that the probability that this takes $(k+1)$ rounds is

$$\phi^k(1-\phi)\,(1-\phi^k)^2+(1-\phi^{k+1})\,\phi^k(1-\phi)\,(1-\phi^k)+(1-\phi^{k+1})^2\,\phi^k(1-\phi),$$

and that this expression reduces to

$$3(1-\phi)\,\phi^k-3(1-\phi^2)\,\phi^{2k}+(1-\phi^3)\,\phi^{3k}.$$

(ii) Deduce from (i) that the p.g.f. for the total number of rounds is

$$G(\xi) = \xi\left(\frac{3(1-\phi)}{1-\phi\xi}-\frac{3(1-\phi^2)}{1-\phi^2\xi}+\frac{(1-\phi^3)}{1-\phi^3\xi}\right)$$

and that the expected number of rounds is $\dfrac{1}{1-\phi^3}+\dfrac{3\phi}{1-\phi^2}$.

(iii) What is the expected value of the total number of inns passed? Compare this with the expected value of the total number passed under the following rules:

Each child drops out as soon as it has collected its first Royal sign and the other children continue in rotation until each has collected a Royal sign.

Show that the ratio of the expected numbers of inns passed under the two rules is

$$3-\frac{2+2\phi+3\phi^2}{(1+\phi)\,(1+\phi+\phi^2)},$$

so that they cannot quite cut the expected number down to a third of its former value by playing to the second set of rules.

(iv) What difference would it make to the results of part (iii) if under a different shortened version of the game they not only dropped out, as in part (iii), but allocated the inns to the children at random and not in rotation?

(v) Simulate ten plays of the game under the first rules with $\phi = \frac{1}{2}$, and compare the mean number of rounds with the expectation calculated in part (ii).

27. In soil specimens from a field infested with eelworm cysts it is found that the number, R, of mature cysts in a specimen may be obtained reasonably well from the following probability function:

$$\text{pr}\,(R = r) = \binom{k+r-1}{k-1} \theta^k (1-\theta)^r,$$

where $r = 0, 1, 2, \ldots$; $k > 0$; and $0 \leqslant \theta \leqslant 1$.

(i) Find the corresponding probability generating function.

(ii) Show that the observations would be consistent with the following model. The field was originally infested uniformly at a rate of k immature cysts per soil-specimen. Each immature cyst had a probability θ of dying without maturing and had a probability $(1-\theta)$ of maturing and producing one more cyst with the same behaviour pattern as its predecessor. The cysts behaved independently of each other.

[The question whether the cysts actually *do* have such a behaviour pattern is of course one which statistics cannot answer. All we can say is that *this* numerical aspect of their behaviour is consistent with this hypothesis. Statistics is no substitute for biology, and the reader would do well to enquire from a biologist whether perhaps other observations about eelworms entirely prevent this model from having any wider relevance.]

(iii) Find the expectation and variance of the number of mature cysts in a specimen. Find also B, ϕ [see Question 21] and comment.

(iv) Simulate for 20 specimens with $k = 6$ and $\theta = \frac{4}{5}$; and compare the resulting mean and variance with the values from part (iii).

(v) Choose k and θ to match the mean and variance of r from the following:

r	0	1	2	3	4	5	6	7
fr $(R = r)$	9	11	10	5	3	1	0	1

28 T. The kth *factorial moment* of a non-negative integral random variable R is defined by

$$\mu'_{(0)} = 1; \quad \mu'_{(k)} = E[r(r-1)\,(r-2) \ldots (r-k+1)] \quad \text{for } k \geqslant 1.$$

(i) Show that:

$$\mu'_{(k)} = k\text{th derivative of } G_R(\xi), \text{ evaluated at } \xi = 1$$

$$\equiv G_R^{(k)}(1), \quad \text{say.}$$

(ii) Express x^3 as a linear combination of $x(x-1)\,(x-2)$, $x(x-1)$ and x; and deduce an expression for the skewness of a random variable in terms of its p.g.f.

(iii) Find the skewness of a binomial random variable with parameters N, θ; and discuss the behaviour for fixed θ as $N \to \infty$.

(iv) Find the skewness of a Poisson variate with parameter μ; and discuss the behaviour as $\mu \to \infty$.

29 F. Using the notation of Section 3, discuss the linearity in N of var (R), and in k of var (W).

4. COMBINATIONS OF RANDOM VARIABLES

4.1 Introduction. There are many problems where the quantity under consideration is the sum of two or more quantities each having the possibility of variation. Thus the weight of a parcel is made up of that of the contents and that of the packaging; a manufactured article may consist of several parts bolted together and the total length be the sum of their lengths; the number of people served at a roadside café is the sum of the contents of several cars and several buses. Sometimes the difference of two quantities is important; whether two medical treatments are effectively the same or not might be decided by the difference of their cure-rates; whether or not a nut fits its bolt depends on the difference of their radii. Sometimes a product or a quotient is concerned; thus the area of a turbine blade and the angle of elevation of a gun depend ultimately on the product and quotient, respectively, of the values of a pair of variables.

To return to the case of the sum of several quantities: perhaps the most far-reaching example of all is the evaluation of a mean. In this process several values of a single random variable are first added together, each value being one of many possible values; (after the addition, of course, the sum is multiplied by a number not dependent on the values of the observations, namely the reciprocal of the number of them.) We will have more to say on this important application in Chapter 11, Section 3.2.

4.2 A numerical case introduced. We next look at the results obtained by adding, subtracting and multiplying values of two random variables, X and Y. We begin with a simple numerical particular case to bring out the ideas involved.

Variable X. Take the score on a single die and code it as follows, (taking the value of X to be the coded score):

If the score is even, record it unchanged; so $X = 2$ or 4 or 6.
If the score is odd, code it as 0; so $X = 0$.
We have:
$$E(X) = (2.\tfrac{1}{6}) + (4.\tfrac{1}{6}) + (6.\tfrac{1}{6}) + (0.\tfrac{1}{2}) = 2,$$
$$\text{var}(X) = (0^2.\tfrac{1}{6}) + (2^2.\tfrac{1}{6}) + (4^2.\tfrac{1}{6}) + (2^2.\tfrac{1}{2}) = \tfrac{16}{3}.$$

Variable Y. Throw two coins and record $Y = 1$ for a pair of heads and $Y = 2$ otherwise.
$$E(Y) = (1.\tfrac{1}{4}) + (2.\tfrac{3}{4}) = \tfrac{7}{4},$$
$$\text{var}(Y) = (\tfrac{3}{4})^2.\tfrac{1}{4} + (\tfrac{1}{4})^2.\tfrac{3}{4} = \tfrac{3}{16}.$$

If we look at the variables X, Y simultaneously we will want to draw up Table 9.2 to show their *joint probability function*; from which we read, for instance, pr $(X = 0, Y = 2) = \frac{9}{24}$. In this case X and Y are independent random variables.

Table 9.2

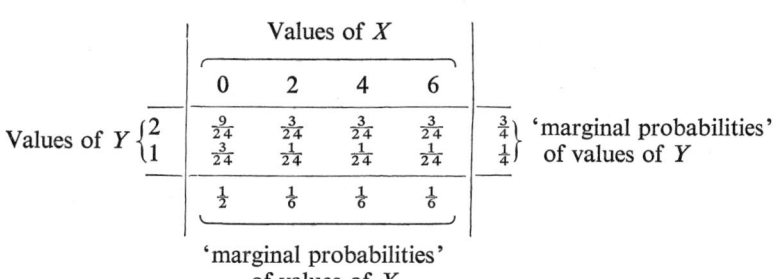

The resemblance of this table to the contingency tables of Chapter 3 should need no emphasizing.

4.3 The numerical case developed. If the values of X, Y are x, y respectively we are concerned with all possible values $t = x+y$, $u = x-y$, $v = xy$ and $w = x/y$, and with their probabilities.

Table 9.3 shows all we need to obtain these probabilities.

Table 9.3

(x, y)	pr $(X = x, Y = y)$	$t =$ $x+y$	$u =$ $x-y$	$v =$ xy	$w =$ x/y
$(0, 1)$	$\frac{3}{24}$	1	-1	0	0
$(2, 1)$	$\frac{1}{24}$	3	1	2	2
$(4, 1)$	$\frac{1}{24}$	5	3	4	4
$(6, 1)$	$\frac{1}{24}$	7	5	6	6
$(0, 2)$	$\frac{9}{24}$	2	-2	0	0
$(2, 2)$	$\frac{3}{24}$	4	0	4	1
$(4, 2)$	$\frac{3}{24}$	6	2	8	2
$(6, 2)$	$\frac{3}{24}$	8	4	12	3

Using Table 9.3 we can obtain such facts as 'that the probability that $v = 0$ is $\frac{3}{24}+\frac{9}{24}$'; and we record these probabilities, multiplied by 24 (to avoid fractions), in further tables. In Table 9.6 for instance the reader will find the record that $24.\,\mathrm{pr}\,(v = 0) = 12$.

Furthermore we include in Tables 9.4, 9.5, 9.6, 9.7 the calculations necessary to find the expectations and variances of the populations of values t, u, v, w. These will be required almost immediately afterwards.

Extracting the relevant information from Table 9.3 we obtain Tables
9.4–7:

Table 9.4

t	$f(t) = 24 \times$ prob.	$tf(t)$	$t^2f(t)$
1	3	3	3
2	9	18	36
3	1	3	9
4	3	12	48
5	1	5	25
6	3	18	108
7	1	7	49
8	3	24	192
	24	90	470

so that
$$E[t] = \tfrac{90}{24} = 3\tfrac{3}{4},$$
$$\text{var}[t] = \tfrac{470}{24} - (\tfrac{15}{4})^2 = \tfrac{265}{48}.$$

Table 9.5

u	$g(u) = 24 \times$ prob.	$ug(u)$	$u^2g(u)$
-2	9	-18	36
-1	3	-3	3
0	3	0	0
1	1	1	1
2	3	6	12
3	1	3	9
4	3	12	48
5	1	5	25
	24	$-21+27$	134
		$= 6$	

so that
$$E[u] = \tfrac{6}{24} = \tfrac{1}{4},$$
$$\text{var}[u] = \tfrac{134}{24} - (\tfrac{1}{4})^2 = \tfrac{265}{48}.$$

Table 9.6

v	$h(v) = 24 \times$ prob.	$vh(v)$	$v^2h(v)$
0	12	0	0
2	1	2	4
4	4	16	64
6	1	6	36
8	3	24	192
12	3	36	432
	24	84	728

so that $$E[v] = \tfrac{84}{24} = 3\tfrac{1}{2},$$

$$\text{var } [v] = \tfrac{728}{24} - (\tfrac{7}{2})^2 = \tfrac{217}{12}.$$

Table 9.7

w	$k(w) = 24 \times \text{prob.}$	$wk(w)$	$w^2k(w)$
0	12	0	0
1	3	3	3
2	4	8	16
3	3	9	27
4	1	4	16
6	1	6	36
	24	30	98

so that $$E[w] = \tfrac{30}{24} = \tfrac{5}{4},$$

$$\text{var } [w] = \tfrac{98}{24} - (\tfrac{5}{4})^2 = \tfrac{121}{48}.$$

If we consider the values t, u, v, w and their respective probabilities, we realize that we have defined four new random variables by forming the sums, differences, products and quotients of all possible pairs of values of the two original random variables, and by alloting to the new values the probabilities of their occurring through the occurrence of the relevant *pairs* of values of X and Y.

4.4   Notation. These new random variables are written $X + Y$, $X - Y$, XY and X/Y; and the reader may think that fairly natural, but he should beware. For if R denotes the score on one die and S denotes the score on another die, then, although R, S have the same probability function as each other, the random variables written $R + S$ and $2R$ are *not* the same. The possible values of the first are obtained by adding together in pairs all possible values of R, S; while the possible values of the second are *defined* as, merely, double the values of R.

The probability functions for $R + S$ and $2R$ are tabulated in Table 9.8 in forms as easily comparable as possible.

Since the phrase 'random variable' is merely another way of talking about a probability function and since the probability functions are different, this table confirms that what we have called $R + S$ and $2R$ must be distinguished from each other.

It happens that $$E(R + S) = 7 = E(2R);$$

but it is also true that

$$\text{var } (R + S) = 2 \cdot \text{var}(R) = 2 \cdot \text{var } (S),$$

whereas $$\text{var } (2R) = 4 \cdot \text{var } (R),$$

as the reader could verify from Table 9.8.

Table 9.8

l	$36 \times \mathrm{pr}\,(R+S = l)$	m	$36 \times \mathrm{pr}\,(2R = m)$
2	1	2	6
3	2		
4	3	4	6
5	4		
6	5	6	6
7	6		
8	5	8	6
9	4		
10	3	10	6
11	2		
12	1	12	6

Reverting to the values t, u, v, w, we go on, by a further extension of notation, so far as to write:

$$T = X+Y; \quad U = X-Y; \quad V = XY; \quad W = X/Y.$$

The meanings of these 'equations' are *defined* by the methods by which the possible values and probabilities associated with t, u, v, w were obtained from those of x, y.

The notations $T = X+Y$, and so on, are fraught with possible misinterpretations as we will now show; but the reader should not despair, for only a little practice seems necessary to get quite used to them.

Let us consider two random variables having the same probability function, as for instance the R and S of this section. They are called *identical random variables*, but we do not usually write this relationship as $R = S$. Furthermore, if we write:

$$L = R+S \quad \text{(so that pr}\,(L = l)\text{ is displayed in Table 9.8),}$$

$$P = RS,$$

and $\quad D = R-S$,

then we do *not* have any of:

$$L = 2R, \qquad P = R^2, \qquad D = 0.$$

That we do *not* have the first is discussed by means of Table 9.8; and the reader can at this stage supply the consistent meanings for RS and R^2, and for $R-S$ and 0, to confirm the rest of the denial. The interpretation of the statement that a random variable is a constant is that it takes the value of the constant with probability 1.

Notice, too, that two random variables can be identical and at the same time independent; indeed this probability is at the heart of the distinction between $R+S$ and $2R$ in this section.

278

Finally, notice that in no sense are the populations defined by the random variables being combined in the same way that a pair of samples of observations were combined to form a new sample in Chapter 4. In fact there is no comparable approach to that of Chapter 4 either meaningful or desirable here.

5. COMBINATORIAL RESULTS AND THEIR PROOFS

5.1 The results. We now pass to a consideration of the parameters, expectation and variance, for each of the new random variables $X+Y$, $X-Y$, XY, X/Y. From the calculations in Tables 9.4, 9.5, 9.6, 9.7 the following results emerge:

$$E[t] = E[x]+E[y], \qquad \text{var } [t] = \text{var } [x]+\text{var } [y],$$

$$E[u] = E[x]-E[y], \qquad \text{var } [u] = \text{var } [x]+\text{var } [y],$$

$$E[v] = E[x].E[y], \qquad \text{no relation between variances for } [v],$$

no relations at all for $[w]$.

These results are particular cases of the following:

If x, y are values of two *independent* random variables, and a and b are constants, and new random variables are formed with values $ax+by$ and xy in the way of Section 4.3 and with probabilities allotted, as there, by a joint probability function, then:

Result I : $\quad E[ax+by] \quad = aE[x]+bE[y]$,

Result II : $\quad \text{var } [ax+by] = a^2 \text{ var } [x]+b^2 \text{ var } [y]$,

Result III: $\quad E[xy] \qquad = E[x].E[y]$.

If, on the other hand, the restriction to independent random variables is removed, then we retain *only*:

Result I: $\quad E[ax+by] = aE[x]+bE[y]$.

In the notation of random variables the results read:

Result I : $\quad E(aX+bY) \quad = aE(X)+bE(Y)$,

Result II : $\quad \text{var } (aX+bY) = a^2 \text{ var } (X)+b^2 \text{ var } (Y)$,

Result III: $\quad E(XY) \qquad = E(X).E(Y)$.

The proofs of Results I and II follow. The proof of Result III is left as an exercise for the reader in Exercise 9D, Question 2.

5.2 Proof of Result I. We will first prove Result I without the restriction of independence.

$$E[ax+by] = \sum\sum_{\text{all } x, y} (ax+by).\text{pr} (X = x \text{ and } Y = y)$$

$$= \sum\sum_{\text{all } x, y} ax.\text{pr} (X = x \text{ and } Y = y)$$

$$+ \sum\sum_{\text{all } x, y} by.\text{pr} (X = x \text{ and } Y = y), \qquad \text{by the linearity of the sums}$$

$$= A_x + A_y, \quad \text{say.}$$

We consider the first of these double sums.

Now

$$\text{pr} (X = x \text{ and } Y = y) = \text{pr} (X = x).\text{pr} (Y = y|X = x)$$

so $\quad A_x = \sum\sum_{\text{all } x, y} ax.\text{pr} (X = x).\text{pr} (Y = y|X = x)$

$$= \sum_{\text{all } x} \{ax.\text{pr} (X = x). \sum_{\text{all } y} \text{pr} (Y = y|X = x)\}, \quad \text{by the linearity of the sums}$$

$$= \sum_{\text{all } x} \{ax.\text{pr} (X = x). 1\}, \quad \text{by the standard property of conditional probabilities}$$

$$= a \sum_{\text{all } x} x.\text{pr} (X = x)$$

$$= a.E[x].$$

A_y is treated similarly and we have

$$E[ax+by] = aE(x)+bE[y].$$

or $\qquad E(aX+bY) = aE(x)+bE(Y).$

The extension to $\qquad E(\sum_r a_r X_r) = \sum_r a_r E(X_r)$

should be obvious to the reader. We may, of course, prove it by Induction.

The proof of Result I when X, Y are independent follows the same lines, but we have the simplification that

$$\text{pr} (Y = y|X = x) = \text{pr} (Y = y).$$

5.3 Proof of Result II. To prove Result II we write $\text{pr} (X = x) = p_x$ and $\text{pr} (Y = y) = p_y$. The condition of independence of X, Y then gives

$$\text{pr} (X = x \text{ and } Y = y) = p_x.p_y.$$

We also write $\qquad E[x] = \mu_x \quad \text{and} \quad E[y] = \mu_y.$

The proof then runs:

$$\text{var } [ax+by] = E[\{(ax+by)-(a\mu_x+b\mu_y)\}^2], \quad \text{by definition and Result I}$$
$$= E[\{a(x-\mu_x)+b(y-\mu_y)\}^2]$$
$$= E[a^2(x-\mu_x)^2+b^2(y-\mu_y)^2+2ab(x-\mu_x)(y-\mu_y)]$$
$$= a^2 E[(x-\mu_x)^2]+b^2 E[(y-\mu_y)^2]$$
$$+2ab \sum_{\text{all } x, y}\sum (x-\mu_x)(y-\mu_y)\,p_x\cdot p_y, \quad \text{by the extension}$$
$$\text{of Result I}$$
$$= a^2 \text{ var } [x]+b^2 \text{ var } [y]+2ab\sum_x\{(x-\mu_x)\,p_x\cdot\sum_y(y-\mu_y)\,p_y\},$$
$$\text{by the linearity of the sums.}$$

Now
$$\sum_y(y-\mu_y)\,p_y = E[y-\mu_y] = 0.$$

So
$$\text{var } [ax+by] = a^2 \text{ var } [x]+b^2 \text{ var } [y].$$

Exercise 9D

1R. To understand why $\sum_{\text{all } x, y}\sum F(x)\,G(y) = \left(\sum_{\text{all } x} F(x)\right)\left(\sum_{\text{all } y} G(y)\right)$, consider

$$F(x) = x \quad \text{for } x = x_1, x_2, x_3$$
$$G(y) = y^2 \quad \text{for } y = y_1, y_2,$$

and write out the six terms involved in the double sum; then factorize them into the form of the product of sums.

2T. Prove that if x, y are values of independent random variables then

$$E[xy] = E[x].E[y].$$

3T. (i) Explain why, with the notation of Section 5.3

$$\text{var } [ax+by] = E[(ax+by)^2]-(a\mu_x+b\mu_y)^2.$$

(ii) Expand the squares on the right-hand side and use Result I to obtain

$$\text{var } [ax+by] = a^2 \text{ var } [x]+b^2 \text{ var } [y].$$

Where have you needed independence of the random variables?

4R. A certain café had a regular booking for lunch every Saturday for 4 coach loads of tourists and 2 minibus loads. The expectations and variances of the numbers of mouths to be fed in a vehicle of each type are given in the following table:

	Expectation	Variance
For each coach	40·2	6·3
For each minibus	9·1	4·4

(i) What can you say about the expectation and variance of the total number to be fed?

(ii) Describe some plausible situations which would break some conditions necessary for your results in part (i), and state which of the results would remain true.

5. A motorway service area deals daily with large numbers, n, of coaches each containing p passengers, each of whom buys b buns at the restaurant. If the expectation and variance of the random variable X are denoted by μ_X and σ_X^2 for $X = N, P, B$, what can you say about the expectation and variance of the number of passengers arriving; of the number of buns sold?

6. You have first throw in a game where the number of squares moved along a line is determined by the throw of a single die.

(i) If n is a predetermined number and *not* the value of a *random* variable, what are the expectation and variance of the number, l, of squares by which you are ahead after you have had $(n+1)$ throws and your opponent has had n?

(ii) Sketch, on a single diagram and labelling them with the values k, the graphs of $E[l] \pm k . \text{stad } [l]$, plotted against n, for $k = 2, 1, \frac{1}{2}, \frac{1}{4}, \frac{1}{8}$.

(iii) Assuming symmetry of the probability function for l, can you usefully employ Chebyshev's Inequality to say anything certain about the probability that you are in fact behind your opponent after you have had your $(n+1)$th throw but before he has?

(iv) Suppose that the probability function for l was symmetrical and that there was a relation (resembling Chebyshev's Inequality) of the form:

$$\text{pr} (\ |l - E[l]| > k . \text{stad } [l] \ ) = p(k; n)$$

where $p(k:n)$ decreased strictly as k increased with fixed n, and as n increased with fixed k.

What could you then say about the probability of being behind your opponent at the time mentioned in parts (i) and (iii)?

[This is a simple case of the classic problem known as *Gamblers' Ruin*.]

7F. [We have only dealt with discrete random variables so far, but we include two examples of the usefulness that the corresponding results will have when continuous random variables have been discussed. In this question assume that Results I, II, III apply to *continuous* random variables.]

(*a*) The mean e.m.f. (in volts) of batteries produced by a company is 22·6 and the standard deviation is 0·03. If four such batteries are connected in series (so that their e.m.f. are added) find the values of mean ± 2 S.D. for the total e.m.f.

(*b*) Certain bolts and washers are made to mean diameter 1 cm. and are packed in pairs at random. If the customer finds that the bolt is bigger than the washer-hole or has diameter more than 0·02 cm too small he rejects it.

The bolts have S.D. (in cm) of their diameters 0·006 and the washers 0·008. Express 0·02 as a multiple of the S.D. of the difference in diameters.

If the probability function of the difference in diameters is approximately Normal, so that about 95 per cent of the values lie within $2 \times$ S.D. from the mean, what proportion of the bolt-and-washer packages will prove acceptable to the customer?

8T. In a sequence of Bernouilli trials the probability of success is θ at each trial; and N is a number chosen before the trials begin.

(*a*) (i) What is the expectation of the number of successes in the first N trials; in the first $(N+1)$ trials?

(ii) What is the difference of these expectations?

(iii) What is the expectation of the difference between the number of successes in the first N trials and the number of successes in the first $(N+1)$ trials?

(iv) What is the expectation of the number of successes on the $(N+1)$th trial?

(v) What is the expectation of the difference between the number of successes in any N trials and the number of successes in another $(N+1)$ trials independent of the first N considered?

(vi) What is the expectation of the number of successes on an arbitrary single trial?

(*b*) Repeat parts (i) to (vi) with 'variance' replacing 'expectation' at each occurrence.

*9 T. In the solution of Question 3, the quantity $E[xy] - \mu_x \mu_y$ occurs.

(i) Show that it equals $E[(x-\mu_x)(y-\mu_y)]$, which is the form in which it appears in the proof of Result II in Section 5.3.

This quantity is called the *covariance* of the random variables X and Y and we will write it cov (X, Y) or cov $[(x, y)]$. It corresponds to the statistic written v_{xy}/N in Exercise 4H, Question 4. (Compare Exercise 4H, Question 5.)

(ii) We have seen that if X, Y are independent then cov $(X, Y) = 0$; show that the converse is false by considering

$$Y = X^2, \quad \text{where pr } (X = x) = \tfrac{1}{5} \quad \text{for} \quad x = -2, -1, 0, 1, 2.$$

(Compare Exercise 4H, Question 7.)

*10 T. [This question develops the theory behind Chapter 4, Section 9, from its beginnings in Exercise 4H, Questions 4, 5.]

Let X, Y be random variables, not necessarily independent, with means μ_x, μ_y and variances σ_x^2, σ_y^2. Let X^* and Y^* be the corresponding standardized variables defined by
$$X^* = (X-\mu_x)/\sigma_x \quad \text{and} \quad Y^* = (Y-\mu_y)/\sigma_y.$$

We define $\rho(X, Y) = \text{cov}(X^*, Y^*)$ and call it the *correlation coefficient* of X and Y. We may also write it as $\rho[(x, y)]$, where x and y are the values of X and Y.

(i) Prove $\rho[(x, y)] = \dfrac{\text{cov }[(x, y)]}{\sigma_x . \sigma_y}$.

(ii) Prove var $[x^* \pm y^*] = \text{var }[x^*] \pm 2 \text{ cov}[(x^*, y^*)] + \text{var }[y^*]$
$$= 2\{1 \pm \rho[(x, y)]\}.$$

(iii) Deduce from (ii) that $|\rho[(x, y)]| \leqslant 1$.

(iv) Deduce also from (ii) that $\rho[(x, y)] = 1$ implies that $x^* - y^*$ is a constant and hence that
$$y = \left(\frac{\sigma_y}{\sigma_x}\right) . x + \text{constant}.$$

(v) In Exercise 4H, Question 4 we obtained the line of regression of y on x for a sample of values (x, y) as $y - \bar{y} = (v_{xy}/v_{xx})(x - \bar{x})$ or $y = (v_{xy}/v_{xx})x + \text{constant}$.
In the notation of this Question this becomes
$$y = \frac{\text{cov }[(x, y)]}{\sigma_x^2} . x + \text{constant}.$$

Show that if $\rho[(x, y)] = 1$ this reduces to

$$y = \left(\frac{\sigma_y}{\sigma_x}\right).x + \text{constant}.$$

[It would of course cause no ambiguity to write $\rho[x, y]$ and cov $[x, y]$ instead of $\rho[(x, y)]$ and cov $[(x, y)]$ and the reader may well prefer to do so. We have preferred the longer notation in these early stages to remind us that what we are calculating is a function of points in a two-dimensional space. In the notation of random variables neither X, Y nor (X, Y) immediately suggest a *point*, and we are free to develop without extra brackets the notations $\rho(X, Y)$ and cov (X, Y) to indicate that we are treating a pair of random variables.]

6. PROBABILITY GENERATING FUNCTIONS (CONTINUED)

Suppose that two independent dichotomous trials are being carried out and that the possibility spaces for the trials are identical (namely: success, failure) but that the probability functions are not necessarily identical.

On the first trial we will take: pr (failure) $= a_0$,

$$\text{pr (success)} = a_1;$$

on the second trial we will take: pr (failure) $= b_0$,

$$\text{pr (success)} = b_1;$$

we will then have $a_0 + a_1 = 1$, $b_0 + b_1 = 1$ but not necessarily $b_0 = a_0$, (and hence, of course, not necessarily $b_1 = a_1$).

The p.g.f. for the numbers of successes are given by:

$$\text{for the first trial:} \qquad G(\xi) = a_0 + a_1 \xi;$$

$$\text{for the second trial:} \quad H(\xi) = b_0 + b_1 \xi.$$

The following table shows the probabilities of various numbers of successes if the two trials are carried out successively, and it is here that we require the independence of the trials.

$$\text{pr (0 successes)} = a_0 b_0,$$

$$\text{pr (1 success)} \quad = a_0 b_1 + a_1 b_0,$$

$$\text{pr (2 successes)} = a_1 b_1.$$

These probabilities are the coefficients of ξ^0, ξ^1, ξ^2 in the expansion of $(a_0 + a_1 \xi)(b_0 + b_1 \xi)$; and this enables us to say that the p.g.f. for the number of successes when both trials are carried is given by K, where

$$K(\xi) = G(\xi).H(\xi).$$

The argument is quite general. If X and Y are integral-valued random variables then we can form p.g.f. as:

$$G_X(\xi) = \Sigma a_i \xi^i, \quad G_Y(\xi) = \Sigma b_j \xi^j,$$

where $a_i = \text{pr} (X = i)$ and $b_j = \text{pr} (Y = j)$.

We calculate the probability that the variable $X + Y$ takes the value r. If X, Y are independent random variables, that is random variables arising from independent trials (and we recollect that this does *not* prevent them from being identical random variables), then the required probability is

$$(a_0 b_r + a_1 b_{r-1} + \dots + a_r b_0),$$

which is the coefficient of ξ^r in the expansion of

$$(a_0 + a_1 \xi + a_2 \xi^2 + \dots + \dots)(b_0 + b_1 \xi + b_2 \xi^2 + \dots + \dots).$$

This product thus gives the p.g.f. for $X + Y$, so that we have in our usual notation:

If X and Y are independent random variables then

$$G_{X+Y}(\xi) = G_X(\xi) \cdot G_Y(\xi);$$

and we can prove by induction that if all the random variables involved are independent then:

$$G_{X+Y+\dots+Z}(\xi) = G_X(\xi) \cdot G_Y(\xi) \cdot \dots \cdot G_Z(\xi).$$

Exercise 9E

1 T. Prove that if G and H are p.g.f. and if $K(\xi) = G(\xi) \cdot H(\xi)$ then

(i) K is a p.g.f.

(ii) $K'(1) = G'(1) + H'(1)$

and

$$K''(1) + K'(1) - \{K'(1)\}^2 = G''(1) + G'(1) - \{G'(1)\}^2 + H''(1) + H'(1) - \{H'(1)\}^2.$$

(iii) Interpret the results of part (ii).

(iv) Show that if two Poisson processes are occurring simultaneously then the *total* number of events (or either kind) occurring in the unit of time is also a Poissonian random variable, and that its parameter is the sum of the parameters of the original processes. [Only a few probability functions have this 'self-reproducing' property.]

(v) Let X be the number of multiples of 3 in 36 throws of a die and Y be the number of multiples of 2 in the same 36 throws; (so that a 'six' is counted once for each). Define $T = X + Y$. Show that X, Y are independent, and that $E(T) = 30$ and var $(T) = 17$. Does T have a binomial probability function?

2. T. X and Y are two independent random variables having Poisson probability function with parameters μ_x and μ_y; $T = X + Y$. We saw in Question 1 (iv) that T is another Poisson variable and that its parameter is $\mu_x + \mu_y$. The parts of this Question are designed to lead to the answer to: 'If it is known that $T = n$, what is the probability that $X = k$?'

285

(i) A simulation of the situation was obtained as follows:

X was the number of occurrences of 00 or 22 in 50 random digit pairs;
Y was the number of occurrences of 77 or 99 in 50 random digit pairs.

Thus approximately X, Y were Poisson variables with $\mu_x = \mu_y = 1$. 30 sets of 50 digits were examined and they gave a joint frequency function for X and T, as shown in the table.

$k = 0$	1	2	3	4	5	6	7	
$n =$								
7	0	0	0	1	0	0	0	0	1
6	0	0	0	0	0	1	0		1
5	0	0	0	0	0	0			0
4	0	1	1	0	0				2
3	0	4	0	2					6
2	0	3	5						8
1	5	3							8
0	4								4
	9	11	6	3	0	1	0	0	30

marginal frequencies of $T = n$

marginal frequencies of $X = k$

Check that the *expected* marginal frequency of $T = 2$ is nearly 8.

(ii) The theoretical approach proceeds as follows. Since T and X are not independent, but Y and X are, we have:

$$\text{pr } (X = k | T = n) = \frac{\text{pr } (X = k \text{ and } T = n)}{\text{pr } (T = n)}$$

$$= \frac{\text{pr } (X = k) . \text{pr } (T = n | X = k)}{\text{pr } (T = n)}$$

$$= \frac{\text{pr } (X = k) . \text{pr } (Y = n - k | X = k)}{\text{pr } (T = n)}$$

$$= \frac{\text{pr } (X = k) . \text{pr } (Y = n - k)}{\text{pr } (T = n)} .$$

Show that this leads to a binomial probability function for X.

(iii) Deduce that if P, Q are Poisson variates then so is $P + Q$, but $P - Q$ is *not*.

(iv) Calculate the expected frequencies of $X = 0, 1, 2$ given that $T = 2$ in the table in part (i).

3T. A random variable X has the following probability function.

$$\text{pr } (X = 0) = 1 - \theta, \quad \text{pr } (X = 1) = \theta.$$

(i) Construct the p.g.f. G_X and find $E(X)$ and var (X).

(ii) If N independent determinations of a value of X are made and these are added together to give a value, r, of a random variable R, *write down*, from the last equation of Section 6, $G_R(\xi)$ in terms of $G_X(\xi)$.

Also *write down* $E(R)$ and var (R). Interpret your results.

(iii) Examine the inverse binomial p.g.f. to see whether it can be constructed in the same sort of way.

(iv) Look again at Exercise 9C, Question 29.

4R. Show that the probability generating function for the score on a throw of $2n$ dice is
$$G : \xi \to \left(\frac{\xi(1 - \xi^6)}{6(1 - \xi)} \right)^{2n}.$$

5R. The p.g.f. for the *winnings* in a certain game of chance is G. An entry fee is then arranged to make the game (classically) fair; what is the p.g.f. for the profit a player now makes?

6. G is a probability generating function and $F(\xi) = G(\xi^n)$, for fixed n.
(i) Prove that
$$F'(1) = n . G'(1),$$
$$F''(1) + F'(1) - (F'(1))^2 = n^2 \{ G''(1) + G'(1) - (G'(1))^2 \}.$$

(ii) Interpret these results in terms of random variables.

7. G is a p.g.f. and $H(\xi) = G(\xi^{-1})$.
(i) Interpret H.
(ii) Use H to prove that $\operatorname{var} (X - Y) = \operatorname{var} (X) + \operatorname{var} (Y)$ for independent random variables X, Y.

8T. We recall from Chapter 8 that if a series of events occurs in a one-dimensional space, called 'time', in such a way that they
(1) are independent,
(2) occur singly,
(3) occur uniformly at a rate of λ per unit time.
Then the probability that r of them occur in the time-interval $0 \leqslant T < t$ is given by $p(r, t)$ where
$$\frac{d}{dt} p(r, t) = \lambda \{ p(r - 1, t) - p(r, t) \}.$$

This holds for all $r \geqslant 0$, if we take $p(-1, t) = 0$.
(i) Multiply this equation by ξ^r and sum the corresponding equations from $r = 0$ to infinity, writing
$$\sum_{r=0}^{\infty} p(r, t) . \xi^r \quad \text{as } g(\xi, t)$$
and show that
$$\frac{d}{dt} g(\xi, t) = \lambda \{ \xi . g(\xi, t) - g(\xi, t) \}.$$

(ii) Deduce that
$$g(\xi, t) = g(\xi, 0) . e^{\lambda(\xi - 1) t}.$$

(iii) Explain why we choose $g(\xi, 0) = 1$; fix t so that $\lambda t = \mu$ (this may be thought of as a scale change for time) and write $g\left(\xi, \frac{\mu}{\lambda} \right) = G(\xi)$ to get the usual Poisson p.g.f.

10

SEQUENCES OF EVENTS— MARKOV MODEL

1. INTRODUCTION

In Chapter 7 we met two types of sequence of trials, the characteristic of the first of them (Bernouillian) being that the probability of each possible event was at each stage independent of the results of any of the preceding trials; the process had no memory. Since then we have investigated various situations where a Bernouilli model with a suitable probability allotment might prove adequate.

At the same time in Chapter 7 we obtained the values of various statistics from a sequence of trials defined in Exercise 3 M, Question 56, where the probabilities of various outcomes were not constant, but depended on the results of previous trials.

This is plainly an important type of probability model, since not all situations can be represented by a system with fixed probabilities. The weather each day is not independent of the weather of the day before; machines tend to get out of adjustment and produce successions of defectives; traffic tends to move in queues and not freely, except possibly on motorways; trucks will drive to different unloading bays at a factory depending on the state of occupancy of each bay; cases of disease occur more frequently once started, because of infection.

2. MARKOV CHAINS

In general if we have a sequence of events $A_1, A_2, ..., A_n, ...$ then we can write:

$$\text{pr} (A_1 \text{ and } A_2 \text{ and } ... \text{ and } A_n)$$
$$= \text{pr} (A_1) . \text{pr} (A_2|A_1) ... \text{pr} (A_r|A_1 \text{ and } A_2 \text{ and } A_3 \text{ and } ... A_{r-1}) ...$$
$$\text{pr} (A_n|A_1 \text{ and } A_2 \text{ and } ... \text{ and } A_{n-1}).$$

The Bernouilli system arises by taking

$$\text{pr} (A_r|A_1 \text{ and } A_2 \text{ and } ... A_{r-1}) = \text{pr}(A_r) \quad \text{for all } r.$$

It is necessary that the probability models available to us should be fairly simple in containing few parameters, otherwise too many parameters have to be estimated from the data and the model loses some of its usefulness.

288

The next simplest model arises by taking the probability of a specified result in a trial to depend on the result of no previous trial except at most the immediately preceding one. In symbols:

$$\text{pr}\,(A_r|A_1 \text{ and } A_2 \text{ and } \ldots \text{ and } A_{r-1}) = \text{pr}\,(A_r|A_{r-1}) \quad \text{for all } r.$$

2.1   Definition. If the probability of a given result in a trial depends on the result of no previous trial except *at most* the immediately preceding one (though it may possibly depend on time) then the situation can be fairly simply handled, as we shall see, and sequences of such results are known as *Markov chains* after their first investigator, A. A. Markov. Markov had been studying dependent trials since 1907 and in 1913 published some statistical work on the novel *Eugene Onegin.*

We will confine ourselves to the simplest cases, where

(i) there is only a finite number of possible outcomes of each trial;

(ii) trials take place at a set of discrete instants, which we will call $t = 1, 2, 3, \ldots$ (the intervals between successive trials may vary in length but we wish only to *count* the trials),

(iii) probabilities are not dependent on time.

Exercise 10 A

1 R. One die is kept in the left hand and another die is passed from hand to hand, changing hands if a score of 4 or less is made with the die (or dice) in the left hand. The sequence of results of the trial is the sequence of L, R indicating whether the transferable die is in the left or right hand. Is this sequence a Markov chain?

2 R. (i) Can the July weather at Puddling Regis be well modelled by a Markov process (Exercise 3F, Question 13); can that at Llandrwnch (Exercise 3F, Question 14)?

(ii) Are the processes of Exercise 3M, Questions 61, 62, Markov processes?

3. Does a model with independent trials seem suitable for the following sequence of results?

[Each result is denoted by a digit.]

Make a table of estimates of pr $((r+1)$th result is $i|r$th result is $j)$ for each ordered pair (i, j) taking $r = 1, 2, \ldots, 64$.

> 2143 2123 4143 4343 2341 4143 2321 2123 2123
> 4323 4343 2341 4341 2123 2143 2323 2

★4. A coin is thrown and the sequence of results recorded by a random variable X which takes values $x_1, x_2, x_3, \ldots$ given by

$$x_t = \begin{cases} 0 \text{ if the result is a tail} \\ 1 \text{ if the result is a head} \end{cases} \quad (t = 1, 2, 3, \ldots).$$

We also define $S_t = \sum_1^t x_r$; $d_t = S_t - S_{t-k}$ for a fixed k; $u_t = \sum_1^t S_r$.

(i) Is the sequence $x_1, x_2, ..., x_t, ...$ a Markov chain?

(ii) Define a trial which leads to the results $S_1, S_2, ..., S_t, ...$ Is the sequence $S_1, S_2, ...$ a Markov chain?

(iii) Define a trial which leads to the results $d_{k+1}, d_{k+2}, ...$ Interpret the values of $\frac{1}{k} d_t$ for $t = k+1, k+2, ...$ Is the sequence $d_{k+1}, d_{k+2}, ...$ a Markov chain for $k = 1$? for $k = 2$?

(iv) Is the sequence $u_1, u_2, ... u_t, ...$ a Markov chain?

5T. Carry out some simulations of the following models for successive events, and interpret their various features as models for the situations quoted.

Choose your own values for r, g, c, d when these are not specified.

Initially r red discs and g green discs are in a bag. A disc is drawn at random. It is replaced, but c discs of its own colour and d discs of the different colour are added to the bag before the next draw.

(i) $c = 0, d > 0$. A model for accidents in a factory. The drawing of a red disc (say) represents the occurrence of an accident.

(ii) $c > 0, d = 0$. A model for an epidemic. ('Polya's model'). The drawing of a red disc (say) represents an occurrence of a case of the disease.

(iii) $c = -1, d = 0$. A model for a certain process of selection. What is it called?

(iv) $c = -1, d = -1$. No obvious practical significance.

In each case is the resulting chain of colours drawn a time-independent Markov chain? In each case is the resulting chain of numbers of red discs left in the bag before each trial a time-independent Markov chain?

2.2 Illustration. We now investigate the kind of situation envisaged in Section 2.1. Consider a rabbit-warren which has three rabbit-holes A, B, C, and a rabbit of indecisive tendencies who wanders from hole to hole every hour. Which hole he is likely to go to next depends *only* upon which hole he is in at the moment, and if he moves at all he moves 'at the hour'.

Suppose Figure 10.1 shows diagrammatically the probabilities of various *transitions.*

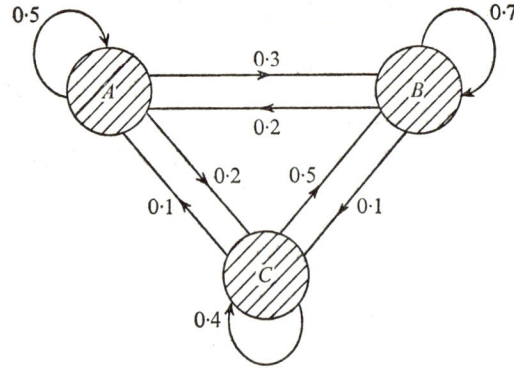

Fig. 10.1

Thus, if the rabbit is in hole B there are probabilities $\begin{pmatrix} 0\cdot2 \\ 0\cdot7 \\ 0\cdot1 \end{pmatrix}$ that he will

next be in holes $\begin{cases} A \\ B \\ C \end{cases}$. Suppose the rabbit is in hole A at a particular moment,

what are the probabilities that he is next in $\begin{cases} A \\ B \\ C \end{cases}$? They are $\begin{pmatrix} 0\cdot5 \\ 0\cdot3 \\ 0\cdot2 \end{pmatrix}$, as inspec-

tion of the figure shows.

2.3 Probability vectors and matrices. What are the probabilities, evaluated at the initial moment (when he is in A, say,) that after his *second* move he will be in A; B; C?

(a) To be in A he must make journeys

$$A \to A \to A \quad \text{or} \quad A \to B \to A \quad \text{or} \quad A \to C \to A.$$

These have probabilities given by

pr (to A from A) $\times$ pr (in A after first move),

pr (to A from B) $\times$ pr (in B after first move),

pr (to A from C) $\times$ pr (in C after first move);

that is: $(0\cdot5 \times 0\cdot5)$, $(0\cdot2 \times 0\cdot3)$, $(0\cdot1 \times 0\cdot2)$.

The sum of these probabilities (they represent *exclusive* possibilities) is what we require and the value can be written

$$(0\cdot5 \quad 0\cdot2 \quad 0\cdot1) \begin{pmatrix} 0\cdot5 \\ 0\cdot3 \\ 0\cdot2 \end{pmatrix}.$$

(b) To find the probability that the rabbit is in B after the second move we evaluate the probabilities for

$$A \to A \to B, \quad A \to B \to B, \quad A \to C \to B;$$

these are $(0\cdot3 \times 0\cdot5)$, $(0\cdot7 \times 0\cdot3)$, $(0\cdot5 \times 0\cdot2)$;

and their sum is $(0\cdot3 \quad 0\cdot7 \quad 0\cdot5) \begin{pmatrix} 0\cdot5 \\ 0\cdot3 \\ 0\cdot2 \end{pmatrix}.$

(c) Similarly to find the probabilities that the rabbit is in C after the second move we evaluate

$$(0\cdot2 \quad 0\cdot1 \quad 0\cdot4) \begin{pmatrix} 0\cdot5 \\ 0\cdot3 \\ 0\cdot2 \end{pmatrix}.$$

All three results may be collected in the statement

$$\begin{pmatrix} \text{pr (hole after 2nd move is } A) \\ \text{pr (hole after 2nd move is } B) \\ \text{pr (hole after 2nd move is } C) \end{pmatrix} = \begin{pmatrix} 0.5 & 0.2 & 0.1 \\ 0.3 & 0.7 & 0.5 \\ 0.2 & 0.1 & 0.4 \end{pmatrix} \begin{pmatrix} 0.5 \\ 0.3 \\ 0.2 \end{pmatrix}.$$

In general, if the probabilities of being in holes A, B, C after r moves are u_r, v_r, w_r and the probability of transition to hole Y from hole X is p_{YX} (note the order of the suffixes) then

$$\begin{pmatrix} u_2 \\ v_2 \\ w_2 \end{pmatrix} = \begin{pmatrix} p_{AA} & p_{AB} & p_{AC} \\ p_{BA} & p_{BB} & p_{BC} \\ p_{CA} & p_{CB} & p_{CC} \end{pmatrix} \begin{pmatrix} u_1 \\ v_1 \\ w_1 \end{pmatrix}.$$

Since the rabbit must be in one of the holes at any moment $u_r + v_r + w_r = 1$ for each r. Since the rabbit must go to or stay in one of the holes at each hour, we also have that $p_{AA} + p_{BA} + p_{CA} = 1$, and similarly for the elements of any other *column* of the matrix.

A vector whose components x all satisfy $0 \leqslant x \leqslant 1$ and also have sum $= 1$ is called a *probability vector*; and a matrix whose elements p_{ij} all satisfy $0 \leqslant p_{ij} \leqslant 1$ and whose columns all have sum 1 is called a *probability matrix* or *stochastic matrix*.

Write $\begin{pmatrix} u_r \\ v_r \\ w_r \end{pmatrix}$ as $\mathbf{u}_r$ and the probability vector representing the initial probabilities as $\mathbf{u}_0$ (it need not be, as here, $\begin{pmatrix} 1 \\ 0 \\ 0 \end{pmatrix}$). Write also the probability matrix as $\mathbf{M}$, then not only do we have $\mathbf{u}_2 = \mathbf{M}\mathbf{u}_1$ as already stated, but also $\mathbf{u}_1 = \mathbf{M}\mathbf{u}_0$, as the reader should verify. From these we deduce $\mathbf{u}_2 = \mathbf{M}^2\mathbf{u}_0$; and the generalization to $\mathbf{u}_r = \mathbf{M}^r\mathbf{u}_0$ is obvious, because of the associativity of matrix multiplication.

The elements of $\mathbf{M}^r$ are also interpretable as transition probabilities because of the associativity of *succession* of transitions, so $\mathbf{M}^r$ is also a probability matrix.

2.4 States and transitions. Instead of concentrating on the rabbit and where it goes, we now concentrate on the warren and how it is occupied by the rabbit. If the rabbit is in hole A we say the warren is in State A. A move of the rabbit is referred to as a transition of the system between states. This allows a more general class of situation to be discussed, where there may be nothing that physically moves from place to place.

292

3. LONG-RUN BEHAVIOUR

3.1 Existence of a limiting behaviour. Does the system (consisting of the inhabited warren) exhibit any tendency towards a definite allotment of probabilities to the various states after a large number of transitions have occurred? *The reader should carry out two or three more steps to observe the tendency.*

To find whether such an allotment of probabilities is approached *whatever* the initial probabilities, we first investigate the matrices $\mathbf{M}^n$, $\mathbf{M}^{n+1}$. Let them be (y_{ij}) and (z_{ij}).

We consider the case when all the elements of $\mathbf{M}$ are positive (not even zero). Then by the column-sum property none is as large as 1 and we have *some* ϵ such that

$$0 < \epsilon < p_{ij} < 1-\epsilon < 1 \quad \text{for all } i, j$$

(and consequently $\frac{1}{2} > \epsilon$).

Now $\mathbf{M}^{n+1} = \mathbf{M}^n . \mathbf{M}$ so

$$\begin{pmatrix} z_{11} & z_{12} & z_{13} \\ z_{21} & z_{22} & z_{23} \\ z_{31} & z_{32} & z_{33} \end{pmatrix} = \begin{pmatrix} y_{11} & y_{12} & y_{13} \\ y_{21} & y_{22} & y_{23} \\ y_{31} & y_{32} & y_{33} \end{pmatrix} \begin{pmatrix} p_{11} & p_{12} & p_{13} \\ p_{21} & p_{22} & p_{23} \\ p_{31} & p_{32} & p_{33} \end{pmatrix}$$

and, for instance,

$$z_{21} = y_{21}p_{11} + y_{22}p_{21} + y_{23}p_{31},$$

but

$$p_{11} + p_{21} + p_{31} = 1,$$

so z_{21} is the co-ordinate on the real line (say an axis of x) of the centroid of masses p_{11}, p_{21}, p_{31} situated at $x = y_{21}, y_{22}, y_{23}$.

To find the extreme possible values of z_{21} we note first that it will lie between the greatest and least values of y_{2j}. We can be more precise.

Let l_2, m_2 be the least and greatest, respectively, of y_{2j}, then by having as much mass at $x = l_2$ as possible, but remembering that at least ϵ is at $x = m_2$ we have

$$(1-\epsilon)l_2 + \epsilon m_2 \leqslant z_{21}$$

as seen in the diagram of Figure 10.2.

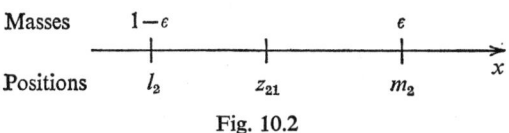

Masses $1-\epsilon$ ϵ

Positions l_2 z_{21} m_2

Fig. 10.2

By reversing the masses we have

$$z_{21} \leqslant \epsilon l_2 + (1-\epsilon)m_2$$

and so

$$(1-\epsilon)l_2 + \epsilon m_2 \leqslant z_{21} \leqslant \epsilon l_2 + (1-\epsilon)m_2$$

or

$$l_2 + h_2\epsilon \leqslant z_{21} \leqslant m_2 - h_2\epsilon, \quad \text{where} \quad h_2 = m_2 - l_2.$$

By the same argument

$$l_2 + h_2\epsilon \leqslant z_{2j} \leqslant m_2 - h_2\epsilon \quad \text{for } j = 2, 3 \text{ also.}$$

We show this diagrammatically in Figure 10.3.

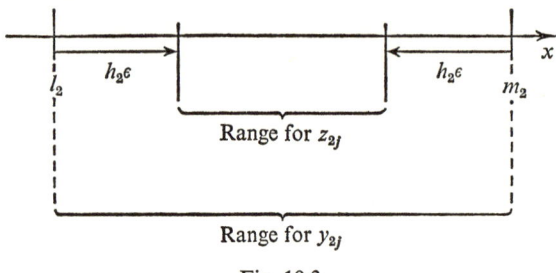

Fig. 10.3

If k_2 is the range for z_{2j} then

$$k_2 \leqslant (1 - 2\epsilon) h_2.$$

But $0 < \epsilon < \frac{1}{2}$ so $0 < 1 - 2\epsilon < 1$, and by repeating the process for the second row of the next power of $\mathbf{M}$ we see that the ranges of variation of the elements of the second rows of successive powers are nested within each other, and tend to zero at least 'geometrically', that is to say at least as fast as the powers of $(1 - 2\epsilon)$. Thus all the elements of the second row have a limit and it is the same limit for each.

Similarly for each other row of the matrix, though the common limit of a row will in general be different from row to row. We may say that a limit of the matrix exists and is a matrix with each row consisting of a repetition of a single element, in other words a matrix whose *columns* are all the same.

Let the common column be $\begin{pmatrix} p_1 \\ p_2 \\ p_3 \end{pmatrix}$

then $\begin{pmatrix} u_n \\ v_n \\ w_n \end{pmatrix} = \mathbf{M}^n \begin{pmatrix} u_0 \\ v_0 \\ w_0 \end{pmatrix}$

$$\rightarrow \begin{pmatrix} p_1 & p_1 & p_1 \\ p_2 & p_2 & p_2 \\ p_3 & p_3 & p_3 \end{pmatrix} \begin{pmatrix} u_0 \\ v_0 \\ w_0 \end{pmatrix}$$

$$= \begin{pmatrix} p_1 (u_0 + v_0 + w_0) \\ p_2 (u_0 + v_0 + w_0) \\ p_3 (u_0 + v_0 + w_0) \end{pmatrix}$$

$$= \begin{pmatrix} p_1 \\ p_2 \\ p_3 \end{pmatrix} \quad \text{since } u_0 + v_0 + w_0 = 1$$

so the limit of $\begin{pmatrix} u_n \\ v_n \\ w_n \end{pmatrix}$ is independent of $\begin{pmatrix} u_0 \\ v_0 \\ w_0 \end{pmatrix}$.

We have proved that if all the elements of a probability matrix are *larger than zero* then the nth power of the matrix 'tends to' a matrix all of whose columns are the same, and that in that case the probability vector tends to just that column whatever the initial probability vector may be.

3.2 Determination of the limiting behaviour. To find the vector $\begin{pmatrix} p_1 \\ p_2 \\ p_3 \end{pmatrix}$ which is the limiting common column and the limiting probability vector, consider that $\mathbf{u}_{n+1} = \mathbf{M}\mathbf{u}_n$ and that $\mathbf{u}_{n+1}$ and $\mathbf{u}_n$ if they have a limit must have the *same* limit, say $\mathbf{u}$, which must therefore satisfy $\mathbf{u} = \mathbf{M}\mathbf{u}$. When we have found such a $\mathbf{u}$ we may put $\begin{pmatrix} p_1 \\ p_2 \\ p_3 \end{pmatrix} = \mathbf{u}$.

To investigate the equation $\mathbf{u} = \mathbf{M}\mathbf{u}$ we rewrite it as

$$\mathbf{I}\mathbf{u} = \mathbf{M}\mathbf{u} \quad \text{or} \quad (\mathbf{M}-\mathbf{I})\,\mathbf{u} = \mathbf{0}.$$

This will have a non-trivial solution for $\mathbf{u}$ provided

$$\det (\mathbf{M}-\mathbf{I}) = 0.$$

That is provided

$$\begin{vmatrix} p_{11}-1 & p_{12} & p_{13} & \cdots & p_{1n} \\ p_{21} & p_{22}-1 & p_{23} & \cdots & \\ \vdots & \vdots & \vdots & & \vdots \\ p_{n1} & \cdots & \cdots & & p_{nn}-1 \end{vmatrix} = 0.$$

(We have dropped the restriction to 3×3 matrices.)

Now the sum of each column of this determinant is zero so the determinant does vanish and a non-trivial solution for $\mathbf{u}$ exists. It may not be unique if some of the p_{ij} are zero.

Example 1. Find the vector(s) $\begin{pmatrix} u \\ v \\ w \end{pmatrix}$ which arises from the rabbit-hole system of Section 2.2, satisfying

$$\begin{pmatrix} 0\cdot5 & 0\cdot2 & 0\cdot1 \\ 0\cdot3 & 0\cdot7 & 0\cdot5 \\ 0\cdot2 & 0\cdot1 & 0\cdot4 \end{pmatrix} \begin{pmatrix} u \\ v \\ w \end{pmatrix} = \begin{pmatrix} u \\ v \\ w \end{pmatrix}.$$

Solution. (An alternative method is indicated in Section 3.4.) We require

$$-0\cdot5u+0\cdot2v+0\cdot1w = 0,$$
$$0\cdot3u-0\cdot3v+0\cdot5w = 0,$$
$$0\cdot2u+0\cdot1v-0\cdot6w = 0,$$

whence

$$-2 \cdot 8u + 1 \cdot 3v = 0 \quad \text{from the 1st and 3rd equations}$$

$$0 \cdot 9u - 1 \cdot 3w = 0 \quad \text{from the 2nd and 3rd equations.}$$

Thus
$$\frac{u}{1 \cdot 3} = \frac{v}{2 \cdot 8} = \frac{w}{0 \cdot 9} = \frac{u+v+w}{5 \cdot 0}$$

gives infinitely many vectors (but all of them are scalar multiples of one another).

We apply the condition $u+v+w = 1$ so that $\begin{pmatrix} u \\ v \\ w \end{pmatrix} = \begin{pmatrix} 0 \cdot 26 \\ 0 \cdot 56 \\ 0 \cdot 18 \end{pmatrix}$ provides us with a suitable vector which is also a probability vector. It is unique.

3.3 Steady state probabilities. If and only if the system has probabilities 0·26, 0·56, 0·18 of being in states A, B, C will it retain fixed probabilities subsequently, but whatever the initial probabilities the probability vector will tend to $\begin{pmatrix} 0 \cdot 26 \\ 0 \cdot 56 \\ 0 \cdot 18 \end{pmatrix}$,

and
$$\begin{pmatrix} 0 \cdot 5 & 0 \cdot 2 & 0 \cdot 1 \\ 0 \cdot 3 & 0 \cdot 7 & 0 \cdot 5 \\ 0 \cdot 2 & 0 \cdot 1 & 0 \cdot 4 \end{pmatrix}^{n} \quad \text{tends to} \quad \begin{pmatrix} 0 \cdot 26 & 0 \cdot 26 & 0 \cdot 26 \\ 0 \cdot 56 & 0 \cdot 56 & 0 \cdot 56 \\ 0 \cdot 18 & 0 \cdot 18 & 0 \cdot 18 \end{pmatrix}$$

as n tends to infinity.

We will call such a vector as $\begin{pmatrix} 0 \cdot 26 \\ 0 \cdot 56 \\ 0 \cdot 18 \end{pmatrix}$ a *steady-state vector* for the system.

It is also called an *equilibrium* vector, and because of the consequences which follow from taking it as the initial vector it is furthermore called a *stationary* vector.

3.4 Alternative determination. An alternative approach to finding **u** and a limiting matrix is *to assume that a limiting matrix exists* and has repeated column $\begin{pmatrix} p_1 \\ p_2 \\ p_3 \end{pmatrix}$, then, since $\mathbf{M} . \mathbf{M}^n = \mathbf{M}^{n+1}$, we take:

$$\mathbf{M} \begin{pmatrix} p_1 & p_1 & p_1 \\ p_2 & p_2 & p_2 \\ p_3 & p_3 & p_3 \end{pmatrix} = \begin{pmatrix} p_1 & p_1 & p_1 \\ p_2 & p_2 & p_2 \\ p_3 & p_3 & p_3 \end{pmatrix}$$

and have $0 \cdot 5p_1 + 0 \cdot 2p_2 + 0 \cdot 1p_3 = p_1$ and two other equations. These equations can be seen to be as before.

The advantage of the solution given in Section 3.2 is that it will lead to certain relations between u, v, w in the case when a non-trivial solution for **u** is not unique, even to within scalar multiples. There is then still something important to say about steady-state vectors although, since the solutions for p_1, p_2, p_3 are also not unique, a limit for the matrix *does not exist*.

3.5 2×2 Probability matrix. In the case of a 2×2 probability matrix all of whose elements are positive, we have the equation

$$\begin{pmatrix} 1-a & b \\ a & 1-b \end{pmatrix} \begin{pmatrix} u \\ v \end{pmatrix} = \begin{pmatrix} u \\ v \end{pmatrix} \quad (0 < a, b < 1)$$

for the steady-state vector $\begin{pmatrix} u \\ v \end{pmatrix}$. Whence

$$\begin{cases} -au + bv = 0, \\ au - bv = 0 \end{cases}$$

or

$$\frac{u}{b} = \frac{v}{a} = \frac{u+v}{a+b}.$$

The required vector is $\dfrac{1}{a+b} \begin{pmatrix} b \\ a \end{pmatrix}$, and the limit of the nth power of the matrix $\begin{pmatrix} 1-a & b \\ a & 1-b \end{pmatrix}$ is $\dfrac{1}{a+b} \begin{pmatrix} b & b \\ a & a \end{pmatrix}$.

4. EIGEN-VECTORS

4.1 General theory. In general an $n \times n$ matrix $\mathbf{M} = (a_{ij})$ (not necessarily stochastic) may admit of a vector **u** and an associated number λ such that

$$\mathbf{Mu} = \lambda \mathbf{u}.$$

Notice that if there is such a pair, say $(\mathbf{u}_1, \lambda_1)$, then $(k\mathbf{u}_1, \lambda_1)$ is also such a pair for any scalar k, since

$$\mathbf{Mu}_1 = \lambda_1 \mathbf{u}_1 \Rightarrow \mathbf{M}(k\mathbf{u}_1) = \lambda_1(k\mathbf{u}_1).$$

We rewrite the equation as

$$\mathbf{Mu} = \lambda \mathbf{Iu}$$

and then as

$$(\mathbf{M} - \lambda \mathbf{I})\,\mathbf{u} = \mathbf{0}.$$

Non-trivial solutions **u** will exist if $\det(\mathbf{M} - \lambda \mathbf{I}) = 0$. This, perhaps unexpectedly, turns our attention from its more natural interest **u** to the number λ and gives an equation of nth degree for λ; namely

$$\begin{vmatrix} a_{11}-\lambda & a_{12} & \cdots & a_{1n} \\ a_{21} & a_{22}\ \lambda & & \\ \vdots & & & \vdots \\ a_{n1} & \cdots & & a_{nn}-\lambda \end{vmatrix} = 0.$$

The roots in λ may be complex, and they may not all be distinct. Each unrepeated root gives rise to a set of vectors (vectors of each set being scalar multiples of one another) obtainable by solving the $(n-1)$ independent linear equations of $(\mathbf{M}-\lambda\mathbf{I})\,\mathbf{u}=\mathbf{0}$ for the $(n-1)$ ratios of the components of $\mathbf{u}$.

The equation for λ (expressed above in determinantal form) is called the *characteristic equation* of the matrix, and its solutions are called *characteristic roots, latent roots* or *eigen-values* for the matrix. The set of roots is called the *spectrum* of the matrix. A vector associated with a root is called a *characteristic vector, latent vector* or *eigen-vector* of the matrix.

We have been dealing so far with column vectors but each eigen-value gives rise also to a row-vector, $\mathbf{v}'$ say, such that $\mathbf{v}'\mathbf{M}=\lambda\mathbf{v}'$. In general the transpose of $\mathbf{u}$, written $\mathbf{u}'$, will differ from $\mathbf{v}'$. When we refer in this chapter to eigen-vectors we will mean eigen column-vectors.

In the case of a probability matrix there is an eigen-value equal to 1, since $\lambda=1$ gives a determinant on the left side of the characteristic equation in which the columns all total to 0. It can be shown that the other eigen-values have modulus not greater than 1 and that any others of modulus 1 are integral roots of 1 (including the possibility that the root $\lambda=1$ may be a multiple root). Only the eigen-vectors associated with the eigen-value 1 give rise to steady-state probability vectors.

4.2 2 × 2 Probability matrices. If we apply the theory of eigen-values to the 2×2 matrix $\begin{pmatrix}1-a & b\\ a & 1-b\end{pmatrix}$ in order to determine whether there are other steady-state vectors, then we find that the characteristic equation is

$$\begin{vmatrix}1-a-\lambda & b\\ a & 1-b-\lambda\end{vmatrix}=0$$

from which $\qquad (1-a-\lambda)\,(1-b-\lambda)-ab=0.$

By inspection one root is 1, as we expect, and the product of the roots is $1-a-b$, so the other root is $1-a-b$.

The second root leads to the pair of equations

$$(1-a)\,u+bv=(1-a-b)\,u,$$
$$au+(1-b)\,v=(1-a-b)\,v,$$

the second being dependent on the first; the solutions being $u=-v$. It is not possible to interpret the resulting vector in terms of probabilities because of the difference in sign of the components.

4.3 Generality. The applications of the theory of eigen-vectors are as widespread as those of linear equations; they extend from deformation of solids to solutions of sets of differential equations governing oscillating systems.

Exercise 10 B

1 R. Interpret the eigen-values and eigen-vectors of the following matrices by illustrating them with diagrams showing the simple geometrical transformations of the plane to which the matrices refer.

$$(i)\ \begin{pmatrix} 2 & 1 \\ 0 & 3 \end{pmatrix}; \quad (ii)\ \begin{pmatrix} 4 & 2 \\ -1 & 1 \end{pmatrix}; \quad (iii)\ \begin{pmatrix} 0 & 3 \\ 2 & 0 \end{pmatrix}; \quad (iv)\ \begin{pmatrix} 2 & 0 \\ 0 & 3 \end{pmatrix}.$$

In each case give an eigen-vector whose modulus is 1, an eigen-vector the sum of whose components is 1 and an eigen-vector whose component of largest modulus has modulus 1.

2. A smooth uniform circular ring of mass M, radius a and centre C hangs smoothly from a point O of itself on a fixed pin, and a bead B of mass m slides freely along the ring. The system is set in oscillation and θ, ϕ are the angular co-ordinates shown in Figure 10.4. It can be proved that

$$(2M+m)\,\ddot{\theta}+m\ddot{\phi}+(M+m)\,n^2\theta = 0,$$
$$\ddot{\theta}+\ \ddot{\phi}+\qquad\quad n^2\phi = 0,$$

where $n^2 = g/a$.

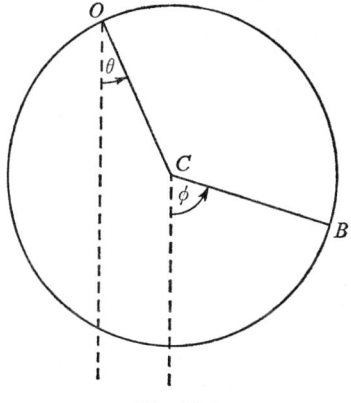

Fig. 10.4

(i) By writing $\begin{pmatrix} \theta \\ \phi \end{pmatrix}$ as **u** and making the substitution $\mathbf{u} = \begin{pmatrix} a \\ b \end{pmatrix} e^{\lambda jt}$, where a, b are constants, show that we require

$$\begin{pmatrix} (M+m)\,n^2-(2M+m)\,\lambda^2 & -m\lambda^2 \\ -\lambda^2 & n^2-\lambda^2 \end{pmatrix} \begin{pmatrix} a \\ b \end{pmatrix} = \mathbf{0}.$$

(ii) Choose the values of λ^2 for this to have non-trivial solutions for $\begin{pmatrix} a \\ b \end{pmatrix}$ and for each value of λ determine $\begin{pmatrix} a \\ b \end{pmatrix}$ to within a scalar multiple.

(iii) If μ^2 is a possible value of λ^2 and $\begin{pmatrix} c \\ d \end{pmatrix}$ is the associated vector $\begin{pmatrix} a \\ b \end{pmatrix}$ then

299

$\begin{pmatrix} \theta \\ \phi \end{pmatrix} = \begin{pmatrix} c \\ d \end{pmatrix} (A e^{\mu jt} + B e^{-\mu jt})$ for arbitrary A, B is called a *normal mode* of oscillation of the system.

Interpret the simpler of the two normal modes for the given system in terms of the geometry of the system.

3. Follow the steps of the following iterative process with approximations and use the method to find an eigen-vector of

$$10^{-1} \begin{pmatrix} 4 & 3 & 1 \\ 3 & 1 & 2 \\ 3 & 6 & 7 \end{pmatrix}.$$

The process:

$$\mathbf{M} = 10^{-1} \begin{pmatrix} 1 & 3 & 5 \\ 4 & 2 & 1 \\ 5 & 5 & 4 \end{pmatrix}$$

$$\Rightarrow \mathbf{M}^2 = 10^{-2} \begin{pmatrix} 38 & 34 & 28 \\ 17 & 21 & 26 \\ 45 & 45 & 46 \end{pmatrix} \simeq 10^{-1} \begin{pmatrix} 4 & 3 & 3 \\ 2 & 2 & 3 \\ 4 & 5 & 4 \end{pmatrix}$$

$$\Rightarrow \mathbf{M}^4 \simeq 10^{-2} \begin{pmatrix} 34 & 33 & 33 \\ 24 & 25 & 24 \\ 42 & 42 & 43 \end{pmatrix}.$$

$$\Rightarrow \text{An eigen-vector is nearly } \begin{pmatrix} 0.33 \\ 0.24 \\ 0.43 \end{pmatrix}.$$

Check
$$10^{-1} \begin{pmatrix} 1 & 3 & 5 \\ 4 & 2 & 1 \\ 5 & 5 & 4 \end{pmatrix} \begin{pmatrix} 0.33 \\ 0.24 \\ 0.43 \end{pmatrix} \simeq \begin{pmatrix} 0.32 \\ 0.22 \\ 0.46 \end{pmatrix}.$$

4 R. Find exactly the eigen-vectors corresponding to eigen-value 1 for each of the five probability matrices of Question 3.

5. Obtain the stationary probabilities of the Markov chain of Exercise 3 M, Question 56 (*a*) and Chapter 7, Section 5.1. What is the estimate from Chapter 7?

6 R. In my grocer's I buy one of two varieties of instant coffee. If, from my previous purchase, I have a coupon for a price reduction on brand 'Hotpot' then I certainly buy Hotpot, otherwise, since 'Cofcup' give away plastic swimming pools, I spin a coin to decide between them. On 50 per cent of occasions that I buy Hotpot I lose the coupon. Find the long-run probability that I buy Hotpot.

7. (i) Show that if in a plant-breeding experiment the members of each generation of plants of a particular species are always crossed with plants of aa-genotype then the long term effect will be to make all the plants of type aa.

(ii) Show that if the plants are always crossed with a known heterozygote (Aa-genotype) then the proportions of the three genotypes will tend towards 1:2:1.

[See also Exercise 10 E, Question 4 (iv).]

8. Mr Mallet, Miss Hoop and I were stranded on a desert island during a cruise in the Bahamas, and we whiled away the time playing croquet together in pairs, the winner of each game next playing the person just left out. It was now a very small club, but our performances remained consistently as in the happier times of Puddling Regis (See Exercise 3 M, Question 12). The weather was, however, much more reliable and we nominally played 60 hours a week, though we spent on average 1 hour 10 minutes per day changing over and, of course, we didn't play on Sundays.

Show that Miss Hoop had 12 hours of spectating a week, that Mallet had 20 and that I had 21.

9 T. (This Question refers to the rabbit problem of Section 2.2.) Use random numbers with digits suitably allotted for each probability and start with 8 rabbits in hole A, 0 in hole B and 2 in hole C. Allow the rabbits to diffuse through the warren independently of each other for half-a-dozen moves each and compare the final states with that predicted by the model.

10 T. Imagine the volume of a vessel divided into a large number, k, of parts, the occupancy of each representing a state for a single particle moving in the vessel. Let there be N particles filling the vessel and moving independently of each other in such a way that for each particle the stationary probability vector for the states has all its components equal to $1/k$.

(i) Consider a fixed value of i; if there are n_i particles in the ith part, what is the expected value of n_i?

(ii) Use a Poisson approximation to find the variance of n_i.

(iii) Show that the coefficient of variation of n_i is $100(k/N)^{\frac{1}{2}}$ per cent. Evaluate this for $k = 10^6$, $N = 10^{20}$.

[This question illustrates in a small way the type of result important in modern physics where the behaviour of individual members of a large assemblage (say of electrons in a metal or of molecules in a gas cloud) may be unpredictable in detail, being clearly only probabilistically known, but where the behaviour of the assemblage appears to be almost known by deterministic laws.]

11. Modify the rabbit hole problem and its solution in Question 9 as follows.
Let there be only 2 holes and let the matrix of transition probabilities be

$$\begin{pmatrix} 0{\cdot}8 & 0{\cdot}3 \\ 0{\cdot}2 & 0{\cdot}7 \end{pmatrix}.$$

Release 5 rabbits in each hole and by simulation with random numbers allow diffusion to take place for 5 transitions.

Repeat 5 times and take the mean of your vectors as an estimate of the relevant eigen vector for the matrix.

[This amounts to a crude Monte Carlo method for finding the eigen-vectors of a matrix, which might originally occur in a problem having no connexion with probabilities. It would need minor modifications if the sum of each column was not 1.]

12.

From ...		A	B	C	D	E	F
To	A	0	$\frac{1}{5}$	0	0	0	0
	B	1	0	$\frac{2}{5}$	0	0	0
	C	0	$\frac{4}{5}$	0	$\frac{3}{5}$	0	0
	D	0	0	$\frac{3}{5}$	0	$\frac{4}{5}$	0
	E	0	0	0	$\frac{2}{5}$	0	1
	F	0	0	0	0	$\frac{1}{5}$	0

$.$

This gives a statistical model (known as the Ehrenfest model) of a system with diffusion but with an elastic force tending to concentrate particles at the centre, the force increasing with distance from the centre.

Find the steady-state probability vector. Should you have foreseen this result?

13. To discuss the lengths of runs of successes in a sequence of Bernouilli trials we can say that the system of results is in state E_0 at the nth trial if the nth trial results in a failure but is in state E_r ($r = 1, 2, 3$) if the last failure occurred at trial number $n - r$. (We conventionally take a zeroth (non-existent) trial as a failure, to start the sequence.) To maintain finiteness we might say the system is in state E_4 if the last success occurred at trial k where $k \leqslant n - 4$.

Taking non-multiplies of 3 as success (A), and multiples of 3 as failure (B) on a fair die the following results (excluding trial 0) occurred:

ABBAAAAAAABBAABAAABBA.

This leads to the succession of states (including the conventional initial state)

$(E_0)\ E_1 E_0 E_0 E_1 E_2 E_3 E_4 E_4 E_4 E_4 E_0 E_0 E_1 E_2 E_0 E_1 E_2 E_3 E_0 E_0 E_1.$

Obtain the transition matrix and the steady-state vector from the model.

What is the relevance of all this to Table 7.3 and to Chapter 7, Section 7.2?

5. CLASSES OF STATES

5.1 Introduction. Consider the matrix

$$\mathbf{M} = \begin{pmatrix} 0\cdot 8 & 0\cdot 3 & 0 & 0 \\ 0\cdot 2 & 0\cdot 7 & 0 & 0 \\ 0 & 0 & 0\cdot 5 & 0\cdot 6 \\ 0 & 0 & 0\cdot 5 & 0\cdot 4 \end{pmatrix}.$$

The steady-state vector for

$$\mathbf{K} = \begin{pmatrix} 0\cdot 8 & 0\cdot 3 \\ 0\cdot 2 & 0\cdot 7 \end{pmatrix} \quad \text{is} \quad \begin{pmatrix} \frac{3}{5} \\ \frac{2}{5} \end{pmatrix}$$

and for

$$\mathbf{L} = \begin{pmatrix} 0\cdot 5 & 0\cdot 6 \\ 0\cdot 5 & 0\cdot 4 \end{pmatrix} \quad \text{is} \quad \begin{pmatrix} \frac{6}{11} \\ \frac{5}{11} \end{pmatrix}$$

and multiplication shows that

$$\mathbf{v}_1 = \begin{pmatrix} \frac{3}{5} \\ \frac{2}{5} \\ 0 \\ 0 \end{pmatrix} \quad \text{and} \quad \mathbf{v}_2 = \begin{pmatrix} 0 \\ 0 \\ \frac{6}{11} \\ \frac{5}{11} \end{pmatrix}$$

are both steady-state vectors for $\mathbf{M}$; a little further multiplication shows more than this: namely, that $p\mathbf{v}_1 + q\mathbf{v}_2$ is a steady-state vector for all non-negative p, q satisfying $p + q = 1$. (The eigen-value 1 is a double root.)

The matrix examined in Section 2 possessed one and only one steady-state vector. The case above represents a situation where states A and B

can be reached from each other as can states C and D, but passage between
the first set and the second set is impossible.

The states A, B, C, D are divided into *classes* by the equivalence relation:

'X, Y can be reached from each other

or X is the same state as Y'.

The example illustrated in Figure 10·5 shows that the relation 'X and Y
can be reached from each other' may not be reflexive, and so is not an
equivalence relation. A cannot be reached *from* A.

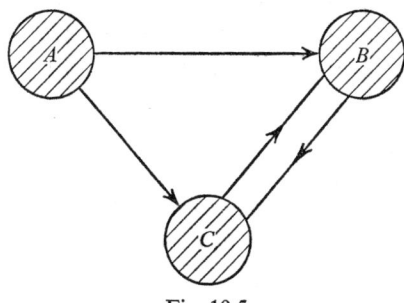

Fig. 10.5

The equivalence classes of the proper equivalence relation are, in this
illustration, $\{A\}$ and $\{B, C\}$.

Exercise 10 C

1R. Examine the effect of the following transition matrix on a particle which
moves to and from each of five positions, A, B, C, D, E.

$$
\begin{array}{cc}
\text{From} \quad \ldots & \begin{array}{ccccc} A & B & C & D & E \end{array} \\
\begin{array}{c} \text{To} \quad A \\ B \\ C \\ D \\ E \end{array} &
\begin{pmatrix}
1 & p & 0 & 0 & 0 \\
0 & 0 & p & 0 & 0 \\
0 & 1-p & 0 & p & 0 \\
0 & 0 & 1-p & 0 & 0 \\
0 & 0 & 0 & 1-p & 1
\end{pmatrix}.
\end{array}
$$

Give two possible steady-state vectors; and the general steady-state vector.

[The motion of the particle is sometimes described as a *random walk*. The
behaviour of positions A and E leads them to be called *absorbing barriers*.]

What are the classes of states?

2R. Discuss the random walk described by the following matrix.

$$
\begin{array}{cc}
\text{From} \quad \ldots & \begin{array}{ccccc} A & B & C & D & E \end{array} \\
\begin{array}{c} \text{To} \quad A \\ B \\ C \\ D \\ E \end{array} &
\begin{pmatrix}
1 & \tfrac{1}{3}p & 0 & 0 & 0 \\
0 & \tfrac{2}{3}p & p & 0 & 0 \\
0 & 1-p & 0 & p & 0 \\
0 & 0 & 1-p & 1-p & 0 \\
0 & 0 & 0 & 0 & 1
\end{pmatrix}.
\end{array}
$$

[The behaviour of A is described by saying that there is an *elastic barrier* between A and B, and that of E by saying that there is a *reflecting barrier* between D and E.]

Give two possible steady-state probability vectors.

What are the classes of states?

5.2 Transient and closed classes. Consider the transition matrix

$$\begin{pmatrix} \times & . & . & . & \times & . & . \\ . & . & . & \times & . & . & . \\ . & . & \times & \times & \times & . & . \\ . & \times & \times & . & . & . & . \\ . & . & . & . & \times & \times & \times \\ . & . & . & . & \times & \times & . \\ . & . & . & . & . & \times & . \end{pmatrix},$$

where $\times$ denotes a positive (non-zero) element and . denotes a zero. This may be partitioned as follows in a way that relates to Figure 10.6 showing the possible transitions.

From ...	E_1	F_1	F_2	F_3	G_1	G_2	G_3
To E_1	$\times$	.	.	.	$\times$	.	.
F_1	.	.	.	$\times$	.	.	.
F_2	.	.	$\times$	$\times$	$\times$	.	.
F_3	.	$\times$	$\times$	.	.	.	.
G_1	.	.	.	.	$\times$	$\times$	$\times$
G_2	.	.	.	.	$\times$	$\times$	.
G_3	.	.	.	.	.	$\times$	.

The states fall into classes which may be diagrammatically represented as Figure 10.7.

The class G of states is called a *transient* class since in the long run it will be vacated and once it is vacated there is no return to it. The states of G will have zeros in any steady-state vector there may be. Classes E and F are called *closed* since once they are entered there is no leaving them. State E_1, being the only state in a closed class, is called *absorbing*.

The reader is warned that writers differ widely over their nomenclature in this matter and that one of the standard works (W. Feller: *An Introduction to Probability Theory and its Applications*, vol. I, Wiley 1965) differs in nomenclature between editions!

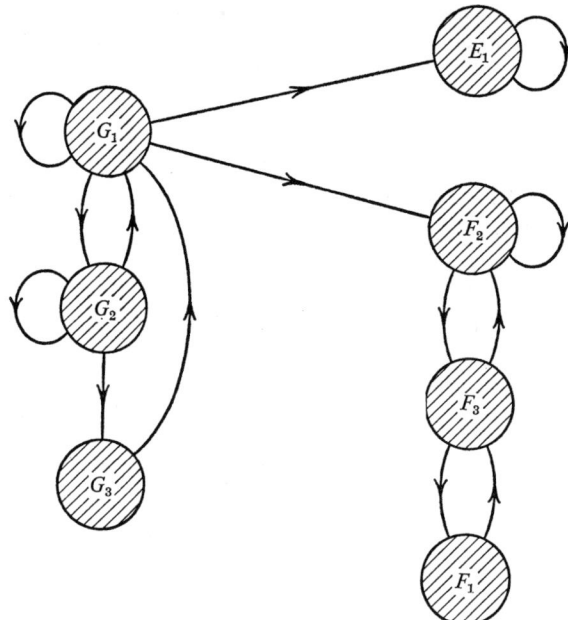

Fig. 10.6

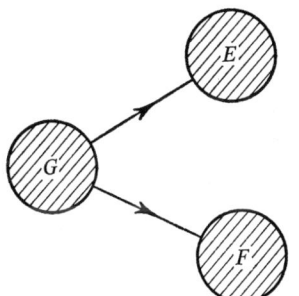

Fig. 10.7

Exercise 10 D

1 R. Consider the following transition matrix

$$
\begin{array}{c}
\quad\quad E_1 \; E_2 \; E_3 \; E_4 \; E_5 \; E_6 \\
\begin{array}{c}
E_1 \\ E_2 \\ E_3 \\ E_4 \\ E_5 \\ E_6
\end{array}
\left(
\begin{array}{cccccc}
\times & . & . & \times & \times & . \\
\times & . & . & . & . & \times \\
\times & . & \times & . & . & . \\
. & . & . & . & \times & . \\
. & . & . & . & \times & . \\
. & x & . & . & . & .
\end{array}
\right)
\end{array}.
$$

Draw a diagram to show the possible transitions and rearrange the orders of columns and rows to enable the matrix to be partitioned as was the matrix in Section 5.2. Which states form transient classes and which closed? Are any states absorbing?

(*Warning*. There are 5 classes, not 3.)

5.3 Refinements of classification. If any state whatever of a system can be reached from any other then the chain is called *irreducible*. Another way of saying this is to say that if the only closed class contains all the states then the chain is irreducible.

To be able to get from one state to another requires a non-zero element in the appropriate position in some power of the transition matrix, but there may be no *single* power of the transformation matrix which contains non-zero elements *in all positions*. For instance, in the diagram of Figure 10.8 it is possible to get from the subset of states $\{A, C\}$ to the subset $\{B, D\}$ or *vice versa* in an *odd* number of transitions, but an *even* number will return the system to the subset of states from which it started.

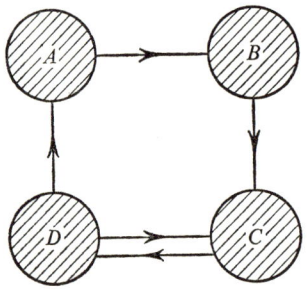

Fig. 10.8

Any state can be reached from any other so this chain is irreducible; in other words, all the states are in the same closed class. This class is said to be *periodic*, of period 2.

If there is a single power of the matrix containing only non-zero elements then the chain is certainly irreducible; it is further called *regular*. Suppose that the kth power is the smallest such power of the matrix and that $n (> k)$ transitions have taken place, then whatever the initial state there is now no state ruled out as impossible for the present state. We have seen that the nth power of a matrix of only non-zero elements tends to a matrix all of whose elements are the same and positive. Thus the nth power of the matrix of a regular chain tends to such a limit. It is then simple to determine the steady state by iteration.

It is apparent that a matrix which has no zeros on the main diagonal and none on the neighbouring diagonals will represent a system in which any state can be reached from any other and will thus give an irreducible chain. It can be shown that in fact it gives a regular chain.

Exercise 10E

1 R. Construct a diagram like Figure 10.6 to show a system of states having an absorbing state, another closed class (this one being periodic with period 3) and three other states falling into two transient classes. There is to be exactly *one* state *from* which any other state, including itself, can be reached ultimately. [If state y can be reached from state x ultimately, y is said to be *consequent* to x.] Label the states and give a correspondingly labelled scheme for a partitioned matrix for the system using × to denote non-zero elements.

2. 4 red discs are in 1 bag and 4 green discs in another. The number of red discs in the first bag is the state of the system. At each transition a disc is taken from each bag and placed in the other. Is the chain regular? Find the steady-state probability vector and verify that in the steady state the probability of being in state r is the probability of drawing exactly r red discs when 4 discs are drawn from a bag of 4 red and 4 green discs; so that we have shuffled effectively.

3 R. Consider the Markov sequence defined by

$$\mathbf{M} = \begin{pmatrix} 0 & 0 & 1 & 0 \\ 0 & 0 & 0 & 1 \\ \frac{3}{4} & \frac{1}{2} & 0 & 0 \\ \frac{1}{4} & \frac{1}{2} & 0 & 0 \end{pmatrix}.$$

(i) How many classes are there? Find the steady-state vector(s).

(ii) Evaluate $\mathbf{N} = \mathbf{M}^2$. Show that there are two closed classes in the Markov system defined by $\mathbf{N}$, and find the steady-state vector(s).

(iii) If the initial probability vector is

$$\mathbf{a} = \begin{pmatrix} \frac{1}{5} \\ \frac{1}{5} \\ \frac{2}{5} \\ \frac{1}{5} \end{pmatrix},$$

find for each of the closed classes of the N-chain the probability of ending in that class.

(iv) Use the results of (iii) to find the steady-state probability vector resulting from the initial vector **a**.

[The chain defined by N is periodic of period 2.]

4. Classify the states corresponding to the given matrices using the following scheme of analysis (which is *not exhaustive*):

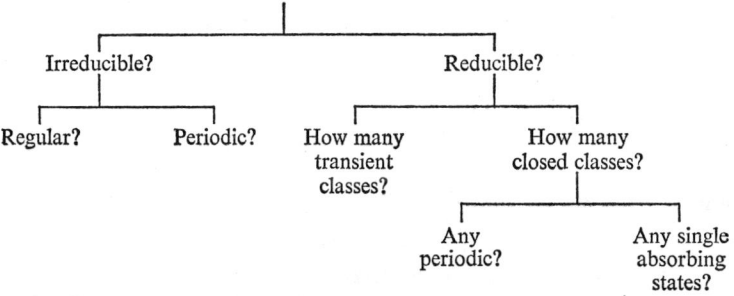

Draw the diagrams to illustrate the possible transitions.

Also describe the behaviours as the number of transitions tends to infinity.

(i) $\begin{pmatrix} 0 & \frac{1}{2} & \frac{1}{2} \\ \frac{1}{2} & 0 & \frac{1}{2} \\ \frac{1}{2} & \frac{1}{2} & 0 \end{pmatrix}.$
(ii) $\begin{pmatrix} 0 & 0 & \frac{1}{2} & 0 \\ 0 & 0 & \frac{1}{2} & 0 \\ 0 & 0 & 0 & 1 \\ 1 & 1 & 0 & 0 \end{pmatrix}.$
(iii) $\begin{pmatrix} \frac{1}{2} & \frac{1}{4} & \frac{1}{2} & 0 & 0 \\ 0 & \frac{1}{2} & 0 & 0 & 0 \\ \frac{1}{2} & \frac{1}{4} & \frac{1}{2} & 0 & 0 \\ 0 & 0 & 0 & \frac{1}{2} & \frac{1}{2} \\ 0 & 0 & 0 & \frac{1}{2} & \frac{1}{2} \end{pmatrix}.$

(iv) The matrices for Exercise 10B, Question 7.
(v) The matrix for Exercise 10B, Question 13.

*5T. Let $\mathbf{M}$ be an $n \times n$ probability matrix with n distinct eigen-vectors $\mathbf{u}_1$, $\mathbf{u}_2$, ..., $\mathbf{u}_n$ corresponding to n distinct eigen-values $1, \lambda_2, ..., \lambda_n$. Choose $\mathbf{u}_1$ to be a probability vector. Let $\mathbf{S}$ be the matrix $(\mathbf{u}_1, \mathbf{u}_2, ..., \mathbf{u}_n)$, that is, whose columns are the eigen-vectors. Then it can be shown that

$$\mathbf{M} = \mathbf{S}. \begin{pmatrix} 1 & 0 & . & & 0 \\ 0 & \lambda_2 & . & & . \\ . & . & . & & . \\ . & . & & . & 0 \\ 0 & . & . & 0 & \lambda_n \end{pmatrix} .\mathbf{S}^{-1},$$

and that $|\lambda_i| \le 1$ for $2 \le i \le n$.
(i) Prove that

$$\mathbf{M}^r = \mathbf{S}. \begin{pmatrix} 1 & 0 & . & . & 0 \\ 0 & \lambda_2^r & . & . & . \\ . & . & . & & . \\ . & . & & . & 0 \\ 0 & . & . & 0 & \lambda_n^r \end{pmatrix} .\mathbf{S}^{-1}.$$

(ii) Verify the given form for $\mathbf{M}$ when $\mathbf{M} = \begin{pmatrix} \frac{1}{2} & \frac{3}{4} \\ \frac{1}{2} & \frac{1}{4} \end{pmatrix}$ and show that differences between the terms of $\mathbf{M}^6$ and the limit as $r \to \infty$ of $\mathbf{M}^r$ do not exceed $\frac{3}{20480}$.

(iii) State a sufficient condition on the eigen-values of a probability matrix for its rth power to tend to a limit.

*6T. Construct a 2×2 probability matrix with eigen-values $1, -1$. Verify the above form for $\mathbf{M}$ and that $\mathbf{M}^r$ does not tend to a limit. Discuss the nature of the class(es) of states of the system.

*7T. The matrix $\begin{pmatrix} 0 & 0 & 1 \\ 1 & 0 & 0 \\ 0 & 1 & 0 \end{pmatrix}$ defines a system with a single (periodic) class (period 3). What are the eigen-values?

6. BEHAVIOUR BEFORE THE LIMITING SITUATION

Consider the system given by Figure 10.9 with matrix:

$$\begin{pmatrix} 1-(b+c) & 0 & 0 \\ b & p & 0 \\ c & 1-p & 1 \end{pmatrix}.$$

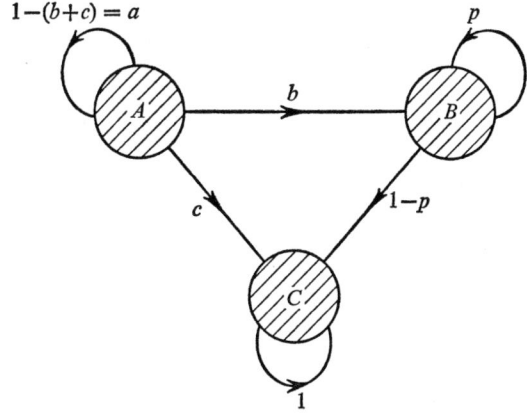

Fig. 10.9

There are three classes:

> that containing state A is transient;
> that containing state B is transient;
> that containing state C is closed.

C is an absorbing state.

We set ourselves various problems.

6.1   Problem 1. If the initial probability vector is $\begin{pmatrix} 1 \\ 0 \\ 0 \end{pmatrix}$ so that the system is in state A, what is the probability that it will remain for r time units in state A? (*It remains for* 1 *unit if it leaves at the first transition.*) What is the expected length of stay in A?

Solution. It is important not to confuse the *length of stay* with the *waiting time* of Chapter 7. *To remain for* r *time units it must leave on the* r*th transition.*

The result is expressed by

$$\text{pr (remains for } r \text{ units)} = \{1-(b+c)\}^{r-1}(b+c).$$

309

The expected length of stay is

$$\sum_{r=1}^{\infty} (b+c)\, ra^{r-1}, \qquad \text{where} \quad a = 1-(b+c),$$

$$= \frac{b+c}{(1-a)^2} = \frac{1}{b+c}.$$

6.2 Problem 2. If the system starts in A, as before, what is the probability that it will ever reach B? And what is the expected time to reach B?

First solution (for first part). Let x be the probability that the system ultimately reaches state B from state A; *this is independent of the time at which we start in state A,* hence the occurrence of the x in the tree-diagram Figure 10.10.

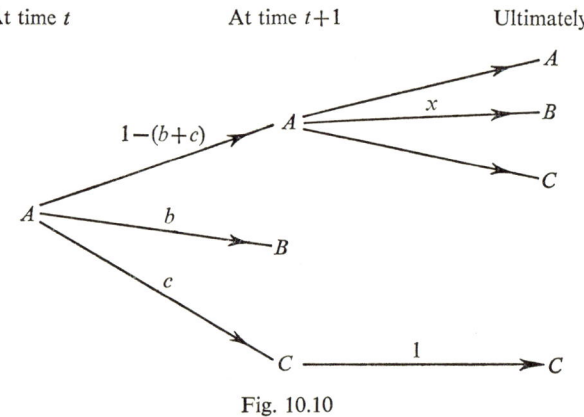

At time t At time $t+1$ Ultimately

Fig. 10.10

We thus have the probability $\{1-(b+c)\}\,x+b$ shown on the branches; but this is itself equal to x, giving

$$\{1-(b+c)\}\,x+b = x,$$

whence $\qquad x = \dfrac{b}{b+c}, \qquad$ *as we might have foreseen.*

This method of attack does not yield the expected time to B.

Second solution. The probability of reaching B on the rth transition is $\{1-(b+c)\}^{r-1}\, b$.

The required probability is

$$\sum_{r=1}^{\infty} \{1-(b+c)\}^{r-1}\, b = \frac{b}{1-(1-(b+c))}$$

$$= \frac{b}{b+c}.$$

For those systems that reach state B at all, the expected time to B is

$$\sum_{r=1}^{\infty} b.ra^{r-1}, \quad \text{where} \quad a = 1-(b+c),$$

$$= \frac{b}{(1-a)^2} = \frac{b}{(b+c)^2}.$$

Taking into account the non-zero probability, $c/(c+b)$, of never reaching B, the expected time to B might be said to be 'infinite'.

6.3 Problem 3. If the system starts in A, as before, what is the probability that it will reach C after r units exactly?

Solution. Let u_r be the probability of reaching C from A (not necessarily directly) for the first time after r units. Let v_r be the probability of reaching C from B for the first time after r units.

We have the tree diagram, Figure 10.11.

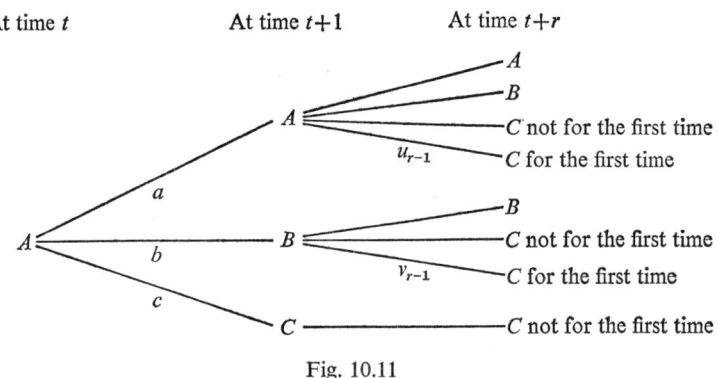

Fig. 10.11

Thus
$$\begin{cases} u_r = a.u_{r-1}+b.v_{r-1} & (r \geqslant 2), \\ u_1 = c. \end{cases}$$

By a similar argument $\quad v_{r-1} = p.v_{r-2};$

whence $\quad v_{r-1} = p^{r-2}(1-p), \quad$ since $v_1 = 1-p.$

Taking these together $\quad u_r = au_{r-1}+bp^{r-2}(1-p)$

and $\quad u_{r-1} = au_{r-2}+bp^{r-3}(1-p)$

$$\vdots \qquad \vdots \qquad \vdots$$

$$u_3 = au_2 + bp(1-p)$$

$$u_2 = au_1 + b(1-p).$$

Multiplying by $1, a, a^2, ..., a^{r-2}$ in turn and adding we have:

$$u_r = a^{r-1}u_1+b(1-p)[p^{r-2}+p^{r-3}a+...+1.a^{r-2}]$$

311

so $$u_r = a^{r-1}.c + b(1-p).\left(\frac{p^{r-1} - a^{r-1}}{p-a}\right) \quad \text{for } r \geqslant 2,$$

and this formula gives correct result for $r = 1$, also.

6.4 Problem 4. What is the expected time to reach C (after which the chain is of no further interest)? We give four solutions, of differing degrees of interest, elegance and importance.

First Solution (using the evaluated probabilities).

T_{CA}, the expected time to C from A is given by

$$\sum_{r=1}^{\infty} r.u_r = \sum_{r=1}^{\infty} \left[c.ra^{r-1} + \frac{b(1-p)}{p-a}(rp^{r-1} - ra^{r-1}) \right]$$

$$= \frac{c}{(1-a)^2} + \frac{b(1-p)}{p-a}\left[\frac{1}{(1-p)^2} - \frac{1}{(1-a)^2}\right]$$

$$= \frac{c}{(1-a)^2} + \frac{b(1-p)}{p-a} \cdot \frac{(p-a)\,[(1-p)+(1-a)]}{(1-p)^2\,(1-a)^2}$$

$$= \frac{c(1-p) + b(1-p) + b(1-a)}{(1-a)^2\,(1-p)}$$

$$= \frac{(1-p)\,(1-a) + b(1-a)}{(1-a)^2\,(1-p)}$$

$$= \frac{1 + b/q}{1-a}, \quad \text{where } q = 1-p.$$

Second Solution (using the recurrence relations for u_r, v_r but not solving them).

The result could have been obtained as follows:

$$u_r = au_{r-1} + bv_{r-1} \quad (r \geqslant 2).$$

Multiply by r and sum from $r = 2$ to ∞;

then $$\sum_{2}^{\infty} ru_r = a.\sum_{2}^{\infty} ru_{r-1} + b.\sum_{2}^{\infty} rv_{r-1},$$

or $$T_{CA} - 1.u_1 = a.\sum_{2}^{\infty}[(r-1)u_{r-1} + u_{r-1}] + b.\sum_{2}^{\infty}[(r-1)v_{r-1} + v_{r-1}],$$

whence $$T_{CA} - c = a[T_{CA} + 1] + b[T_{CB} + 1].$$

To find T_{CB}: $$v_r = p.v_{r-1}, \quad \text{for } r \geqslant 2,$$

so $$\sum_{r=2}^{\infty} rv_r = p.\sum_{r=2}^{\infty} rv_{r-1}$$

or $$T_{CB} - 1.(1-p) = p.\sum_{r=2}^{\infty}[(r-1)v_{r-1} + v_{r-1}], \quad \text{since } v_1 = 1-p,$$

or $$T_{CB} - (1-p) = p.[T_{CB} + 1]$$

whence $$T_{CB} = 1/(1-p), \quad \text{and the result follows.}$$

Third solution (using probability generating functions).

Multiply the recurrence relation for u_r by ξ^r to get

$$\xi^r . u_r = a\xi^r . u_{r-1} + b . \xi^r . v_{r-1} \quad (r \geqslant 2),$$

whence
$$\sum_{r=2}^{\infty} \xi^r u_r = a\xi \sum_{r=2}^{\infty} \xi^{r-1} u_{r-1} + b . \xi \sum_{r=2}^{\infty} \xi^{r-1} v_{r-1};$$

then using $u_1 = c$, $u_0 = v_0 = 0$ and writing the generating functions for u_r and v_r as U, V we have

$$U(\xi) - c\xi = a\xi . U(\xi) + b\xi . V(\xi).$$

Multiplying $v_{r-1} = pv_{r-2}$ by ξ^{r-1} and adding for $r \geqslant 2$

$$\sum_{r=2}^{\infty} \xi^{r-1} v_{r-1} = p . \xi \sum_{r=2}^{\infty} \xi^{r-2} v_{r-2},$$

then using $v_1 = (1-p)$; $v_0 = 0$

$$V(\xi) - (1-p) \xi = p\xi . V(\xi).$$

Differentiating these results with respect to ξ we have:

$$U'(\xi) - c = a\xi U'(\xi) + aU(\xi) + b\xi V'(\xi) + bV(\xi),$$

$$V'(\xi) - (1-p) = p\xi V'(\xi) + pV(\xi),$$

putting $\xi = 1$ gives
$$T_{CA} - c = aT_{CA} + a + bT_{CB} + b,$$

$$T_{CB} - (1-p) = p . T_{CB} + p$$

from which the results follow as before. This method would obviously yield the variances of the time to C from A, by consideration of $U''(\xi)$.

Fourth solution. The second and third solutions will be seen to justify the following argument, which runs that 'the mean time to C from A (say T_{CA}) is independent of how long the system has been at A and similarly the mean time to C from B (say T_{CB}) is independent of how long the system has been at B'.

Furthermore, there are probabilities a, b, c that after one move from A the system will be in A, B, C respectively from which the mean times to C are T_{CA}, T_{CB} and 0. Thus

$$T_{CA} = a(1 + T_{CA}) + b . (1 + T_{CB}) + c(1 + 0).$$

Similarly
$$T_{CB} = p(1 + T_{CB}) + (1-p) . (1 + 0).$$

From there we find
$$T_{CB} = 1/(1-p)$$
as already shown, and
$$T_{CA} = (1 + b/q)/(1 - a)$$
as already shown.

(This fourth solution is perhaps the hardest to justify directly, *but is, by far, the simplest to apply once it is understood.*)

Exercise 10 F

1 R. Consider the Markov sequence defined by

$$\begin{pmatrix} \tfrac{1}{2} & \tfrac{1}{2} & 0 \\ \tfrac{1}{2} & 0 & 1 \\ 0 & \tfrac{1}{2} & 0 \end{pmatrix}.$$

(i) What are the closed classes? What (if any) are the transient classes?

(ii) Is it regular? Find the steady-state vector(s).

(iii) Find the expected passage time from each state to the first subsequent arrival in each other state, and the expected duration in each state upon arrival in it.

(iv) Starting from state B and using random number tables simulate 100 transitions and evaluate the mean times corresponding to the calculated expectations. (Remember that $BCBA$ yields 3 and 1 as passage times from B to A.)

2 R. Repeat parts (iii) and (iv) of Question 1, using the rabbit matrix from Section 2.

3. Find the expected passage time from each of states B, C, D to state A in the random walk of Exercise 10 C, Question 2.

Find whether or not $T_{BD} = T_{BC} + T_{CD}$.

4 T. Discuss the formulation of a problem similar to Question 3 but concerning the random walk of Exercise 10 C, Question 1.

5. Solve problems (B), (C), (D), (E) of Exercise 3 M, Question 56 for the chains resulting from rules (a), (b), (c) of that question.

Compare any relevant theoretical results with the observed results in Tables 7.2–7.6.

6. (i) Construct a matrix with the elastic force properties of the Ehrenfest model of Exercise 10 B, Question 12, but with 7 states and find the expected passage times from each state to the central one.

(ii) Express the expected passage times as fractions of the expected passage times from the extreme states to the centre.

(iii) What would be the corresponding fractions if the particle moved in a deterministic manner under the elastic law, moving freely from rest at the extreme?

7 T. (i) Find the *recurrence* time for each state of the rabbit matrix of Section 2.1 —that is, the expected time from each state to the first recurrence of that state.

(ii) Show that if these times are T_{AA}, T_{BB}, T_{CC} then

$$T_{AA} : T_{BB} : T_{CC} = u^{-1} : v^{-1} : w^{-1},$$

where u, v, w are the steady-state probabilities of each state.

(iii) What difficulties arise in attempting an analogous result for the problem of Section 6?

(iv) If the record of a sequence of states called E_1, E_2, ..., E_i, ..., E_n contains, in some order, r_1, r_2, ..., r_i, ..., r_n of each state respectively, and if the steady-state probability of E_i is estimated by $u_i = r_i/\Sigma r_i$, explain in simple terms why the observed mean recurrence times for the states must be nearly u_i^{-1}, *whether the chain is Markovian or not.*

8. In Exercise 10A, Question 3 the sequence was constructed from Random Number Tables using the following transition matrix:

From ...	1	2	3	4
To 1	0	$\frac{1}{2}$	0	$\frac{1}{2}$
2	$\frac{1}{2}$	0	$\frac{1}{2}$	0
3	0	$\frac{1}{2}$	0	$\frac{1}{2}$
4	$\frac{1}{2}$	0	$\frac{1}{2}$	0

Discuss the nature of the classes of states and hence of the chain. Calculate various parameters of interest and compare their values with observations from Exercise 10A, Question 3.

9. Express the equations for the queue problem of Exercise 3M, Question 54 in matrix form.

10T. A sequence $\{u_r\}$ of digits 0, 1, 2 is constructed as follows:
To obtain u_{r+1}, write $v_r \equiv u_{r-1} + u_r$ (modulo 3) and use the table:

u	pr $(u_{r+1} = u \mid v_r = 0)$	pr $(u_{r+1} = u \mid v_r = 1)$	pr $(u_{r+1} = u \mid v_r = 2)$
0	$\frac{1}{2}$	$\frac{1}{4}$	$\frac{1}{4}$
1	$\frac{1}{4}$	$\frac{1}{2}$	$\frac{1}{4}$
2	$\frac{1}{4}$	$\frac{1}{4}$	$\frac{1}{2}$

The following is a chain of 201 such digits starting arbitrarily with 00:

```
000 1 2 0 2 1 0 22 0000 2 1 2 000 2 0 2 0 2222
1 2 0 2 111 2 1 0 11 2 00 22 11 2 000 1 0 2 0
2 0 222 00 22 1 0 1 00000 2 1 22 0 22 1 2 0 11
2 0 1 2 00 1 2 1 2 0 2 1 0 2 0 2 0 1 0 222 1
0 11 222 1 2 0 2 000 111 22 1 0000 11 0 22 11 2
1 00 22 0 222 11 22 1 0 11 0 11 2 0 2 00 2 1
000 1 2 1 0 1 2 00 111 0 1 0 1 0 1111 2 0 2 11
0 1 0 111
```

(i) Verify that for the first 101 members (giving 100 transitions) the transition matrix is *observed* as

From ...	0	1	2
To 0	0·41	0·26	0·45
1	0·18	0·22	0·29
2	0·41	0·52	0·26

(ii) Now the chain is *not* Markovian, but you are to verify that a Markov model with the above transition matrix would predict just the observed numbers of 0, 1, 2 in the first 100 digits (excluding the initial 0) from which it was obtained.

Hence show that a model with *independent* digits (having suitable probabilities) and this Markov model would make the same predictions about the numbers of 0, 1, 2 in the second 100 digits. Compare these predictions with the observed values for the second 100 digits.

(iii) Confirm the values in the following table:

Based on	Mean transition times between digits		
	From 0 to 1	From 1 to 1	From 2 to 1
Model with independence	4·35	4·35	4·35
Observation of 2nd to 101st digits	4·41	4·13	3·70
Markov model from 1st 100 transitions	4·55	4·35	4·11
Observation of 102nd to 201st digits	2·42	2·45	2·50

[Note the slight improvement in predicted mean transition times obtained in the first 100 digits from using a Markov model (as opposed to one with independence) but that the model is not sufficiently flexible to be useful outside the data from which it was constructed, the chain not being Markovian. Models involving the history of the chain before the immediately preceding trial are analytically troublesome.]

11

PROPERTIES OF SAMPLES

1. GENERAL OUTLINES

1.1 Two central problems. We have now laid the foundations of a knowledge of probability and of descriptive statistics and have seen how to bring these branches together in the discussion of what are called random variables. We have all the necessary techniques to use the idea of a random variable to provide us with models of real situations.

Two of the central problems of statistics are first *the estimation of parameters*, which basically means the choice of a suitable value for each parameter to specify our model (this is usually called *point-estimation*); and secondly *the testing of hypotheses*, which in the simplest cases means deciding whether or not to reject a specified model, as being unsuitable to describe a situation for which it was not specifically designed. These problems are later linked, in that we can choose a *range* of values to allot to a parameter, by selecting a type of model and leaving the parameter values undecided until we have tested whether the resulting specific model would have to be rejected. This leaves us with a range of unrejected values of the parameter and is known as *interval-estimation*.

1.2 Method of argument. It is perhaps worth emphasizing, even at this early stage, that the keynote is *rejection* of various hypotheses and that the grounds for rejection are that the hypothesis would lead with only low probability to the observed results. In mathematics we can argue '$A \Rightarrow$ not-B, but B; so not-A'. In statistics we argue 'pr $(B|A)$ is small, but B; so we choose to reject A'. This by no means *proves* not-A. In mathematics we cannot argue '$A \Rightarrow C$, but C; so A' and in statistics we do not argue 'pr $(C|A)$ is large, but C; so accept A' or, rather, if we do use this form of words we only mean 'pr $(C|A)$ is large, but C; so we do not reject A on *these* grounds'.

1.3 The field narrowed. In this chapter we begin a study of estimation. As we have already indicated the problem arises in the following sort of way: we have a population of values of a random variable about which we do not know very much and we wish to find suitable values for the various parameters.

Let us, at first, concentrate on the mean and variance of some measurement. Thus a manufacturer of pistons will want to know the mean and variance of their diameters so that he can be sure they will fit their cylinders properly. Again, botanists find that one of the few simple characteristics that differentiate between various species of Sorbus (Whitebeam) is their length of leaf; what is required in the identification table is the range of about 'mean $\pm 2 \times$ S.D.' which may give about a 95 per cent coverage (as we saw in Chapter 6). If we are studying the hospital queues of Chapter 1, Section 7 we will want to know, amongst other things, the mean and variance of the length of interview; of the time by which those who are early arrive early; and of the number of people who arrive late. Makers of light bulbs want to know the mean life of their bulbs and many even want to know whether a new process, while perhaps not increasing the mean life, has reduced the variance.

1.4 The need for sampling. In all these cases we want the values of parameters in a population because we would like to argue from the population to the future observed measurements, which will form a sample from the population.

In the case of pistons it might well be that every item actually is measured as it comes off the line. Is it possible to *conceive* of us actually measuring all the leaves of all Whitebeams in any year, and thus determining the values required for use in succeeding years? Even if it is possible to conceive such a massive operation, it is scarcely desirable to undertake it. In the case of the hospital we want to know suitable values long before the hospital closes down and the final records can be studied. In the case of the light bulbs, however, there would be none left for sale if all were tested until they failed.

It is obvious that we do not in practice find the population parameters and argue from them, and it is no surprise to hear that we take a sample (in the ordinary usage of that word) and make a guess about the population, basing our guess on the relevant statistics of the sample. The method seems, however, complicated if we think about it: the route runs from sample to hypothetical population to other samples.

The key to the investigation is the relation of population to sample, and the reader has already used the appropriate method of argument in Exercise 7D, Questions 14, 17, 18, 20.

2. A PARTICULAR PROBLEM

2.1 Outline of problem. Before considering fairly realistic but more complicated situations, we will consider a problem in which the type of probability function of the parent population is a very simple one—namely rectangular.

We are set the following problem. A challenger says 'I am thinking of a certain population of consecutive integers, X, from which an integer can be selected at random in such a way that all integers in the population have equal chances of occurring at each selection, whether they have been selected before or not. Can you guess the mean of my population? I am prepared to give you a random sample of the integers.' On our asking for such a sample he turns to Table 8 (Random Sampling Numbers) of *Cambridge Elementary Statistical Tables* and reads out the first ten numbers from his population that occur. They are:

$$29 \quad 42 \quad 23 \quad 38 \quad 42 \quad 35 \quad 29 \quad 24 \quad 23 \quad 25.$$

He does not show us the page, so we do not know which numbers occurred in the table but were not in his population, nor do we know the lengths of the intervals in the table (made up of numbers not in his population) between the numbers given in the sample.

2.2   Outline of solution. We may proceed in various ways:

(*A*) Since the probability function for the population is symmetrical (indeed, rectangular) the mean of the extremes, α, β, of the population is the mean μ of the population. The mean of the extremes of the sample is $\frac{1}{2}(23+42)$; that is 32·5. This would give a rapid estimate of μ.

(*B*) The population is multimodal. All x are modes. The modes of the sample are 23, 29, 42. These have little to recommend them as estimates of μ, as they might equally well be anywhere between the extremes, the probability function being rectangular.

(*C*) The median of the population is equal to its mean (with the obvious definition of median, in this parent population, as a number such that a sample-member is equally likely to be greater than or less than it). The median of the sample is 29.

(*D*) The mean of the sample, m, seems a sound estimate of μ. It is 31.

After looking at these various estimates we have to take a plunge and make a selection of one of them. Mathematics has taken us as far as it can, and the rest is luck and judgement. Many will wish to use the mean of the sample as an estimator and give as their estimate the value of the mean of the sample, in this case 31.

In fact this sample was produced by a random choice from the numbers from 23 to 43 inclusive, so that the population-mean is 33. In this particular situation, with this particular sample, the mean of the extremes of the sample gives a closer estimate of the population mean than does the mean of the sample. How much weight to put on this fact we consider in the next section.

2.3   Samples of increasing sizes. It is apparent that in the problem outlined in Section 2.1 the extremes of the population are useful since, as

319

we increase the size of the sample, they are as likely to occur as any other value of X, *but* once they have both occurred then a further increase in the size of the sample will not affect the mean of the extremes of the sample, which by then will be the actual population mean.

We illustrate this by taking the first samples of sizes $N = 1, 2, ..., 10$ that would have occurred had we stopped short before a sample of size 11 had been given.

(Size) N	Sample	(Mean) m	(Mean of extremes) $\frac{1}{2}(a+b)$
1	(29)	29·0	29
2	(29, 42)	35·5	35·5
3	(29, 42, 23)	31·3	32·5
4	(29, 42, 23, 38)	33·0	32·5
5	(29, 42, 23, 38, 42)	34·8	32·5
6	(29, 42, 23, 38, 42, 35)	34·8	32·5
7	(29, 42, 23, 38, 42, 35, 29)	34·0	32·5
8	(29, 42, 23, 38, 42, 35, 29, 24)	32·8	32·5
9	(29, 42, 23, 38, 42, 35, 39, 24, 23)	31·7	32·5
10	(29, 42, 23, 38, 42, 35, 29, 24, 23, 25)	31·0	32·5

By taking larger and larger samples beyond this size we shall continually alter m but the only possible alteration in $\frac{1}{2}(a+b)$, the mean of the extremes of the sample, is when '43' occurs in the sample. This will give $\frac{1}{2}(a+b) = 33$, which is the value of μ.

A graph of m and of $\frac{1}{2}(a+b)$ against N will be helpful. (See Figure 11.1).

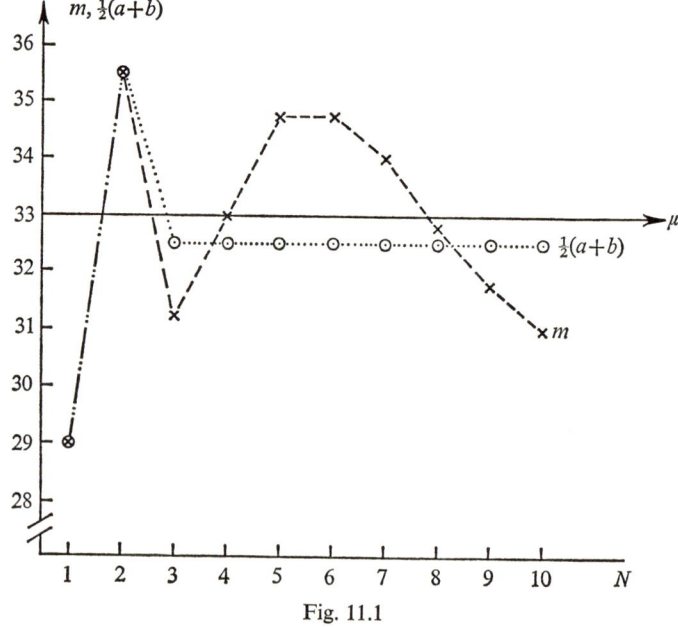

Fig. 11.1

It resembles Figures 2.1 and 2.2 and, we feel, not at all superficially; they all indicate the approach of a statistic value to some kind of limiting value.

It is true that m will approach μ as N increases, as we have indicated in Chapter 9, Section 1, but it is evident that in this case $\frac{1}{2}(a+b)$ approaches μ more quickly. This is a special feature of this particular type of parent population, namely one with a rectangular probability function. It is *not* the usual state of affairs.

In general it is the nearness of m to μ that is the most manageable attack and we will consider this approach more fully in Section 3.

Exercise 11 A

Keep the solutions

1 R. (*a*) Throw a die 36 times, and arrange the list of results into 18 consecutive pairs. Record the 'score' in each pair. You now have a sequence of 18 members of a population with a triangular probability function. Repeat the procedure of Section 2.3 and observe graphically the approaches of m and $\frac{1}{2}(a+b)$ to μ. (*Hint:* Calculate with a slide-rule on the cumulative total.)

(*b*) Using same 36 throws treat them as 36 single throws and then as 12 throws of a trio of dice and in each case repeat the processes of Section 2.3 and observe the approaches of m and $\frac{1}{2}(a+b)$ to μ.

2. A die is thrown and the results are used as follows to produce a skew probability function:

> If the score is even it is recorded unchanged.
> If the score is odd it is recorded as 5.

Discuss the use of the mean of extremes of a sample to estimate the population mean.

3. (*a*) Devise a symmetrical probability function such that samples from the corresponding population are very much more successfully used to estimate the population mean when the mean of the extremes of the sample is taken than when the mean of the sample is taken.

(*b*) Two given populations are fairly similar in many of their parameters; which parameters would you want to compare in the two populations to give some idea of the relative usefulness of the mean of the extremes as opposed to the mean of a sample from each population to estimate the mean of each population?

2.4 Samples of a fixed size. In real life we cannot go on and on adjoining members to our sample; time and money tell us to stop.

Suppose we were only able to have a sample of given limited size, and for instance were instructed to estimate μ from the value of m in the single sample of size 10 in Section 2.1. Our graph would only be as Figure 11.2.

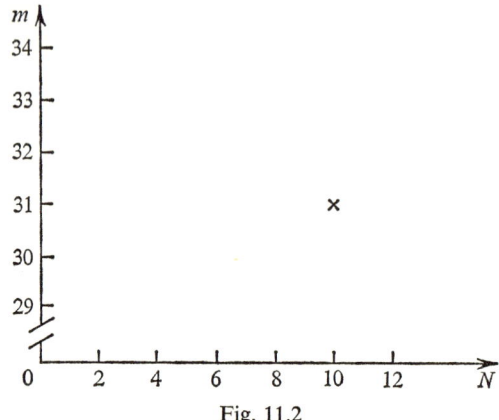

Fig. 11.2

However could we estimate μ? It would seem a reasonable relaxation to be allowed to follow out some other sequences of values of m and obtain means of various samples of size 10.

We should then have a diagram looking like Figure 11.3.

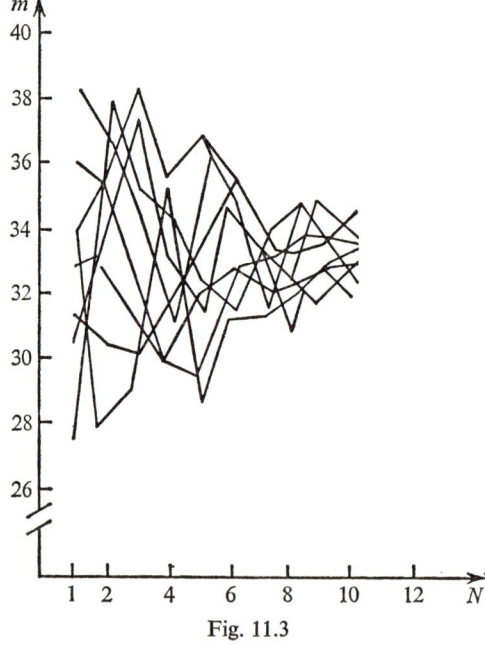

Fig. 11.3

By omitting all the routes to the values of m for samples of size 10 we could build up a collection of such values of m into a graph looking like Figure 11.4.

322

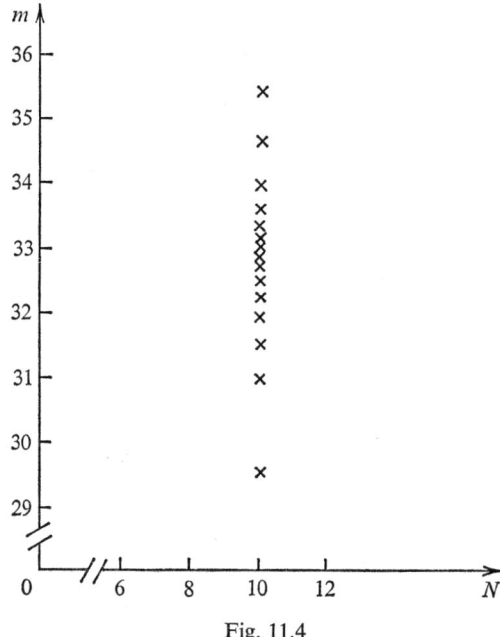

Fig. 11.4

In Figure 11.4 the rough location of the main bulk of crosses would suggest that μ is about 33 and the degree of concentration of the crosses would give us some idea of the reliability of this estimate.

It is our business in this chapter to discuss where the bulk occurs and how concentrated it is.

Since the situation of Figure 11.2, we have acquired extra information by repeated sampling, but it is remarkable how much information can be extracted from a single sample if we do not merely distil the information it contains down to a single statistic, for instance, in this case the mean. This will emerge later in this chapter.

With all this in view we change the conditions of the problem away from being allowed samples of larger and larger size, back even from being allowed several samples of a fixed size, to being given *a single sample of a fixed size*. To see how to use such a single sample we must *imagine* the possibility of studying all the possible samples of that fixed size.

It would be absurd to suppose that every sample of size 10 for instance, agreed with the one in Section 2.1 in having its mean $m = 31 \cdot 0$. In fact, from the 21 different members of the parent population, 21^{10} different samples are possible (remember replacement occurs), allowing for order. We will make sure by our selection methods that all these *ordered* samples are equally likely to be drawn and we then have a simple method of

323

allotting suitable values to the probabilities of various events. This approach is known as *simple random sampling* (*with replacement*). Other sampling methods exist; they allow more refined statements to be made and the reader who wants to pursue them later cannot do better than read A. Stuart: *Basic Ideas of Scientific Sampling* [Griffin's Statistical Monographs and Courses, 1962].

We thus have the notion of a set of *all possible* samples of a given size. Each sample has a mean and from the probability function for the various samples we obtain a probability function for various values of the mean. A new probability function has emerged in the problem. Alternatively we can say we have a population [*m*], (of values of means), defined by giving the various values their appropriate probabilities.

In general the probability function of the population of mean of samples of a fixed size depends on two things:

(i) the probability function for the parent population;

(ii) the fixed size of the samples chosen.

We could generalize the problem by considering some other statistics of the samples, say the variance, and obtaining a population of variances from samples of size 10 from the parent population. It would then be important to distinguish between the variance of the population of means and the mean of the population of variances! There is no reason why these should be the same, and indeed they are only equal if the samples are of size 2.

Exercise 11 B

1F. (i) In a sample of size 2 prove that the relation between the variance S^2, and the range, r, is $S^2 = \frac{1}{4}r^2$.

(ii) Complete the following table which shows, as number pairs, the values of (sum, square of range) for each possible ordered sample of size $N = 2$ drawn from the numbers 1 to 6 with replacement.

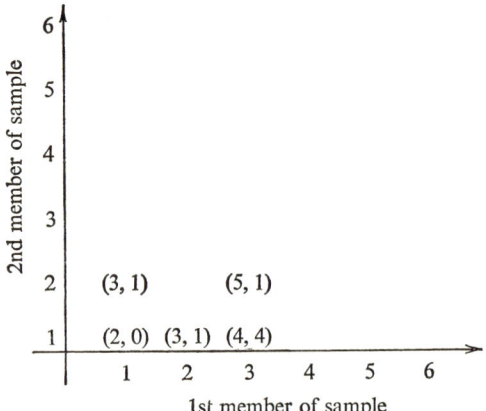

(iii) Draw up the probability functions for the populations [m], [v] of means, variances of samples of size 2 from the *uniform parent population* of integers 1, 2, ..., 6, (whose members are equally likely) and represent them graphically using isolated points (joined by dotted lines to taste).

(iv) Obtain $E[m]$, var [m], $E[v]$, var [v] and note the (*coincidental*) equality of var [m] and $E[v]$. (The coincidence occurs if and only if $N = 2$, whatever the parent population, as we shall see later.)

(v) Draw a scatter diagram with values m on one axis and values v on the other, marking each point with the value of 36 times the probability associated with the pair (m, v). Discuss whether the mean and variance of the samples from this uniform population are independent random variables or not, by first judging in a commonsense way from this diagram and secondly calculating $\rho[(m, v)]$ as defined in Exercise 9D, Question 10.

3. POPULATION OF MEANS OF SAMPLES OF FIXED SIZE

3.1   Mean of such a population. We now consider in more generality one of the four parameters calculated by the reader in Exercise 11B, namely $E[m]$.

We do not attempt complete generality yet, but for a little longer will keep to our policy of considering simple parent populations in order to bring out the *principles* involved.

We will consider the parent population [x] given by Table 11.1.

Table 11.1

x	pr ($X = x$)
0	0·2
1	0·2
2	0·2
3	0·2
4	0·2

That is to say we have a rectangular probability function; or an infinite population with equal probability of selection at each stage; or a finite population with replacement between selections and with equal probabilities of each selection.

Admittedly this population possesses symmetry, which is rather a special property, but it is not a parent population already much concentrated at its centre, so the behaviour of means of samples in bunching at the centre is not given a flying start.

We shall take samples—for instance (0, 2, 0) or (3, 4, 0, 1).

325

Now there are:　　　　5 samples of size 1

25 samples of size 2

125 samples of size 3

625 samples of size 4,

where we have counted (0, 2, 0) and (2, 0, 0) once each, for instance, since this corresponds to the particular allotment of suitable probabilities which we wish to study.

It is not a very long process to analyse these samples and find their totals and hence means. We obtain as follows the probability functions for the populations of means of samples of various sizes.

If the total of a sample is denoted by Y then we will write

$g_N(y)$ = frequency with which $Y = y$ in the 5^N different ordered samples of size N.

The result is Table 11.2.

Table 11.2

y	$g_1(y)$	$g_2(y)$	$g_3(y)$	$g_4(y)$
0	1	1	1	1
1	1	2	3	4
2	1	3	6	10
3	1	4	10	20
4	1	5	15	35
5	—	4	18	52
6	—	3	19	68
7	—	2	18	80
8	—	1	15	85
9	—	—	10	80
10	—	—	6	68
11	—	—	3	52
12	—	—	1	35
13	—	—	—	20
14	—	—	—	10
15	—	—	—	4
16	—	—	—	1

(The reader might care to produce the column for $g_5(y)$.) We now convert the total, y, of a sample into the mean, m, of the sample, by dividing by the size of the sample. We also convert the frequency $g_N(y)$ into a probability $p_N(m)$ by dividing the frequency by 5^N. Since the resulting table has symmetry about the value $m = 2$ the bottom half of the table is merely indicated. We have omitted the columns for samples of size 3 so that the reader can supply them as part of Exercise 11C. The result is Table 11.3.

Table 11.3

Table of probabilities of various means for various sizes of sample

m	$p_1(m)$	$p_2(m)$	$p_4(m)$
0	0·2	0·04	0·0016
0·25			0·0064
0·50		0·08	0·0160
0·75			0·0320
1·00	0·2	0·12	0·0560
1·25			0·0832
1·50		0·16	0·1088
1·75			0·1280
2·00	0·2	0·20	0·1360
.	.	.	.
.	.	.	.
.	.	.	.
4·00	0·2	0·04	0·0016

What does it mean to say: $p_1(2·00) = 0·2$

or: $p_4(2·00) = 0·1360$?

$p_1(2·00) = 0·2$ means that if a random sample of size 1 is taken, then the probability that its mean is 2·00 is 0·2.

$p_4(2·00) = 0·1360$ means that if a random sample of size 4 is taken, then the probability that its mean is 2·00 is 0·1360.

At first sight it rather looks as if we are less likely to get a satisfactorily central value for the mean (namely 2·00) if we take a larger sample. But consider further. Since there are only 5 possible means when a sample of size 1 is drawn and 17 possible means when a sample of size 4 is drawn, we might expect the probability of getting a particular mean (if it is available at all in the first case) to be higher in that first case than in the second. In the second, however, we could get a mean nearly equal to 2 whereas in the first case we could get no nearer than 1 or 3 if we failed to get 2. This resolves the paradox and we see that the situation is ripe for the application of a histogram technique, as developed in Chapter 5 for frequencies and frequency densities.

We will use different class intervals in each column, choosing the class-marks at the values m which have non-zero probability and keeping all the intervals equal in any one column. By dividing the relevant probability by the length of the class interval we shall obtain an 'averaged' probability density over the interval, which on a diagram will be suggestive of what is actually going on among the numbers.

In Table 11.4 all this has been done and the resulting averaged probability density functions are called $\phi_1(m)$, $\phi_2(m)$, $\phi_4(m)$.

Table 11.4

m	$\phi_1(m)$	$\phi_2(m)$	$\phi_4(m)$
0	0·2	0·08	0·006
0·25			0·026
0·50		0·16	0·064
0·75			0·128
1·00	0·2	0·24	0·224
1·25			0·333
1·50		0·32	0·435
1·75			0·512
2·00	0·2	0·40	0·548
.	.	.	.
.	.	.	.
.	.	.	.
4·00	0·2	0·08	0·006

We next plot a graph using the numbers as follows. For example, $\phi_4(0\cdot75) = 0\cdot128$ indicates that we are to draw a rectangle of base 0·25 (which is the class-interval for m in that column), and height 0·128, centred at $m = 0\cdot75$.

This is done in Figure 11.5.

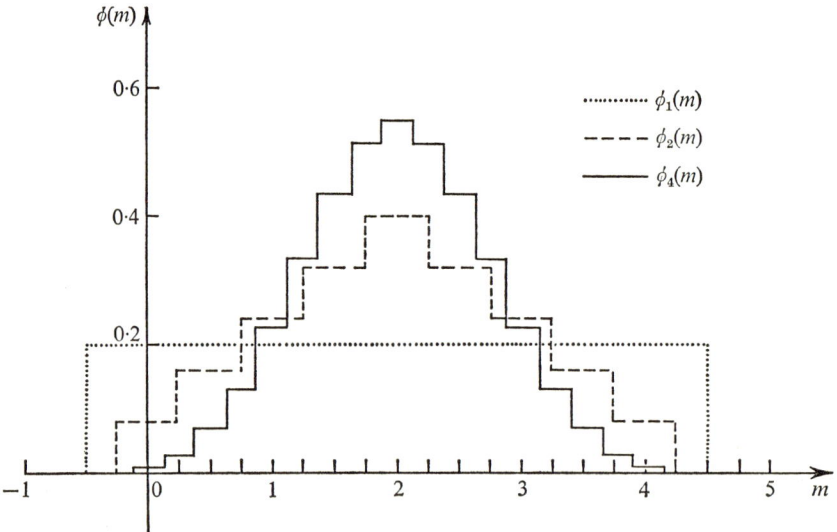

Fig. 11.5. Graphs of averaged probability density for populations of means of samples of size 1, 2, 4.

What does it mean to read from the graphs that

$$\phi_1(1\cdot8) = 0\cdot2 \qquad \phi_2(1\cdot8) = 0\cdot4 \qquad \phi_4(1\cdot8) = 0\cdot512 ?$$

It means that there are probabilities

	$0\cdot2 \times 1$	$0\cdot4 \times 0\cdot5$	$0\cdot512 \times 0\cdot25,$
or	$0\cdot2$	$0\cdot2$	$0\cdot128,$

that means from samples of sizes

1	2	4

will lie in the ranges

$1\cdot5$ to $2\cdot5$	$1\cdot75$ to $2\cdot25$	$1\cdot675$ to $1\cdot925,$

these being the extremes of the bases of the rectangles on the graphs (their vertical sides having been omitted for clarity).

It is necessary to revert to ranges in this way because the probability densities have only been rather artificially obtained from such ranges. The reason for changing to averaged probability densities at all is to point the way forward to the possibility of considering random variables which may take any value in an interval (that is to say continuous random variables).

In general the graphs display the facts that

(i) for any given size of sample, the population of means of samples has *its* mean at the mean of the parent population;

(ii) as the sample size increases, so the population of means of samples becomes more concentrated near the mean of the parent population.

We will prove that (i) is true whatever the parent population may be, and that (ii) is true under very general conditions, namely merely that the parent population has a finite variance at all.

These provide us with what is essentially the Law of Large Numbers, and are the basis for using a sample rather than an individual when examining a population.

Before answering the question suggested by (ii) above—'How concentrated?'—we insert Exercise 11 C and the proof of (i).

Exercise 11 C

1 R. Produce the column of values of $\phi_3(m)$ in the manner of Section 3.1.

2. Throw a die a hundred times and record the scores *in sequence*. (64 further throws beyond those of Exercise 11 A, Question 1 will serve.)

Carry out the following mappings to get a skew frequency function.

$$\text{even numbers} \to 2$$
$$3 \to 3$$
$$1, 5 \to 6$$

(Alternatively: use random number tables to generate the same probability function.)

The resulting population is regarded as the parent population for this experiment. Group the values into 20 consecutive samples of size 5 and calculate the mean of each such sample. Adjoin consecutive pairs of samples to get 10 samples of size 10. Calculate their means. Take intervals of 1·5 to 2·5; 2·5 to 3·5 and so on and draw the histograms for the following three quantities:

(i) The relative frequency density of the parent population. (This gives its averaged probability density function.)

(ii) The relative frequency density for each of the two populations of observed means of samples. (If a sample has mean = 3·5, for instance, add $\frac{1}{2}$ to each neighbouring frequency.)

3 F. Suppose you say that you will obtain with given probability a sample (from the population of Section 3.1) of given size whose mean is within a given range of a given value. How do the given quantities affect one another? Answer by considering the following situations in the light of Table 11.3.

(i) The given probability is $66\frac{2}{3}$ per cent. The given value is 1·5. What is the least interval (in the form $1·5 \pm h$) that you dare risk to have a $66\frac{2}{3}$ per cent chance of its covering your actual mean? Consider samples of sizes 1, 2, 3, 4 and list the corresponding intervals. Comment in general on the effect of increasing the sample size.

(ii) The given value is again 1·5 and the range is given as $1·5 \pm 0·76$. What is the greatest probability that you dare risk to have the chance indicated by it of covering your actual mean? Consider samples of sizes 1, 2, 3, 4 and list the corresponding probabilities. Comment in general on increasing the sample size and compare with part (i).

(iii) Repeat (ii) with given value 2·0 and range $2·0 \pm 0·76$. Compare with (ii).

3.2 General result on mean of population of means. The numerical work of Section 3.1 has suggested that if a random variable M is defined by drawing (with replacement) samples of a fixed size from a parent population with mean μ and taking a value of M to be m, the mean of a sample, then $E[m] = \mu$.

In this Section we will prove this result in the general case. We take up the hint dropped in Chapter 9, Section 4.1 that the theory of expected values of combinations of random variables would have one of its main applications here.

Now in the calculation of the mean of a sample whose members have values $x_1, x_2, ..., x_N$ (*possibly with repetitions*) we evaluate

$$m = \frac{1}{N}(x_1 + x_2 + ... + x_N).$$

We make what may not seem a very significant change by writing this as

$$m = \frac{1}{N}x_1 + \frac{1}{N}x_2 + ... + \frac{1}{N}x_N.$$

Each value $x_1, x_2, ..., x_N$ is a value of the random variable which defines the parent population. This is a special case of the situation in which (in

330

the notation of Chapter 9) the generalization of $E[ax+by] = aE[x]+bE[y]$ operates.

This generalization runs:

$$E(a_1 X_1+a_2 X_2+...+a_N X_N) = a_1 E(X_1)+a_2 E(X_2)+...+a_N E(X_N).$$

All we do is:

(i) to take the random variables X_1, X_2, ..., X_N as identical, all identical to the random variable X which defines our parent population (in fact all we *need* is that the X_i should have the same expected value);

(ii) to take $a_1 = a_2 =...= a_N = \dfrac{1}{N}$.

We obtain:

$$E[m] = E\left[\frac{1}{N}x_1+\frac{1}{N}x_2+...+\frac{1}{N}x_N\right]$$

$$= \frac{1}{N}E(X_1)+\frac{1}{N}E(X_2)+...+\frac{1}{N}E(X_N)$$

$$= \frac{1}{N}\cdot\mu+\frac{1}{N}\mu...+\frac{1}{N}\mu$$

$$= \mu.$$

It is important to realize that we have made no assumptions whatever about the parent population, except that it should have a (finite) mean. It certainly does not, for instance, have to be symmetrical, however reasonable such a demand might have seemed. There *is* a strong requirement, however, and it lies in the method of choosing the members of the sample—so that they will individually have expected value μ and will not show *selection bias*.

For instance we may choose two members at a time from the population and then discard one of them by a method which in no way depends on the values chosen—we might take a coin and spin it, or always take the one that just occurred first in the selection of the pair; but we may *not* choose two members at a time and then discard, say, the smaller of them; that process would alter the expected value of the mean of the numbers retained.

3.3   Variance of the population of means. At the end of Section 3.1 we suggested that the next question about the population of means of samples of fixed size from the parent population defined in Table 11.1 was 'How concentrated?'. It is that question we now answer.

We calculate the variance of the population of means of samples of size 4. We write the variance as σ_4^2.

m	$m-2$	$(m-2)^2$	$625 \times p_4(m)$	$625 \times (m-2)^2 . p_4(m)$
0	-2	4	1	4
$\frac{1}{4}$	$-1\frac{3}{4}$	$\frac{49}{16}$	4	$12\frac{1}{4}$
$\frac{1}{2}$	$-1\frac{1}{2}$	$\frac{9}{4}$	10	$22\frac{1}{2}$
$\frac{3}{4}$	$-1\frac{1}{4}$	$\frac{25}{16}$	20	$31\frac{1}{4}$
1	-1	1	35	35
$1\frac{1}{4}$	$-\frac{3}{4}$	$\frac{9}{16}$	52	$29\frac{1}{4}$
$1\frac{1}{2}$	$-\frac{1}{2}$	$\frac{1}{4}$	68	17
$1\frac{3}{4}$	$-\frac{1}{4}$	$\frac{1}{16}$	80	5
				$\overline{\overline{156\frac{1}{4}}}$
2	0	0	85	$\overline{0}$

and since the population is symmetrical about $m = 2$ we see that

$$\sigma_4^2 = \frac{312\frac{1}{2}}{625} = \tfrac{1}{2}, \quad \text{exactly.}$$

Exercise 11 D

Keep the Solutions

1 *T*. Calculate σ_2^2 and σ_3^2 in the manner of Section 3.3.
   What relation do you observe between σ_N^2 and the variance of the parent population?

3.4 General result on variance of population of means. The reader should have noted after Exercise 11 D that, for the given parent population of Table 11.1 and for $N = 1, 2, 3, 4$, we have:

$$\sigma_N^2 = \frac{1}{N}\sigma^2.$$

To prove this in the general case we apply another of the results of Chapter 9, Section 5.1, namely (in the notation of that Chapter)

$$\text{var}\,[ax+by] = a^2\,\text{var}\,[x]+b^2\,\text{var}\,[y].$$

We generalize to the N random variables X_1, X_2, ..., X_N and we take the multipliers to be $a_1, a_2, ..., a_N$, so that:

$$\text{var}\,(a_1 X_1+a_2 X_2+...+a_N X_N) = a_1^2\,\text{var}\,(X_1)+a_2^2\,\text{var}\,(X_2)+...+a_N^2\,\text{var}\,(X_N).$$

This is true if the random variables X_1, X_2, ..., X_N are independent. Independence does not, of course, prevent the variables from being 'identical' (which, we recall, only means that they have the same probability function) so we take:
   (i) all the variables to be X, the variable of the parent population,
   (ii) all the multipliers to be $1/N$.

We obtain:

$$\text{var}\,[m] = \text{var}\left(\frac{1}{N}X_1 + \frac{1}{N}X_2 + \ldots + \frac{1}{N}X_N\right)$$

$$= \frac{1}{N^2}\text{var}\,(X_1) + \frac{1}{N^2}\text{var}\,(X_2) + \ldots + \frac{1}{N_2}\text{var}\,(X_N)$$

$$= \frac{1}{N^2}\cdot\sigma^2 + \frac{1}{N^2}\cdot\sigma^2 + \ldots + \frac{1}{N^2}\cdot\sigma^2$$

$$= N\left(\frac{1}{N^2}\sigma^2\right)$$

$$= \sigma^2/N.$$

4. THE LAW OF LARGE NUMBERS

4.1 Summary on population of means.

The situation. A certain parent population is under consideration. In symbols: A random variable X is defined by a function p such that $\text{pr}\,(X = x) = p(x)$.

Conditions on the parent population. Only that the variance exists.
   In symbols: (i) $\mu \equiv E[x] \equiv \Sigma xp(x)$ exists;
                       (ii) $\sigma^2 \equiv \text{var}\,[x] \equiv E[(x-\mu)^2] \equiv \Sigma(x-\mu)^2\,p(x)$ exists.
(In fact this second condition being satisfied ensures that the first is.)
   No other restriction is placed on p at all.

Procedure. A *random* sample of previously fixed size is drawn (with replacement) from the parent population and its mean calculated.
   In symbols: The size of sample is fixed beforehand at N. The sample is $x_1, x_2, x_3, \ldots, x_N$ (with possible repetitions) and $m = \bar{x} = (1/N)\Sigma x_i$.
   Notice that we are not to decide the sample size on any characteristic of the sample observed during the drawing of the sample—the value of N is *not* to be itself a random variable.

Result of the procedure. The mean of the sample defines a new random variable whose value is the value of this mean.
   In symbols: the possible values of m define a population, $[m]$, of values of a random variable M, say. This population has its own probability function, say q such that $q(m) \equiv \text{pr}\,(M = m)$.

Theorem. This population of means of samples has the same 'centre' as the parent population but is more 'concentrated', the concentration depending on and increasing with the size of the sample.
   Precisely, and in symbols:

$$E[m] = \mu \quad\text{and}\quad \text{var}\,[m] = \sigma^2/N.$$

or $\qquad\qquad E[\bar{x}] = E[x] \quad\text{and}\quad \text{var}\,[\bar{x}] = \frac{1}{N}\,\text{var}\,[x].$

Diagrams illustrating the theorem. The arrows indicate an interval of 1 S.D. on each side of the mean, respectively in each graph.

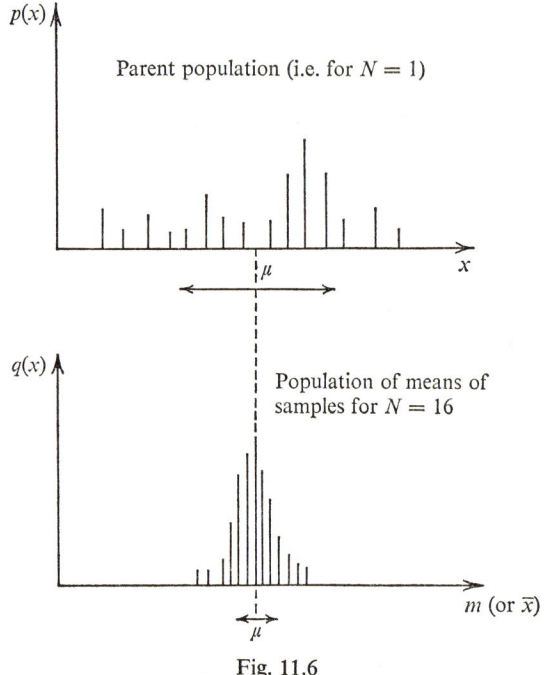

Fig. 11.6

Comments. Notice that we do not (at this stage) make any statements about the form of q apart from giving the values of two of its parameters. Other parameters of q depend on the form of p and on N. It is an extremely noteworthy fact, however, that for large enough N the form of q becomes almost independent of the form of p; q becomes more and more near the so-called 'Normal' form as N increases. Clearly this loose wording will need tightening later; one aspect of this being that so far we have only considered *discrete* random variables X.

4.2 The Weak Law of Large Numbers. We apply Chebyshev's Inequality to the population of means, m_N, of samples of size N from a parent population with mean μ and variance σ^2. We note that

$$\text{stad } [m_N] = \sigma/\sqrt{N} = \sigma N^{-\frac{1}{2}}.$$

We have $\qquad \text{pr}\,(|m_N - \mu| > \lambda\sigma N^{-\frac{1}{2}}) < 1/\lambda^2$

or $\qquad\qquad \text{pr}\,(|m_N - \mu| > \epsilon) < \sigma^2/N\epsilon^2$

or $\qquad\qquad \text{pr}\,(|m_N - \mu| > \epsilon) \to 0 \quad \text{as} \quad N \to \infty,$

334

or given any ϵ, $\eta > 0$ there exists a number k such that for any $N \geqslant k$ we have

$$\text{pr}\,(|m_N - \mu| > \epsilon) < \eta.$$

Results of this type will obviously be valuable in the problem of interval-estimation for the following reason.

What the 'law' or theorem tells us is that if we are sampling from a population of any sort that has a (finite) variance and if we take a large enough random sample, then there will be very few occasions when the mean of our sample differs seriously from the mean of the population.

4.3   Two consequences. One point of interest is the way that, for a given probability η of a discrepancy, the range of discrepancy, 2ϵ, falls at first rapidly for increasing N and then less rapidly. Thus, by increasing N from 2 to 5 we reduce $\epsilon = \lambda\sigma N^{-\frac{1}{2}}$ from $0 \cdot 71\lambda\sigma$ to $0 \cdot 45\lambda\sigma$, a reduction of 37 per cent; but by increasing N by a further 3 we only reduce $\lambda\sigma N^{-\frac{1}{2}}$ from $0 \cdot 45$ to $0 \cdot 37\lambda\sigma$, a further reduction of only about 16 per cent. In fact to reduce the range by a factor of 2 at any stage we need a four-fold increase in sample size and, for instance, increasing the sample size from 30 to, even, 40 is scarcely worth the increased cost. (The reader should obtain the resulting percentage reduction in ϵ to check his grasp of this point.)

Similarly for a given range of discrepancy, 2ϵ, the probability, η, of being outside it falls disappointingly slowly once an even quite modest value of N is reached.

A second point of interest is that, with the random sampling (with replacement) that we are considering, the proportion of sample size to size of parent population is irrelevant. This has a consequence surprising to many people.

A survey of perhaps 2000 people is used to determine some national characteristic in a group of 50,000,000 people, and the man-in-the-street exclaims at the temerity of statisticians in basing their conclusions on so negligible a number of people. Our results have shown, however, that in sampling with replacement the only matters of relevance are a carefully randomized method of selection of the sample members, and the *absolute* size of the sample. Now a national survey is not usually conducted by sampling with replacement, and in sampling without replacement the relative size of the sample is relevant. Its importance, however, is small when the relative size is small, and with increasing size of parent population and fixed size of sample it does of course become smaller. We can think of our national survey as, in a certain sense, sampling without replacement from an 'almost infinite' population, and so, effectively, sampling *with* replacement. For this reason the absolute and not the relative size of the sample is the matter of importance.

We will merely quote the result for the variance of the population of means of samples where the sampling is without replacement. If the size of the parent population is N_0 and its variance is σ^2, and if the size of the sample is N, then

$$\text{var (mean)} = \left(\frac{N_0 - N}{N_0 - 1}\right) \cdot \left(\frac{\sigma^2}{N}\right) = \left(1 - \frac{N-1}{N_0 - 1}\right) \cdot \left(\frac{\sigma^2}{N}\right).$$

To particularize, if we wished to survey the constituency of, say, Wolverhampton North-East (50,000 voters) and our standards of reliability required a sample of 2,000 people then the standard deviation of the mean of our sample would be $\sqrt{\left(\frac{48,000}{49,999}\right)} \cdot \frac{\sigma}{\sqrt{(2000)}}$ and the factor $\sqrt{\left(\frac{48,000}{49,999}\right)}$ is almost negligible. (Its value is 0·98.)

To survey the whole country of 50,000,000 persons to the same standard of reliability would require a sample (properly randomized for the whole country this time) of size N given by

$$\left(\frac{50,000,000 - N}{49,999,999}\right) \frac{1}{N} = \frac{48,000}{49,999} \cdot \frac{1}{2000}.$$

Whence $N = 2083$ approximately. By sticking to $N = 2000$ we should not come noticeably to grief.

4.4   Convergence in probability. The behaviour of the sequence of random variables of Section 4.2 of which typical values were $m_1, m_2, m_3, \ldots, m_N, \ldots$ suggests the following definition.

Let $Y_1, Y_2, Y_3, \ldots, Y_r, \ldots$ be a sequence of random variables with typical values $y_1, y_2, y_3, \ldots, y_r, \ldots$. If a number l exists such that, given any $\epsilon, \eta > 0$, there exists a number K (generally dependent on ϵ, η) such that

$$\text{pr}\left(|y_r - l| > \epsilon\right) < \eta \quad \text{for all } r > K,$$

then we say the sequence of random variables *converges in probability* to l.

We will not use the idea, formally, very much but we mention it as an example of the way in which parts of statistics can be put on a quite firm basis of definitions and theorems, like a branch of *pure* mathematics. It is over the rules for decisions as to whether or not we will reject a hypothesis that we still have choice and where the subject takes on the aspect of *applied* mathematics. The numerical consequences of the selected rules are of course again material for pure mathematics.

5. APPLICATION TO BERNOUILLI TRIALS

We may think of a sequence of Bernouilli trials as a method of determining a succession of independent values of a random variable X, where only two values are possible, 0 or 1, and the probability function of the parent population is

$$\text{pr}(X = 1) = \pi, \quad \text{pr}(X = 0) = 1 - \pi.$$

In the form of words where the occurrence of $X = 1$ is called 'a success' and of $X = 0$ is called 'a failure' we see that we have pr (success) $= \pi$. If we then have r successes in N trials we have a proportion $r/N = p$, say, of successes. But if N is fixed beforehand so that we do *not* stop at a time determined by the trials themselves, then the N trials may be thought of as the drawing of a sample of size N from the parent population. In that case the mean value of X in the N trials, being calculated as

$$(1.r+0.(N-r))/N,$$

is r/N or p.

Thus, applying the results of Section 4.1 we have:

$$E[p] = E(X), \quad \text{var}\,[p] = \frac{1}{N}\,\text{var}\,(X).$$

Now
$$E(X) = 1.\pi+0.(1-\pi) = \pi$$

and
$$\begin{aligned}
\text{var}\,(X) &= E(X^2)-(E(X))^2 \\
&= \{1^2.\pi+0^2.(1-\pi)\}-(\pi)^2 \\
&= \pi(1-\pi),
\end{aligned}$$

so
$$E[p] = \pi \quad \text{and} \quad \text{var}\,[p] = \pi(1-\pi)/N.$$

This is one of the more elegant ways of obtaining these standard relations between the proportion, p, of successes in N Bernouilli trials and the probability π of success at each trial. The application of the result of Section 4.2 to a binomial parent population is called Bernouilli's Theorem.

6. POPULATION OF VARIANCES OF SAMPLES OF FIXED SIZE

6.1   Introduction. We now have two reasons for wanting to know about the variance of a population. First because it may itself be of interest; thus if we are filling containers with some liquid and wish to be sure that we do not put too little in, then we will have to aim to overfill slightly because of the variability inherent in any real process, and if we could measure and then reduce the variability we could achieve a saving by overfilling less. Or again, it may be very uneconomical to replace a worn part on some complicated mechanism containing several of them, or in something like a blast furnace which depends for profitability on long uninterrupted running; and in these cases we should like all the replacements to be done at one time and, since that time must be no longer than the life of the weakest of them, we should want all the parts for replacement to be at nearly the same state of wear, to avoid wasting the unexpired life of the stronger ones.

Secondly, the variance of the parent population may be of interest because we wish to obtain, using σ^2/N, an idea of the reliability of an estimate of the mean of the population.

A study of the variance of samples from a population seems likely to throw light on the variance of the parent population and it is such a study we undertake in the next section.

6.2 Mean of population of variances of samples. As in other problems in this chapter, we begin with a simple numerical investigation. We consider a parent population already much studied, namely that defined by

$$\text{pr}\,(X = 1) = \tfrac{1}{2}, \quad \text{pr}\,(X = 0) = \tfrac{1}{2}.$$

For this population $\mu = \tfrac{1}{2}$, $\sigma^2 = \tfrac{1}{4}$.

We look at the populations of variances of samples of various sizes (the size fixed *before* the sample is drawn).

Samples of size 1. These are (0) and (1), their probabilities are each $\tfrac{1}{2}$, their means are 0 and 1 and their variances zero.

We tabulate as follows:

Sample	Prob.	m	S^2		S^2	Prob.
(0)	$\tfrac{1}{2}$	0	0	and	0	1
(1)	$\tfrac{1}{2}$	1	0			

we see that $E[S^2] = 0$. So far this is trivial but it does lead us to expect that $E[S^2] < \sigma^2$, since each sample has its variance calculated from its *own* mean (*not* the mean of the parent population) so that even if the sample is concentrated far from the mean of the parent population its variance is not necessarily large.

We continue with larger samples; for each given size of sample the first table is of the samples and the second shows the probability function of the population of variances of the samples.

Samples of size 2.

Sample	Prob.	m	S^2		S^2	Prob.
(0, 0)	$\tfrac{1}{4}$	0	0	and	0	$\tfrac{1}{2}$
(0, 1)	$\tfrac{1}{2}$	$\tfrac{1}{2}$	$\tfrac{1}{4}$		$\tfrac{1}{4}$	$\tfrac{1}{2}$
(1, 1)	$\tfrac{1}{4}$	1	0			

$$E[S^2] = 0 . \tfrac{1}{2} + \tfrac{1}{4} . \tfrac{1}{2} = \tfrac{1}{8}$$
$$\Rightarrow E[S^2] = \tfrac{1}{2}\sigma^2 < \sigma^2.$$

Samples of size 3.

Sample	Prob.	m	S^2		S^2	Prob.
(0, 0, 0)	$\tfrac{1}{8}$	0	0	and	0	$\tfrac{1}{4}$
(0, 0, 1)	$\tfrac{3}{8}$	$\tfrac{1}{3}$	$\tfrac{2}{9}$		$\tfrac{2}{9}$	$\tfrac{3}{4}$
(0, 1, 1)	$\tfrac{3}{8}$	$\tfrac{2}{3}$	$\tfrac{2}{9}$			
(1, 1, 1)	$\tfrac{1}{8}$	1	0			

$$E[S^2] = 0 . \tfrac{1}{4} + \tfrac{2}{9} . \tfrac{3}{4} = \tfrac{1}{6} = \tfrac{2}{3} . \tfrac{1}{4}$$
$$\Rightarrow E[S^2] = \tfrac{2}{3}\sigma^2 < \sigma^2.$$

It is the case, and we prove it later, that for any parent population whatever, provided only that it has finite variance σ^2, if we take samples of size N and calculate their variances we shall get a population of variances whose mean is

$$\frac{N-1}{N}\sigma^2;$$

in symbols, we shall get

$$E[S^2] = \frac{N-1}{N}\sigma^2.$$

This result is the basis for the remark at the end of Section 2.4 about var $[m]$ and $E[S^2]$ only being equal if $N = 2$, for we can now write

$$\text{var}\,[m] = E[S^2]$$

$$\Rightarrow \frac{\sigma^2}{N} = \frac{N-1}{N}\sigma^2$$

$$\Rightarrow N(N-2)\,\sigma^2 = 0$$

$$\Rightarrow N = 2 \quad \text{or trivial alternatives.}$$

Some beginners find that when they have no reference book at hand they become confused between the two results:

$$\text{var}\,[m] = \frac{1}{N}.\sigma^2 \quad \text{and} \quad E[S^2] = \frac{N-1}{N}.\sigma^2.$$

The obvious check as to which is which is surely that with increasing sample size, N, the mean of the sample, m, has less chance of variation, so that var $[m]$ decreases; but also with increasing sample size the variance of the sample is more likely to be nearly the population variance, so that $E[S^2]$ becomes nearly σ^2.

Furthermore we have the check that with $N = 1$ the sample is a mere individual member; the mean of the sample has all the variability of the population itself and the variance of the sample is expected to be zero.

A second point which can puzzle beginners is that if we have a parent population of the weights of, say, 6 people and the population has variance σ^2, why do we find that the expected variance of samples of size 6 is only $\frac{5}{6}\sigma^2$, instead of σ^2, when, as they argue, 'we surely have the whole population in a sample of size 6'. The unmasking of this fallacy is left to the reader.

6.3 Further numerical example.

To find the probability function for the population of variances of samples of size 3 drawn at random, with replacement, from a parent population of values of a random variable X given by:

$$\text{pr}\,(X = -a) = \text{pr}\,(X = a) = \tfrac{1}{4}, \quad \text{pr}\,(X = 0) = \tfrac{1}{2}.$$

This parent population gives $E(X) = 0$, $\sigma^2 \equiv \text{var}(X) = a^2/2$. If we draw samples of size 3 we have the following:

Unordered sample	$64 \times$ prob.	Mean	Variance/a^2
$(-a, -a, -a)$	$1 \times 1 \times 1 \times 1 = 1$	$-a$	0
$(-a, -a, 0)$	$3 \times 1 \times 1 \times 2 = 6$	$-\frac{2}{3}a$	$\frac{1}{3}(\frac{1}{9}+\frac{1}{9}+\frac{4}{9}) = \frac{2}{9}$
$(-a, -a, a)$	$3 \times 1 \times 1 \times 1 = 3$	$-\frac{1}{3}a$	$\frac{1}{3}(\frac{4}{9}+\frac{4}{9}+\frac{16}{9}) = \frac{8}{9}$
$(-a, 0, 0)$	$3 \times 1 \times 2 \times 2 = 12$	$-\frac{1}{3}a$	$\frac{1}{3}(\frac{4}{9}+\frac{1}{9}+\frac{1}{9}) = \frac{2}{9}$
$(-a, 0, a)$	$6 \times 1 \times 2 \times 1 = 12$	0	$\frac{1}{3}(1+0+1) = \frac{6}{9}$
$(-a, a, a)$	$3 \times 1 \times 1 \times 1 = 3$	$\frac{1}{3}a$	$\frac{1}{3}(\frac{16}{9}+\frac{4}{9}+\frac{4}{9}) = \frac{8}{9}$
$(0, 0, 0)$	$1 \times 2 \times 2 \times 2 = 8$	0	0
$(0, 0, a)$	$3 \times 2 \times 2 \times 1 = 12$	$\frac{1}{3}a$	$\frac{1}{3}(\frac{1}{9}+\frac{1}{9}+\frac{4}{9}) = \frac{2}{9}$
$(0, a, a)$	$3 \times 2 \times 1 \times 1 = 6$	$\frac{2}{3}a$	$\frac{1}{3}(\frac{4}{9}+\frac{1}{9}+\frac{1}{9}) = \frac{2}{9}$
(a, a, a)	$1 \times 1 \times 1 \times 1 = 1$	a	0

Check: total $= 64$

From which arises:

S^2	$64 \times \text{pr (variance} = S^2)$
0	10
$\frac{2}{9}a^2$	36
$\frac{6}{9}a^2$	12
$\frac{8}{9}a^2$	6

so that

$$E[S^2] = \left((0 \times 10) + (\tfrac{2}{9} \times 36) + (\tfrac{6}{9} \times 12) + (\tfrac{8}{9} \times 6)\right) \frac{a^2}{64}$$

$$= \frac{64}{3} \cdot \frac{a^2}{64} = \tfrac{2}{3} \cdot \frac{a^2}{2} = \tfrac{2}{3}\sigma^2.$$

(Thus verifying again the general result, and throwing light on the fallacy indicated at the end of Section 6.2.)

Exercise 11 E

1 R. After the manner of Section 6.2 find the probability functions for populations of variances of samples of fixed sizes from the parent populations below, and the given sample sizes. In each case find $E[S^2]$ and var $[S^2]$ and draw a diagram to illustrate the probability function.

(a) Parent population: pr $(X = a) = $ pr $(X = -a) = \frac{1}{2}$;
Sample size: 4.

(b) Parent population: pr $(Y = -3a) = $ pr $(Y = 3a) = \frac{1}{8}$;
pr $(Y = -a) = $ pr $(Y = a) = \frac{3}{8}$;
Sample size: 2.

2 R. If the reader has not already done so he should now work Exercise 6 M, Question 9. In any case that question will repay being contemplated again at this stage, and having the relationships between the various statistics and parameters checked in the light of this chapter.

6.4 General result on the mean of population of variances.

The numerical work of Sections 6.2 and 6.3 has suggested that if a random variable V is defined by drawing (with replacement) samples of fixed size N from a parent population of values of X, with mean μ and variance σ^2, and taking a value of V to be v, the variance of a sample, then

$$E[v] = \frac{N-1}{N} \cdot \sigma^2.$$

In this Section we prove this result in the general case.

In the calculation of v we evaluate

$$v = \frac{1}{N}\left((x_1-m)^2+(x_2-m)^2+\ldots+(x_N-m)^2\right),$$

where the mean of the sample is m, and the x_i may include repetitions. We have often used the rewritten form of this as follows:

$$v = \frac{1}{N}\left((x_1-k)^2+(x_2-k)^2+\ldots+(x_N-k)^2\right)-(m-k)^2,$$

where k is arbitrary.

For our present purposes we choose $k = \mu$, and get:

$$v = \left(\frac{1}{N}(x_1-\mu)^2+\frac{1}{N}(x_2-\mu)^2+\ldots+\frac{1}{N}(x_N-\mu)^2\right)-(m-\mu)^2.$$

Now the $(x_i-\mu)^2$ and $(m-\mu)^2$ are all values of random variables. Those for $i = 1, \ldots, N$ are ultimately obtained from identical random variables, each identical to the random variable X, which is the parent population variable. The value $(m-\mu)^2$ is ultimately obtained from the random variable with values m (the mean of the sample).

We make use of the facts that $\mu = E[x_i]$ and $\mu = E[m]$ as follows: we write

$$E[(x_i-\mu)^2] = E[(x_i-E[x_i])^2] \equiv \text{var}\,[x_i] = \sigma^2$$

and

$$E[(m-\mu)^2] = E[(m-E[m])^2] \equiv \text{var}\,[m] = \sigma^2/N.$$

Then, by the general result for the expected value of a linear combination of random variables, we have:

$$E[v] = \left(\frac{1}{N}\cdot\sigma^2+\frac{1}{N}\cdot\sigma^2+\ldots+\frac{1}{N}\cdot\sigma^2\right)-(\sigma^2/N)$$

$$= \left(\frac{N-1}{N}\right)\sigma^2.$$

We note that not all of $x_1, x_2, \ldots, x_N, m$ are values of independent variables, because of the linear relation between the x_i and m, but that the result for the expected value of a sum did not depend on the variables being independent, as the reader can verify by turning back to Chapter 9, Section 5.2.

341

6.5 Variance of the population of variances. We have now obtained values in the general case for each of the quantities $E[m]$, var $[m]$, $E[v]$; and we pause to remark that we will not pursue at all far var $[v]$.

Part of the difficulty lies in the non-independence of the random variables with values $(x_i - \mu)^2$ (for $i = 1, 2, ..., N$) and $(m - \mu)^2$. The Result II, on variances, in Chapter 9, Section 5.1 requires the random variables to be independent.

All we will do is to satisfy the reader's curiosity by quoting:

$$\text{var } [S^2] = N^{-3}(N-1)\{(N-1)\,\mu_4 - (N-3)\,\sigma^4\},$$

where $\mu_4 = E((X-\mu)^4)$ in the parent population.

It would be a very profound insight that produced this result from simple numerical cases.

Exercise 11 F

1. Verify the result of Section 6.5 on the parent population and samples of Sections 6.2 and 6.3.

2. Verify the result of Section 6.5 on the parent population and samples of Exercise 11 E, Question 1.

6.6 Form of the probability function for variances of samples. Figure 11.7 shows the probability diagram for the population of variances of samples of size 3 examined in Section 6.3.

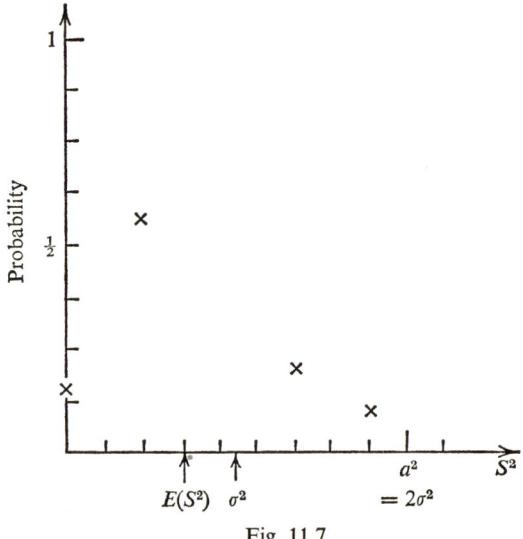

Fig. 11.7

The S^2 axis has been graduated in $\frac{1}{9}a^2$ units and the probability axis in $\frac{1}{8}$ units for greater convenience in each case.

The shape is typical for such populations: We get a positively-skewed, even sometimes a positively-J-shaped diagram, having the mean at $S^2 = \left(\dfrac{N-1}{N}\right)\sigma^2$.

The diagram arising from Exercise 6M, Question 9 provides another example, though in this case the diagram is positively-J-shaped. We postpone further discussion of this until a study of continuous random variables is made.

12

POINT-ESTIMATION

1. INTRODUCTION

We look at the problem set in Exercise 3M, Question 25 and quote 'You draw a random sample of 4 discs from a bag containing 6 discs each of which is either red or green and the sample consists of 3 red discs and 1 green. What numbers of red and green discs in the original bag gave the highest probability of this sample being drawn?'

We construct a table showing down the page a number-pair to indicate the various possible states of the original bag and across the page a number-pair to indicate the various possible samples, in each case the number of red discs being quoted first. Each pair has a single number beside it. This is the ratio of red discs to the total number of discs. It may be thought of as a value of ρ, a parameter uniquely defining the original state of the bag (assuming the number of discs, 6, was fixed); or as a value of r, a statistic defining the sample similarly. Our problem is now to set up a system which will give us grounds for estimating sensibly the value of the parameter ρ from the value of the statistic r. See Table 12.1.

Table 12.1

	$(0, 4)$ $r = 0$	$(1, 3)$ $r = 0.25$	$(2, 2)$ $r = 0.5$	$(3, 1)$ $r = 0.75$	$(4, 0)$ $r = 1$
$(6, 0)\,\rho = 1.00$	—	—	—	—	$\frac{15}{15}$
$(5, 1)\,\rho = 0.83$	—	—	—	$\frac{10}{15}$	$\frac{5}{15}$
$(4, 2)\,\rho = 0.67$	—	—	$\frac{6}{15}$	$\frac{8}{15}$	$\frac{1}{15}$
$(3, 3)\,\rho = 0.5$	—	$\frac{3}{15}$	$\frac{9}{15}$	$\frac{3}{15}$	—
$(2, 4)\,\rho = 0.33$	$\frac{1}{15}$	$\frac{8}{15}$	$\frac{6}{15}$	—	—
$(1, 5)\,\rho = 0.17$	$\frac{5}{15}$	$\frac{10}{15}$	—	—	—
$(0, 6)\,\rho = 0$	$\frac{15}{15}$	—	—	—	—

In the body of the table are entered numbers which are to be read as follows: In any row they are the conditional probabilities of the occurrence of the sample whose column they are in, given that the parent population is the one whose row they are in. That is to say: given $\rho = 0.67$ (so that the original bag had 4 red and 2 green discs) then the probability of a sample (of size 4) having 3 red and 1 green discs is $\frac{8}{15}$. The sum of the numbers in any row is unity since they are the probabilities of exclusive events which exhaust the possibilities for the relevant value of ρ.

344

But what if we read the table the other way? The set of numbers in a column is *not* a set of probabilities. For instance if $r = 0.75$ so that the sample contained 3 red discs then we cannot say that the probability that there were originally 4 red discs in the bag is $\frac{8}{15}$. The number of discs originally in the bag is *not* a random variable. There either were or were not 4 red discs in the bag originally, the probability of there being 4 of them is either 1 or 0; the element of doubt arises because we do not know which. The element of doubt about which sample *will* be drawn at any stage arises because it has not yet *been* drawn, not because we merely do not know.

Remember that the way we use the word 'probability' (with our frequency-based 'probabilities') depends ultimately on a possible indefinite repetition of the experiment. We cannot sensibly talk about arranging the indefinite repetition of an experiment where bags of varied and unknown constitution are selected at random from some collection (itself selected from what?) of bags from which exactly 3 red discs have been drawn. The final proof that the numbers in a column do not form a set of probabilities is that they do not sum to 1.

2. LIKELIHOOD

2.1 Definition and use. Considered as columns, the numbers are, however, useful and we call them *likelihoods* and use the symbol $L(\rho)$. What we can say is that if the sample had 3 red discs then the maximum value of $L(\rho)$ occurs when $\rho = 0.83$. This would be one reason for betting on $\rho = 0.83$, if we found we were forced to bet. Is it a compelling reason? That is a matter of taste; it depends on the better.

The study of the probability function for the population of means of samples, or of any other statistics of samples, is a matter of pure mathematics and could be undertaken for its own sake. It involves no value-judgements, no mention of estimates and, *a fortiori*, no judgements as to whether estimates are 'reliable' or 'satisfactory'. We can still pick any number we like as an estimate of the population parameter. However rational or irrational our choice, and whether we succeed in hitting the true parameter value or not, we can quote our figure as an *estimate*.

The reason we might use here to make the suggested choice seem credible is that for each value of the population parameter there is a calculable probability that the sample which will occur will have such-and-such a value of the statistic, and there is a unique sample for each value of the statistic.

For instance, if $r = 0.75$ then $\rho = 0.83$, 0.67 and 0.5 are all *possible* values of ρ; the probabilities associated with $r = 0.75$ are

$$\text{pr}\,(r = 0.75\,|\,\rho = 0.83) = \tfrac{10}{15},$$
$$\text{pr}\,(r = 0.75\,|\,\rho = 0.67) = \tfrac{8}{15},$$
$$\text{pr}\,(r = 0.75\,|\,\rho = 0.5) \;\; = \tfrac{3}{15}.$$

We could choose to estimate ρ as 0·83 because any other value of ρ would make the occurrence of $r = 0·75$ an event of smaller probability, and if the probabilities have been suitably allotted in the first place, then events whose calculated probabilities are small will seldom occur. In other words, we feel that if we observed many samples (each drawn from a possibly different mixture of 6 discs) and if each time that we found $r = 0·75$ we guessed $\rho = 0·83$, then we should be wrong fewer times than if we consistently chose any other single value of ρ to be associated with $r = 0·75$.

Of course if we said '$r = 0·75$ so I believe that $\rho = 0·83$ or $0·67$' we should be wrong even fewer times. But that is another matter, and it opens the wide and fascinating field of *interval-estimation*. We are dealing here with what is called *point-estimation*, and will delay the consideration of interval-estimation.

To summarize: we could *choose* to *estimate* the population parameter as that value whose row has the biggest of the likelihoods in the column referring to the value of the statistic actually observed. This is the famous 'Principle of Maximum Likelihood' as applied to this situation in which there is only one parameter. It leads, in this case, to Table 12.2.

Table 12.2

Value of r	Estimate of ρ
0	0
0·25	0·17
0·5	0·5
0·75	0·83
1·0	1·0

It is not a matter of pure mathematics. It is a matter of taste. It was suggested by R. A. Fisher in 1922. Estimates selected by the principle have a number of simple properties. Other criteria exist for selecting estimates. In many cases the estimates agree with those selected by this principle.

It is very important to be clear where mathematics begins and ends in this matter of estimates. It is the non-mathematical content that makes *Statistics* a different discipline from *Mathematics*.

We recapitulate: A certain criterion of estimation is selected on non-mathematical grounds. The mathematical *consequences* of the criterion lead to certain estimates. Different criteria may lead to different estimates.

2.2 *A priori* assumptions. Any attempt to gain practical experience of this estimating procedure by actually using a collection of six discs (for instance one of a pair of people using six coins in a pocket and calling copper ones red and silver ones green) immediately runs into the difficulty that the person who is about to do the estimating does not *start* with the belief that all constitutions of the collection of six discs or coins are equally likely, and—whether the reader has noticed it yet or not—this is an underlying assumption of the discussion so far. For instance; $\frac{8}{15}$ in the column (Table 12.1) headed $r = 0.75$ is not the probability that $r = 0.75$, but is only, as already remarked, pr $(r = 0.75 | \rho = 0.67)$. We can only find pr $(r = 0.75)$ if we know pr $(\rho = 0.67)$; and if we know the probabilities of all the values of ρ before the experiment begins—know them *a priori* as we say—then we should be foolish not to take them into account in our estimating procedure. Even if we merely had strong views about the values of the *a priori* probabilities of values of ρ, a good estimating procedure might take account of them.

In the case of all values of ρ being equi-probable we have that each of the probabilities is $\frac{1}{7}$, so that to obtain the probabilities of values of r we multiply *all* the table entries by $\frac{1}{7}$. This means that their *grand* total is now 1. It does not change their relative sizes so that the Principle of Maximum Likelihood still chooses for us in the same way.

3. BAYESIAN METHOD

3.1 Outline of method. Suppose, however, that the bagful of six discs was made up by selecting the discs at random from an enormous pool of equal numbers of red and green discs so that we had, by the well-worn binomial result:

ρ	*a priori* probability
1	$\frac{1}{64}$
0.83	$\frac{6}{64}$
0.67	$\frac{15}{64}$
0.5	$\frac{20}{64}$
0.33	$\frac{15}{64}$
0.17	$\frac{6}{64}$
0	$\frac{1}{64}$

Then Table 12.3 can be read to give the actual probabilities of the various possible number-pairs (ρ, r), and not merely the conditional probabilities for values of r given the values of ρ. The table indicates the probabilities multiplied by 64×15.

Table 12.3

	$r = 0$	$r = 0.25$	$r = 0.5$	$r = 0.75$	$r = 1.0$
$\rho = 1.0$	—	—	—	—	15
$\rho = 0.83$	—	—	—	60	30
$\rho = 0.67$	—	—	90	120	15
$\rho = 0.5$	—	60	180	60	—
$\rho = 0.33$	15	120	90	—	—
$\rho = 0.17$	30	60	—	—	—
$\rho = 0$	15	—	—	—	—

The reader should check that he fully understands this table. For instance:

$$\text{pr} \, (\rho = 0.67 \text{ and } r = 0.75)$$
$$= \text{pr} \, (\rho = 0.67) \times \text{pr} \, (r = 0.75 \,|\, \rho = 0.67)$$
$$= \frac{15}{64} \times \frac{8}{15}$$
$$= \frac{120}{64 \times 15}.$$

We now have a situation that we have met before and we can actually obtain such quantities as $\text{pr} \, (\rho = 0.67 \,|\, r = 0.75)$. The appropriate technique is the application of Bayes' Theorem, and the reader solved a problem in estimation by its use in Exercise 3C, Question 10, and Exercise 3M, Questions 11, 12. The theorem gives, for instance:

$$\text{pr} \, (\rho = 0.83 \,|\, r = 0.75) = 60/(60 + 120 + 60) = \tfrac{1}{4},$$
$$\text{pr} \, (\rho = 0.67 \,|\, r = 0.75) = 120/(60 + 120 + 60) = \tfrac{1}{2},$$
$$\text{pr} \, (\rho = 0.5 \;|\, r = 0.75) = 60/(60 + 120 + 60) = \tfrac{1}{4}.$$

The values $\tfrac{1}{4}$, $\tfrac{1}{2}$, $\tfrac{1}{4}$ are called the *a posteriori* probabilities of ρ when the value $r = 0.75$ has been observed. Notice that the *a posteriori* values depend on the allotment of *a priori* probabilities to values of ρ; because if, for instance, equal *a priori* probabilities are allotted, then the *a posteriori* probabilities of these three values of ρ (given that $r = 0.75$ has been observed) are $\tfrac{10}{21}$, $\tfrac{8}{21}$, $\tfrac{3}{21}$. This change is the usual case in such situations, but Section 9 takes up this point again.

To return to the table based on the Binomial allotment of *a priori* probabilities, we see that a reasonable estimating process would be to select that value of ρ which has greatest *a posteriori* probability. This leads to Table 12.4.

Table 12.4

Value of r	A Bayesian estimate of ρ
0.0	0.17
0.25	0.33
0.5	0.5
0.75	0.67
1.0	0.83

We have placed an *indefinite* article before 'Bayesian' to remind us that the Bayesian results differ for different allotments of *a priori* probabilities. This dependence is not the great drawback in real-life problems that it might seem. We nearly always have some degrees of belief about the various values of ρ and if these could be formalized into a theory of probabilities we could use them. This is the road to 'subjective probability'.

348

3.2 Bias. Notice that this particular Bayesian estimating procedure never leads us to estimate ρ as 0 or 1. The *a priori* ideas about these extreme values have had the quite reasonable effect that, even if $r = 0$ is observed, the method prefers to suppose that this arose from a low value of r from a bag with one red disc, than to suppose that the 'very improbable' bag with no red discs is the one we are faced with. The method has a bias against choosing $\rho = 0$ or $\rho = 1$ *even when, unknown to us, ρ actually has the value* 0 *or* 1.

This 'bias' has nothing to do with what we called *selection bias* in Chapter 11, Section 3.2, for in this investigation the sampling is properly random.

In this case, however, the bias is so strong that whenever $\rho = 0$ or 1 we shall *certainly* be wrong in our estimate. This shock may be lessened by noting that, for each particular estimating procedure for selecting a single value of ρ, there are too few values of r in this *discrete* problem for a $(1 - 1)$ correspondence between $\{r\}$ and $\{\rho\}$, so that some values of ρ will never be selected.

We might well ask 'Shall we be "on average" right when $\rho = 0.83$, for instance?' To answer this, we must first analyse what we could mean by 'on average'. We shall then also see that we have, by a rather round about route, a definition for a term 'bias'.

4. UNBIASEDNESS

4.1 Estimators as random variables. We can think about the estimates of ρ in the following abstract way: Given a sample, we have obtained from it a certain number, called the estimate of ρ (the method is to calculate r and consult an appropriate table to find the entry associated with r); a mapping from a sample to a number defines a statistic. It is important to realize that although we talk about estimators for ρ we are in fact handling functions of r. We can use a notation which brings out this fact. In general $\hat{\rho}$ will be used as a functional symbol to denote a function whose domain is the set of all possible samples and whose images are the numbers we intend to use to estimate ρ; that the numbers are to be used as estimates is indicated by the 'hat'. (If other parameters are involved we might use functions $\hat{\theta}$, $\widehat{\sigma^2}$ and so on—read 'theta-hat', 'sigma-squared-hat' and so on.) The numbers can be good or bad as estimates of ρ—the use of the symbol ρ as part of the symbol $\hat{\rho}$ is not to prejudge that.

In this simple problem, and indeed in many problems, the sample can be entirely described in terms of a single statistic, in this case r, and we will write $\hat{\rho}(r)$ for the estimate, suffixing the ρ to indicate the estimating procedure involved, if there may be more than one.

In the Table 12.5 the relation '$r \to$ estimate of ρ' is tabulated for various procedures.

Table 12.5

Estimating method and estimate

Value of r	Maximum likelihood $\hat{\rho}_L(r)$	Bayesian (binomial *a priori*) $\hat{\rho}_B(r)$	Direct correspondence $\hat{\rho}_C(r)$
0·0	0·0	0·17	0·0
0·25	0·17	0·33	0·25
0·5	0·5	0·5	0·5
0·75	0·83	0·67	0·75
1·0	1·0	0·83	1·0

The first two columns of estimates are from Tables 12.2 and 12.4; the third column of estimates is based on the principle that we estimate the proportion in the population (that is in the bag) directly, by the proportion in the sample; a method which has intuitive appeal (even though, in this case, it, too, is *certain* to give the wrong answer sometimes).

As we have already remarked any function of r, such as $\hat{\rho}(r)$, is merely another statistic and if we are interested in its probability function we often call such a statistic a random variable. We could discuss whether our estimating procedure was 'on average' correct about the value of ρ by finding, for each value of ρ, the expected value of the random variable $\hat{\rho}(r)$, and comparing this expected value with the particular value of ρ from which we started. This may seem more complicated in these words than in the symbols which follow in the next Section.

4.2 Expected value of estimator. *For each value of ρ we will calculate:*

$$\sum_{\text{all values of } r_i} \hat{\rho}(r_i) . \text{pr} \, (r = r_i | \rho \text{ has the named value}),$$

and we will then compare our answer with the value of ρ.

A careful look at the formula above will show that the answer obtained will depend on (reading from left to right in the formula)

(i) the particular estimating method selected, that is to say on the form of $\hat{\rho}$; for instance we might choose $\hat{\rho}(r_i) = 1$ for all r_i if that seemed worth studying;

(ii) the values of pr $(r = r_i | \rho$ has the named value); these are defined in terms of probability theory and the problem, and, after the initial probabilities defined by the problem are allotted, we have no choice here;

(iii) the value of ρ in use at the time; once the problem and estimating procedures are selected the value of ρ is the only variable, and we will consider each value in turn, or obtain a formula applicable to every value.

This look at the formula shows that we have indeed a function of ρ and that comparison of the answer with the value of ρ is a meaningful course of action.

4.3 Details in a particular case. A typical calculation is set out below. We take the maximum likelihood estimator as the statistic whose expected value is being calculated, and, among all the values of ρ, we deal with $\rho = \frac{1}{3}$ first; the conditional probabilities are obtained from Table 12.1; and the values of $\hat{\rho}_L(r)$ from Table 12.5, after conversion from decimal form.

Expected value of estimator

$$= \hat{\rho}_L(0).\mathrm{pr}\ (r = 0|\rho = \tfrac{1}{3}) + \hat{\rho}_L(\tfrac{1}{4}).\mathrm{pr}\ (r = \tfrac{1}{4}|\rho = \tfrac{1}{3})$$

$$+\hat{\rho}_L(\tfrac{1}{2}).\mathrm{pr}\ (r = \tfrac{1}{2}|\rho = \tfrac{1}{3}) + 0 + 0$$

$$= (0\times\tfrac{1}{15})+(\tfrac{1}{6}\times\tfrac{8}{15})+(\tfrac{1}{2}\times\tfrac{6}{15})$$

$$= \tfrac{26}{90}.$$

We note that the value is not equal to $\frac{1}{3}$, which is the value of ρ concerned here; so that in this sense we will *not* be on average correct when $\rho = \frac{1}{3}$.

For the maximum likelihood estimates we can concisely calculate all the expected values we require (that is: one for each value of ρ) by the following matrix product. (The matrix is used here *merely as a shorthand*; there is no matrix algebra involved.)

$$
\begin{pmatrix}
0 & 0 & 0 & 0 & \frac{15}{15} \\
0 & 0 & 0 & \frac{10}{15} & \frac{5}{15} \\
0 & 0 & \frac{6}{15} & \frac{8}{15} & \frac{1}{15} \\
0 & \frac{3}{15} & \frac{9}{15} & \frac{3}{15} & 0 \\
\frac{1}{15} & \frac{8}{15} & \frac{6}{15} & 0 & 0 \\
\frac{5}{15} & \frac{10}{15} & 0 & 0 & 0 \\
\frac{15}{15} & 0 & 0 & 0 & 0
\end{pmatrix}
\begin{pmatrix}
0 \\ \frac{1}{6} \\ \frac{3}{6} \\ \frac{5}{6} \\ \frac{6}{6}
\end{pmatrix}
=
\begin{pmatrix}
1 \\ \frac{80}{90} \\ \frac{64}{90} \\ \frac{45}{90} \\ \frac{26}{90} \\ \frac{10}{90} \\ 0
\end{pmatrix}.
$$

The reader should identify the result previously written out in full and thus discover how the calculation has been laid out. The rectangular matrix gives the conditional probabilities; it is post-multiplied by the vector of the *estimates* and the result is the vector of the *expected values of the estimator*. We see that in general the individual expected values are *not* equal to the parameter values from which they arose, so that we shall not always be 'on average' correct. If we knew the parameter value already we should be able to say whether the estimates were 'on average' correct (as they are for $\rho = 0, \frac{1}{2}, 1$), but if we knew the parameter value already we should be wasting our time in estimating it!

4.4 Contrast with another particular case. The calculations with the correspondence-principle estimates are as follows:

$$
\begin{pmatrix}
0 & 0 & 0 & 0 & \frac{15}{15} \\
0 & 0 & 0 & \frac{10}{15} & \frac{5}{15} \\
0 & 0 & \frac{6}{15} & \frac{8}{15} & \frac{1}{15} \\
0 & \frac{3}{15} & \frac{9}{15} & \frac{3}{15} & 0 \\
\frac{1}{15} & \frac{8}{15} & \frac{6}{15} & 0 & 0 \\
\frac{5}{15} & \frac{10}{15} & 0 & 0 & 0 \\
\frac{15}{15} & 0 & 0 & 0 & 0
\end{pmatrix}
\begin{pmatrix}
0 \\ \frac{1}{4} \\ \frac{2}{4} \\ \frac{3}{4} \\ \frac{4}{4}
\end{pmatrix}
=
\begin{pmatrix}
\frac{60}{60} \\ \frac{50}{60} \\ \frac{40}{60} \\ \frac{30}{60} \\ \frac{20}{60} \\ \frac{10}{60} \\ 0
\end{pmatrix}.
$$

We have the beautiful result that for each value of the parameter the expected value of this estimator is equal to the parameter-value. This means that *without knowing what the parameter value is* we can say that 'on average' this estimate will be correct.

4.5 Definition of unbiasedness. We make the following definition: If an estimating procedure (which in the last resort means a function, called the estimator) is such that for *each* value of the parameter the estimates satisfy

$$E[\text{estimates}] = \text{parameter-value},$$

then we call the estimating procedure (or estimator) *unbiased*. We often loosely call an estimate obtained from an unbiased procedure an unbiased estimate, but strictly speaking it is the procedure or estimator which is unbiased. Compare the situation when describing a game as 'fair'. The adjective really applies to the game as a set of rules and not to a particular play of the game (although there is often no harm in loosely applying the adjective 'fair' to a particular play). The fact that we can call a game 'fair' even when we have lost a particular play is quite a useful analogy. It reminds us that, just as a fair game may or may not result in a draw, and just as we cannot tell from a particular play of a game whether the game is fair or not, so we cannot tell whether a particular estimate derived from an unbiased estimator will actually equal the parameter value or not.

The distinction we are frequently making between *estimator* and *estimate* can be exemplified by a sentence such as: 'Using the mean of the sample as an estimator for the mean of the parent population, the estimate was 0·7; but using the median of the sample as an estimator for the mean of the parent population, the estimate was 0·6.' The distinction is simply that between a random variable and a value of it, or between a statistic and a value of it.

We notice from our numerical example that the maximum likelihood estimator in a problem is not necessarily unbiased. The correspondence estimator in this particular problem is unbiased, but we shall see that there are problems where the correspondence estimator is not unbiased.

Notice that if the equality of expected value and parameter-value breaks down for even a single value of the parameter, then the estimator is *biased*. This is obviously a sensible usage, since in an actual problem of estimation we do not know the value of the parameter and hence could not make use of the fact that for some values of the parameter we might have

$$E[\text{estimates}] = \text{parameter-value}.$$

Finally we remark that the arithmetic was as heavy as it was because we had no algebraic formulation at hand for this problem. In general the whole matter is done simultaneously for all values of the parameter by handling it algebraically. The summations involved often provide *uses* for all those chapters in algebra books on summation of series!

5. ESTIMATORS FOR MEAN AND VARIANCE

5.1 General results. We have already seen (in Chapter 11) that if a sample of size N is taken at random (and with replacement) from a parent population whose mean and variance are μ and σ^2 and the mean and variance of the sample are m and S^2, then

$$E[m] = \mu; \qquad E[S^2] = \left(\frac{N-1}{N}\right).\sigma^2.$$

In the language of the present chapter *the mean of a sample is an unbiased estimator of the population mean* and *the variance of a sample is a biased estimator of the population-variance*.

*5.2 Further discussion on Section 4.1. If we look again at the correspondence-estimator for the proportion of red discs in the bag of Section 1, we can see why we might have anticipated that it would be unbiased, in the following way.

At the ith draw define a random variable Y_i by $Y_i = 1$ or 0, according as the disc is red or green. Then $\text{pr}(Y_i = 1) = \rho$ *for each value of i*, even though, for instance, $\text{pr}(Y_1 = 1 \text{ and } Y_2 = 1) \neq \rho^2$, because of the non-independence; this is the message of Exercise 3M, Question 26 (iv), whose importance may have escaped the reader at the time. We also see that r (defined as the ratio of the number of red discs to the number of all discs in the sample) is the mean of the Y_i-values obtained by regarding the drawing of the sample in this way; we will rename it m. We can interpret ρ as the expected value of the Y_i-variables, since it can be calculated from the parameters of the parent population, arising from the six discs, in the appropriate way; we will rename it μ. The reader should convince himself of these statements, as the ideas are perhaps rather subtle; compare Chapter 11, Section 5, where, however, the variables were independent.

We are now ready to show that even when the sampling is done without replacement we still have $E[m] = \mu$: the necessary result about expectations (Result I in Chapter 9, Section 5.1) does not require the variables to be independent. We have shown above that our Y_i, though not independent, are identical (that is: have the same probability function) and so have a common expected value, here called μ. Thus we do have $E[m] = \mu$, that is to say $E[r] = p$. Indeed we even have var $[m] < \{(N-1)/N\}$ var (Y_i), so that the values m are more closely clustered round their expected value μ, than they would have been if the sampling had been with replacement.

The different value of var $[m]$ should not surprise us if we remember that var $[m]$ is a quantity which depends on the values m *and on their probabilities* (as the square-bracketed notation is intended to remind us); so that when we change the method of getting values m we can expect to change var $[m]$. Furthermore the proof that var $[m] = \{(N-1)/N\}.\sigma^2$ breaks down here, since we no longer have independence, so we still have no cause to be alarmed at a change. That there actually is a change, and that the change is one of *reduction*, when we sample without replacement is to be anticipated for the following reason. If the sampling is with replacement then a possible sample consists of, for instance, the most extreme member of the population taken N times, this makes a larger contribution to $\Sigma(m - E[m])^2$ than is available at all if the sampling is without replacement.

5.3   Unbiased estimator for variance. If we wish to have an unbiased estimator for the population-variance then we only need to use the statistic s^2 defined by $s^2 = \left(\dfrac{N}{N-1}\right).S^2$. Notice the internationally-recommended notational distinction with 'capital' and 'small' letters, and do not confuse it with the distinction between a random variable and its values.

It is easy to see that s^2 is an unbiased estimator, for:

$$E[s^2] = \frac{N}{N-1}.E[S^2]$$

$$= \frac{N}{N-1}.\left(\frac{N-1}{N}.\sigma^2\right)$$

$$= \sigma^2.$$

Now since
$$s^2 = \frac{N}{N-1}.S^2,$$

we have
$$s^2 = \frac{N}{N-1}\left(\frac{1}{N}\Sigma(x_i - m)^2 f(x_i)\right)$$

$$= \frac{1}{N-1}\Sigma(x_i - m)^2 f(x_i).$$

This is the justification for the rule: 'To estimate the variance of the parent population by using a sample, proceed as if calculating the variance of the sample but divide by $N-1$ instead of N'. This process is sometimes called applying 'Bessel's Correction'. It is inapplicable to the formula:

$$S^2 = \frac{1}{N}\Sigma x_i^2 f(x_i) - m^2.$$

The most convenient form for a direct calculation of s^2 from the observations in a sample is:

$$s^2 = \frac{1}{N-1}\left(\sum_1^N t_j^2 - \frac{1}{N}\left(\sum_1^N t_j\right)^2\right),$$

where the t_j are the observations reduced to some convenient origin, and *possibly with repetitions*. It does not seem necessary at this late stage to burden the reader with the full notational horrors of

$$s^2 = \frac{1}{N-1}\left(\sum_1^n (x_i-k)^2 f(x_i) - \frac{1}{N}\left(\sum_1^n (x_i-k) f(x_i)\right)^2\right),$$

since he should be able by now to make the necessary adjustments for repetitions of readings and the resulting frequency function. We also leave to the reader the verification that the formula does, as asserted, evaluate s^2.

5.4 A different definition of variance of a sample. The need in many applications to have an *unbiased* estimator for the variance of a population leads many writers to define the variance *of a sample* as the sum of the squares of the deviations from its mean, divided by $N-1$ instead of by N.

Now if the size of the sample is 1 then the sample obviously gives no information about the variance of the parent population, and the modified definition given above (of the variance *of a sample*) has the attractive property that it does not even define the variance of a sample in this case. The whole approach of the present text, however, to descriptive statistics has been to emphasize their descriptive nature and to develop *various* methods of using them for estimation. In the early stages of the subject when the student first meets 'variance', but before he can deal fully with estimation, it is not easy to justify the $N-1$ divisor, and, furthermore, any attempt to do so prejudges the decision to use *unbiased* estimators as opposed to any others.

As we shall see in Exercise 12B, Question 27, the largest member (in its obvious sense) of a sample from a uniform parent population is a biased estimator for the largest member of the population; but that does not seem, to the present author, an adequate reason for defining inconsistently, in a book concerned with *theory* rather than *practice*, the largest member of a sample and the largest member of a population.

The upshot of all this is that in consulting a statistics text which the reader has not followed through from the first page, it is very important to find out which definition of the variance of a sample is being used.

355

With increasing N the difference diminishes, of course. For $N = 20$ it is only 5 per cent in the variance and about $2\frac{1}{2}$ per cent in the standard deviation; and this is well within the fluctuations due to *sampling*, as expressed by the variance of the variance itself.

5.5   Degrees of freedom. The alternative definition of variance of a sample is the first example we have met of a very widespread feature of more advanced statistics. It often happens that the integer most intimately connected with an estimation problem is not the total number of observations available but the number of observations which could have their values arbitrarily chosen in succession *after* certain statistics of the observations have been given their observed values.

In this case, if we are calculating the value of $\sum\limits_{1}^{N}(t_i - \bar{t})^2$ preparatory to measuring the 'variability', and if we suppose the mean, $\bar{t}$, to have its value allotted already, then only $(N-1)$ of the t_j can be given arbitrary values because we have a linear relation $\Sigma t_j = N\bar{t}$.

Some people may prefer to look at the matter in the following way. If we are trying to give a measure of 'variability', by any method not solely employing ranks, then we are using values of the deviations $d_j = (t_j - \bar{t})$ and these automatically satisfy $\sum\limits_{1}^{N} d_j = 0$, so that they are not linearly independent, and only $(N-1)$ of them could have values allotted to them arbitrarily.

$(N-1)$ is called the number of degrees of freedom of the sample with given mean. If the sample is regarded as a point in N-dimensional space, then $(N-1)$ is the dimension of the sub-space to which the linear relation $\Sigma t_j = N\bar{t}$ restricts us. It can happen in more complicated problems that there are $k\,(> 1)$ linear relations between the t_j, and in that case the number of degrees of freedom in the problem is given by $\nu = N - k\,(\geqslant 1)$.

Many people form the view that it is *because* the number of degrees of freedom of the sample with given mean is $(N-1)$ that we divide by $(N-1)$ to obtain an unbiased estimator for the population-variance; but statements about degrees of freedom are only convenient ways of talking about what we actually do, and what we actually do is determined by mathematical theorems like

$$E[S^2] = \left(\frac{N}{N-1}\right).\sigma^2.$$

5.6   Example 1. In Chapter 4, Section 7.2 two populations $[y]$, $[z]$ were quoted:

$$[y] = (78, 81, 82, 85, 87, 87, 88);$$
$$[z] = (63, 64, 67, 70, 71, 71, 72, 73, 73, 73, 73).$$

If these populations are now regarded as samples of values of a couple of random variables, find unbiased estimates of the expectations and variances of the random variables.

Solution. (i) *Estimates from* [y]. Use new origin $y = 85$; $N = 7$.

Sum of deviations $= -7 -4-3+0+2+2+3 = -7$.

Sum of (deviations)$^2 = 49+16+9+0+4+4+9 = 91$.

Unbiased estimate of mean $= 85+(-\frac{7}{7}) = 84$.

Unbiased estimate of variance $= \frac{1}{6}(91-(-7)^2/7) = \frac{84}{6} = 14$.

(ii) *Estimate from* [z]. Use new origin $z = 70$; $N = 11$.

Sum of deviations
$$= -7 -6-3+0+1+1+2+3+3+3+3 = 0.$$

Sum of (deviations)^2
$$= 49+36+9+0+1+1+4+9+9+9+9 = 136.$$

Unbiased estimate of mean $= 70+(0/11) = 70$.

Unbiased estimate of variance $= \frac{1}{10}(136-(0^2)/7) = 13 \cdot 6$.

Comment. Now the variances of the samples themselves were calculated in Chapter 4, Section 7.3 as var $[y] = 12 \cdot 0$ and var $[z] = 12 \cdot 4$. We now deduce that the smaller variance of $[y]$ may possibly have been due to its smaller size, since we now suspect that it may have been drawn from a parent-population of *larger* variance.

In fact the samples are so small and the variance-ratio so nearly equals unity that the differences are not significant; we should be unwise to take any action based on such a ratio; it is well within the limits of sampling variability as expressed by the variance of the populations of variances.

If the variances of the parent populations are taken as equal, say as σ^2 (one parent population might be related to the other by a simple translation along the number line), then the problem arises as to how to combine the information from the samples to obtain an estimate of the value of σ^2. This problem is discussed in Section 8.3.

6. OBTAINING UNBIASED ESTIMATORS

6.1 A method. The method by which, from $E[S^2] = \left(\dfrac{N}{N-1}\right) . \sigma^2$, we obtained an unbiased estimator for σ^2 can be generalized.

Suppose we want to obtain an unbiased estimator for a parameter θ. We first try to spot a useful statistic (this part of the process cannot be finitely programmed, but requires ingenuity, experience and luck) and then obtain its expected value. The expected value will depend on the probability function for the statistic and this will involve the unknown value of θ; thus the expected value of the statistic will be a function of θ. If it is a *linear* function, all is well; for, if u is the statistic concerned, we have:

$$E[u] = a+b\theta \quad \Rightarrow \quad E[(u-a)/b] = \theta;$$

and if we define $v = (u-a)/b$, then we have $E[v] = \theta$; so that v is an unbiased estimator for θ.

If $E[u]$ is *not* a linear function of θ then we must try another statistic to start from. The reader should have been accustomed from Exercise 4C, Question 3, onwards, to the link between linearity and means.

Now means are connected with summation and summation is closely linked to integration. In dealing with continuous parent populations we shall see that integration plays the role of summation. The difficulties involved in obtaining unbiased estimators are comparable to (and indeed arise from) the difficulties involved in integration of functions.

Thus

$$\int g(x).dx = G(x) + \text{constant}$$

$$\Rightarrow \int g(ax+b).dx = \frac{1}{a}G(ax+b) + \text{constant},$$

but we cannot draw any conclusion about $\int g(x^2).dx$.

It is interesting to notice that the Method of Maximum Likelihood involves *differentiation*. Now differentiation can be finitely programmed; for instance:

$$\frac{d}{dx}G(x) = g(x) \Rightarrow \frac{d}{dx}G(ax+b) = a.g(ax+b),$$

but also implies

$$\frac{d}{dx}G(x^2) = g(x^2).2x.$$

The property of maximum likelihood estimators analogous to the first line is that we can specify a finite routine which will find them.

The second line gives a perhaps unanticipated bonus. Whereas on the one hand $E[g(u)] \not\equiv g(E[u])$ so that if u is an unbiased estimator for θ we do not in general have $g(u)$ is an unbiased estimator for $g(\theta)$, yet on the other hand we *do* have that if v is a maximum likelihood estimator for θ then $g(v)$ is a maximum likelihood estimator for $g(\theta)$. This is the property of *invariance* taken up again in Section 10.

6.2   A common error. A common error among beginners is to say:

$$'E[u] = a+b\theta \quad \Rightarrow \quad u = a+bE[\theta] \quad \Rightarrow \quad (u-a)/b = E[\theta]$$

$$\Rightarrow \quad (u-a)/b \text{ is an unbiased estimator for } \theta.'$$

This bizarre array of symbols depends for its effect upon a pun. Roughly speaking the reader is saying 'The mean of my sample is 2·4 so I *expect* the mean of the population is about 2·4, in fact the *expected* value of the mean of the population is 2·4'. This of course is not the *technical usage* of 'expected value'; the actual error in the symbolic argument is in the first 'implication' where θ has been treated as a value of a random variable on the right-hand side and as a *non-random* variable on the left, while u has been treated the other way round. The fact that the conclusion is correct is an added hazard.

Exercise 12 A

1 R. Find whether the given Bayesian estimator of Section 3 is unbiased in the problem of Section 1.

2 T. Produce a table as in Section 1 for taking samples (without replacement) of 3 discs from the bag. Verify that in this case the maximum likelihood estimator for ρ is given by

$$\hat{\rho}_L(r) = r,$$

thus showing that, even with same parent population, the maximum likelihood estimator function may depend on the size of the sample.

3 R. A certain bag contains 4 discs which may be red or green. $\rho = $ (number of red discs in bag)/4. A sample of size 5 is taken, with replacement, from the bag and $r = $ (number of red discs in sample)/5.

(i) Draw up a table like that in Section 1 showing the probability of each possible value of r given each possible value of ρ.

(ii) Make a table to show the value of the maximum likelihood estimate of ρ given each value of r.

4 T. A certain number N is chosen and fixed and then a sequence of N Bernouilli trials is to be carried out, the probability of success at each trial being θ. The probability that the sequence includes r successes is $\binom{N}{r} . (1 - \theta)^{N-r} \theta^r$.

When N, θ are fixed we will call this quantity $P(r; \theta)$ to emphasize its nature as a probability; and when N, r are fixed we will call it $L(\theta; r)$ to emphasize its nature as a likelihood.

(i) Keeping N, θ fixed, write down $E[r,]$ and using the properties proved in Exercise 9 B, Question 1, give an unbiased estimator for θ.

(ii) Keeping N, r fixed, find the value of θ which maximizes $L(\theta; r)$, and hence find the maximum likelihood estimator for θ, and whether it is unbiased.

5 T. A certain number r is chosen and fixed and then a sequence of Bernouilli trials is to be carried out and stopped at the rth success, the probability of success at each stage being θ. The probability that the sequence includes w failures is

$$\binom{r+w-1}{r-1} (1 - \theta)^w \theta^r.$$

When r, θ are fixed we will call this quantity $P(w; \theta)$ to emphasize its nature as a probability; and when r, w are fixed we will call it $L(\theta; w)$ to emphasize its nature as a likelihood.

(i) Keeping r, θ fixed write down $E[w]$ and use the properties of Exercise 9 B, Question 1 to construct a statistic to serve as an unbiased estimator for as simple a function of θ as you can. Compare your result with that of Question 4, part (i).

(ii) Keeping r, w fixed find the value of θ which maximizes $L(\theta, w)$, and hence find the maximum likelihood estimator for θ, and whether it is unbiased. Compare your result with that of Question 4, part (ii).

6T. We have seen that the expected value of the waiting time, W_1, to the first success in a sequence of Bernouilli trials is given, with our usual meaning for θ, by $E(W_1) = (1-\theta)/\theta$; from this we can deduce that $\theta = \dfrac{1}{1+E(W_1)}$.

A beginner *mistakenly* believes that this implies that $1/(1+W_1)$ will provide an unbiased estimator for θ.

(i) What is $E(1/(1+W_1))$?

(ii) Show that if the waiting time to the second success is W_2 then $1/(1+W_2)$ *does* provide an unbiased estimator for θ.

(iii) To show further that, for fixed $k, k > 1$, if the kth success occurs on the rth trial then $\dfrac{k-1}{r-1}$ is an unbiased estimator for θ, proceed as follows:

$$\text{pr }(k\text{th success is on }r\text{th trial}) = \text{pr }(k-1 \text{ successes in } r-1 \text{ trials}).\theta$$

so
$$E\left[\frac{k-1}{r-1}\right] \text{ reduces to } \sum_{r=k}^{\infty} \binom{r-2}{r-k} (1-\theta)^{r-k} \theta^k,$$

but
$$\binom{i}{j} = (-1)^j \binom{-(i-j+1)}{j}, \dots \text{ and so on.}\dots$$

7. A sequence of Bernouilli trials, with probability θ of success at each trial, is continued until the first success occurs. Let this be on the Sth trial.

(i) Find $E[s]$ and prove that

$$E\left[\frac{1}{s}\right]. = \frac{\theta}{1-\theta} \log \left(\frac{1}{\theta}\right).$$

(ii) The Bernouilli trial consists of observing a time interval in which random events occur and the length of the interval is such that the expected number of events occurring in such an interval is 1.

Success occurs in the Bernouilli trial if and only if no events occur in the interval. What is the value of θ?

(iii) Show that for the value of θ obtained in part (ii),

$$E\left[\frac{1}{s}\right] = \frac{1}{E[s-1]} \neq \frac{1}{E[s]}.$$

(iv) What general statement could you make about unbiased estimators, based on part (iii)?

8T. (i) t is an unbiased estimator for a parameter θ, and $t_j, j = 1, 2, \dots, N$, are values of t obtained from N experiments. Prove that $\dfrac{1}{N}\Sigma t_j$ is an unbiased estimate of θ.

(ii) Find the conditions on the set of numbers λ_j if $\Sigma \lambda_j t_j$ is also to be an unbiased estimate of θ.

[This demonstrates the important property that if we have a number of estimates derived from unbiased estimators and we take a weighted mean of their values we are using a method which provides another unbiased estimator, whatever may be the weightings.]

9. A certain dichotomous trial has constant probability of success θ. A dozen sequences, $A, B, \dots, L$, of the trial are carried out and the following particulars recorded.

Code letter of sequence	Ratio of successes to total number of trials for first 10 trials	Ratio of successes to total number of trials for first 4 successes
A	1 in 10	4 in 16
B	6 in 10	4 in 6
C	3 in 10	4 in 11
D	3 in 10	4 in 11
E	1 in 10	4 in 15
F	5 in 10	4 in 8
G	7 in 10	4 in 4
H	5 in 10	4 in 9
I	8 in 10	4 in 5
J	3 in 10	4 in 14
K	2 in 10	4 in 15
L	6 in 10	4 in 8

Use the results of Questions 4, 5, 6, 8 to do the following:

(a) Obtain separately from each column of results a maximum likelihood estimate of θ.

(b) Obtain, one from each column of results, unbiased estimates of θ, and from the second column an unbiased estimate of $1/\theta$.

10. By considering the unbiasedness property (proved in Question 4) of r/N as an estimator of θ in a binomial population, show that, in the estimation problem of Section 1, if the 6 discs in the bag were earlier drawn from a large collection of discs of any proportion of red and green whatever, then the ratio of red to total number of discs in a sample drawn from the bag without replacement is an almost unbiased estimator of ρ. (The use of the word 'almost' is only to remind us that selecting without replacement from a finite collection of discs, however large, cannot be quite Bernouillian.)

11 R. The following are two pairs of samples.

$A1$ 16, 15, 18, 16, 16, 16, 17, 15, 19, 17, 15, 16.
$A2$ 11, 14, 17, 15, 16, 20, 11, 18, 15, 13, 20, 16, 19, 12, 14, 20.
$B1$ 126, 110, 107, 103, 105, 112, 118, 107.
$B2$ 102, 118, 115, 105, 105, 100.

(i) Make unbiased estimates of the means and variances of the four parent populations from which they were drawn. Find also the ratio of the larger to the smaller of the estimated variances of the parent populations of A_1 and A_2; and repeat for B_1 and B_2.

(ii) In fact the means, μ, and variances, v, of the parent populations from which these samples were taken at random are given in mild disguise by the solutions of the following pairs of equations.

$$A1 \quad \begin{cases} \mu - 4v = 9 \\ 2\mu + 3v = 40 \end{cases} \qquad A2 \quad \begin{cases} \mu + 2v = 32 \\ \mu - 2v = -1 \end{cases}$$

$$B1 \quad \begin{cases} \mu - v = 10 \\ 10\mu - 11v = 0 \end{cases} \qquad B2 \quad \begin{cases} 2\mu - v = 120 \\ -4\mu + 6v = 120 \end{cases}$$

361

The reader should resist the temptation to solve these equations until he has made his estimates. Having made the estimates, however, he is to find for each sample the value of the variance-ratio $\left(\text{that is } \dfrac{\text{larger variance}}{\text{smaller variance}}\right)$ when the variances are the estimated and actual population variances.

[Samples *B* were taken from approximately Normal parent populations and for samples of these sizes from Normal parent populations it can be shown that the three ratios calculated (in parts (i) and (ii)) will have values exceeding 4·9 (even if the actual population variances were equal), 2·0 and 2·2 in about 10 per cent of cases; the ratios observed in this question are thus well within limits of reasonable chance fluctuations.]

12F. (This question forms an important introduction to the rest of the chapter.)

A population $[t]$ has mean θ. A sample of size k is drawn with members $t_1, t_2, \ldots, t_k$ (in their random order of drawing). There may be repetitions among the values $t_1, t_2, \ldots, t_k$ since the sampling is with replacement.

(i) Show that, whatever may be the probability function of the parent population $[t]$, each of the following statistics u, v, w, x provides an unbiased estimator for θ. That is to say, prove that $E[u] = E[v] = E[w] = E[x] = \theta$.

    (*a*) $u = t_1$.

    (*b*) $v = \Sigma\lambda_j t_j/\Sigma\lambda_j$ for any numbers λ_j whatever.

    (*c*) w is formed, as the readings arrive, as a sort of running average in the following way:

The first two readings have their mean m_2 computed, then the mean $\frac{1}{2}(m_2 + t_3)$ is computed and so on.

Defined inductively we have:

$$\begin{cases} m_0 = t_1, \\ m_j = \tfrac{1}{2}(m_{j-1} + t_j) & (j=1, 2, \ldots, k). \end{cases}$$

The definition of w is m_k.

    (*d*) x is the mean of the sample.

(ii) If the variance of the population $[t]$ is β^2, calculate var $[u]$, var $[v]$, var $[w]$, var $[x]$.

[A full appreciation of part (i) (*a*) is very important. It reminds us that to take any single member of a population is to use an unbiased estimator of the mean of the population. Part (ii) reminds us that the variance of this estimator is the very variance of the population itself. *Members of populations are often usefully thought of as samples of size* 1.]

13. Use the data collected in Exercise 1A, Question 2 to provide a table of conditional probabilities by allotting to the conditional probabilities values directly proportional to the corresponding relative frequencies.

(i) Make a table of the maximum likelihood estimates (of the true rank) from each value of the guessed rank.

(ii) Suppose that the items to be estimated were not lines on cards but rods of lengths 5·0, 5·3, 5·6, ..., 7·7 cm, cut from ten rods each of length 21 cm in such a way that the frequencies were given by:

Length/cm	5·0	5·3	5·6	5·9	6·2	6·5	6·8	7·1	7·4	7·7
Frequency	8	3	3	3	3	3	3	2	2	2

One of these rods is then selected at random so that the *a priori* probabilities are proportional to the frequencies in this table.

Assume further that the conditional probabilities are unaffected by this alteration and make a table of the resulting Bayesian estimates for the true rank given the guessed rank.

(iii) For each estimating procedure make a table of the expected value of the estimating statistic (that is: guessed rank) for each value of the parameter (that is: true rank).

7. FURTHER COMPARISONS BETWEEN ESTIMATORS

7.1 Unbiased estimator not unique. Exercise 12A, Question 12 has shown that there may be several quite different unbiased estimators. That is to say a particular single sample of size N may provide several estimates of the same population parameter all by unbiased procedures. We do not expect the *estimates* to be equal; we do not know which if any is correct on this particular occasion, but only that each has arisen from an unbiased estimating procedure. How do we develop a criterion for selecting one unbiased *procedure* as preferable to another?

7.2 Examination of criterion of unbiasedness. It will help to look again, perhaps more critically, at the reason for choosing an unbiased estimator at all. It is that if a great many samples are taken from a population with fixed but unknown parameter, and each sample gives rise to a value of the estimating statistic, and the *mean* of all the resulting estimates is calculated, then the answer will be near the value of the parameter, with the usual sort of chance fluctuations that are inevitable in, indeed are the essence of, statistical problems.

The reader should not imagine that in real life he is going to be able to take an indefinite number of samples before basing a decision on them. If he did get the chance of a couple of samples, sure to be from the same population and of sizes say N_1 and N_2, he would do better to treat them as a single sample of size $(N_1 + N_2)$ as we shall see; but there comes a moment when he runs out of time, or money, and will have to stop amalgamating extra samples into his super-sample. What advantage comes then from using an estimator which, *if* we were allowed indefinite repetitions, *would* be 'closer on average'?

First: it might be that every year a decision has to be taken and, while each year there will be some error, it would be a worthwhile goal to try to get the errors to cancel rather than to accumulate. Of course if the losses are not directly proportional to the errors, then we should like to minimize some other function than the mean (and hence sum) of the errors (in economics some quite complicated functions might be appropriate), and much present-day research on estimating procedures is concerned with what are called *loss functions*. Direct cancellation of errors, that is to say use of the *mean* for its own sake, is thus not a very strong reason.

Secondly: Chebyshev's Inequality tells us that, whatever probability function the estimating statistic may have, the probability of a value of the statistic occurring more than k times its standard deviation *from its mean* is at most $1/k^2$. Of two possible estimators, then, both having the same variance, the unbiased one, that is the one whose mean is the mean of the parent population, *may* give an estimate more likely to be near the mean of the parent population than an estimate from the biased estimator. But even in this situation curious things could happen. For instance in Exercise 4E, Question 6 there are three populations, of which we now take the first unchanged, ignore the second, and transform the third by subtracting $\frac{1}{2}$ from all the values of x, calling the result z.

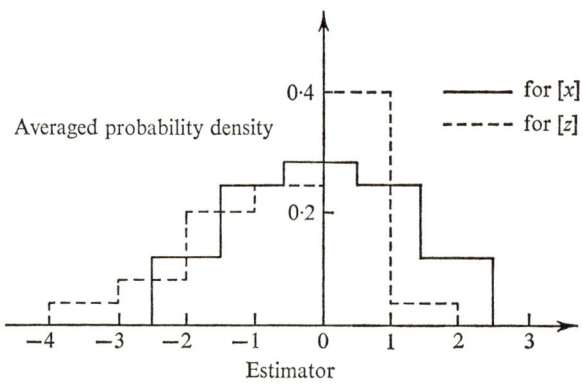

Fig. 12.1

If $[x]$ and $[z]$ are taken as the populations of two estimating statistics in a problem where the parameter-value is zero, then $[x]$ gives an unbiased estimator and $[z]$ does not, and both populations of statistics have the same variance. The probability functions are illustrated in Figure 12.1.

Assuming uniform spread within each 'class' we have:

Range of values of statistic	pr (x lies in range)	pr (z lies in range)
$-\frac{1}{2}$ to $\frac{1}{2}$	0·28	0·32
-1 to 1	0·52	0·64
$-1\frac{1}{2}$ to $1\frac{1}{2}$	0·76	0·76
-2 to 2	0·88	0·88
$-2\frac{1}{2}$ to $2\frac{1}{2}$	1·00	0·94

The example is, of course, artificial, but it should make the point that unbiasedness is not the only thing to seek. Notice that the choice of z here as the estimating statistic would mean we were usually nearer the parameter value than we should be with x, but that we could suffer, on rare occasions, from values wildly out, in a way that does not occur with x. As in life, we see that general rules may well exist but circumstances alter cases. This is not to decry the value of any particular general rule, and we will sometimes regard unbiasedness as a desirable characteristic in itself, but we must turn our attention more to the question of the variance of the population of estimating statistics supplied by a particular estimator.

So far, it is when the mean, and not for instance the median, is being used as a measure of the 'centre' or 'position' of a population that unbiased estimators seem to have some special status; and the mean as we know from Chapter 4 is an arbitrary choice of statistic for this purpose. Our third reason, however, was also hinted at in Chapter 4. Nothing can detract from the fact that the algebra of expected values is really very simple; and it is this *convenience* which is so much in favour of unbiased estimators. An estimator whose properties we could not develop would be quite useless for any serious work.

In a similar way the strong point in favour of maximum likelihood estimators is not the nearly magical effects of having the word 'likelihood' in their name, but the analytical simplifications which *they* introduce; analytical simplicities which differ from those of unbiased estimators.

7.3 Variance of an estimator.
We now look more closely at the question of the variance of the estimating statistic. This question was raised in Section 7.2. We develop a different numerical example to illustrate matters: one based on the problem with which Chapter 11 began, namely the estimation of the mean of a uniform population.

The reasons for choosing a uniform parent population are two.

First: the construction of the entire population of values of the statistic is fairly easy.

But secondly: a uniform parent population is about as unfavourable a start as we can take, if bunching of estimates at the centre is what we are trying to illustrate. This is because any population which starts bunched may reasonably be imagined to lead more quickly to bunching of the estimates when we start taking samples. This is a general rule which does not allow for the possibility that a population may have too large a proportion of its members larger than any value we may name beforehand, while having most of its members bunched in the centre. One such population was discussed by Cauchy and will be mentioned again later. The variance of Cauchy's population is 'infinite' and no size of sample from it, however large, produces a population of means with a probability function different from that of the parent population!

We return to less bizarre populations, and the population chosen for an

365

illustration has members denoted by X and the probability function shown in Table 12·6.

Table 12.6

x	$\operatorname{pr}(X = x)$
-9	$\frac{1}{4}$
-3	$\frac{1}{4}$
$+3$	$\frac{1}{4}$
$+9$	$\frac{1}{4}$

(The curious choice of values of X is simply to avoid fractions later.) We will take samples of size 3: they are listed in Table 12·7 together with three of their statistics: there are 64 different samples, if order is taken into account, and they reduce to 20 different unordered samples. As always, we make the assumption that if the model represents a real collection of items then the number of items of any type and the probability of the associated value of X which identifies the type must be proportional; and further that each item must have an equal chance of selection in the sample at each stage and be returned to the collection before the next selection of an item.

Table 12.7

Unordered sample	$64 \times$ prob.	Mean of extremes	Median	Mean
$-9, -9, -9$	1	-9	-9	-9
$-9, -9, -3$	3	-6	-9	-7
$-9, -9, +3$	3	-3	-9	-5
$-9, -9, +9$	3	0	-9	-3
$-9, -3, -3$	3	-6	-3	-5
$-9, -3, +3$	6	-3	-3	-3
$-9, -3, +9$	6	0	-3	-1
$-9, +3, +3$	3	-3	3	-1
$-9, +3, +9$	6	0	3	1
$-9, +9, +9$	3	0	9	3
$-3, -3, -3$	1	-3	-3	-3
$-3, -3, +3$	3	0	-3	-1
$-3, -3, +9$	3	3	-3	1
$-3, +3, +3$	3	0	3	1
$-3, +3, +9$	6	3	3	3
$-3, +9, +9$	3	3	9	5
$+3, +3, +3$	1	3	3	3
$+3, +3, +9$	3	6	3	5
$+3, +9, +9$	3	6	9	7
$+9, +9, +9$	1	9	9	9

These requirements (*random sampling with replacement*) underlie all the theory of statistics developed here. Quite different results hold for sampling without replacement; though, of course, the differences become more negligible as the size of the parent population increases relative to the size of the sample.

From Table 12.7 we construct Tables 12.8 to 12.10, the tables of probability of the various statistics. (These tables define the functions called by many writers the *sampling distributions* of the mean of the extremes, of the median and of the mean, respectively; we have avoided this phraseology here, but the reader will frequently meet it in the literature and it is most important to understand what is meant by it.) We quote the values of two of the parameters of each population of statistic values—namely the expected value and variance.

Table 12.8

Mean of extremes	64 × prob.	64 × averaged prob. density
−9	1	$\frac{1}{3}$
−6	6	2
−3	13	$4\frac{1}{3}$
0	24	8
3	13	$4\frac{1}{3}$
6	6	2
9	1	$\frac{1}{3}$

We have:

E(mean of extremes) $= 0$; var (mean of extremes) $= \frac{828}{64} = 12\cdot9$.

Table 12.9

Median	64 × prob.	64 × averaged prob. density
−9	10	$1\frac{2}{3}$
−3	22	$3\frac{2}{3}$
3	22	$3\frac{2}{3}$
9	10	$1\frac{2}{3}$

We have: E(median) $= 0$; var (median) $= \frac{2016}{64} = 31\cdot5$.

Table 12.10

Mean	64 × prob.	64 × averaged prob. density
−9	1	$\frac{1}{2}$
−7	3	$1\frac{1}{2}$
−5	6	3
−3	10	5
−1	12	6
1	12	6
3	10	5
5	6	3
7	3	$1\frac{1}{2}$
9	1	$\frac{1}{2}$

We have: E(mean) $= 0$; var (mean) $= \frac{960}{64} = 15\cdot0$.

As a check on the last calculation we see that var (mean) = $\frac{1}{3}$ × variance of parent population as it must with samples of size 3.

These results are shown as histograms in Figure 12.2. The probability densities have been averaged over intervals of length 3, 6 and 2 respectively. For simplicity only half of each graph is shown.

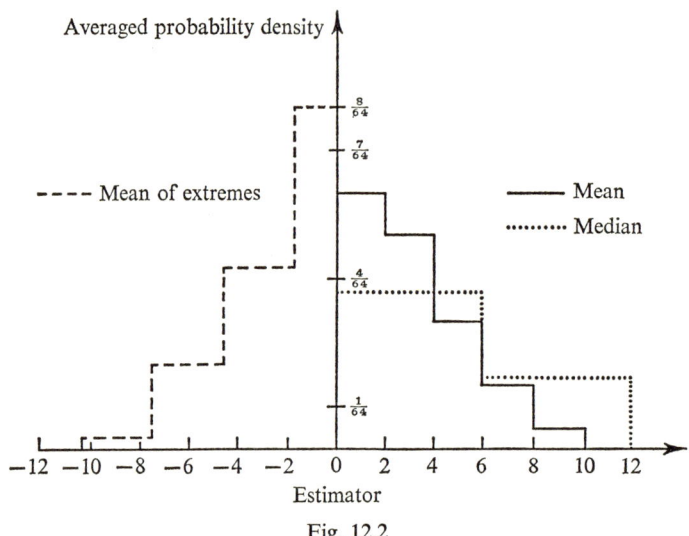

Fig. 12.2

We return to Tables 12.8, 12.9 and 12.10, and note that the probability of getting a value of the estimating statistic within various ranges is as shown in Table 12.11.

Table 12.11

Range	64 × prob. for mean of extremes	64 × prob. for median	64 × prob. for mean
0 to 0	24	0	0
−1 1	24	0	24
−2 2	24	0	24
−3 3	50	44	44
−4 4	50	44	44
−5 5	50	44	56
−6 6	62	44	56
−7 7	62	44	62
−8 8	62	44	62
−9 9	64	64	64

The advantage over the mean possessed by the mean of extremes is not large but it should be apparent that of two unbiased estimators the one that gives the lower variance of statistics is generally preferable. This is the criterion of *minimum variance*. There are cases where it too leads to an unsatisfactory estimator (see Exercise 12B, Question 6), but it is certainly worth our attention next.

8. MINIMUM VARIANCE

At the end of the last section we mentioned the principle of choosing an estimator to have the lowest variance among estimators of that class. We look at the application of this principle to the estimators dealt with in Exercise 12A, Question 12.

8.1 Two trivial cases. We deal first with a couple of trivial but intriguing cases. In Exercise 12A, Question 12 part (i)(*a*) we see that the variance does not decrease with increasing sample size; as is obvious from first principles since no notice is taken of any reading after the first. The method is equivalent, of course, to taking the reading from any predetermined position in the sample, or indeed to taking a single reading chosen at random after the sample has been drawn.

Part (i)(*c*) of the question shows a rather more favourable state of affairs. This time the weighting factor for a reading is dependent on the position of the reading in the sample as it is ordered by the times at which the readings are taken.

We have $\quad \text{var}\,[w] = ((\tfrac{1}{4})^{k-1} + (\tfrac{1}{4})^{k-1} + (\tfrac{1}{4})^{k-2} + \ldots + \tfrac{1}{4})\,\beta^2$

$$= \left((\tfrac{1}{4})^{k-1} + (\tfrac{1}{4})\left(\frac{1 - (\tfrac{1}{4})^{k-1}}{1 - (\tfrac{1}{4})}\right)\right).\beta^2$$

$$= (\tfrac{1}{3} + \tfrac{2}{3}.(\tfrac{1}{4})^{k-1}).\beta^2.$$

Thus the variance of this estimator does decrease with increasing sample size but does not tend to zero; it always exceeds $\tfrac{1}{3}\beta^2$, which is the variance of the population of means of samples of size 3. It follows that taking a sample of size 3 and using its mean as an estimator is preferable to this method.

8.2 Consistency. If an estimator T for a parameter θ satisfies $E(T) \to \theta$ and var $(T) \to 0$ as the sample size, N, tends to infinity, then the estimator is called *consistent*. This means that the values of the estimator are ever more closely clustered round this mean as the sample size increases, and that furthermore this mean is either equal to the parameter value (in which case of course the consistent statistic happens to be unbiased) or gets ever closer to the parameter value as $N \to \infty$ and can be made as close as we please by taking N large enough. In this second case the bias can be made negligible by taking large enough samples. It is more important that a statistic should be consistent even if biased than that it should be merely unbiased but fail to have var $(T) \to 0$.

The reader has the opportunity of examining the concept of consistency more closely in Exercise 12B, Question 9.

8.3   An important case of minimum variance. We now consider part (i)(*b*) of Exercise 12A, Question 12. We first of all point out that although the question was framed in terms of a parent population [*t*], with mean θ and variance β^2, from which a sample was drawn to estimate θ, the situation can be very much more general than that.

What we were *really* dealing with was a sample of k values, one each of k independent estimators $T_1, T_2, ..., T_k$, under the following conditions:

(*a*) Each estimator satisfied $E(T) = \theta$ (so that they were unbiased estimators of θ),

(*b*) the variance of each estimator was β^2.

At no stage did we require that the estimators should agree in more than these two parameters. We were then asked to investigate other estimators with values derived from the values of $T_1, T_2, ..., T_k$.

We now look at the derived estimator v in the light of the principle of minimum variance.

We have $v = \Sigma\lambda_j t_j / \Sigma\lambda_j$, and we first write $\lambda_j/\Sigma\lambda_j = (1/k) + \delta_j$, so that δ_j measures the discrepancy of the 'weighting factor' $\lambda_j/\Sigma\lambda_j$ from strict proportion. It follows that $\Sigma\delta_j = \Sigma(\lambda_j/\Sigma\lambda_j) - (k/k) = 0$.

Then
$$\text{var}(v) = \Sigma\left(\frac{1}{k} + \delta_j\right)^2 . \text{var}[t_j]$$

$$= \Sigma\left(\frac{1}{k^2} + \frac{2\delta_j}{k} + \delta_j^2\right) . \beta^2$$

$$= \frac{k\beta^2}{k^2} + \frac{2\beta^2}{k} . \Sigma\delta_j + \beta^2\Sigma\delta_j^2$$

$$= \frac{\beta^2}{k} + \beta^2\Sigma\delta_j^2, \quad \text{since } \Sigma\delta_j = 0,$$

$$\geq \beta^2/k, \quad \text{with equality only if } all \ \delta_j = 0.$$

This shows that if a number, k, of independent unbiased estimators, T_j, for a parameter all have the same variance, then the linear combination $\frac{1}{k}\Sigma T_j$ has a smaller variance than any other linear combination of them. The value of this minimum variance being $(1/k)$ times the common variance of the estimators. Now the value of $\frac{1}{k}\Sigma T_j$ is merely the mean value of the original estimates, so, despite all our efforts to deal in terms of general

370

unbiased estimators T_j for a parameter θ, we see that 'taking the mean' creeps into the problem by way of taking the mean of the estimates.

Three special cases of this situation are of supreme importance, and we give them a section to themselves.

8.4   Pooled estimates from equal-sized samples.

I. Take the T_j to be single members of a population whose mean, θ, we wish to estimate. The variance of the population is σ^2, say. We verify that the T_j satisfy our conditions:

(a) They each have $E(T_j) = \theta$: by definition of θ.

(b) They all have the same variance: it is in fact, by definition, the variance of the population being studied, so $\beta^2 = \sigma^2$.

It follows that the mean of a random sample of k independently drawn members of a population is an unbiased estimator of the population mean having smaller variance than any other linear combination of the individual observations. Its variance is $\beta^2/k = \sigma^2/k$.

This is merely the result already proved in Exercise 12A, Question 12, part (i)(d), and so verifies the assertion at the start of Section 8.3 about the process of generalizing the result of part (i)(b) of the question.

II. Take the T_j to be the means of k samples of equal size, say N, from a population whose mean θ we wish to estimate. The variance of the population is σ^2, say. We verify that the T_j satisfy our conditions:

(a) They each have $E(T_j) = \theta$: by the result in Chapter 11, Section 3.2.

(b) They all have the same variance by symmetry: it is in fact, by the result in Chapter 11, Section 3.4, given by $\beta^2 = \sigma^2/N$.

It follows that the means of k independent samples of equal sizes should be treated as if they were k single observations; this leads to amalgamating them into a single sample and calculating

$$\text{its mean} = \Sigma(\text{means of samples})/k$$

$$= \Sigma\left(\frac{\text{sum of sample members}}{N}\right)\Big/k$$

$$= \frac{\text{sum of all members of all samples}}{Nk}$$

$$= \frac{\text{sum of all observations}}{\text{total number of observations}}.$$

This unbiased estimator has the smallest variance for estimators of its linear pattern. The variance of this estimator is

$$(\sigma^2/N)/k = \sigma^2/Nk.$$

This equals (population variance)/(total number of observations). These results will be shown later to be true even when the restriction to samples of equal size is lifted. They are thus of great importance.

III. Take the T_j to be $\{N/(N-1)\} \times$ (variance of sample) for k samples of equal size N drawn from populations *whose means are irrelevant* but whose variances and fourth moments are equal, and whose common variance, θ, we wish to estimate. We verify that the T_j satisfy our conditions:

(a) $$E(T_j) = \frac{N}{N-1} E(\text{variance of samples of size } N)$$

$$= \frac{N}{N-1} \cdot \frac{N-1}{N} \cdot \theta$$

$$= \theta.$$

(b) All the T_j have the same variance since variances are unaffected by any translation necessary to bring the means of the various populations into coincidence: it is in fact given by

$$\beta^2 = \left(\frac{N}{N-1}\right)^2 \cdot \left(\frac{N-1}{N^3} \{(N-1)\,\mu_4 - (N-3)\,\theta^2\}\right),$$

referring to Chapter 11, Section 6.5. μ_4 is the common fourth moment of the populations.

It follows that the variances of k independent samples of size N from such populations should be used to form the statistic

$$\frac{1}{k}\Sigma\frac{N}{N-1}\,(\text{variance of sample}).$$

Now:

$N \times$ (variance of sample) = sum of squares of deviations of members of sample from mean of sample,

so the statistic is:

$$\frac{1}{k}\Sigma\frac{1}{N-1}\,(\text{sum of squares of deviation of members of sample from mean of sample})$$

which is:

$$\frac{1}{Nk-k}\,(\text{total sum of (squares of deviations of members of sample from sample's } own \text{ mean)}).$$

But Nk is the total number of observations and k is the number of samples under consideration, so we finally make the following statement:

If samples of equal size are drawn from populations whose means are irrelevant, but which have a common variance and a common fourth moment, and if we estimate this common variance in the manner described below, then we shall have an estimator which is unbiased and which has a

372

smaller variance than has any other linear combination of the squares of
the deviations. The manner referred to is to use the statistic:

> {total sum of (squares of deviations of members of sample from
> sample's own mean)} divided by {total number of observations
> minus the number of samples employed}.

We shall see in Exercise 12 B, Question 21 that this estimator retains the
properties of unbiasedness and minimum variance for estimators of its
form even when the sizes of the samples are unequal. This makes it a very
important one.

Notice, incidentally, that if the means of the k samples are thought of as
being allotted first, then the divisor is the number of degrees of freedom of
the observations. This way of describing what we have written may prove
helpful to the reader in reminding him of the form of the statistic.

Finally we remark that the condition on the fourth moments is not a
painful one since the hypothesis that we usually make is that the populations are identical apart from their means; and furthermore many populations that occur in practice are defined, apart from position, by only one
more parameter, so that once their variances are taken to be equal their
higher moments are also equal.

8.5   Efficiency. If two estimators are unbiased then we have seen that
we will prefer the one with smaller variance; in this way the inverses of the
variances give some measure of the extent to which we prefer one estimator
to another. The ratio of the inverses of the variances or, what is the same
thing, the inverse ratio of the variances of two unbiased estimators is
called the *relative efficiency* of the estimators.

Thus in the example dealt with in Section 7.3 we have:

efficiency of median relative to mean $= \frac{64}{2016}/\frac{64}{960} = 48\%$;

efficiency of median relative to mean of extremes $= \frac{64}{2016}/\frac{64}{828} = 41\%$;

efficiency of mean relative to mean of extremes $= \frac{64}{960}/\frac{64}{828} = 86\%$.

The reason for calling this ratio a *relative efficiency* is as follows. If we
take samples of size N from a parent population whose variance is σ^2 then
the variance of the population of means of samples is given by

$$\mathrm{var}\,[m] = \sigma^2/N.$$

Now suppose that the population of some other estimating statistic of
position, say t, had, for this size of sample, a variance var $[t]$. This would
be the same variance as that of the population of means of samples of
size N' given by $\sigma^2/N' = \mathrm{var}\,[t]$. From this we have:

$$N' = \frac{\sigma^2}{\mathrm{var}\,[t]} = N.\frac{\mathrm{var}\,[m]}{\mathrm{var}\,[t]}.$$

The ratio of the 'effective' size, N', of the sample when the second statistic is calculated to the actual size, N, is $\dfrac{\text{var}\,[m]}{\text{var}\,[t]}$. In a loose manner of speaking the statistic only uses a fraction var $[m]$/var $[t]$ of the data in the sample from which it was calculated. By an extension of this idea we define the relative efficiency of any two unbiased estimators as the inverse ratio of the variances of the populations of their values. Broadly speaking if the first of two estimators has an efficiency of 50 per cent relative to a second estimator, then twice as large a sample will have to be taken for the first statistic to get about the same sort of likely closeness to the parameter value as the second statistic gets. In some problems there exists an unbiased estimator whose variance can be proved to be smaller than (or equal to) that of any other unbiased estimator, and efficiencies relative to this estimator are often called *efficiencies* without the adjective 'relative'; an *efficiency* is therefore bound not to exceed 100 per cent and it is customary for *relative efficiencies* to be quoted relative to the estimator of smaller variance and so to be less than 100 per cent too. It is a pity that the *inverse* of the efficiency was not the quantity defined by early writers as it would save inverting twice on the route from variances *via* efficiencies to altered sample sizes!

If one estimator is biased and another is unbiased their relative efficiency is defined in a slightly modified way, into which we will not enter here, but which is discussed in Section 11, Example 3.

9. SUFFICIENCY

Suppose we are investigating a parameter θ in a population and that we have a sample from the population; sometimes the situation holds that there is a statistic (whose value is determined by the sample) such that all the information which the sample supplies about the parameter value is expressible by the value of the statistic. When the parameter and statistic are related in this way the statistic is called a *sufficient* estimator for the parameter. Not all parameters have sufficient estimators; when a parameter does admit of a sufficient estimator then that estimator is obviously a desirable one to use.

All this can be put into symbolic form in statements about the forms of the various probability functions, but to illustrate the ideas, let us consider a sequence of Bernouilli trials. We will characterize them by a random variable X, such that

$$\text{pr}\,(X = 1) = \theta, \quad \text{pr}\,(X = 0) = 1 - \theta.$$

The parameter θ is the only parameter in the problem. We will show that if a sample (let us say of size N) of values of X is given then the statistic whose value is the mean of the values of X in the sample is a sufficient estimator for θ. In ordinary words, to which we may return with relief, if we know the proportion of successes in a run of N Bernouilli trials then no other information about the sample will enable us to improve our estimate of θ.

For instance, if we had observed a sequence of 10 trials, and among them there had been only 1 success then we would estimate the probability of success as $\frac{1}{10}$. If next we carried out 1 more trial and observed another success we would estimate the probability of success as $\frac{2}{11}$. Now we might argue that, because of the discrete nature of the process, either one or no successes was bound to occur on the 11th trial, so that the sudden jump in the estimate was misleading. We would like to retain the information that the first 10 trials only contained 1 success and to make use of this information in our new estimate. One suspects that many people would do this. They are in effect saying 'Given that the first 11 trials contained 2 successes, the information that the last of the trials showed a success is useful information'. They might even bolster their argument by saying 'If the probability of success was $\frac{1}{10}$ then the expected waiting time between trials is 9 so that we can expect a jump in the estimate about every 9 or 10 trials, and to be swept off our feet and double our estimate because of *one* more success is rash'.

We will prove, however, that given the proportion of successes, $\frac{2}{11}$, for the first 11 trials the conditional probability that the 11th trial shows a success is not a function of θ at all. For $k \leqslant N$, given that the proportion of success in the first N trials is r/N, the conditional probability that the kth trial shows a success is not a function of θ. This will mean that information about whether the kth trial actually *was* a success cannot then help us to decide the value of θ; any more than can information about my neighbour's cat or about any other matter which does not involve θ.

To find pr (kth trial is a success | r successes in N trials) we have to overcome the difficulty that the N trials *include* the kth trial, so that arguments that require independence are not yet available. Notice how we proceed (the first equality is a mere definition);

pr (kth trial is a success | r successes in N trials)

$$= \frac{\text{pr } (k\text{th trial is a success and there are } r \text{ successes in } all \ N \text{ trials})}{\text{pr (there are } r \text{ successes in } all \ N \text{ trials})}$$

$$= \frac{\text{pr } (k\text{th trial is a success and there are } r-1 \text{ successes in the } other \ N-1 \text{ trials})}{\text{pr (there are } r \text{ successes in } all \ N \text{ trials})}$$

$$= \frac{\text{pr } (k\text{th trial is a success}) \times \text{pr (there are } r-1 \text{ successes in the } other \ N-1 \text{ trials})}{\text{pr (there are } r \text{ successes in } all \ N \text{ trials})}.$$

375

The last step depends on having so framed the events that we had independent events to deal with (the independence of the events ultimately used is part of the structure of *Bernouilli* trials).

It follows that required probability $= \dfrac{\theta \cdot \binom{N-1}{r-1} (1-\theta)^{N-r} \theta^{r-1}}{\binom{N}{r} (1-\theta)^{N-r} \cdot \theta^r} = \dfrac{r}{N}.$

We have shown, in terms of the estimation problem above, what are the consequences of this independence of θ. These consequences are not usually regarded as obvious by those who have not met them before, so we give an alternative method of looking at the problem.

Another way of stating the conclusion in more specific terms is to say that if I am spinning a fair coin and after 6 turns tell you that I have observed heads 4 times, then any bet you *now* place upon what *was* the result of the first trial should use the odds 2:1 and not 1:1, since the probability that the first trial was a head *given that I have had* 4 *heads in the first* 6 *throws* is $\frac{2}{3}$, and not $\frac{1}{2}$. In effect you are to say 'I don't care whether you call the coin *fair*, so long as it is the same coin all the time and the trials are independent; the proof of the pudding is in the eating'.

This way of looking at it makes it seem a highly satisfactory result; but it is the *proof* and not the satisfactory nature which makes it true of probabilities as we have defined them. We look a little further into the matter and contrast it with another problem.

Consider Exercise 3C, Question 10, where we first met Bayesian inference. 'A pair of fair coins is first thrown to decide whether next to throw one die or two dice. A number of dice equal to the number of heads will be thrown (and *both* coins will be rethrown if a double tail occurs).

(i) What is the probability of a score of 5 given that two dice were thrown?

(ii) What is the probability that two dice were thrown given that a score of 5 was obtained?

(iii) Explain....'

In this problem the *a posteriori* probability that two dice were thrown depends on the *a priori* probabilities of one or two dice (namely $\frac{2}{3}$ and $\frac{1}{3}$) and not merely on the value of the statistic, that is on the value of the score.

This is the usual state of affairs. The situation with the Bernouilli trials is unusual; we cannot usually say 'I don't care what the *a priori* probabilities were'.

We now bring the two games into closer contrast. We formalize the coins and dice problem in such a way that the probability that we employ more than one (that is to say a second) die is θ. And we contrast that with

376

a situation where two throws of a possibly biased coin are used to produce the Bernouilli sequence. In these games we respectively calculate for $t = 5$ and $t = 1$ the values of pr (2nd score is zero | total score is t).

The problem of Exercise 3 C, Question 10 now reads:

Two random devices are to be used, the first produces scores with the following probability function:

Score	Prob.	Score	Prob.
1	$\frac{1}{6}$	4	$\frac{1}{6}$
2	$\frac{1}{6}$	5	$\frac{1}{6}$
3	$\frac{1}{6}$	6	$\frac{1}{6}$

The second produces scores with the following probability function:

Score	Prob.	Score	Prob.
0	$1-\theta$	4	$\frac{1}{6}\theta$
1	$\frac{1}{6}\theta$	5	$\frac{1}{6}\theta$
2	$\frac{1}{6}\theta$	6	$\frac{1}{6}\theta$
3	$\frac{1}{6}\theta$		

The random devices are used once each in turn and the total score is announced as 5; what is *now* the probability that the score on the second device was 0? (This is the *a posteriori* probability that *one* die was used: the probability that two were used is the complement of it.)

The solution, fully formal, runs:

pr (2nd score is zero | total score is 5)

$$= \frac{\text{pr (2nd score is zero)} \cdot \text{pr (total score is 5} | \text{2nd score is zero)}}{\sum\limits_{k=0}^{6} \text{pr (2nd score} = k) \cdot \text{pr (total score is 5} | \text{2nd score} = k)}$$

$$= \frac{\frac{1}{6}(1-\theta)}{\frac{1}{6}(1-\theta) + 4 \cdot \frac{1}{6}\theta \cdot \frac{1}{6}}$$

$$= \frac{3(1-\theta)}{3-\theta}.$$

Thus under the rules of the game where $(1-\theta) =$ probability that only one die is used $= \frac{2}{3}$, we have pr (2nd score is zero | total score is 5) $= \frac{6}{8} = \frac{3}{4}$.

Thus the *a posteriori* probability that two dice were used is $\frac{1}{4}$; and the reader has probably already reached this answer in Exercise 3 C by looking in an unsophisticated way at the possibility space of scores. The vital point however is that this *a posteriori* probability depends on θ. The statistic 'score' is *not* a sufficient estimator for θ.

The corresponding analysis of a similar Bernouilli problem is now given.

The throwing of two coins (whether biased or not) can be formalized as follows in parallel to the above:

Two random devices are used, the first produces scores with the following probability function

Score	Prob.
0	$1-\theta$
1	θ

and so does the second.

Bayes' Theorem gives:

pr (2nd score is zero | total score is 1)

$$= \frac{\text{pr (2nd score is zero).pr (total score is 1 | 2nd score is zero)}}{\{\text{pr (2nd score is zero).pr (total score is 1 | 2nd score is zero)}}$$
$$+ \text{pr (2nd score is 1).pr (total score is 1 | 2nd score is 1)}\}$$

$$= \frac{(1-\theta).\theta}{(1-\theta).\theta + \theta.(1-\theta)}$$

$$= \tfrac{1}{2}.$$

This says that if two coins have been thrown, then, *however wildly biased they may have been*, the observation of 1 success in 2 trials tells us, about the second trial, only that the *a posteriori* probability that it showed a success is $\tfrac{1}{2}$. This is fairly acceptable, as we have said, but we re-emphasize that any *a priori* information about how biased the coin might be is of no use at all in *estimating* how biased it actually is once we have calculated the value of the statistic r/N, which in this case is $\tfrac{1}{2}$.

The idea of a sufficient estimator, as a statistic which contains all the information in the sample that could help towards estimating the parameter concerned, is really quite a simple one. Results about sufficiency or otherwise of estimators can be rather heavy mathematically. An important property of maximum likelihood estimators is that if a parameter has a sufficient estimator then the maximum likelihood estimator is that sufficient one. This removes some of the worries about the *a priori* assumptions mentioned in Section 2.2, because, as we have seen, the *a priori* probabilities are irrelevant to the behaviour of a sufficient estimator.

10. CONFLICT OF PROPERTIES

It will be apparent to the reader that, given a probability model and a parameter, it is not usually possible to achieve in a single estimator for the parameter all the properties listed so far as desirable. We very much want *consistency*. Being of *minimum variance* is helpful, of course; but if we have consistency then the variance can be reduced as far as we please by taking large enough samples. There is, though, a snag here: large samples are in real life expensive or may even be, in a particular problem, impossible, and if impossible then consistency is a useless luxury; it is then that the

378

relative efficiency in smaller samples is important. If the estimator is consistent then bias too can be reduced by taking larger samples (with the same snag). Incidentally the extent of bias does not have a special name.

Unbiasedness of itself attains its importance because a 'centre' of the population of statistic values then lies at the parameter value—but by remembering, as we saw in Chapter 4, that the centre chosen might have been, say, the median and that then some other statistic would have been sought, we keep unbiasedness in perspective. That, however, does not reduce the algebraic simplicities which it introduces. The simplicities are dependent on the linearity at the heart of the property, and this has the defect of its virtue: that if T is an unbiased estimator for θ then in general $g(T)$ is not an unbiased estimator for $g(\theta)$. It is of course *an* estimator, but its properties may not be easy to determine. The estimator is said to lack *invariance*, and this is a handicap in a commonsense way.

A parameter may not admit of a *sufficient statistic* but if it does then that statistic has a strong claim on our attention.

A matter that we have not yet raised is that of *knowing the probability function* for a statistic, beyond its mean and variance. This knowledge is needed if we are to make numerical statements about the 'reliability' of our estimate. Statistics which have a Normal probability function are then particularly convenient because so much is known about that function. In some cases we can transform a statistic to obtain this property, by a simple transformation like $t' = \log t$ or $z = \tanh^{-1} r$.

10.1. Various methods of obtaining estimators. The use of *Bayes' Theorem* depends only on the definition of probability and not on introducing any new principles.

Calculating the expected value of a *relevant-looking statistic* is often possible and has commonsense appeal, though the appeal is tarnished if we think about the lack of invariance that will follow. If the expected value turns out to be a linear function of the parameter under investigation, then we have a method of constructing an unbiased estimator. If it does not and we still want an unbiased estimator, then we start somewhere else.

The *method of maximum likelihood*, introduced by Fisher in papers published in 1922 and 1925, has a certain commonsense appeal about 'likelihood', though we must be careful not to get carried away by a name. (It is worth restating that a distinction must be kept between a set of likelihoods and a set of probabilities; likelihoods say nothing about repeated trials, but the probabilities we have used in this text do. Furthermore we talk about the probability that a certain statistic-value *will* occur in a sample given the parameter value, but it is perhaps significant that we do not talk about the likelihood that a certain parameter value *will* occur given the statistic value in the sample, but merely about the likelihood that the parameter-value *was* such and such, given the statistic value. When we appear to talk about the probability that the result of an early trial *was* such-and-such given the result of a subsequent trial, we are of course using an extended meaning of probability (conditional probability), and we had to justify this use by verifying that the axioms were satisfied. We also showed that a frequency meaning could be given to it in the restricted set of trials in which both the earlier and subsequent trials have the quoted results.)

After this digression we return to the properties of the maximum likelihood estimators and of the method of obtaining them. They have many attractive properties. They are always consistent, and in large samples they are the most efficient of a very wide class; they need not however be unbiased. They have the property of invariance, that if T is a maximum likelihood estimator for θ then $g(T)$ is a maximum likelihood estimator for $g(\theta)$. If a sufficient estimator exists for the parameter then it is the maximum likelihood estimator. Finally, under very wide conditions, the larger the sample size the more nearly do the probability functions of the estimators approach Normal form. As to the method of obtaining them; once the probability of observing the relevant sample has been obtained it is regarded as a function of the (previously given) parameter and the rest is by well-worn analytical techniques for maximization. It is not after all *only* in the first-sight appeal of maximizing that the virtues of the method lie.

It very often happens that the data we are faced with are drawn from a population about which we know nothing much, except that it spawned this random sample. We cannot obtain the probability functions of various statistics and all our methods fail. At other times the probability function of the parent population may be known and the probability functions of various statistics may be known approximately for large samples but not for small samples. In all these cases we can use those statistics like the median and interquartile range which depend only on *rank*. A problem tackled by this sort of method was suggested in Exercise 1A, Question 8, and the method justified in Exercise 7E, Question 3. These statistics have properties which often depend only on the binomial theorem for a knowledge of their behaviour. We buy this knowledge at the cost of efficiency. Something quite as general as an order-statistic may not be the most efficient statistic for a particular problem but it may be the only one whose properties we can develop. If we expect to lead a varied life it would be nice to possess waterproof boots and dancing shoes and bed-room slippers, but if we do not possess them we can get by with a single pair of shoes. Methods using order-statistics are sometimes called 'non-parametric' and sometimes 'distribution-free'.

10.2. Diagrammatic summary. We show the sort of probability diagram that statistics have when they possess, or lack, some of the properties we have discussed. If the random variables concerned are discrete the probability diagrams are of course discontinuous. We have drawn smoothed curves of the sort that will be important when continuous random variables are discussed.

Figure 12.3 shows the probability diagrams, plotted together on the same axes, for the populations of three statistics u_1, u_2, u_3, calculated from samples of a fixed size, and each being used as an estimator for a parameter whose value is θ. The value of θ is also marked on the statistic-axis.

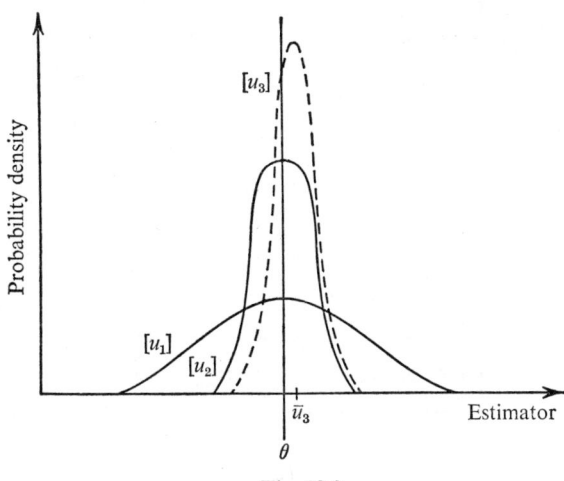

Fig. 12.3

(i) u_1, u_2 are unbiased since $E[u] = \theta$ in each case. u_3 is biased.

(ii) We cannot easily show consistency since to do so requires us to show the many probability curves for samples of different sizes.

(iii) u_1 is less efficient than u_2, which is less efficient than u_3. (The efficiency of biased estimation is taken up in Section 11, Example 3). u_3 would be preferable to u_1 as an estimator because of the far greater efficiency and might well be preferable to u_2 since although it is biased it is more efficient. The consistency or otherwise of the estimators might alter this situation, of course.

11. TWO WORKED EXAMPLES

Example 2. A sample of size 2 is taken from a uniform population of the integers 1 to N inclusive. The members are x, y and the value of a random variable Z is the larger of these. N is a fixed number; it is not a random variable.

381

Find an estimator for N, nearly unbiased for large enough N and based on Z; compare it with an estimator based on the mean of the sample.

Solution. We consider x, y as the values of two identical independent random variables X, Y. This bivariate situation is represented in Figure 12.4. We have

$$\mathrm{pr}\,(Z \leqslant z) = \mathrm{pr}\,((x, y) \in A) = z^2/N^2; \quad \mathrm{pr}\,(Z \leqslant z-1) = (z-1)^2/N^2$$

so
$$\mathrm{pr}\,(Z = z) = (z^2 - (z-1)^2)/N^2 = (2z-1)/N^2.$$

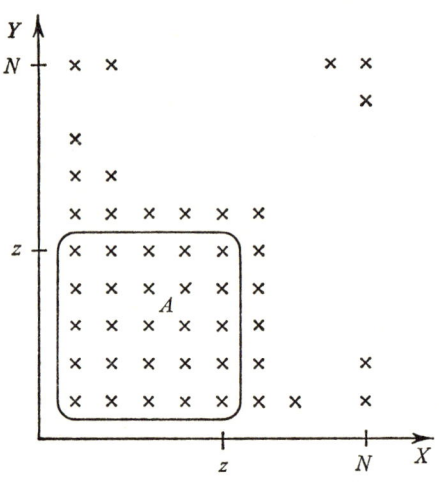

Fig. 12.4

This can also be seen from the second diagram, Figure 12.5, where

$$\mathrm{pr}\,(Z = z) = \mathrm{pr}\,((x, y) \in B) = (z+z-1)/N^2.$$

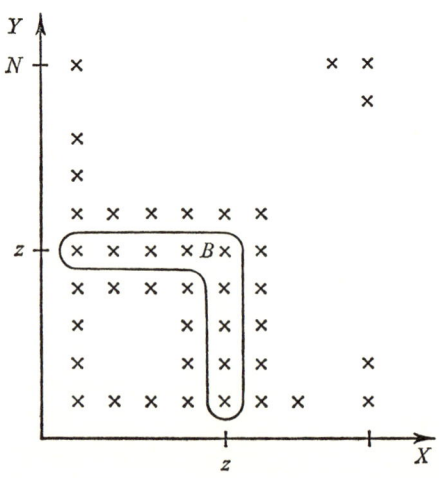

Fig. 12.5

First:
$$E(Z) = \frac{1}{N^2} \sum_1^N z(2z-1)$$

$$= \frac{1}{N^2} \left\{ \frac{2N(N+1)(2N+1)}{6} - \frac{N(N+1)}{2} \right\}$$

$$= \frac{(N+1)(4N-1)}{6N}$$

$$\sim \tfrac{2}{3}N, \text{ for large } N.$$

Thus $3Z/2$ is an estimator with the properties sought.

Now
$$\mathrm{var}\,(Z) = E(Z^2) - (E(Z))^2$$

but
$$E(Z^2) = \frac{1}{N^2} \sum_1^N z^2(2z-1)$$

$$= \frac{1}{N^2} \left\{ \frac{2N^2(N+1)^2}{4} - \frac{N(N+1)(2N+1)}{6} \right\}$$

$$= \frac{(N+1)(3N^2+N-1)}{6N},$$

so
$$\mathrm{var}\,(Z) = \frac{(N+1)(3N^2+N-1)}{6N} - \frac{(N+1)^2(4N-1)^2}{36N^2}$$

$$= \frac{(N-1)(N+1)(2N^2+1)}{36N^2}.$$

Secondly:
$$E(\tfrac{1}{2}(X+Y)) = E(\text{mean of sample})$$

$$= \text{mean of population}$$

$$= \tfrac{1}{2}(N+1),$$

so
$$E(X+Y-1) = N,$$

and $(X+Y-1)$ is an unbiased estimator for N.

Now
$$\mathrm{var}\,(X) = E(X^2) - (E(X))^2$$

$$= \frac{1}{N} \left(\frac{N(N+1)(2N+1)}{6} \right) - \left(\frac{N+1}{2} \right)^2$$

$$= \frac{(N+1)(N-1)}{12},$$

so
$$\mathrm{var}\,(X+Y-1) = \frac{(N+1)(N-1)}{6}.$$

To compare the estimators we note that $X+Y-1$ is unbiased, and if we assume that N is large enough for the bias in $3Z/2$ to be negligible we have:
The efficiency of $(X+Y-1)$ relative to $3Z/2$

$$= \text{var}\,(3Z/2)/\text{var}\,(X+Y-1)$$

$$= \frac{9}{4}\cdot\frac{(N-1)\,(N+1)\,(2N^2+1)}{36N^2}\cdot\frac{6}{(N+1)\,(N-1)}$$

$$= \frac{3}{4}\left(1+\frac{1}{2N^2}\right)$$

$$\sim 75\%, \quad \text{for large } N.$$

Example 3. To prove that Z in the previous Example is the maximum likelihood estimator for N, and to find its efficiency relative to the two unbiased estimators.

Solution. We have $\quad \text{pr}\,(Z=z) = (2z-1)/N.$

For fixed z this is the likelihood function of N. It has its largest value when N is least; but N cannot be less than z since z is one of the numbers $1, 2, 3, \ldots, N$, so the value of N which maximizes the likelihood is z. Thus Z is the maximum likelihood estimator for N.

If we wish to make a sensible definition of relative efficiency for possibly-biased estimators T_1 and T_2 of a parameter θ we say:

$$\text{Efficiency of } T_1 \text{ relative to } T_2 = \frac{E((T_2-\theta)^2)}{E((T_1-\theta)^2)}.$$

In the case of unbiased estimators this reduces to the previous definition, since $\theta = E(T)$.
In the case of Z we require $E((Z-N)^2)$, not $E((Z-\tfrac{2}{3}N)^2)$.
We have

$$E((Z-N)^2) = \text{var}\,(Z)+(E(Z)-N)^2$$

$$\sim \frac{N^2}{18}+(\tfrac{2}{3}N-N)^2, \quad \text{for large } N$$

$$= \frac{N^2}{6}.$$

Thus

$$\text{Efficiency of } Z \text{ relative to } (X+Y-1) \sim \frac{N^2}{6}\Big/\frac{N^2}{6}, \quad \text{for large } N$$

$$= 100\%.$$

$$\text{Efficiency of } Z \text{ relative to } \frac{3Z}{2} \sim \frac{N^2}{8}\Big/\frac{N^2}{6}, \quad \text{for large } N.$$

The maximum likelihood estimator in this problem has little to recommend it as far as unbiasedness and efficiency are concerned.

384

Exercise 12B

1R. The following table shows all the samples of size 3 from the uniform population of integers $(1, 2, 3, 4, 5, 6)$. Beside each sample is the quantity: $216 \times$ pr (the sample is drawn at random, with replacement).

111	1	222	1	333	1	444	1	555	1	666	1
112	3	223	3	334	3	445	3	556	3		
113	3	224	3	335	3	446	3	566	3		
114	3	225	3	336	3	455	3				
115	3	226	3	344	3	456	6				
116	3	233	3	345	6	466	3				
122	3	234	6	346	6						
123	6	235	6	355	3						
124	6	236	6	356	6						
125	6	244	3	366	3						
126	6	245	6								
133	3	246	6								
134	6	255	3								
135	6	256	6								
136	6	266	3								
144	3										
145	6										
146	6										
155	3										
156	6										
166	3										

We write the general sample of size 3 as abc, $(a \leqslant b \leqslant c)$.

(i) Consider the statistics

$$t = \text{median of sample} = b,$$
$$m = \text{mean of sample} = \tfrac{1}{3}(a+b+c),$$
$$u = \text{mean of extremes of sample} = \tfrac{1}{2}(a+c).$$

It is obvious, by symmetry, that these are unbiased estimators for the population mean. Find their variances, and arrange them in order of increasing efficiency.

(ii) u appears to be a linear function of the members of a sample, how can it have a smaller variance than m?

(iii) $m = \tfrac{1}{3}(2u+t)$, why do we *not* have

$$\text{var}[m] = \tfrac{4}{9}\text{var}[u] + \tfrac{1}{9}\text{var}[t]?$$

2R. See Chapter 11, Section 6.3 for a useful table for this question. A sample of size 3 is drawn from a binomial parent population with the following probability function: $\text{pr}(X = -a+b) = \text{pr}(X = a+b) = \tfrac{1}{4}$; $\text{pr}(X = b) = \tfrac{1}{2}$.

A value of one of the following statistics can be made available but the actual values of X in the sample are not known.

$$T_1 = \text{mean of the extremes of the sample},$$
$$T_2 = \text{median of the sample},$$
$$T_3 = \text{mean of the sample}.$$

(i) Are the statistics T_r ($r=1, 2, 3$) unbiased estimators for b?

(ii) Find $\text{var}(T_r)$ ($r=1, 2, 3$).

(iii) On the grounds of maximum efficiency which statistic is the one whose value you would wish to know?

3 R. The following three samples were drawn at random from parent populations of the types mentioned. The parameters are concealed in the simple equations, in order to prevent accidental sighting among the answers.

Make unbiased estimates of the mean and variance of the parent populations from which the samples were drawn.

If your estimating statistic for the mean is the mean of the sample, then give an estimate of its variance.

(i) *Rectangular population*

$$\mu = k + \tfrac{1}{2}, \quad \sigma^2 = l - \frac{1}{l}, \quad \text{where} \quad \begin{pmatrix} 1 & -1 \\ -1 & 2 \end{pmatrix}\begin{pmatrix} k \\ l \end{pmatrix} = \begin{pmatrix} 3 \\ 9 \end{pmatrix};$$

sample = (14, 12, 11, 15, 10, 21, 11, 19, 20, 12, 21, 17, 13, 21, 10, 21, 12, 12, 18, 18).

(ii) *Triangular population*

$$\mu = t^2 + \frac{1}{t}, \quad \sigma^2 = t^2 - 1 + \frac{1}{t^2}, \quad \text{where } t \text{ is the positive root of}$$

$$(x \quad 1)\begin{pmatrix} 1 & 1 \\ 1 & -8 \end{pmatrix}\begin{pmatrix} x \\ 1 \end{pmatrix} = (0);$$

sample = (2, 8, 8, 3, 6, 6, 5, 6, 2, 4, 4, 1, 4, 8, 4, 4, 7, 5, 4, 4).

(iii) *Normal population.* (Assume that the results for discrete populations hold here too.)

$$\mu = 20p, \quad \sigma^2 = 20q, \quad \text{where } q = p^2 + 1 = 3p - 1 > 4;$$

sample = (30, 62, 16, 34, 46, 46, 43, 49, 27, 29, 36, 26, 34, 57, 29, 36, 51, 21, 39, 26).

4. (In this question assume that the necessary results hold for continuous parent populations, as they do for discrete ones.)

Two small objects have weights x gm and y gm. A physicist obtains the sum and difference of their as yet unknown weights by putting them in the same pan and balancing them, and then putting them in opposite pans and balancing them. Let the measured sum and difference of the unknown weights be u gm and v gm respectively. The physicist calculates the values of x and y from $\tfrac{1}{2}(u+v)$ and $\tfrac{1}{2}(u-v)$.

Let the presence of experimental errors in weighings imply that an indefinitely large set of weighings of a fixed object gives rise to a population of weights having standard deviation σ gm and that this is independent of the weight (within a reasonable range).

Show that the variance of the calculated value of x is equal to the variance of the mean of two separate weighings of x by itself; and hence that if the weights were to be determined within these same limits by weighing the objects separately, then twice as many weighings would be needed. [See the footnote to Question 18.]

5 R. A bag contains β discs each labelled with one of the numbers 1, 2, ..., β. A single disc is drawn from the bag and R denotes the number on it.

(i) Find pr $(R = r)$ for $r = 1, 2, ..., \beta$.

(ii) Find $E[r]$, $E[r^2]$, var $[r]$.

(iii) Define a new number t, a linear function of r, such that $E[t] = \beta$, and show that var $[t] = (\beta^2 - 1)/3$.

386

(iv) Let a sample of N values of t have mean m, show that m is an unbiased estimator for β and find var $[m]$. (The disc is replaced between each drawing.)

(v) Verify the results of Example 2, as special cases.

6T. (i) What is the variance of the estimator suggested in Exercise 12A, Question 6 part (ii).

(ii) Is the number of successes on the first trial an unbiased estimator for θ and if so what is its variance?

(iii) Which is the more efficient of these two estimators and which would you prefer? (*Remember:* you are not being offered a choice between a *sample* of values of each of these estimators but between a *single value* of each.)

7. Discuss the following proposal:

'$\dfrac{1}{N+1} \displaystyle\sum_{i=1}^{N} x_i$ has variance less than that of $\bar{x}$ and so is more reliable; it is biased as an estimator of μ but we will multiply by $\dfrac{N+1}{N}$ to compensate for that; and anyway for large enough N the bias is negligible.'

8. What is the relationship between the various properties of estimators and

(i) the relative courses of the lines in Figure 11.1,

(ii) the shape of the mass of lines in Figure 11.3 (give a rough description of the way the width contracts as n increases).

(iii) the position of the bulk of the points in figure 11.4?

9. Give as straightforward an example as you can of a statistic to satisfy each of the following descriptions. Name the parameter being estimated in each case.

(i) Consistent and unbiased.

(ii) Consistent but biased.

(iii) Not consistent but unbiased.

(iv) Not consistent and furthermore biased.

10. In Chapter 11, Section 6.3 there is a table of variances of samples of size 3 from a certain parent population. There is no change in the table if we take the population to be that of Question 2, namely:

$$\text{pr}\,(X = -a+b) = \text{pr}\,(X = a+b) = \tfrac{1}{4}; \quad \text{pr}\,(X = b) = \tfrac{1}{2}.$$

The table is then:

S^2	$64 \times$ probability
0	10
$4\sigma^2/9$	36
$12\sigma^2/9$	12
$16\sigma^2/9$	6

where we have written the values of S^2 in terms of σ^2 rather than of a^2; (this is merely so that we can discuss estimating σ^2 rather than a^2).

If the statistic $t \equiv S^2$ is to be used to estimate the variance of the parent population,

(i) find λ if λt is an unbiased estimator for σ^2, that is to say find λ if $E[\lambda t] = \sigma^2$.

(ii) Show that if t is regarded as fixed then the probability table may be rewritten as

σ^2	Likelihood
$9t/16$	$6/64$
$9t/12$	$12/64$
$9t/4$	$36/64$

Find the expression for σ^2 in terms of t which maximizes the likelihood. This is the maximum likelihood estimator for σ^2. Writing it as $\widehat{\sigma_L^2}(t)$, find $E[\widehat{\sigma_L^2}(t)]$ and hence whether it is biased.

(iii) Find the variances of the estimators of (i) and (ii) and thus find the more efficient of them and the efficiency of the other relative to it.

11. The previous question reminds us that the estimators obtained under various general methods do not always coincide. A less particular example follows:

When a sample of size N is drawn from a Normal population and the mean and variance of the sample are m, S^2, then it can be proved that the estimators $\widehat{\mu_L}(m, S^2)$ and $\widehat{\sigma_L^2}(m, S^2)$ which, taken simultaneously, maximize the likelihood of the given sample are:

$$\widehat{\mu_L}(m, S^2) = m \quad \text{and} \quad \widehat{\sigma_L^2}(m, S^2) = S^2.$$

These contrast with unbiased estimators of least variance which we have seen to be

$$\widehat{\mu_U}(m, S^2) = m, \quad \widehat{\sigma_U^2}(m, S^2) = \frac{N}{N-1} S^2.$$

Suppose that you had only 5 members of a Normal population whose mean and variance you wished to estimate; use each successive 5 non-overlapping members of the sample in Question 3 (iii) to study the effects of the estimating procedures above on the estimate of the second parameter of the pair.

12. A sample $x_1, x_2, \ldots, x_N$ (with possible repetitions) is drawn from a parent population of *known mean, μ*. Show that an unbiased estimate of the variance of the population is $\Sigma(x_i - \mu)^2/N$. For some simple choice of parent population and of N, compare the variance of this estimate with that of the estimate useful if μ is not known, namely $\Sigma(x_i - \bar{x})^2/(N-1)$.

13T. Use properties of unbiased estimators and of maximum likelihood estimators quoted in the chapter, together with the results at the start of Question 11 to show that $\sqrt{\left(\dfrac{N}{N-1} S^2\right)}$ is not an unbiased estimator of the S.D. of a normal population, but that m and S are simultaneous maximum likelihood estimators for the mean and S.D. of a Normal population.

[The search for an unbiased estimator of the S.D. of a Normal population is rather fruitless since the algebraic simplicities of using the square root of an unbiased estimator for the variance are enormous. It is in fact the variance and not the S.D. which occurs in the basic probability functions.]

14R. Independent statistics T_1, T_2 have means θ_1, θ_2 and variances σ_1^2, σ_2^2.
(i) Find the expectation and variance of the statistic given by $T = k_1 T_1 + k_2 T_2$.

(ii) If T_1 and T_2 are both unbiased estimators for a parameter θ of a certain population, find values of k_1 and k_2 which make T an unbiased estimator for θ having minimum variance among estimators formed in this way.

(iii) Prove that the variance of T is then given by

$$\frac{1}{\text{var}(T)} = \frac{1}{\sigma_1^2} + \frac{1}{\sigma_2^2}.$$

15. If x_1 and x_2 are two independent unbiased estimates of a parameter, having variances s_1^2 and s_2^2 respectively,

(i) Prove that the choices $a_r = \dfrac{1}{s_r^2} \bigg/ \left(\dfrac{1}{s_1^2} + \dfrac{1}{s_2^2}\right)$ for $r = 1, 2$ give $a_1 x_1 + a_2 x_2$ as

an unbiased estimate of x from a method giving smaller variance than that of any other choice of the a_r.

(ii) Tidy up the formulae of part (i) by writing $1/u_r^2$ for the variances and generalize to four independent unbiased estimates.

[Once again the inverse of the variance is a useful quantity.]

(iii) A certain physical constant is determined by four experimenters, working independently, as follows:

$$2 \cdot 03 \pm 0 \cdot 08, \quad 2 \cdot 20 \pm 0 \cdot 05, \quad 2 \cdot 25 \pm 0 \cdot 10, \quad 2 \cdot 12 \pm 0 \cdot 04$$

(the figure after $\pm$ sign denoting the standard deviation of that experimenter's population of estimates.)

Find the value of a single estimator which is a linear combination of the estimators whose values these are, and which has least variance among such estimators. Estimate also the value of the standard deviation of this combined estimator.

16 T. A sample of size N_1 is taken from a population with mean μ and variance σ^2 and the mean of the sample is x. Another sample, but of size N_2, is taken and it has mean y.

(i) Show that the statistic $z = \frac{1}{2}(x+y)$ is an unbiased estimator for μ having variance $\dfrac{1}{4}\left(\dfrac{1}{N_1} + \dfrac{1}{N_2}\right)\sigma^2$, and that the statistic $m = (N_1 x + N_2 y)/(N_1 + N_2)$ is an unbiased estimator for μ having variance $\sigma^2/(N_1 + N_2)$.

(ii) Show that the second variance never exceeds the first and discuss subsection (II) of Section 8.4 in the light of this.

(iii) To what do the results of part (i) reduce if $N_1 = N_2 = N$, say?

(iv) What is the relation of m in part (i) to the statistic in Question 15 part (i)?

17 T. With the notation of Question 16, $t_\lambda = \lambda x + (1-\lambda)y$ is an unbiased estimator for μ. Prove that to obtain the estimator of this type with minimum variance we must take $\lambda = N_1/(N_1 + N_2)$.

18. In Exercise 1A, Question 1 three methods of finding a measure for the thickness of the leaves of a book are given. The reader has now acquired the theory necessary to discuss the relative merits of the first two, provided he will assume that the results of the theory still hold when the random variables are continuous. The third method involves regression lines and a theory about their reliability. This we have not yet developed.

Assume that the population $[x]$ of thicknesses (in cm) of leaves has mean μ and variance σ^2; assume that the measurement processes introduce an error δ cm into a measurement, but that there is no systematic bias of measurement, so that, $E[\delta] = 0$; write var $[\delta] = \epsilon^2$ and assume, finally, that ϵ is independent of the actual measurement made, within the range of measurements under consideration.

With this notation we can write that the actual measurement made of the thickness (in cm) of N leaves is given by

$$t = \left\{ \sum_1^N x_i \right\} + \delta.$$

The problem is then reduced to the discussion of how best to use the values of t from samples of x of various sizes, N, to estimate μ with minimum variance.

[Question 4 and the present question illustrate the important role that can be played by theoretical statistics in apparently quite ordinary problems of measurement. This is a small part of its role in the Design of Experiments. As remarked in Chapter 1, Section 9 the statistician is too often merely called in to interpret data that he would never have suggested collecting if he had known, before the experiment, what it was that the experiment was intended but ill-designed to do.]

19. (i) T_1 is an unbiased estimator for a certain parameter. It has a known variance σ_1^2. T_2 is another unbiased estimator, independent of T_1, but it has unknown variance σ_2^2. What condition on σ_2^2 is sufficient for $\frac{1}{2}(T_1 + T_2)$ to have a smaller variance than T_1?

(ii) You have taken a sample of size N_0 of values of a statistic T to estimate a certain parameter θ. A friend offers you the result of his labours in the same field and it consists of the mean of a sample of values of the same statistic. He cannot remember the size, N, of the sample upon which it was based. What is the value of N which must be exceeded if you are to improve your estimate by merely averaging your estimate and his after the manner of part (i)?

$\star$**20.** Your friend, in the last question, says he will be able to trace his records and find the value of N. You form the impression that all values of N from 1 to N_0, inclusive, are equally likely. Show that in that case the expected value of the multiplier you will apply to his mean in forming the minimum variance combined-estimate is

$$\lambda = 1 - \sum_{r=1}^{N_0} \frac{1}{r+N_0}; \quad \text{and that this satisfies } 1 - \log 2 < \lambda < 1 - \log\left(\frac{2N_0+1}{N_0+1}\right).$$

21T. Two populations of equal variance and fourth moment but differing possibly in their means are given, the common variance being σ^2. A sample of size N_1 and variance S_1^2 is drawn from the first and a sample of size N_2 and variance S_2^2 is drawn from the second; the samples being independent and each drawn with replacement.

(i) Write down $\qquad\qquad E[S_1^2] \quad \text{and} \quad E[S_2^2].$

(ii) Find a relation between λ, μ so that

$$E[\lambda S_1^2 + \mu S_2^2] = \sigma^2.$$

(iii) Show that

$$\lambda_0 = \frac{N_1}{N_1+N_2-2}\left\{1 + \frac{\delta}{N_1-1}\right\},$$

$$\mu_0 = \frac{N_2}{N_1+N_2-2}\left\{1 - \frac{\delta}{N_2-1}\right\}$$

satisfy the relation of (ii).

(iv) If the parent populations are normal, then the variance of the population of variances of samples of size N is $2\sigma^4(N-1)/N^2$. Show that then we have:

$$\text{var} [\lambda_0 S_1^2 + \mu_0 S_2^2] = \frac{2\sigma^4}{(N_1+N_2-2)} \{1+\delta^2/(N_1-1)(N_2-1)\}.$$

(v) Deduce that $\dfrac{N_1 S_1^2 + N_2 S_2^2}{N_1+N_2-2}$ is an estimator for σ^2 that is (always) unbiased, and has (for Normal parent populations) least variance of all estimators obtained by this linear combination method.

(vi) Show that if p_i and q_i are respectively the sum and the sum of squares of members of the samples, $(i=1, 2)$, then the estimator may be written:

$$\frac{\sum\limits_{1}^{2}(q_i-p_i^2/N_i)}{\left(\sum\limits_{1}^{2}N_i\right)-2}.$$

[The general form for k samples is then

$$\text{pooled estimate} = \frac{(\text{sum of squares of deviations of sample members from their own means})}{(\text{total size of combined samples})-(\text{number of samples})}.$$

Once again the restriction to equality of fourth moments in the parent population is not severe since the parent populations are ordinarily identical in the hypothesis to be tested.]

22 T. Show that the form of estimate in Question 21 part (v) can be rearranged in the following way:

The number of degrees of freedom of a sample of size N is defined by $\nu = N-1$; s^2 is the usual unbiased estimate of the variance of the population given by

$$s^2 = \frac{N}{N-1}S^2,$$

then the pooled estimate is

$$\frac{\nu_1 s_1^2 + \nu_2 s_2^2}{\nu_1+\nu_2}.$$

[In fact this generalizes to $\Sigma\nu_i s_i^2/\Sigma\nu_1$. It is of course the simplicity of this form in terms of ordinary weighted means of unbiased estimates that gives s^2 and ν their attractions as statistics of the sample, compared with S^2 and N. As already remarked, many books develop the whole subject of variance in terms of s^2 and ν.]

23 R. (i) It is believed that the following samples were drawn from populations differing only, if at all, in their means. On this assumption estimate the common variance of the populations.

(3 2 5 3 5 3)
(6 4 2 7 3 5)
(7 4 4 3 8 7)
(3 2 3 8 4 4)

(ii) Estimate the variance of the population of means of these samples.

24 T. If we have k different varieties of wheat that we wish to test (to see whether their yields, for instance, are different), we might plant a plot of each and study the k yields. It would however be very difficult to ensure that *other* condition in the plots, for instance drainage, mineral content, pest infestation and so forth, were the same. We could proceed as in Chapter 2, Section 9.2, which the reader should now turn to and read again (in that section $k = 4$). We would then have k possibly different samples (created by the varieties of wheat) in each of which are k members (the different members being produced by we hope only random fluctuations from mean drainage etc., the fluctuations have equal *variances*).

A more general type of problem with four samples each of five members is outlined in Exercise 4 F, Question 2.

The analysis of such problems is the object of this question. We make the Null Hypothesis that there are no significant differences anywhere and we then have a model from which to draw conclusions which can be tested against the data.

Consider the following array denoting k samples each of size l, with the statistics given in the headed columns.

	Sum	Sum of squares	Correction term	Mean	Variance
Sample 1 $x_{11}x_{12} \ldots x_{1l}$	p_1	q_1	p_1^2/l	m_1	v_1
Sample 2 $x_{21}x_{22} \ldots x_{2l}$	p_2	q_2	p_2^2/l	m_2	v_2
$\vdots$	$\vdots$	$\vdots$	$\vdots$	$\vdots$	$\vdots$
Sample k $x_{k1}x_{k2} \ldots x_{kl}$	p_k	q_k	p_k^2/l	m_k	v_k
Sums =	T	S	C		

Let the mean of the values m_i be m and the variance of the values m_i be b^2.

Assume in what follows that the x_{ij} are all drawn at random from a single population with variance σ^2.

(i) Show that
$$E[b^2] = \frac{k-1}{k} \cdot \left(\frac{\sigma^2}{l}\right).$$

Deduce that an unbiased estimate of σ^2 is

$$\frac{1}{k-1} \times l \times \text{(sum of squares of the deviations of the } m_i \text{ from } m).$$

Show also that

l(sum of squares of the deviations of the m_i from m)

$$= l\left((m_1^2 + m_2^2 + . + m_k^2) - \frac{(m_1 + m_2 + . + m_k)^2}{k}\right)$$

$$= \Sigma(p_i^2/l) - \frac{(\Sigma p_i)^2}{kl}$$

$$= C - \frac{T^2}{N}, \quad \text{where } N = kl = \text{total numbers of observations,}$$

so that first unbiased estimate is $\dfrac{1}{k-1}\left(C - \dfrac{T^2}{N}\right)$.

This is called the *estimated variance between classes* (or *samples*.)

(ii) Use the standard result (proved in Section 8.4) about an unbiased estimate for the variance of a population based on the variances of samples of equal size to show that

$$\text{a second unbiased estimate for } \sigma^2 = \frac{1}{N-k}.\sum_i \left(q_i - \frac{p_i^2}{l}\right)$$

$$= \frac{1}{N-k}\,(S-C).$$

This is called the *estimated variance within classes* (or *samples.*)

(iii) By linearly combining the two unbiased estimates already obtained, in a way that will give another unbiased estimate, show that

$$\text{a third unbiased estimate} = \frac{1}{N-1}\left(S - \frac{T^2}{N}\right).$$

Verify, from the general result about unbiased estimates of σ^2 based on N values from a population, that this is the unbiased estimate of this sort with least variance. It is called the *estimated variance overall*.

[In the branch of statistics called *analysis of variance* the first two of these estimates are compared to see if they differ significantly. If they do, then the hypothesis that all the N observations x_{ij} were drawn at random from a single population is rejected. It may be that the means of the samples differed significantly or that the variances of the samples did, but it is then worth investigating by further experiments such things as the differences in the varieties of wheat or the effects of drainage and so on.]

25R. Assume now that all the observations of Question 23 were drawn at random from a single population (and so in manner which was independent of the sample they are in). On this assumption obtain from the data the three unbiased estimators (of Question 24) for the variance of their parent population.

Find the ratio, $\dfrac{\text{variance between classes}}{\text{variance within classes}}$ (or its inverse, whichever exceeds unity).

[The process of testing whether the estimates in fact differ significantly or not is not taken up here. It is their ratio as calculated above that we test—using tables called *Tables of the F-Distribution*.]

26R. Carry out the processes of Question 25 on each of the sets of data of Exercise 4G, Question 6.

27. For the bag and discs of Question 5 a single disc is taken, the value of R recorded and the disc replaced. This is done N times.

(i) For $b \neq 1$, prove that the probability that any x of the discs have $R = b$ and all the rest have $R < b$ is $\dbinom{N}{x}\left(\dfrac{1}{\beta}\right)^x\left(\dfrac{b-1}{\beta}\right)^{N-x}$.

(ii) By writing this probability as $\left(\dfrac{b}{\beta}\right)^N.\dbinom{N}{x}\left(\dfrac{b-1}{b}\right)^{N-x}\left(\dfrac{1}{b}\right)^x$ and summing from $x = 1$ to $x = N$ show, for $b \neq 1$, that

$$L(\beta) \equiv \text{pr (the largest value of } R \text{ is } b) = \left(\frac{b}{\beta}\right)^N\left(1 - \left(\frac{b-1}{b}\right)^N\right)$$

$$= \left(\frac{b}{\beta}\right)^N - \left(\frac{b-1}{\beta}\right)^N$$

(and verify that this is also the correct result for $b = 1$). Obtain the result also by a generalization of the method of Example 2.

(iii) If b is regarded as fixed, the value of β which maximizes $L(\beta)$ is called the maximum likelihood estimate of β. Show why the maximum likelihood estimate is b.

(iv) Show that

$$E[b] = \beta^{-N}\{\beta^{N+1} - ((\beta-1)^N + (\beta-2)^N + \ldots + 1^N)\};$$

and by using

$$\sum_{p=1}^{p=q} p^k = \frac{1}{k+1} \cdot q^{k+1} + (\text{polynomial in } q \text{ of degree } k)$$

show that

$$E[b] \sim \frac{N}{N+1} \cdot \beta, \quad \text{for large } \beta.$$

Verify the corresponding result of Example 2 (where the notation is very different).

(v) Define

$$u = \frac{N+1}{N} \cdot b,$$

and show that, for large β, u is nearly unbiased and

$$\text{var}[u] \sim \beta^2/N(N+2), \quad \text{for large } \beta.$$

Verify that u is *consistent* estimator for β.

(vi) For large values of N, compare var $[u]$ with the variance of the statistic m (from Question 5 part (iv)) based on N values of t, and discuss whether u is preferable to m as an estimator.

(vii) Use random number tables to simulate about thirty situations in which a sample of size 10 is available from a range $1 \leqslant r \leqslant \beta$, and compare the behaviour of the two estimators.

28 T. In the animal-culling survey of Exercise 7D, Question 30 it is decided to assume that the discrepancies between the surveys, while large, have no concrete grounds for being regarded as significant. We are to combine the surveys and use the probability model below:

	Unpolluted	Polluted	
Valuable	$\alpha\beta$	$(1-\alpha)\beta$	β
Not valuable	$\alpha(1-\beta)$	$(1-\alpha)(1-\beta)$	$1-\beta$
	α	$1-\alpha$	

(i) Using the multinomial theorem of Chapter 7 and stating all the hypotheses of independence that you assume, show that the probability of the observed figures is:

$$\frac{35!\,10!}{2!\,8!\,3!\,22!\,3!\,7!} \cdot (\alpha\beta)^5\,\alpha^8(1-\beta)^8\,(1-\alpha)^3\,\beta^3(1-\alpha)^{22}\,(1-\beta)^{22}\,(1-\alpha\beta)^7.$$

(ii) If the numbers observed are regarded as fixed, this may be written as a function $L(\alpha, \beta)$ when it is called a likelihood (as we have seen). Show that

$$\log L(\alpha, \beta) = 13 \log \alpha + 8 \log \beta + 25 \log (1-\alpha) + 30 \log (1-\beta)$$
$$+ 7 \log (1-\alpha\beta) + \text{constant}.$$

(iii)   Treat α as fixed while you differentiate with respect to β. Interchange the roles of α, β and differentiate again. If these derivatives are set equal to zero and the resulting equations solved for α, β we have pairs of values α, β which may simultaneously maximize the likelihood of the observed sample. If a pair does, it provides the maximum likelihood estimates of α, β.

(iv)   Show the equations to be solved are a cubic in α and a cubic in β.

(v)   Check that $\alpha = \frac{1}{3}$ is a value which together with a suitable value for β would make the derivatives vanish. Find the corresponding suitable value for β.

(vi)   Assuming a model in which the values of α, β are equal to these estimates, and in which 45 animals are concerned, draw up a table showing the expected numbers in each category. (It can easily be proved that, even in a multinomial problem, the expected number of animals in a category with associated probability θ is $N\theta$, where N is the total number of animals available.)

(vii)   Assume a binomial probability function for the number of valuable unpolluted animals, and under the same hypothesis about the values of α, β find the variance of the number of valuable unpolluted animals observed.

(viii)   As a rough check on the reasonableness of the agreement between model and observation, show that the discrepancy between observed and expected numbers of valuable unpolluted animals is well within the $2 \times$ S.D. limits.

[This problem has shown the way in which the maximum likelihood method operates in a problem with more than one parameter. It is typically used to determine a value-set (for the set of parameters) which maximizes the likelihood of the entire sample.

Since likelihoods start life as members of a set of probabilities, and probabilities are typically formed by multiplications, it often happens that likelihoods involve the *product* of many expressions. The logarithm of the likelihood, called loglikelihood, is thus a useful quantity to consider since the loglikelihoods from several samples will be additive. In point of fact it is the derivative of the loglikelihood that we want. This is called the *likelihood score*. When it is a function of a single parameter one might write it $Sc(\theta)$; the method then consists in finding the value of θ which satisfies $Sc(\theta) = 0$.]

29.   A sequence of Bernouilli trials has probability θ of success at each trial. N is a fixed number, and N trials yield r successes.

(i)   Show that r/N is an unbiased estimator for θ and find var $[r/N]$.

(ii)   Use Chebyshev's Inequality to show that

$$\text{pr}\left(\left|\frac{r}{N} - \theta\right| \leqslant \epsilon\right) > 1 - \frac{\theta(1-\theta)}{N.\epsilon^2}$$

and hence that this probability exceeds $1 - (4N\epsilon^2)^{-1}$, whatever may be the value of θ.

(iii)   If r/N is to be used as an estimator for θ and if we want to quote an interval of length 0·01 each way from its midpoint and have at least 95 per cent probability that it covers the value of θ, how large does the deduction from Chebyshev's Inequality indicate that we should take N?

[Remember that Chebyshev's Inequality is very weak; smaller values of N can in fact be used.]

(iv)   Referring to the data of Table 2.3 on drawing pins, we see that, in 2,848 throws of plastic pins, the pin fell point-down 1230 times, giving an overall proportion of 0·432. If the probability of a plastic pin falling point-down was in fact 0·410 what does Chebyshev's Inequality give for a maximum probability of a *discrepancy* of the observed size or more in the proportion? Does the observed discrepancy now alarm you?

(v) If the probability of a brass pin falling point-down was in fact equal to the observed proportion of plastic pins falling point-down, show that about $98\frac{1}{3}$ per cent is the minimum probability (calculated from Chebyshev's Inequality and assuming nothing about the probability function for values of r/N) that the proportion of brass pins falling point-down would *exceed* the observed proportion after 32 boxfuls, namely $379/(37 \times 32)$.

Interpret this probability in terms of repetitions of the experiment.

(vi) In Chapter 2, Section 2.3 it is stated that the proportion for 32 boxfuls of brass pins, although nearer 0·41 than was the proportion for 16 boxfuls, makes it 'harder to envisage' that the long term rate for brass pins is really 0·41. Discuss that statement.

30. In the study of a certain small nocturnal animal a cage is primed with three decoy animals and is left out for $3\frac{1}{2}$ hours each night. At the end of each hour a door at the end of the entry tunnel is opened inwards by a time-mechanism for a short period, and either an animal comes in or one escapes. There is only room in the tunnel, and time, for at most one of these events to occur, but the inquisitive nature of the animals at the door (which is the only place they can see in or out of the cage) ensures that one of them actually does occur. At the end of the third hour the animals seem to behave rather differently, perhaps because they are becoming accustomed to the cage; and anyway, if there are by then five animals in the cage, the door cannot usually open because the animals already in are blocking it.

The following simple probabilistic model is suggested:

At the end of each of the first two hours:

$$\text{pr (animal enters)} = \theta, \quad \text{pr (animal escapes)} = 1 - \theta.$$

At the end of the third hour:

(a) If less than 5 animals in cage: as above, but $\theta = \frac{1}{2}$.

(b) If 5 animals in cage: nothing occurs.

(i) Use a die to simulate the situation with $\theta = \frac{5}{6}$ and carry out the equivalent of 20 nights trapping. Repeat with $\theta = \frac{1}{6}$.

(ii) Let N denote the number of animals finally in the cage and obtain from theory

$$p_n(\theta) \equiv \text{pr } (N=n) \quad \text{for} \quad n = 0, 1, 2, 3, 4, 5.$$

Sketch the graphs of $p_n(\theta)$ against θ for $n = 0, 2, 4, 5$.

(iii) Find the expected value and variance of N, each as a function of θ.

(iv) Construct a statistic S as a linear function of N such that $E[s] = \theta$. If $\bar{s}$ is the mean of several values of S from successive nights, then is it an unbiased estimate of θ? Find var $[s]$.

(v) A scoring system is devised as follows to define another statistic T, with values t, in terms of the values, n, of N.

n	0	2	4	5
t	$-a$	a	$1-a$	1

Find $E[t]$, $E[t^2]$, var $[t]$, each as a function of θ and a. Verify that T is an unbiased estimator of θ, and hence show that $\bar{t}$ is also an unbiased estimator.

(vi) Take $a = 0$ and plot var $[t]$ against θ. Repeat for $a = \frac{1}{2}$.

(vii) If θ is known to be more than $\frac{1}{2}$, suggest on theoretical grounds the better of the two estimators of part (vi). Test each estimator with the data of your simulated run for each value of θ.

31. (This question is a continuation of Question 30.)

(i) Using the number of animals in the cage after only two openings of the door construct an unbiased estimator for θ and showing by considering its variance that the third opening 'only confuses the issue', explaining what this means in this context.

[This would suggest a redesigning of the experiment, saving one decoy animal and late hours.]

(ii) Let there be Q animals in the cage after the second opening, and for each q in turn define θ_q as the value of θ which maximizes pr $(Q = q)$; then θ_q is the maximum likelihood estimate of θ for $Q = q$.

Find the maximum likelihood estimates of θ for each q and compare them with the unbiased estimates of part (i).

(iii) Show that the maximum likelihood estimator for θ using the numbers in the cage after three openings may be biased and that the bias can be 50 per cent.

[It is the ease with which maximum likelihood estimators can be found in quite general situations that is one of their strong recommendations.]

32. To obtain a simple deterministic model of the action of a certain insecticide, it is assumed that when the insecticide is applied with a dose of strength x to an insect then the probability, θ, that the insect dies satisfies an equation:

$$\frac{d\theta}{dx} = \frac{b\theta(1-\theta)}{x}$$

for some positive constant b.

(i) Explain why this model has certain reasonable characteristics and show that it leads to

$$\log\left(\frac{\theta}{1-\theta}\right) = a+b\log x$$

for suitable a.

(ii) Experiments are going to be carried out to determine a, b. In applying the insecticide to a fixed large number N of insects the following probability model is used. The probability of death is the same for all the insects and the deaths are independent of each other, so that the number killed, k, has a binomial probability function and k/N is nearly θ.

In order to avoid difficulties when either none or all of the insects are killed it is proposed not to calculate the value of the statistic $t = \log\left(\frac{k/N}{1-k/N}\right)$ but of the statistic y given by

$$y = \log\left(\frac{k+\frac{1}{2}}{(N-k)+\frac{1}{2}}\right),$$

where $(k+\frac{1}{2})$ has been written for k in the definition of t. Compare the behaviours of t and y for $k = 0, \frac{1}{2}N, N$.

By writing $k = N\theta+h$, show that y may be written as

$$\log\left(\theta+\frac{1}{2N}\right) - \log\left(1-\theta+\frac{1}{2N}\right) + \frac{h}{N} \cdot \frac{1+1/N}{\left(\theta+\frac{1}{2N}\right)\left(1-\theta+\frac{1}{2N}\right)} + \text{higher power of } \frac{h}{N}$$

and find the term in $(h/N)^2$.

(iii) What are the values of $E[h]$, $E[h^2]$, var $[h]$ obtained from the Binomial probability function for k.

(iv) Neglecting all terms in $(h/N)^3$ and higher powers show that

$$E[y] = \log\left(\left(\theta+\frac{1}{2N}\right)\Big/\left(1-\theta+\frac{1}{2N}\right)\right)$$

$$+\theta(1-\theta)(2N)^{-1}\left(\left(\theta+\frac{1}{2N}\right)^{-2}+\left(1-\theta+\frac{1}{2N}\right)^{-2}\right);$$

and neglecting all terms in $(h/N)^2$ and higher powers show that

$$\text{var}[y] = \frac{\theta(1-\theta)}{\left(\theta+\frac{1}{2N}\right)^2\left(1-\theta+\frac{1}{2N}\right)^2}\cdot\frac{1}{N}\left(1+\frac{1}{N}\right)^2.$$

(v) Show that, for large N, y is nearly an unbiased estimator for $\log(\theta(1-\theta)^{-1})$ and that var $[y] \to 0$ as $N \to \infty$.

[If an estimator t based on a sample of size N has the properties that $E[t] \to \theta$ and var $[t] \to 0$ as $N \to \infty$, then it is called a *consistent* estimator for θ.]

SOLUTIONS TO EXERCISES

Exercise 1 A (p. 13)

1. (a) Method (A): Note the effect of widely differing thicknesses of bunches with possibly widely differing calculated averages, even if no mistakes are made. Too much 'weight' is given to these bunches, and these are most likely to have unreliable averages.

Method (B): This eliminates the effect noted for (A). Note, however, the effect of a possible mistake of measurement on counting in a bunch, or of an unnoticed gross non-uniformity of paper.

Method (C): This eliminates the effects noted for (A) and (B).
(b) Take equal numbers of leaves in each bunch; (A) and (B) are then identical.
(c) Difficulty of measuring small thicknesses with small proportional error. Effects of pressure in squeezing the bunches.
(d) The following is a possible analogy, shown schematically:

book	collection of animals
leaves	individual animals
same paper	same species or type
(not too...different paper)	(not too different in habitat or composed of merely similar species)
bunches of leaves	samples of animals
thickness of bunch	total weight of sample

Method (C) is appropriate since the size of sample could be as large as convenient and could differ when convenient. Furthermore, the uniformity of paper in a book can be almost certainly assumed because of the care of the publishers, but non-uniformity cannot be assumed in nature; and in Method (C), differences unnoticed otherwise would be detected by the fact that different lines would appear on the graph paper.

4. None of the modes can be deduced from others.

5. $N \equiv 0$ [mod 4].

q_1 is any number between the readings with labels $\frac{1}{4}N$ and $\frac{1}{4}N+1$
q_3 is any number between the readings with labels $(N+1)-(\frac{1}{4}N+1)$ and $(N+1)-\frac{1}{4}N$.

$N \equiv 1$ [mod 4].

q_1 is any number between the readings with labels $\frac{1}{4}(N-1)$ and $\frac{1}{4}(N-1)+1$
q_3 is any number between the readings with labels $\frac{3}{4}N+\frac{1}{4}$ and $\frac{3}{4}N+1\frac{1}{4}$.

These results should be used as hints for the other cases. The medians are dealt with in Section 3.

6.

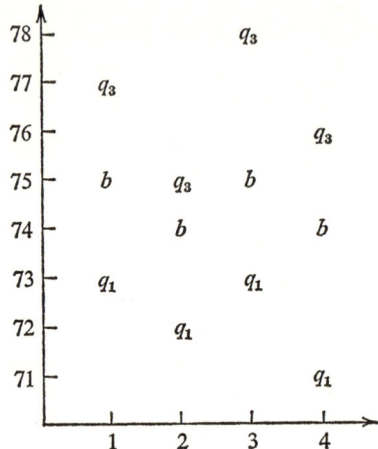

(i) There seems to be a tendency for the scores to be lower (growing familiarity) except in Round 3 (bad weather?)

(ii) Cannot be answered by this analysis, in which rounds and not players are being treated.

(iii) There seems to be no bunching of q_3 with b more than of q_1 with b. Such bunching might be expected if the list was truncated.

(iv) The mean of the quartiles does not seem particularly above or below the median.

(v) Cannot be answered by this analysis.

9. 'The strength of a chain is its weakest link.'

10. x = Number of inches per month   0 1 2 3 4 48

   $f(x)$ = Number of months with this number of
                inches: at P.R.   0 0 4 5 3 0

   $g(x)$ = Number of months with this number of
                inches: at Ll.   1 3 3 0 4 1

Higher 'averages':

   Using median for 'average': P.R. with median = 3 in.
   Using mode for 'average': Ll. with mode   = 4 in.
   Using mean for 'average': Ll. with mean   = 6·1 in.

11. The 'average' using mode is 13 yr; using median is 14 yr. The range between the quartiles ('Interquartile range') is $(15-13)$ yr = 2 yr.

400

Exercise 3 A (p. 58)

1.

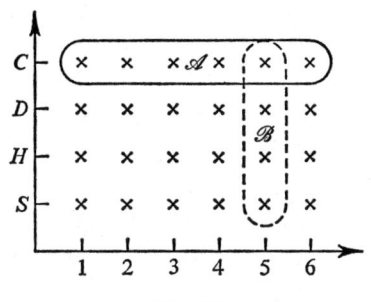

Fig. S.1

3.

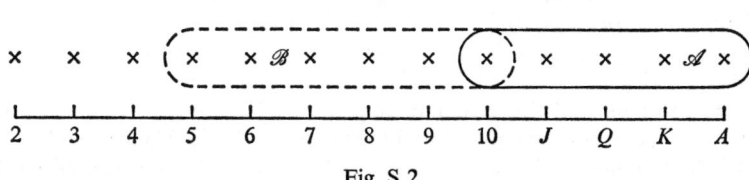

Fig. S.2

5.

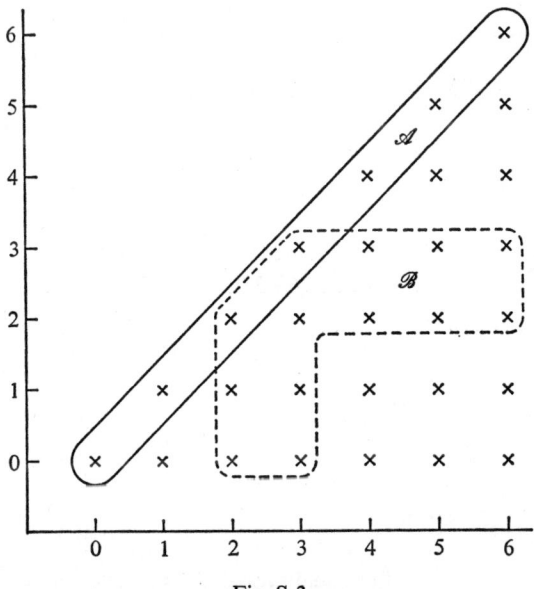

Fig. S.3

1, 3, 5. The following results are read from Figures S. 1, S. 2, S. 3.

	Qu. 1	Qu. 3	Qu. 5
pr (A)	$\frac{1}{4}$	$\frac{5}{13}$	$\frac{1}{4}$
pr (B)	$\frac{1}{6}$	$\frac{6}{13}$	$\frac{13}{28}$
pr (A')	$\frac{3}{4}$	$\frac{8}{13}$	$\frac{3}{4}$
pr (B')	$\frac{5}{6}$	$\frac{7}{13}$	$\frac{15}{28}$
pr (A and B)	$\frac{1}{24}$	$\frac{1}{13}$	$\frac{1}{14}$
pr (A and B')	$\frac{5}{24}$	$\frac{4}{13}$	$\frac{5}{28}$
pr (A' and B)	$\frac{1}{8}$	$\frac{5}{13}$	$\frac{11}{28}$
pr (A' and B')	$\frac{5}{8}$	$\frac{3}{13}$	$\frac{5}{14}$
pr (A and A)	$\frac{1}{4}$	$\frac{5}{13}$	$\frac{1}{4}$
pr (A and A')	0	0	0
pr (A or B)	$\frac{3}{8}$	$\frac{10}{13}$	$\frac{9}{14}$
pr (A or B')	$\frac{7}{8}$	$\frac{8}{13}$	$\frac{17}{28}$
pr (A' or B)	$\frac{19}{24}$	$\frac{9}{13}$	$\frac{23}{28}$
pr (A' or B')	$\frac{23}{24}$	$\frac{12}{13}$	$\frac{13}{14}$
pr (B or B)	$\frac{1}{6}$	$\frac{6}{13}$	$\frac{13}{28}$
pr (B or B')	1	1	1

5. Bose–Einstein statistics give the same probability model because of the indistinguishability of the photons and the possibility of double occupancy. The ends of a domino correspond to the photons; the numbers on the ends to the 'states'.

7. (i) B tries to do something twice as difficult as A tries, but he gets twice as many throws; nevertheless B 'wastes' a six if he throws a six on each of two throws which happen to be part of a single turn: only one success is counted for them.

(ii) $2 \times \frac{1}{6} \times \frac{1}{6} = \frac{1}{18}$.

(iii) $\frac{6}{36} = \frac{1}{6}$.

(iv) $(2 \times \frac{1}{6} \times \frac{5}{6}) + (\frac{1}{6} \times \frac{1}{6}) = \frac{11}{36}$.

9. The frequency ratios are:

$$plastic: \quad \frac{244 + 274 + 230 + 241 + 241}{16(36 + 39 + 33 + 34 + 36 + 37)} = 0.415,$$

$$\frac{104 + 62}{433} = 0.384.$$

$$brass: \quad \frac{186}{16 \times 185} = 0.063$$

$$\frac{38}{433} = 0.088.$$

There seem no reasons to allot equal probabilities.

Exercise 3B (p. 65)

1, 3, 5, 7.

	Qu. 1	Qu. 3	Qu. 5	Qu. 7
pr $(B\|A)$	$\frac{1}{6}$	$\frac{1}{4}$	$\frac{2}{7}$	$\frac{10}{15}$
pr $(A\|B)$	$\frac{1}{4}$	$\frac{1}{6}$	$\frac{2}{13}$	$\frac{10}{18}$
pr $(B\|A')$	$\frac{1}{6}$	$\frac{5}{8}$	$\frac{11}{21}$	$\frac{8}{9}$
pr $(A\|B')$	$\frac{1}{4}$	$\frac{4}{7}$	$\frac{5}{15}$	$\frac{5}{6}$
pr $(B'\|A)$	$\frac{5}{6}$	$\frac{4}{5}$	$\frac{5}{7}$	$\frac{5}{15}$
pr $(A'\|B)$	$\frac{3}{4}$	$\frac{5}{6}$	$\frac{11}{13}$	$\frac{8}{18}$
pr $(B\|A')$	$\frac{5}{6}$	$\frac{3}{8}$	$\frac{10}{21}$	$\frac{1}{9}$
pr $(A\|B')$	$\frac{3}{4}$	$\frac{3}{7}$	$\frac{10}{15}$	$\frac{1}{6}$

7.

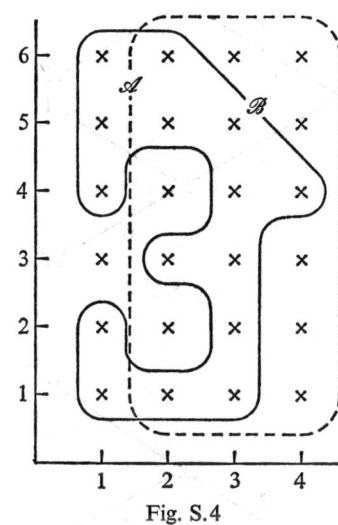

Fig. S.4

Exercise 3C (p. 70)

1. (i)

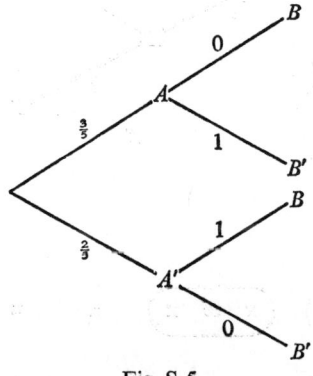

Fig. S.5

(ii) $0, \frac{3}{5}, \frac{2}{5}, 0.$

(iii)

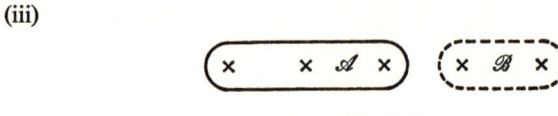

Fig. S.6

(iv)

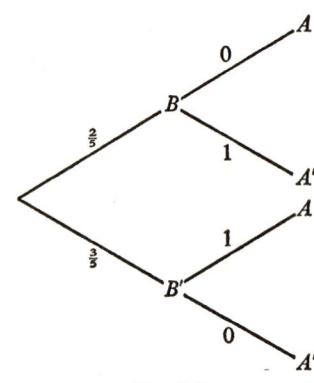

Fig. S.7

3. (i)

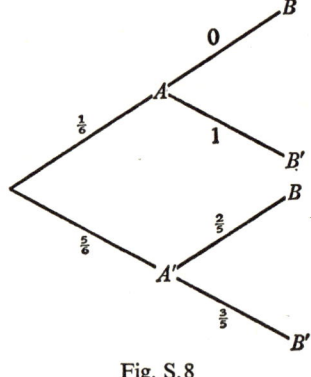

Fig. S.8

(ii) $0, \frac{1}{6}, \frac{1}{3}, \frac{1}{2}.$

(iii)

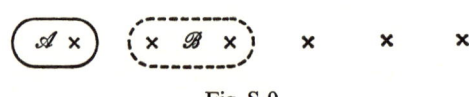

Fig. S.9

(iv)

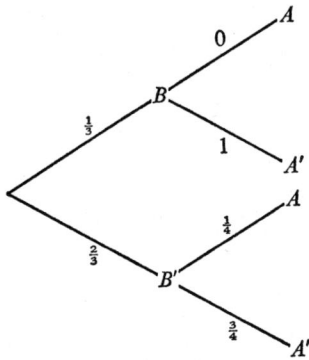

Fig. S.10

5. (i)

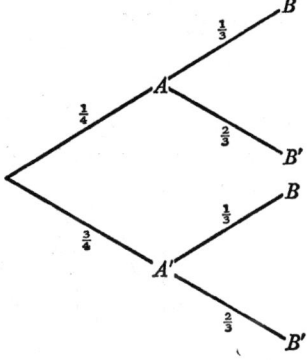

Fig. S.11

(ii) $\frac{1}{12}, \frac{1}{6}, \frac{1}{4}, \frac{1}{2}$.

(iii)

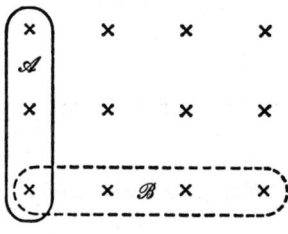

Fig. S.12

(iv)

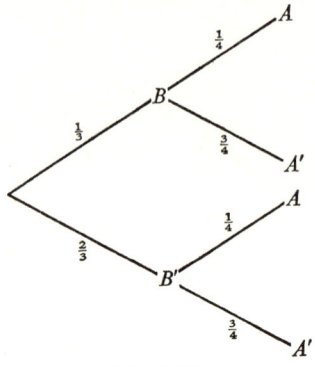

Fig. S.13

7. pr (I shall get a hat) $= \frac{3}{4} \times \frac{1}{3} \times \frac{1}{2} = \frac{1}{8}$,
   pr (I shan't) $\quad = \frac{7}{8}$.

8. (i) fairy $\frac{6}{13}$, witch $\frac{6}{15}$; (ii) fairy $\frac{2}{13}$, witch $\frac{2}{10}$.

9.

	Situation			pr (you win)	
You choose wheel A and your opponent chooses B				$\frac{7}{32} + \frac{25}{32} \cdot \frac{16}{32} = \frac{624}{1024}$	
,,	,,	A	,, ,, C	$\frac{7}{32} + \frac{25}{32} \cdot \frac{7}{32} = \frac{399}{1024} < \frac{400}{1024}$	
,,	,,	B	,, ,, A	$\frac{16}{32} \cdot \frac{25}{32} + 0 = \frac{400}{1024}$	
,,	,,	B	,, ,, C	$\frac{16}{32} + \frac{16}{32} \cdot \frac{7}{32} = \frac{624}{1024}$	
,,	,,	C	,, ,, A	$\frac{25}{32} \cdot \frac{25}{32} + 0 = \frac{625}{1024} > \frac{624}{1024}$	
,,	,,	C	,, ,, B	$\frac{25}{32} \cdot \frac{16}{32} + 0 = \frac{400}{1024} < \frac{1}{2}$	

If playing an intelligent opponent
   You choose C and resign yourself to losing.

If your opponent chooses at random
   You choose C and hope that his choice falls on A.

Exercise 3 D (p. 74)

1, 3, 5, 7. A *not* independent of B.

9. pr $(A|B) = $ pr (A). A proof is given at the end of Section 6.6.

Exercise 3 E (p. 81)

1. $\frac{192}{1000}$; $\frac{8}{1000}$; $\frac{192}{992}$; $\frac{1}{2}$.
   No pairs are independent.

3. The contingency table is:

	Stung	Not stung	
Sunburned	2	6	8
Not sunburned	4	12	16
Total	6	18	24

5. (i) $\frac{1}{3} \times \frac{1}{4} \times \frac{3}{5} = \frac{1}{20}$.

(ii) $\frac{2}{3} \times \frac{3}{4} \times \frac{3}{5} = \frac{3}{10}$.

(iii) $\frac{1}{4}$.

(iv) pr (I have ticket, passport, cheques) $= (1 - \frac{1}{3}) \times (1 - \frac{1}{4}) \times \frac{3}{5}$, and this exceeds the probability of each other possibility.

7. (a) (i) $(\frac{1}{2})^6$;   (ii) $(\frac{1}{2})^7$;   (iii) $(\frac{1}{2})^7/(\frac{1}{2})^6 = \frac{1}{2}$.

(b) The 'Law of Averages' says 'it will try to average out by giving a tail on the seventh throw'; or words to that effect.

(c) That the coin was double-headed; the probability would be estimated as 1.

9. pr $(A|A) = 1 \neq 0 = $ pr $(A|A')$

$\Rightarrow A$ is not statistically independent of A;

$\Rightarrow$ the relation is not reflexive.

Consider the events:

$\qquad A = $ a six is thrown on the first of two dice;

$\qquad B = $ a total of 7 is thrown on the two dice;

$\qquad C = $ a one is thrown on the first of two dice.

Then A is statistically independent of B, and B is statistically independent of C but A is not statistically independent of C. Thus the relation is not transitive.

11. (i) True: pr $(A) = \frac{2}{4} = \frac{1}{2} = $ pr $(A|B)$.

(ii) True: by the truth of (i) and the symmetry between B and C.

(iii) False: pr $(B \text{ and } C) = \frac{1}{4} \neq \frac{1}{2} = $ pr $(B \text{ and } C|A)$.

(iv) True: pr $(B \text{ and } C) = \frac{1}{4} = (\frac{1}{2})^2 = $ pr $(B).$ pr (C).

(v) False: pr $(A \text{ and } B \text{ and } C) = \frac{1}{4} \neq (\frac{1}{2})^3 = $ pr $(A).$ pr $(B).$ pr (C).

13. The answers are as for Question 11, but the numerical values of the probabilities differ.

Exercise 3 F (p. 84)

1. $\frac{9}{24}$; $\frac{21}{24}$; $\frac{23}{24}$.   (iii) is true.

3. $\frac{10}{13}$; $\frac{8}{13}$; $\frac{12}{13}$.   None is true.

5. $\frac{18}{28}$; $\frac{17}{28}$; $\frac{26}{28}$.   None is true.

7. (a) It is possible if and only if one of the events is impossible.

Proof. A and B are exclusive $\Rightarrow$ pr $(A \text{ and } B) = 0$.

But pr $(A).$ pr $(B) \neq 0$ unless either or both of pr $(A) = 0$, pr $(B) = 0$.

Thus pr $(A \text{ and } B) \neq $ pr $(A).$ pr (B) unless one of A, B is impossible.

On the other hand B, say, is impossible $\Rightarrow$ pr $(B) = 0$ and $(A \text{ and } B)$ impossible

$\qquad\qquad\qquad\qquad\qquad \Rightarrow$ pr $(A).$ pr $(B) = 0 = $ pr $(A \text{ and } B)$

$\qquad\qquad\qquad\qquad\qquad \Rightarrow A$ and B independent.

(*b*) It is possible if and only if one of the events is itself exhaustive.

Proof. A and B independent ⇒ A′ and B′ independent

$$⇒ pr (A' \text{ and } B') = pr (A').pr (B').$$

But pr $(A').pr (B') \neq 0$ unless either or both of pr $(A') = 0$, pr $(B') = 0$.
Thus pr $(A' \text{ and } B') \neq 0$ unless one at least of A, B is separately exhaustive.
On the other hand B, say, is exhaustive ⇒ pr (A) = pr $(A|B)$

$$⇒ A \text{ and } B \text{ are independent.}$$

(*c*) It is possible since A and A′ are exclusive and exhaustive.

9. $1-(1-a)(1-b) = 1-(1-a-b+ab) = a+b-ab.$

$$\begin{aligned}
pr (A \text{ or } B) &= pr (A)+pr (B)-pr (A \text{ and } B) \\
&= pr (A)+pr (B)-pr (A) pr (B) && \text{by independence} \\
&= 1-(1-pr (A))(1-pr (B)) && \text{by result above} \\
&= 1-pr (A').pr (B') \\
&= 1-pr (A' \text{ and } B') && \text{by independence.}
\end{aligned}$$

11. The body of the table is

pr $(A).pr (B|A)$	pr $(A)+pr (B)-pr (A).pr (B|A)$
pr $(A).pr (B)$	pr $(A)+pr (B)-pr (A).pr (B)$
0	pr $(A)+pr (B)$
pr $(A).pr (B|A)$	1

13. (i) 2 is the number most likely to occur, because

$$pr (0 \text{ fine days}) = (\tfrac{1}{4})^2,$$
$$< pr (1 \text{ fine day}) = 2 \times \tfrac{3}{4} \times \tfrac{1}{4},$$
$$< pr (2 \text{ fine days}) = (\tfrac{3}{4})^2.$$

(ii) 2 is the number most likely to occur. This is a further consequence of the independence.

(iii) 3 is the number most likely to occur, because

$$pr (0 \text{ fine days}) = (\tfrac{1}{4})^4,$$
$$< pr (1 \text{ fine day}) = 4 \times \tfrac{3}{4} \times (\tfrac{1}{4})^3,$$
$$< pr (2 \text{ fine days}) = 6 \times (\tfrac{3}{4})^2 \times (\tfrac{1}{4})^2,$$
$$< pr (3 \text{ fine days}) = 4 \times (\tfrac{3}{4})^3 \times (\tfrac{1}{4})$$
$$> pr (4 \text{ fine days}) = (\tfrac{3}{4})^4.$$

Notice that this is *not* the sum of the most likely numbers in each of the pairs of days.

(iv) 2 and 3 are equally likely to occur, because

$$pr (0 \text{ fine days}) = (\tfrac{1}{4})^3,$$
$$< pr (1 \text{ fine day}) = 3 \times (\tfrac{3}{4}) \times (\tfrac{1}{4})^2,$$
$$< pr (2 \text{ fine days}) = 3 \times (\tfrac{3}{4})^2 \times (\tfrac{1}{4}),$$
$$= pr (3 \text{ fine days}) = (\tfrac{3}{4})^3.$$

Exercise 3M (p 91)

1. (i) $pq(1-r)$;   (ii) $(1-p)q(1-r)$;   (iii) $1-(1-p)(1-q)(1-r)$;
  (iv) $pq(1-r)+p(1-q)r+(1-p)qr$.

3. pr (large gold coin, under witch's instructions) $= \dfrac{g}{s+g}\cdot\dfrac{x_2}{x_2+y_2}$,

   pr (large gold coin, under fairy's instructions) $= \dfrac{gx_2}{s(x_1+y_1)+g(x_2+y_2)}$.

   Equality $\Leftrightarrow x_2+y_2 = x_1+y_1$ or $s = 0$,
            $\Leftrightarrow$ packets all contain same number of coins, or there are no
              silver coins.

5. (i) pr (valuables get through) $= xy\cdot\tfrac{4}{5}+x(1-y)+(1-x)y+(1-x)(1-y)\cdot\tfrac{3}{5}$.
  (ii)

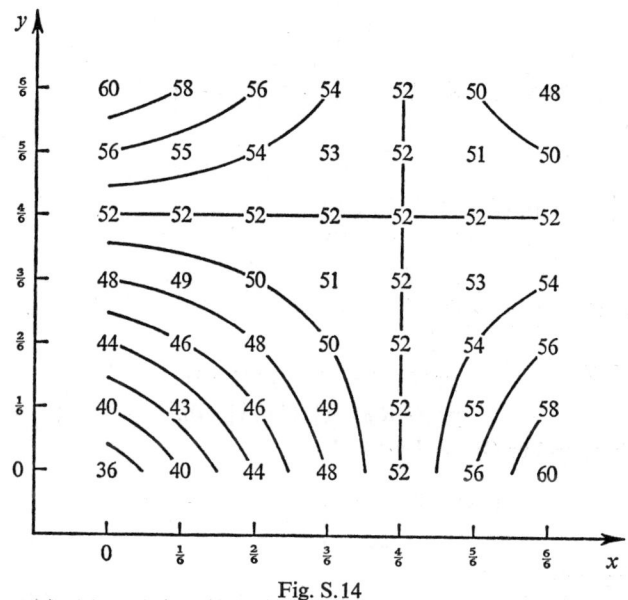

Fig. S.14

(iii) For x fixed, z is first degree in y, so the graph is linear.

$\qquad x < \tfrac{2}{3} \Rightarrow -3(x-\tfrac{2}{3}) > 0$

$\qquad\qquad \Rightarrow z$ is least when $(y-\tfrac{2}{3})$ is least

$\qquad\qquad \Rightarrow z$ is least when $y = 0$

$\qquad\qquad \Rightarrow$ minimum value of z is $4\tfrac{1}{3}-3(x-\tfrac{2}{3})\cdot-\tfrac{2}{3} = \dfrac{3+2x}{5}$.

$\qquad x > \tfrac{2}{3} \Rightarrow -3(x-\tfrac{2}{3}) < 0$

$\qquad\qquad \Rightarrow z$ is least when $(y-\tfrac{2}{3})$ is greatest

$\qquad\qquad \Rightarrow z$ is least when $y = 1$

$\qquad\qquad \Rightarrow$ minimum value of z is $4\tfrac{1}{3}-3(x-\tfrac{2}{3})\cdot\tfrac{1}{3} = \dfrac{5-x}{5}$.

409

$$x = \tfrac{2}{3} \Rightarrow z \text{ is fixed, and } z = \frac{4\tfrac{1}{3}}{5}.$$

$$m(x) = \begin{cases} \dfrac{3+2x}{5} & x \leqslant \tfrac{2}{3}, \\[2mm] \dfrac{5-x}{5} & x > \tfrac{2}{3}, \text{ furthermore } x = \tfrac{2}{3} \text{ here also gives } m(\tfrac{2}{3}). \end{cases}$$

By choosing $x = \tfrac{2}{3}$ we maximize $m(x)$ at $\dfrac{4\tfrac{1}{3}}{5} = \tfrac{13}{15}$.

The probability the valuables get through is $\tfrac{13}{15}$.

(iv) $\tfrac{2}{3}$; $\tfrac{13}{15}$; the values giving the saddle-point.

(v) If the valuables always went town route, the robbers would always ambush town route.

The optimum strategy cuts the *losses* by a proportion equal to

$$\frac{(1-\tfrac{4}{5})-(1-\tfrac{13}{15})}{(1-\tfrac{4}{5})} = 17\%.$$

7. pr (machine works) $= \tfrac{1}{3} \times \tfrac{1}{2} \times \tfrac{5}{6} + (\tfrac{4}{5} \times \tfrac{1}{3} \times \tfrac{5}{6}) \times 2 = \tfrac{19}{36}.$

9. $\dfrac{\theta}{\theta + (1-\theta)/p}.$

11. pr $(U|W) = $ pr $(U$ and $W)/$pr (W)

$$= \frac{(\tfrac{1}{5} \times \tfrac{1}{2} \times 1) + (\tfrac{4}{5} \times \tfrac{1}{2} \times \tfrac{2}{3})}{(\tfrac{1}{5} \times \tfrac{1}{2}) + (\tfrac{4}{5} \times \tfrac{1}{2})},$$

pr $(U|W') = \dfrac{(\tfrac{1}{5} \times \tfrac{1}{2} \times \tfrac{1}{3}) + (\tfrac{4}{5} \times \tfrac{1}{2} \times 0)}{\tfrac{1}{2}}.$

More directly

$$\text{pr } (U|R) = (\tfrac{1}{2} \times 1) + (\tfrac{1}{2} \times \tfrac{1}{3}),$$
$$\text{pr } (U|R') = (\tfrac{1}{2} \times \tfrac{2}{3}) + (\tfrac{1}{2} \times 0).$$

Using these and pr $(W) = \tfrac{1}{2}$ we find

$$\text{pr } (W|U) = \frac{\tfrac{1}{2} \times \tfrac{11}{15}}{(\tfrac{1}{2} \times \tfrac{11}{15}) + (\tfrac{1}{2} \times \tfrac{1}{15})} = \tfrac{11}{12}.$$

Also
$$\text{pr } (R|U) = \tfrac{1}{3}.$$

The odds can be 11:1.

It is certain that his wife has been at him so I can accept any odds.

13. The product of the matrix and vector gives

$$\begin{pmatrix} \text{pr } (B|A) . \text{pr } (A) + \text{pr } (B|A') . \text{pr } (A') \\ \text{pr } (B'|A) . \text{pr } (A) + \text{pr } (B'|A') . \text{pr } (A') \end{pmatrix}$$

and the result follows.

$$\begin{pmatrix} a & b \\ c & d \end{pmatrix} = ad - bc = (1-c)(1-b) - bc = 1 - (b+c),$$

$$1 - (b+c) \geqslant -1 \quad \text{since } b, c \leqslant 1,$$
$$1 - (b+c) \leqslant 1 \quad \text{since } b, c \geqslant 0,$$
$$1 - (b+c) = -1 \Rightarrow b = 1, \ c = 1$$
$$\Rightarrow A, B \text{ are exclusive, exhaustive,}$$

$$1-(b+c) = 0 \quad \Rightarrow b = 1-c$$
$$\Rightarrow \text{pr } (B|A') = 1 - \text{pr } (B'|A)$$
$$\Rightarrow \text{pr } (B|A') = \text{pr } (B|A)$$
$$\Rightarrow A \text{ and } B \text{ are independent,}$$
$$1-(b+c) = 1 \quad \Rightarrow b = 0, c = 0$$
$$\Rightarrow A, B \text{ are the same event.}$$

There are two independent elements:

$$\begin{pmatrix} 1-c & b \\ c & 1-b \end{pmatrix}$$

square of the matrix $= \begin{pmatrix} (1-c)^2+bc & (2-(b+c))\,b \\ (2-(b+c))\,c & bc+(1-b)^2 \end{pmatrix}$

which is of the form $\begin{pmatrix} 1-\theta & \phi \\ \theta & 1-\phi \end{pmatrix}.$

15. (i) pr (total is 6) $= 2 \times \frac{1}{6} \times \frac{1}{6} + 2 \times \frac{1}{6} \times K/3 + \left(\dfrac{1-K}{3}\right)^2$

$$= \frac{5+4(K-\frac{1}{2})^2}{36}.$$

This is maximized by $K = 0$ or 1.

(ii) pr (total is 7) $= \dfrac{3-4(K-\frac{1}{2})^2}{18}.$

This is maximized by $K = \frac{1}{2}$.

(iii) Probability maximized by $K = 0$ or 1.

17. pr (gene is a) $= \theta \Rightarrow$ pr (genes are different) $= 2\theta(1-\theta)$
$$= \tfrac{1}{2} - 2(\theta - \tfrac{1}{2})^2.$$

The results follow.

18. The answers to this question are deducible from those to Question 19.

19. (i) Frequency table showing constitution of bulked collection.

	K	L	M
type AA	$\frac{1}{4} \times 12$	$\frac{4}{9} \times 144$	$\frac{1}{100} \times 100$
Aa	$\frac{1}{2} \times 12$	$\frac{4}{9} \times 144$	$\frac{18}{100} \times 100$
aa	$\frac{1}{4} \times 12$	$\frac{1}{9} \times 144$	$\frac{81}{100} \times 100$

pr (genotype is AA) $= (3+64+1)/256 = \frac{68}{256}$,

pr (genotype is Aa) $= (6+64+18)/256 = \frac{88}{256}$,

pr (genotype is aa) $= (3+16+81)/256 = \frac{100}{256}$.

pr (gene is a) $= \frac{100}{256} \times 1 + \frac{88}{256} \times \frac{1}{2} = \frac{9}{16}$,

pr (gene is A) $= \frac{68}{256} \times 1 + \frac{88}{256} \times \frac{1}{2} = \frac{7}{16}$.

To justify the alternative method we note:

$$\text{pr (gene is a)} = (2 \times 100 + 88)/512 = \tfrac{9}{16}, \text{ as before,}$$

$$\text{pr (gene is A)} = (2 \times 68 + 88)/512 = \tfrac{7}{16}, \text{ as before.}$$

(ii) hypothetical value of pr (gene is a) $= \sqrt{(\tfrac{100}{256})} = \tfrac{10}{16}$,
   hypothetical pr (genotype is AA) $= \tfrac{36}{256}$,
   hypothetical pr (genotype is Aa) $= \tfrac{120}{256}$,
   hypothetical proportion of AA- or Aa-types to aa-type $= 156:100 \simeq 61:39$.

(iii) pr (genotype is AA) $= \tfrac{49}{256}$,
   pr (genotype is Aa) $= \tfrac{126}{256}$,
   pr (genotype is aa) $= \tfrac{81}{256}$.

Expected proportion of two types $= 175:81 \simeq 68:32$.

21. Table of probabilities

		Genotypes			Genes	
		BB	Bb	bb	B	b
First generation	Adult	0	1	0	$\tfrac{1}{2}$	$\tfrac{1}{2}$
Children	At birth	$\tfrac{1}{4}$	$\tfrac{1}{2}$	$\tfrac{1}{4}$		
	Adult	0	$\tfrac{2}{3}$	$\tfrac{1}{3}$	$\tfrac{1}{3}$	$\tfrac{2}{3}$
Grandchildren	At birth	$\tfrac{1}{9}$	$\tfrac{4}{9}$	$\tfrac{4}{9}$		
	Adult	0	$\tfrac{1}{2}$	$\tfrac{1}{2}$	$\tfrac{1}{4}$	$\tfrac{3}{4}$
Great grandchildren	At birth	$\tfrac{1}{16}$	$\tfrac{6}{16}$	$\tfrac{9}{16}$		
	Adult	0	$\tfrac{2}{5}$	$\tfrac{3}{5}$	$\tfrac{1}{5}$	$\tfrac{4}{5}$

23.

	(i) Wing colour	(ii) True-breeding	(iv) Probabilities required (see below)
CCII	white	white	$\tfrac{1}{16}$
CCIi	white	—	$\tfrac{2}{16}$
CCii	black	black	$\tfrac{1}{16}$
CcII	white	white	$\tfrac{2}{16}$
CcIi	white	—	$\tfrac{4}{16}$
Ccii	black	—	$\tfrac{2}{16}$
ccII	white	white	$\tfrac{1}{16}$
ccIi	white	white	$\tfrac{2}{16}$
ccii	white	white	$\tfrac{1}{16}$

(iii) The following table shows the possible parental gene contributions on the edge of the table, and the genotype of the resulting offspring in the body of the table. (The parents are true-breeding white-winged.)

	CI	cI	ci
CI	CCII	CcII	CcIi
cI	CcII	ccII	ccIi
ci	CcIi	ccIi	ccii

The result follows.

The required probabilities are tabulated above. They are derived from:

		Gene-contribution from other parent			
		CI	Ci	cI	ci
Gene-con-	CI	CCII	CCIi	CcII	CcIi
tribution	Ci	CCIi	CCii	CcIi	Ccii
from one	cI	CcII	CcIi	ccII	ccIi
parent	ci	CcIi	Ccii	ccIi	ccii

The various possible parental gene contributions are equiprobable, thus the full table of probabilities. The expected proportion is $3:13$

(v) The frequency-ratios for the genotypes BB, Bb, bb are $1:2:1$. The frequency-ratios for the appearance types (technically called *phenotypes*) is black-winged: white-winged as $1:3 \neq 3:13$.

[Large numbers would be needed before a significant difference could be established.]

(vi) (a) The crossing of true-breeding white-winged can produce not-true-breeding white-winged under first theory but cannot do so under second theory.

(b) There can be not-true-breeding black-winged under first theory, but there cannot be under second theory.

25. (i) (a) $4 \times \dfrac{4 \times 3 \times 2 \times 2}{6 \times 5 \times 4 \times 3} = \frac{8}{15}$;

(b) $4 \times \dfrac{3 \times 2 \times 1 \times 3}{6 \times 5 \times 4 \times 3} = \frac{1}{5}$.

(ii) With 5 red and 1 green disc the probability is $\frac{10}{15}$, and this is the largest value obtainable.

27. (i) There are 10 cracked tumblers.

$$\text{pr (buyer accepts)} = \frac{40 \times 39 \times 38 \times 37 \times 36}{50 \times 49 \times 48 \times 47 \times 46} = 0\cdot31;$$

(ii) pr (buyer rejects) $= 5 \times \dfrac{49 \times 48 \times 47 \times 46 \times 1}{50 \times 49 \times 48 \times 47 \times 46} = 0\cdot1.$

29.

$$\begin{pmatrix} \frac{20}{36} & \frac{5}{36} & \frac{10}{36} & \frac{1}{36} \\ \frac{20}{36} & \frac{5}{36} & \frac{10}{36} & \frac{1}{36} \\ 0 & 0 & \frac{5}{6} & \frac{1}{6} \\ 0 & 0 & 0 & 1 \end{pmatrix}.$$

$\sum\limits_{j \geqslant i} a_{ij} = $ probability that resulting hand is of at least as high a type.

31. $\frac{9}{24}$ [significantly expressible as $1 - \dfrac{1}{1!} + \dfrac{1}{2!} - \dfrac{1}{3!} + \dfrac{1}{4!}$. See W. W. Rouse Ball: *Mathematical Recreations and Essays*, Macmillan, 11th Edition, 1947, p. 46.]

33. $b(n, N) = N!/N^n(N-n)!$

$n = 1$	$\dfrac{n}{N} = 0.1$	0.2	0.3	0.4	0.5
$N = 10$	1	1	0.9	0.72	0.504	0.302
$N = 50$	1	0.814	0.382	0.096	0.012	—
$N = 100$	1	0.643	0.134	0.008	—	—

$k(10) \simeq 0.402$; $k(50) \simeq 0.175$; $k(100) \simeq 0.125$; $k(1000) = 0.0375$.

35. $1-(\frac{3}{4})^3 = \frac{37}{64}$. That the three people are so chosen that their replies are independent. Without this assumption there is no unique solution to the problem.

37.
$$\text{pr (4 bulls)} = 5 \times (\tfrac{4}{5})^4 \times \tfrac{1}{5} = 0.41,$$
$$\text{pr (5 bulls)} = (\tfrac{4}{5})^5 = 0.33,$$
$$\text{pr (less than 4 bulls)} = 1-(0.41+0.33) = 0.26.$$

He is more likely to exceed than to fall short.

39. $\text{pr (4 aces)} = 5 \times (\tfrac{1}{6})^4 \times \tfrac{5}{6} = 3.21 \times 10^{-3}$,,
$$\text{pr (4 faces the same)} = 6 \times 3.21 \times 10^{-3} = 1.93 \times 10^{-2},$$
$$\text{pr (5 faces the same)} = 1 \times (\tfrac{1}{6})^4 = 0.08 \times 10^{-2},$$
$$\text{pr (at least 4 faces the same)} = 2.01 \times 10^{-2}.$$

41. (i) 5000×365 to 1 or 1.8×10^6 to 1;

(ii) $1/(1-(1-\tfrac{1}{5000})^{1/365})$ to 1;

(iii) $1-(1-\tfrac{1}{5000})^{1/365} \simeq \dfrac{1}{5000 \times 365}$ with *proportional* error $\simeq \dfrac{1}{3650000}$.

The proportional error < 0.01 so the result is correct to 2 S.F.

43. (i) $(1-1/K)^n$;

(ii) $(1-1/K)^n + (n/K)(1-1/K)^{n-1}$;

(iii) $1-(1-1/K)^n - (n/K)(1-1/K)^{n-1}$.

The approximations are e^{-2}, $3e^{-2}$, $1-3e^{-2}$ or 0.135, 0.41, 0.59.

45. $\text{pr (0 or 1 or 2 entrants)} = (\tfrac{1}{4})^7 + 7(\tfrac{1}{4})^6 \cdot \tfrac{3}{4} + \dfrac{7 \times 6}{1 \times 2}(\tfrac{1}{4})^5 (\tfrac{3}{4})^2$
$$= 0.013.$$

One would report the figures as they were quoted, 2 out of 7. There are inevitably some fluctuations and their magnitude helps to determine the reliability of any estimates made.

47. (i) $\tfrac{1}{4}$; (ii) $(1-\tfrac{3}{4}) = \tfrac{1}{4}$; (iii) $(1-\tfrac{3}{4})\tfrac{1}{4} = \tfrac{1}{16}$;

(iv) $(1-\tfrac{3}{4})\tfrac{3}{4} = \tfrac{3}{16}$; (v) $\tfrac{1}{4}+(1-\tfrac{3}{4}) \cdot \tfrac{1}{4} = \tfrac{5}{16}$;

(vi) $1-\dfrac{3 \times 5}{16} = \tfrac{1}{16}$; (vii) $(1-\tfrac{15}{16}) \cdot \tfrac{15}{16} = \tfrac{15}{256}$;

(viii) $\lim\limits_{n \to \infty} (\tfrac{1}{16})^n = 0$.

49. $\text{pr (Ann wins at rth round)} = \text{pr (no result for $(r-1)$ rounds)} \times a = \theta_r a$, say.
$\text{pr (Bob wins at rth round)} = \theta_r \times \text{pr (Ann fails at rth round)} \times b = \theta_r(1-a)b$.

We require
$$\sum_{r=1}^{n} \theta_r a \Big/ \sum_{r=1}^{n} \theta_r (1-a)\, b = \frac{a\phi}{(1-a)\, b\phi},$$

where
$$\phi = \sum_{r=1}^{n} \theta_r.$$

51. Write pr (Ann wins)/pr (Bob wins) $= P$.

(i) $P = \dfrac{\frac{2}{5}}{\frac{3}{5} \times \frac{3}{5}} \neq 1$ so the game is not fair.

(ii) $a = \frac{1}{2}, b = \frac{1}{2}$ $P = \dfrac{\frac{1}{2}}{\frac{1}{2} \times \frac{1}{2}}$,   result follows;

$a = \frac{1}{3}, b = \frac{2}{3}$ $P = \dfrac{\frac{1}{3}}{\frac{2}{3} \times \frac{2}{3}} = \frac{3}{4}$,   result follows.

[This means that we cannot correct the overall imbalance towards Ann of 2:1 by making an imbalance 1:2 for each game.]

(iii) $P = \dfrac{u_r/v_r}{(1 - u_r/v_r)\,(1 - u_r/v_r)} = \dfrac{u_r v_r}{(v_r - u_r)^2}$,

$u_r = 1, 1, 2, 3, 5$ for $r = 0, ..., 4$,   with a in its lowest terms,

$v_r = 2, 3, 5, 8, 13$ for $r = 0, ..., 4$,

$P = \dfrac{2}{1^2}, \dfrac{3}{2^2}, \dfrac{10}{3^2}, \dfrac{24}{5^2}, \dfrac{65}{8^2}$, as required.

Let Ann and Bob pick a card in turn. Ann wins on that turn if she picks an honour. If she fails Bob wins if he next draws a non-honour.

(iv) $\dfrac{a}{(1-a)\, b} = 1$; $a+b = 1 \Rightarrow \dfrac{1-b}{b^2} = 1 \Rightarrow b = \dfrac{-1+\sqrt{5}}{2} \ (> 0)$;

$b = 0{\cdot}618$ (calculated) compares with $1 - \dfrac{u_4}{v_4} = \dfrac{8}{13} = 0{\cdot}615$.

53. $\frac{8}{27}; \frac{12}{27}; \frac{6}{27}; \frac{1}{27}; 0.$
The most likely number is 1 (with probability $4 \times (\frac{1}{3}) \times (\frac{2}{3})^3$).

55. (i) $\frac{1}{3}; \frac{2}{3}; \frac{1}{3}; \frac{1}{3}; \frac{1}{3};$
(ii) $\frac{13}{27}; \frac{9}{27}; \frac{4}{27}; \frac{1}{27}.$
During the fourth minute.

57. u_n = pr (system is in state L at nth instant)
   = pr (state L at $(n-1)$ instant).pr (no transition from L)
      $+$pr (state R at $(n-1)$ instant).pr (transition from R)
   = $u_{n-1}.(1-a) + v_{n-1}.b$.

Similarly $v_n = u_{n-1}.a + v_{n-1}.(1-b)$.

(i) Put $n = 1$ in the above;
(ii) Arrange above in matrix form;
(iii) $\mathbf{w}_n = \mathbf{A}.\mathbf{w}_{n-1}$, result follows by induction;

(iv)

n	$\mathbf{w}_n$	$(\mathbf{A})^n$
1	$\frac{1}{4}\begin{pmatrix} 2 \\ 2 \end{pmatrix}$	$\frac{1}{4}\begin{pmatrix} 2 & 1 \\ 2 & 3 \end{pmatrix}$
2	$\frac{1}{4^2}\begin{pmatrix} 6 \\ 10 \end{pmatrix}$	$\frac{1}{4^2}\begin{pmatrix} 6 & 5 \\ 10 & 11 \end{pmatrix}$
3	$\frac{1}{4^3}\begin{pmatrix} 22 \\ 42 \end{pmatrix}$	$\frac{1}{4^3}\begin{pmatrix} 22 & 21 \\ 42 & 43 \end{pmatrix}$
4	$\frac{1}{4^4}\begin{pmatrix} 86 \\ 170 \end{pmatrix}$	$\frac{1}{4^4}\begin{pmatrix} 86 & 85 \\ 170 & 171 \end{pmatrix}$

(v) $\begin{pmatrix} 1-a & b \\ a & 1-b \end{pmatrix}\begin{pmatrix} u_n \\ v_n \end{pmatrix} = \begin{pmatrix} u_n \\ v_n \end{pmatrix} \Rightarrow u_n:v_n:1 = b:a:a+b.$

(vi) Matrix is $\begin{pmatrix} 1-10^{-6} & 2\times10^{-6} \\ 10^{-6} & 1-2\times10^{-6} \end{pmatrix}$.

$$\text{pr (gene is C)} = \frac{2\times10^{-6}}{(1+2)\times10^{-6}} = \tfrac{2}{3}.$$

59. (i) $3\times(\tfrac{2}{1000})\times(\tfrac{998}{1000})^2 \simeq 0.006.$

(ii) $500\times(\tfrac{2}{1000})\times(\tfrac{998}{1000})^{499} \simeq (1-\tfrac{1}{500})^{500}/(1-\tfrac{1}{500}) \simeq e^{-1} \simeq 0.38.$

(iii) $q(0) = (\tfrac{998}{1000})^{1000} \simeq e^{-2} \simeq 0.135;$

$q(1) = 1000\times(\tfrac{2}{1000})\times(\tfrac{998}{1000})^{999} = \tfrac{1000}{1}\times\tfrac{2}{998}\,q(0) \simeq 0.271;$

$$q(2) = \frac{1000\times999}{1\times2}\times(\tfrac{2}{1000})^2\times(\tfrac{998}{1000})^{998} = \tfrac{999}{2}\times\tfrac{2}{998}\,q(1) \simeq 0.271.$$

(iv) $1-(0.135+0.271+0.271) \simeq 0.32.$

61.

(i) Ordered pair	(i) and (ii) Probability	(ii) Number of steps
back, back	p^2	2
back, stay	$p(1-2p)$	1
back, forward	p^2	0
stay, back	$(1-2p)p$	1
stay, stay	$(1-2p)^2$	0
stay, forward	$(1-2p)p$	1
forward, back	p^2	0
forward, stay	$p(1-2p)$	1
forward, forward	p^2	2

(ii) $f(p) = p^2+(1-2p)^2+p^2 = 1-4+6p^2,$

$g(p) = p(1-2p)+2(1-2p)p+p(1-2p) = 4p-8p^2,$

$h(p) = p^2+p^2 = 2p^2.$

(iii) $f(p) = 6(p-\tfrac{1}{3})^2+\tfrac{1}{3}$, which is minimized by $p = \tfrac{1}{3}.$

(iv) $g(p) = \tfrac{1}{2}-8(p-\tfrac{1}{4})^2$, which is maximized by $p = \tfrac{1}{4}.$

(v)

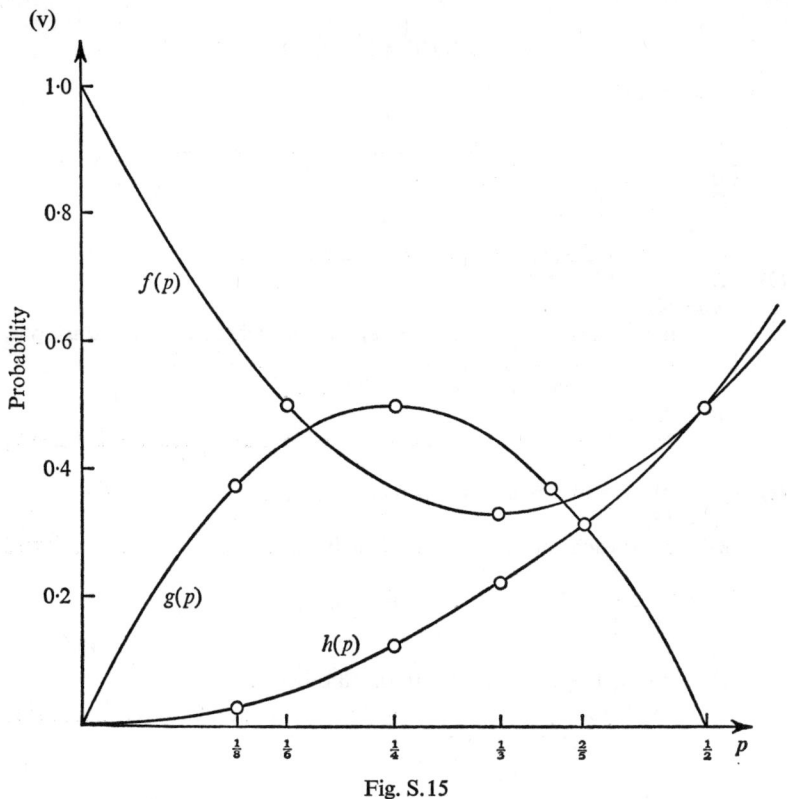

Fig. S.15

We seek the range of values of p between the roots of $f(p) = g(p)$; that is $0.185 < p < 0.387$.

63. Probability $= \dfrac{16 \times (\frac{11}{4} - 1)^2}{(11 - 1)^2} = 0.49$.

65. The events of falling in given parts are not independent.

Required probability $= (1 - 1/n)\,(1 - 2/n) \dots \left(1 - \dfrac{n-1}{n}\right)(1 - n/n) = 0$.

67. By consideration of *HHH, HHT, HTH, THH; HTT, THT, TTH, TTT* we see that:

pr (the third is the same as the pair | there is a pair of heads) $= \frac{1}{4}$

and   pr (the third is the same as the pair | there is a pair of tails) $= \frac{1}{4}$.

It follows that

pr (the third is the same as the pair | there is some pair) $= \frac{1}{4} \neq \frac{1}{4} + \frac{1}{4} = \frac{1}{2}$.

417

Exercise 4A (p. 116)

1. (*a*) $S =$ {plants mentioned},
    $N = 12$,
    m maps each plant onto the number of leaves it has,
    $[x] = (4, 4, 4, 4, 4, 4, 4, 4, 4, 4, 5, 7)$,
    $D = \{4, 5, 7\}$,
    $n = 3$,
    f: $4 \to 10$, $5 \to 1$, $7 \to 1$, $x \to 0$ for all other x.

  (*b*) $S =$ {months in 1964},
    $N = 12$,
    m maps each month onto the number of inches of rainfall during it,
    $[x] = (4\cdot2, 3\cdot4, 1\cdot8, 2\cdot4, 2\cdot6, 2\cdot4, 4\cdot5, 4\cdot2, 2\cdot4, 1\cdot3, 3\cdot8, 3\cdot6)$,
    $D = \{4\cdot2, 3\cdot4, 1\cdot8, 2\cdot4, 2\cdot6, 4\cdot5, 1\cdot3, 3\cdot8, 3\cdot6\}$,
    $n = 9$,
    f: $1\cdot3 \to 1, 1\cdot8 \to 1, 2\cdot4 \to 3, 2\cdot6 \to 1, 3\cdot4 \to 1, 3\cdot6 \to 1, 3\cdot8 \to 1, 4\cdot2 \to 2$,
       $4\cdot5 \to 1$, $x \to 0$ for all other x.

  (*c*) $S =$ {throws of a pair of coins mentioned},
    $N = 15$,
    m maps each throw onto 2 or 1 or 0, recording as the throw showed
       2 or 1 or 0 heads,
    $[x] = (2, 2, 2, 1, 1, 1, 1, 1, 1, 1, 0, 0, 0, 0, 0)$,
    $D = \{0, 1, 2\}$,
    $n = 3$,
    f: $0 \to 5$, $1 \to 7$, $2 \to 3$, $x \to 0$ for all other x.

Note that defining m by the method suggested in Section 2.3 (vi) is unsuitable
since we have not preserved a distinction between (0, 1) and (1, 0)

Exercise 4B (p. 124)

1. (i)

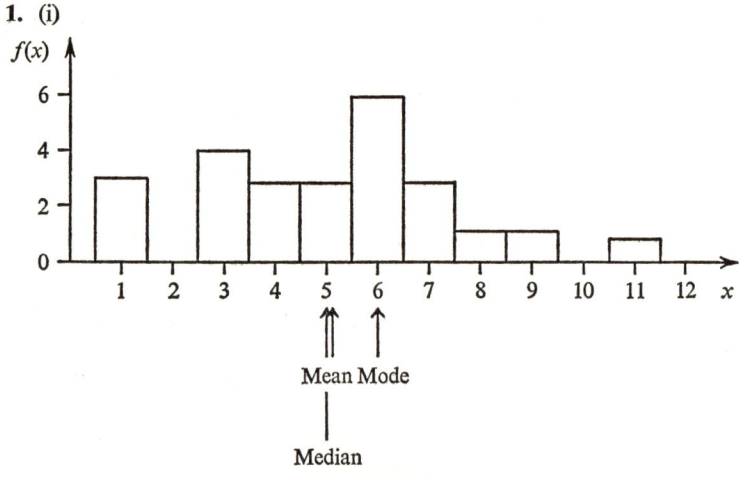

Fig. S.16

mean $[x] = 5\cdot1$ median $[x] = 5$ mode $[x] = 6$.

(ii)

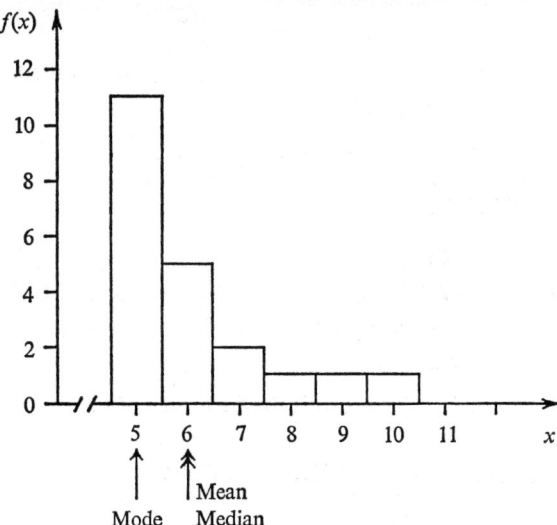

Fig. S.17

mean $[x] = 6$ median $[x] = 6$ mode $[x] = 5$.

3. (i) Stretch in x-direction; y-axis invariant; factor $= 2$.

    (ii) Translation along x-axis; magnitude $= 1$; direction: to the left.

    (iii) The transformation of part (ii) followed by: stretch in x-direction; y-axis invariant; factor $= \frac{1}{2}$.

5. (a) Mean of new sample $= \dfrac{(20 \times 3 \cdot 4) + (2 \times 4 \cdot 5)}{20 + 2} = 3 \cdot 5$.

    (b) $3 \cdot 4 \leqslant$ new median $\leqslant 4 \cdot 5$; we cannot sharpen this.

    (c) New mode is either $3 \cdot 4$ or $4 \cdot 5$; we cannot tell which, or whether both.

7. (a) Let the samples be $[x]$, $[y]$; then $m_1 = \dfrac{1}{N_1} \Sigma x_i$, $m_2 = \dfrac{1}{N_2} \Sigma y_i$;

    New mean $= \dfrac{\Sigma x_i + \Sigma y_i}{N_1 + N_2} = \dfrac{N_1 m_1 + N_2 m_2}{N_1 + N_2}.$

    (b) Samples of sizes N_i and means m_i are combined;

    New mean $= \dfrac{\overset{n}{\underset{1}{\Sigma}} N_i m_i}{\overset{n}{\underset{1}{\Sigma}} N_i}.$

This proves the result of Exercise 1 A, Question 1 (b); for if $N_i = N$, say, for all i, then new mean $= \dfrac{1}{n} \overset{n}{\underset{1}{\Sigma}} m_i =$ mean of the old means.

(c) Mean of the constructed population $= \dfrac{\Sigma x_i f(x_i)}{\Sigma f(x_i)}$, by definition

$$= \dfrac{\Sigma m_i N_i}{\Sigma N_i}, \quad \text{as required.}$$

Exercise 4C (p. 129)

1. (i) 101·08; (ii) 818·75.

3. $\bar{x} = \dfrac{\Sigma x_i f(x_i)}{\Sigma f(x_i)} = \dfrac{\Sigma (ct_i + k) f(x_i)}{\Sigma f(x_i)}$

$$= \dfrac{c\Sigma t_i f(x_i) + k\Sigma f(x_i)}{\Sigma f(x_i)}$$

$$= c\bar{t} + k, \quad \text{since } f(x_i) \text{ is the frequency of } t_i.$$

5. (i) $\dfrac{1}{N}\Sigma(x_i - k) f(x_i) + k = \dfrac{1}{N}\Sigma(x_i f(x_i) - kf(x_i)) + k$

$$= \dfrac{1}{N}\Sigma x_i f(x_i) - \dfrac{1}{N}\Sigma kf(x_i) + k$$

$$= \bar{x} - \dfrac{k}{N}\Sigma f(x_i) + k$$

$$= \bar{x}, \quad \text{since } N = \Sigma f(x_i);$$

(ii) $M_1(k) = 0 \Leftrightarrow \dfrac{1}{N}\Sigma(x_i - k) f(x_i) = 0 \Leftrightarrow \bar{x} - k = 0 \Leftrightarrow k = \bar{x}.$

Exercise 4D (p. 131)

1. $\bar{x} = 7$; mean value of $(x - \bar{x})^2 = \frac{35}{6}$.

3. For $n = 6$: mean value $= \frac{2}{6}(0 + \frac{3}{4} + \frac{3}{4}) = \frac{1}{2}$;
for $n = 9$: mean value $= \frac{1}{9}(0 + 2(0·413 + 0·970 + 0·750 + 0·117))$
$$= 0·50;$$
for $n = 12$: mean value $= \frac{2}{12}(0 + \frac{1}{4} + \frac{3}{4} + 1 + \frac{3}{4} + \frac{1}{4}) = \frac{1}{2}$.

Exercise 4E (p. 137)

1.

	(i)	(ii)
(a)	2	1·4
(b)	8	2·8
(c)	8	2·8
(d)	2	1·4
(e)	8	2·8
(f)	5	2·2
(g)	$6\frac{2}{3}$	2·6

3. For instance, the diagram for the frequency function whose only non-zero values are given by $f(-2) = 1, f(1) = 2$.

5. (a) These are results from experiments with no unique results.

(b) (i)

x	1	2	3	4	5	6
$f(x)$	12	12	12	12	12	12

mean $= 3 \cdot 5$; variance $= \frac{35}{12}$;

(ii)

x	1	1·5	2	2·5	3	3·5	4	4·5	5	5·5	6
$f(x)$	2	4	6	8	10	12	10	8	6	4	2

mean $= 3 \cdot 5$; variance $= \frac{35}{24}$.

(c) The means and variances in part (b) would be unchanged. The answers to part (a) would be likely to be closer to those of part (b) in case (i) and less close in case (ii).

7. (a) Let the common mean be m.

Denote the members of the first population by y_i ($i = 1, ..., N_1$) and the members of the second population by y_i ($i = N_1+1, ..., N_1+N_2$) where repetitions among the y_i are possible.

We have $v_1 = \dfrac{1}{N_1} \sum_{1}^{N_1} (y_i - m)^2; \quad v_2 = \dfrac{1}{N_2} \sum_{N_1+1}^{N_1+N_2} (y_i - m)^2;$

$$v = \frac{1}{N_1+N_2} \sum_{1}^{N_1+N_2} (y_i - m)^2 = \frac{1}{N_1+N_2} \left(\sum_{1}^{N_1} (y_i - m)^2 + \sum_{N_1+1}^{N_1+N_2} (y_i - m)^2 \right)$$

$$= \frac{N_1 v_1 + N_2 v_2}{N_1 + N_2}.$$

(b) Consider the populations:

First population $= (0)$; $N_1 = 1$, mean $= 0$, variance $= 0$.
Second population $= (2)$; $N_2 = 1$, mean $= 2$, variance $= 0$.

Then combined population $= (0, 2)$ and we have:

$$v = 1 \neq \frac{(1 \times 0) + (1 \times 0)}{1 + 1}.$$

It is possible for the relationship to break down.

(c) $S = \sqrt{\left(\dfrac{N_1 S_1^2 + N_2 S_2^2}{N_1 + N_2} \right)}.$

9. This result is proved in Section 10.2.

11. Hints only:
    (a) a translation;
    (b) a translation;
    (c) a reflexion of (b) in $x = 5$;
    (d) a stretch applied to (b) ($x = 0$ invariant);

(e) a stretch applied to (a) ($x = 6$ invariant) *or* a stretch applied to (b) ($x = 0$ invariant), followed by a translation;

(f) a stretch applied to (a) ($x = 0$ invariant).

Exercise 4F (p. 142)

1. (i) $[x'] = [x]$; mean $[x'] = 2\frac{1}{2}$;
    var $[x]$ = var $[x'] = \frac{1}{4}(1^2+2^2+3^2+4^2)-6\frac{1}{4} = 1\frac{1}{4}$; stad $[x] = 1\cdot1$.

    (ii) $[x'] = (-3, -2, -1, 0, 1, 2, 3, 4)$; mean $[x'] = \frac{1}{2}$;
    var $[x] = 5\cdot25$; stad $[x] = 2\cdot3$.

    (iii) $[x'] = (1, 2, 3, ..., (n-1), n)$; mean $[x'] = \frac{1}{2}(n+1)$;

    $$\text{var } [x] = \frac{1}{n} \cdot \frac{n(n+1)(2n+1)}{6} - \frac{(n+1)^2}{4} = \frac{1}{12}(n^2-1);$$

    $$\text{stad } [x] = \sqrt{(\frac{1}{12}(n^2-1))}.$$

    (iv) var $[x] = 296/13^2$; stad $[x] = 1\cdot3$.

    (v)

x	$f(x)$	x'	$x'f(x)$	$(x')^2 f(x)$
69	3	-1	-3	3
70	7	0	0	0
71	6	1	6	6
72	4	2	8	16
	20		$-3+14$	25
			$= 11$	

    $$\text{var } [x] = \tfrac{25}{20} - (\tfrac{11}{20})^2 = 379/20^2; \text{ stad } [x] = 0\cdot97.$$

    (vi) var $[x] = 37900/20^2$; stad $[x] = 9\cdot7$.

Exercise 4G (p. 144)

1.
$$N \times \text{variance} = \Sigma(t_i - \bar{t})^2 f(t_i)$$
$$= \Sigma(t_i^2 - 2t_i\bar{t} + \bar{t}^2) f(t_i)$$
$$= \Sigma t_i^2 f(t_i) - 2\bar{t}\Sigma t_i f(t_i) + \bar{t}^2 \Sigma f(t_i)$$
$$= \Sigma t_1^2 f(t_i) - 2\bar{t}.N\bar{t} + \bar{t}^2.N.$$

Hence,
$$\text{variance} = \frac{1}{N}\Sigma t_i^2 f(t_i) - \bar{t}^2.$$

3. Full answers not given. Check: the sum of the variances is $71\frac{3}{4}$.

5. (i)
$$\text{Mean } [m_i] = \frac{1}{k}\sum_{i=1}^{k} m_i = \frac{1}{k}\sum_{i=1}^{k}\left(\frac{1}{l}\sum_{j=1}^{l} x_{ij}\right)$$
$$= \frac{1}{kl}\Sigma\Sigma x_{ij}$$
$$= m.$$

422

(ii)   Mean $[v_i] = \dfrac{1}{k} \sum\limits_{i=1}^{k} v_i$

$$= \frac{1}{k} \sum_{i=1}^{k} \left\{ \frac{1}{l} \sum_{j=1}^{l} (x_{ij} - m)^2 - (m_i - m)^2 \right\}$$

$$= \frac{1}{kl} \sum_{j} \sum_{j} (x_{ij} - m)^2 - \frac{1}{k} \sum_{i=1}^{k} (m_i - m)^2.$$

But $\text{mean } [m_i] = m \Rightarrow \dfrac{1}{k} \sum\limits_{i=1}^{k} (m_i - m)^2 = \text{var } [m_i]$

and $\dfrac{1}{kl} \sum\limits_{i} \sum\limits_{j} (x_{ij} - m)^2 = \text{var } [x_{ij}].$

So $\text{mean } [v_i] = \text{var } [x_{ij}] - \text{mean } [m_i].$
The result follows.

7. (ii) Follows from the definition of V;

(iii) $v_1 - (m_1 - M)^2$; $v_2 - (m_2 - M)^2$;

(iv) $m_2 = 6\cdot116$; $v_2 = 0\cdot149784$; $M = 6\cdot0472$;

(v) $V = 0\cdot093860152$.

Final result   mean $= 6\cdot12$; S.D. $= 0\cdot31$.

9. (i) $M = m_1 R_1 + m_2 R_2$ is immediate.

We note also that $R_1 + R_2 = 1.$

It follows that $m_1 - M = m_1(1 - R_1) - m_2 R_2$

$$= R_2(m_1 - m_2).$$

Similarly $m_2 - M = R_1(m_2 - m_1).$

Then $\dfrac{N_1(m_1 - M)^2 + N_2(m_2 - M)^2}{N_1 + N_2}$

$$= R_1 . (R_2(m_1 - m_2))^2 + R_2(R_1(m_2 - m_1))^2$$

$$= R_1 R_2(m_1 - m_2)^2, \quad \text{using again } R_1 + R_2 = 1.$$

The result for V follows.

(ii) Since $R_1 R_2(m_1 - m_2)^2$ will only vanish when $m_1 = m_2$, the relation of Exercise 4E, Question 7(b) will always break down when the means are unequal.

Exercise 4H (p. 149)

1. (i) $\bar{x} = \bar{y} = 0$.

(ii) The values are

$$\tfrac{1}{5}(10p^2 - 26p + 18) \text{ or } 2\{(p - \tfrac{13}{10})^2 + \ldots\}$$

and $\tfrac{1}{5}(18r^2 - 26r + 10) \text{ or } \tfrac{18}{5}\{(r - \tfrac{13}{18})^2 + \ldots\}.$

(iii) The required values are $p = \tfrac{13}{10}$ and $r = \tfrac{13}{18}$.

(iv) The required graphs are of $y = 13x/10$ and $x = 13y/18$.

3. The figures are, of course, useless for predicting tobacco consumption. Even the repeated occurrence together of two phenomena does not imply that one *causes* the other—consider the case of two accurate striking clocks. To investigate causation we would need to examine the effect of altering one of the phenomena to see whether the other would respond.

The statistical situation is well discussed by R. A. Fisher in *Smoking: The Cancer Controversy* [Oliver and Boyd, 1959]. Remember however that much evidence has been gathered since 1959 on such things as the effects on the cancer-incidence in a group of doctors of *altering* the tobacco consumption.

5.
$$\text{cov}\,[(x, y)] = \frac{1}{N}\Sigma(x_i - \bar{x})\,(y_i - \bar{y})$$

$$= \frac{1}{N}\Sigma(x_i y_i - \bar{x} y_i - \bar{y} x_i + \bar{x}\bar{y})$$

$$= \frac{1}{N}\Sigma x_i y_i - \bar{x}\frac{1}{N}\Sigma y_i - \bar{y}\frac{1}{N}\Sigma x_i + \bar{x}\bar{y}.\frac{N}{N}$$

$$= \frac{1}{N}\Sigma x_i y_i - \bar{x}\bar{y} - \bar{y}\bar{x} + \bar{x}\bar{y}$$

$$= \frac{1}{N}\Sigma x_i y_i - \bar{x}\bar{y}.$$

7. From the methods of Question 1 the lines are $y = 2$ and $x = 0$.

Exercise 4I (p. 153)

1.

K	J	Y	A	Z	B	M	U	C	V	S
16	0	—	0	—	0	1·375	4	1·890625	2·109375	1·4524
	1	3	3	9	9					
	2	0	3	0	9					
	3	1	4	1	10					
	4	0	4	0	10					
	5	0	4	0	10					
	6	0	4	0	10					
	7	1	5	1	11					
	8	1	6	1	12					
	9	5	11	25	37					
	10	1	12	1	38					
	11	4	16	16	54					
	12	2	18	4	58					
	13	1	19	1	59					
	14	2	21	4	63					
	15	0	21	0	63					
	16	1	22	1	64					

3. The numbers are stored on a tape alternately x- and y-coordinates.

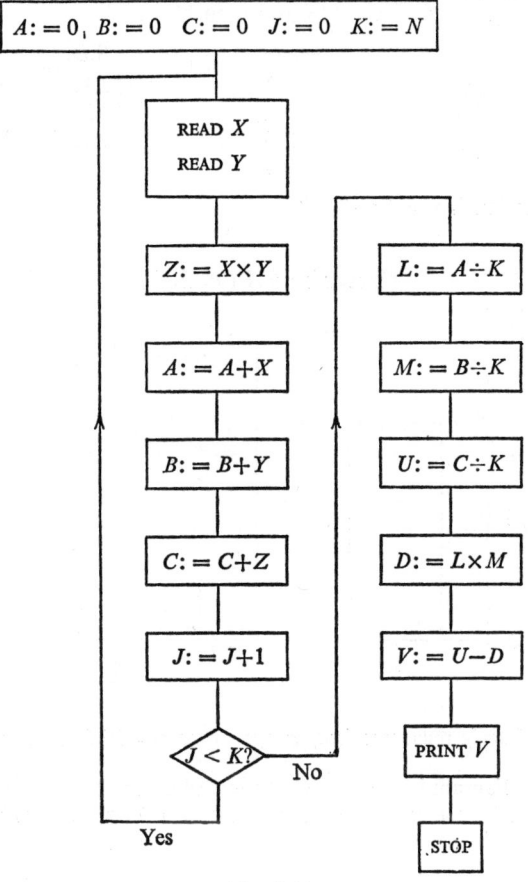

Fig. S.18

Exercise 4J (p. 158)

1. (i)

	Variance	S.D.
(a), (b), (c)	12·0	3·46
(d)	4 × 12·0	2 × 3·46
(e)	9 × 12·0	3 × 3·46
(f)	$\frac{1}{100}$ × 12·0	0·346

(ii) $\frac{1}{12}(n^2-1)\cdot\left(\dfrac{1}{n-1}\right)^2 = \dfrac{1}{12}\left(\dfrac{n+1}{n-1}\right),$

$\dfrac{1}{12}\cdot\left(\dfrac{n+1}{n-1}\right)\cdot\left(\dfrac{n-1}{n}\right)^2 = \dfrac{1}{12}\left(1-\dfrac{1}{n^2}\right).$

2. [The answers to this Question, and *not* to Question 3, are given.]
The S.D. are given to four S.F. as a better check for the reader, and not because it is suggested that the data would warrant such precision.

	Mean	S.D.
(a)	16·3	12·32
(b)	163	10·95
(c)	2·6	0·4320
(d)	0·4	0·3838

The transformation is $y = (x - 3\frac{1}{2})/8$.

Exercise 5 A (p. 168)

1. (i) (a)

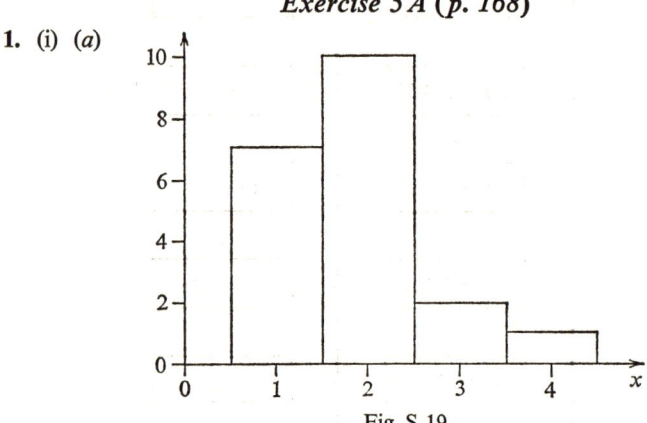

Fig. S.19

Vertical scale measures frequency;   Mean = 1·85; Median = 2.

(b) The diagram is the same, but the vertical scale measures journeys per hour-of-length-of-journey. Approximate mean = 1·85; approximate median $= 1·5 + \frac{3}{10} \times 1 = 1·8$.

(ii) (a)

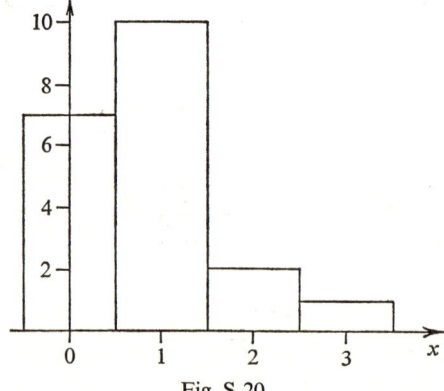

Fig. S.20

Vertical scale measures frequency;   Mean = 0·85; Median = 1.

(b)

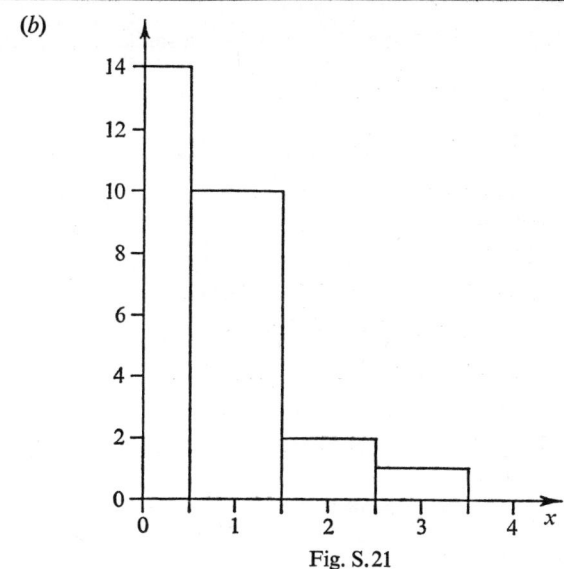

Fig. S.21

Vertical scale measures journeys per hour-of-length-of-journey;

$$\text{Approximate mean} = \frac{(7 \times \frac{1}{4}) + (10 \times 1) + (2 \times 2) + (1 \times 3)}{20} = 0{\cdot}94;$$

Approximate median $= 0{\cdot}8$.

3. Mean height at which bar fell $= (170 - \frac{21}{90} \times 2{\cdot}5)$ cm;

Approximate mean maximum height reached by jumpers $= (170 - \frac{21}{90} \times 2{\cdot}5$
$$- \tfrac{1}{2} \times 2{\cdot}5) \text{ cm}$$
$$= 168{\cdot}2 \text{ cm}.$$

5. Approximate mean from (i) $15\frac{1}{2}$; from (ii) $15{\cdot}0$.

Comparisons: (a) fixed, while (b) varies.

By moving the classes to the right from (i) to (ii) we have lowered the approximating mean; largely because of the effect of $h(21) = 3$.

(b) fixed, while (a) varies.

By shortening the length of interval from Example 1 to part (i) of this Question, we have, for no obvious reason, raised the approximating mean.

In all these cases the approximations are larger than the true mean.

7. Let h be the largest class-interval.

In formally we may proceed as follows: All the readings are within $\frac{1}{2}h$ from their nearest classmark, so that the 'mean' calculated by replacing each reading by its nearest classmark will be within $\frac{1}{2}h$ of the true mean.

Alternatively we may argue: Let μ be the 'mean' calculated from the table of grouped data. By adding $\frac{1}{2}h$ to each classmark and using these new numbers in place of the class marks we would calculate a quantity $\mu + h/2$; but each of the values of (classmark $+ \frac{1}{2}h$) will exceed all the readings in the class from which it arose, and by at most h, so the value of $\mu + h/2$ will exceed the true mean by at most h.

Let the true mean be m.

In symbols: $\mu + \dfrac{h}{2} \leqslant m + h$, whence $\mu - \dfrac{h}{2} \leqslant m$.

By subtracting $h/2$ from each classmark we obtain in a similar way $m \leqslant \mu + \dfrac{h}{2}$.

Whence
$$\mu - \frac{h}{2} \leqslant m \leqslant \mu + \frac{h}{2}$$

or
$$|\mu - m| \leqslant \frac{h}{2}, \quad \text{the required result.}$$

Finally, we may proceed quite formally.

Let the ith class mark be X_i and let the frequency of the ith class be q_i. Write $N = \sum_i q_i =$ total frequency.

Let $x_{ij}(j = 1, \ldots q_i)$ be the members of the ith class.

Then
$$X_i - \tfrac{1}{2}h \leqslant x_{ij} \leqslant X_i + \tfrac{1}{2}h,$$

and so
$$(X_i - \tfrac{1}{2}h)\, q_i \leqslant \sum_{j=1}^{q_i} x_{ij} \leqslant (X_i + \tfrac{1}{2}h)\, q_i.$$

Furthermore
$$\frac{1}{N}\sum_i (X_i - \tfrac{1}{2}h)\, q_i \leqslant \frac{1}{N}\sum_i \sum_{j=1}^{q_i} x_{ij} \leqslant \frac{1}{N}\sum_i (X_i + \tfrac{1}{2}h)\, q_i.$$

It follows that
$$\mu - \tfrac{1}{2}h\frac{\Sigma q_i}{N} \leqslant m \leqslant \mu + \tfrac{1}{2}h.\frac{\Sigma q_i}{N},$$

which gives the required result.

9.

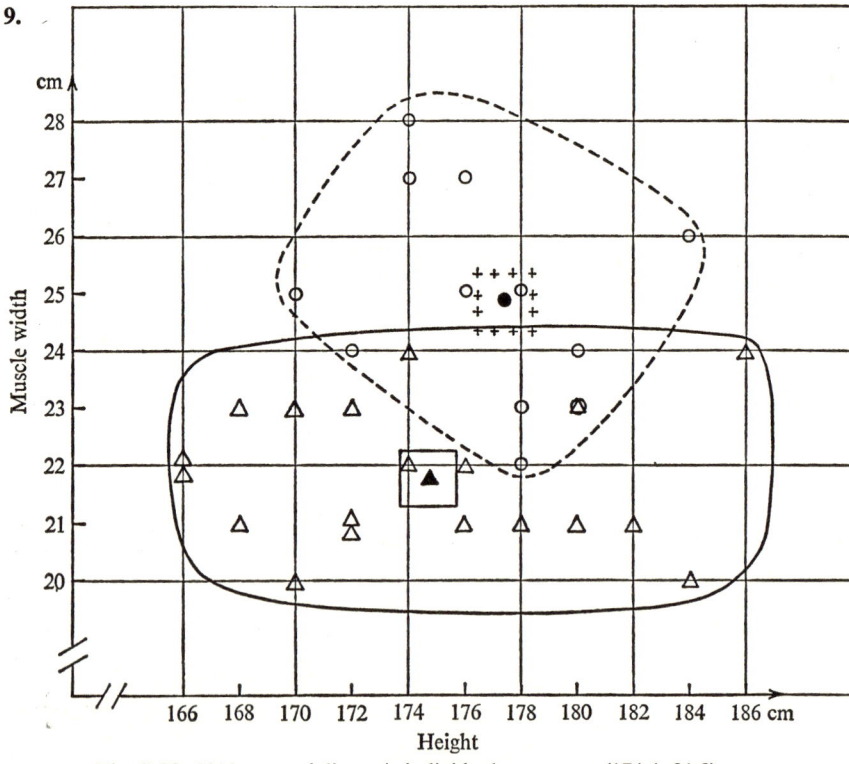

Fig. S.22. 5000 m specialists: △ individuals, ▲ mean (174·4, 21·8);
100 m specialists: ○ individuals, ● mean (176·7, 24·9).

Exercise 5B (p. 172)

1. (i) Approximation to mean is $\frac{435}{48} = 9·06$ in each case.
Approximation to variance are (a) 37, (b) 26.

(ii) If any population of two *distinct* members is grouped at their mean then the approximation to the mean is the true mean, but the 'approximation' to the variance is zero, which is not the original variance.

Exercise 5C (p. 173)

1. ...populations.........table....

3. ...frequency function....

5. ...values of the variable....

7. ...tables....

9. ...density diagrams....

11. ...frequency density function....

13. In this sentence we would replace 'frequency distributions' by 'finite populations'.

Exercise 6A (p. 178)

1. (i)

Pop.	S.D.	Proportion
1st	0·535	100%
2nd	$\frac{1}{2}$	$\frac{3}{4} = 75\%$
3rd	0·472	78%

(ii) $f(-1) = 1$; $f(0) = 16$; $f(1) = 1$.

3. Choose mean as origin, standard deviation as unit.

5. (i) (a)

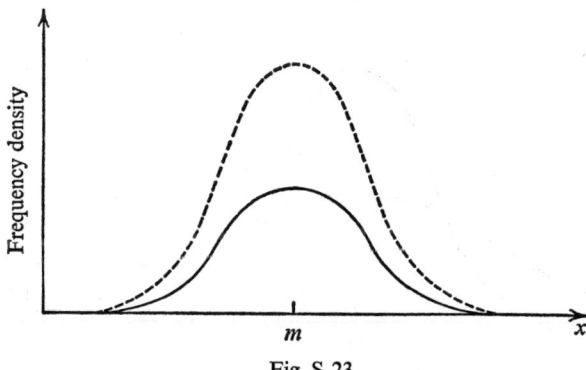

Fig. S.23

(b)

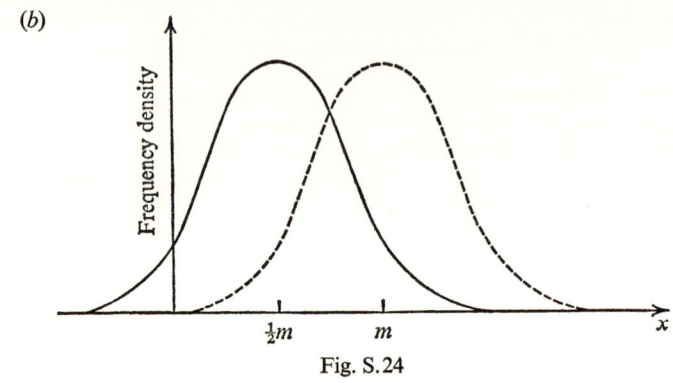

Fig. S.24

(c)

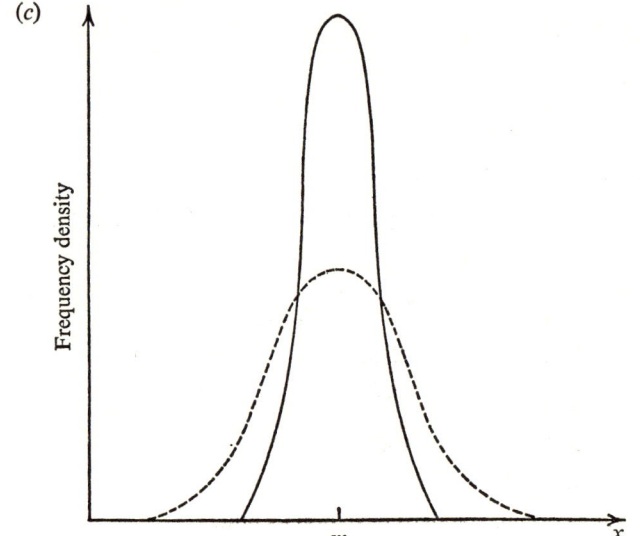

Fig. S.25

(d)

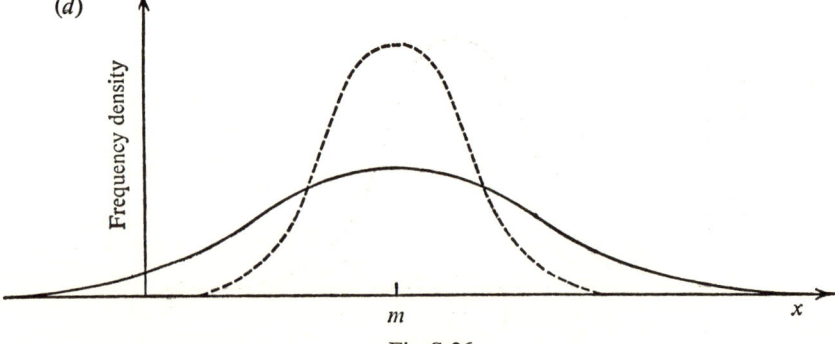

Fig. S.26

(ii) (a) A stretch: $f(x) = 0$ invariant; factor $\frac{1}{2}$.

     (b) A translation to the left, of magnitude $\frac{1}{2}$m.

     (c) A stretch: $f(x) = 0$ invariant; factor 2;
       with a stretch: $x = m$ invariant; factor $\frac{1}{2}$.

     (d) A stretch: $f(x) = 0$ invariant; factor $\frac{1}{2}$;
       with a stretch: $x = m$ invariant;   factor 2.

7. (i)

	Mean	S.D.	Coeff. of var.
(a)	7	$\sqrt{2}$	20
	70	$5\sqrt{2}$	2×20
(b)	7	$\sqrt{2}$	20
	30	$\sqrt{2}$	$\frac{7}{30} \times 20$
(c)	1	$\sqrt{2}$	140
	2	$\sqrt{2}$	$\frac{1}{2} \times 140$
(d)	7	$\sqrt{2}$	20
	280	$\sqrt{2}$	$\frac{1}{40} \times 20$
(e)	86	$9\sqrt{2}$	15
	$\frac{5}{9}(86-32)$	$\frac{5}{9} \times 9\sqrt{2}$	$\dfrac{86}{86-32} \times 15$
(f)	141	$\sqrt{2}$	1·003
	70	$\frac{1}{2}\sqrt{2}$	1·01
(g)	7	$\sqrt{2}$	20
	51	$13\sqrt{2}$	36

Parts (c) and (d) remind us that when the zero of measurement is arbitrary the magnitude of the mean can be altered without altering the standard deviation. This makes the coefficient of variation useless in such cases as occur in (b), (c), (d), (e). It is a helpful statistic in (a), (f), (g).

(ii) The transformations are those involving only a multiplication of the values of the variable by a fixed number—such as occur with change of units.

Exercise 6 B (p. 183)

1.

Pop.	S^2	S	$m-2S$	$m-S$	m	$m+S$	$m+2S$
1	1·3	1·14	−0·3	0·86	2	3·14	4·3
2	$1·3+(1·5)^2$	1·88	1·7	3·62	5·5	7·38	9·3
3	$1·3+(2·5)^2$	2·74	0	2·76	5·5	8·24	11
4	2·3	1·51	1·0	2·48	4	5·52	7·0

Proportions within given ranges:

Pop.	Inner range	Outer range
1	$\dfrac{57·6+50+25·6}{200} = 67\%$	$\dfrac{90+50+40+8}{200} = 94\%$
2	$\dfrac{2(90+44)}{400} = 67\%$	$\dfrac{2(90+50+40+8)}{400} = 94\%$
3	$\dfrac{2(20+40+37)}{400} = 48\%$	100%
4	$\dfrac{0·8+90+10+1·0}{200} = 51\%$	$\dfrac{200-5}{200} = 97\frac{1}{2}\%$

Exercise 6M (*p. 185*)

1. Mean = 0·61. Variance = 0·61 (unrounded result is 0·6079).

3.

	Mean	S.D.
Authorized (or King James) Version	4·06	1·92
New English Bible	3·88	1·75

It is doubtful whether the differences in means is significant, taking into account the variability indicated by the standard deviations. A fuller discussion of this type of problem requires techniques for the testing of hypotheses.

5. The effects of such a transformation are exhibited.
The logarithms selected have base = 4.

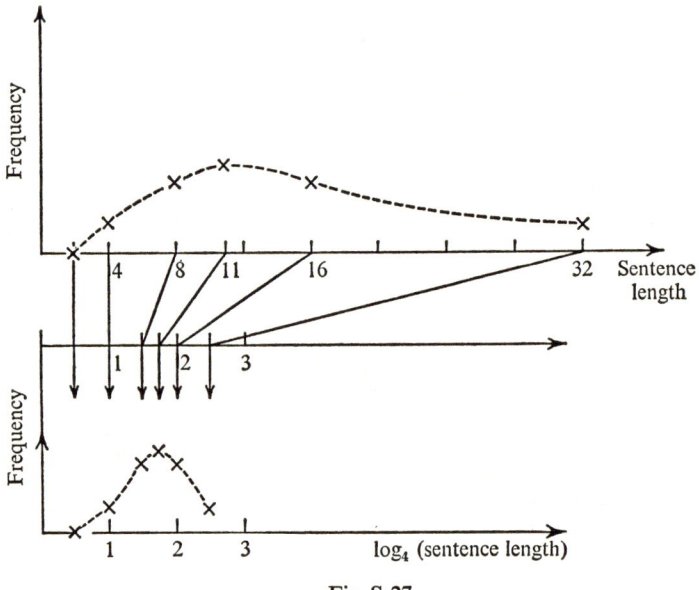

Fig. S.27

7. The first and fifth *sextiles* are 'one sixth of the way in from each end'.
First sextile = $1·6 + \frac{11}{23}(1·8 - 1·6) = 1·7$.
Fifth sextile = 2·2.
If 68 per cent of a population lies between $m - S$ and $m + S$, then about 17 per cent (or $\frac{1}{6}$) lies to the left of $m - S$; thus $m - S$ is nearly the first sextile. Similarly $m + S$ is nearly the fifth sextile. An approximation to the standard deviation could be obtained by halving the difference between these sextiles.
In this problem we might take standard deviation $\simeq \frac{1}{2}(2·2 - 1·7) = 0·25$.

The approximation obtained from Example 1 is 0·249.

9. (i) The data needed for the calculations and to draw the histogram are as shown below:

Interval	Interval length	Class-mark	Frequency	Frequency density
0–0·05	0·05	0·025	9	180
0·05–0·25	0·2	0·15	11	55
0·25–0·75	0·5	0·5	13	26
(0–0·75	0·75	0·375	33	44)
0·75–2·75	2	1·75	25	12·5
2·75–5·25	2·5	4	17	6·8
5·25–10·25	5	7·75	14	2·8
10·25–15·25	5	12·75	6	1·2
15·25–25·25	10	20·25	5	0·5

We also have

y	$\frac{1}{2}$	1	2	4	8	16
$\phi(y)$	26·6	17·7	11·0	6·0	2·6	0·7

(ii) $\bar{y} \simeq 4·06$; var $[y] \simeq 25·7$.

(iii) The proportion is 83 per cent.

The variance of a sample is calculated from the deviations from its own mean, thus even a sample of two fairly extreme values will not have a particularly large variance if both values are on the same side of the population mean. About half the samples with extreme values will show this effect. There is thus a preponderance of samples with variances smaller than the variance of the parent population.

If the mean square deviation of the values of the sample from the mean of the *parent* population is calculated, this effect disappears and the mean of the resulting population of statistics is the variance of the parent population.

The variance of the sample is said to be a *biased estimator* of the variance of the parent population, while the second statistic mentioned above is said to be an *unbiased estimator*. The matter is treated at length in Chapters 11 and 12.

11.

	m	S
Males	2·00	0·77
Females (all)	1·50	1·01
Females (1st group)	0·67	0·79
Females (2nd group)	2·00	0·77

The genetic constitutions of colonies of *Maniola jurtina*, of which the spots are symptoms, have been extensively studied by W. H. Dowdeswell and E. B. Ford.

13. No answers meaningful.

15. No answers meaningful.

17.

	Skewness	Kurtosis
f	0	2·08
g	0	3·94
h	−0·69	2·78
k	0	—

The warning delivered by the skewness of k also sounds in Question 18.

19. It is a further advantage of the third layout of column headings in Chapter 4, Section 10.3 that, by appending one further column to those needed anyway to calculate the skewness, we can employ the identity:

$$\Sigma x_i^2(x_i+1)\,f(x_i) = \Sigma x_i^3 f(x_i) + \Sigma x_i^2 f(x_i).$$

Exercise 7A (p. 193)

1. Attributes: poker dice, liar dice (in both games the attributes are thought of as *ordered*, but they are not calculated with), crown and anchor.
   Numbers: ludo, snakes-and-ladders, 'Monopoly'.

3. (i) No. The words are classified into more than two categories.
   (ii) Yes. Each word either is or is not of length 2 letters. If we record a success for each word of 2 letters we are carrying out a dichotomous trial and the answer we want is the number of successes.
   (iii) No. Each letter either is or is not letter 'e'; but the record of the experiment demands more than the number of successes in finding 'e'. The *words* are classified into more than two categories.
   (iv) Yes. Each word either does or does not contain exactly 2 letters 'e'. If we record a success for each word with 2 letters 'e' we are carrying out a dichotomous trial and the answer we want is the number of successes.

Exercise 7B (p. 202)

1. No answers meaningful.

Exercise 7C (p. 211)

1.

	(a)	(b)	(c)
	0001	0011	0111
	0010	0101	1011
	0100	1001	1101
	1000	0110	1110
		1010	
		1100	

16 arrangements are possible.

3. 2^5 is the number of different arrangements in a line of 5 objects any one of which may be either a 'zero' or a 'one'.

5. 2^4 is the number of subsets of a set of 4 objects; the empty set and the full set of 4 objects being counted.
   The subsets of $\{A, B, C, D\}$ are:

$$\varnothing$$

$$\{A\}, \{B\}, \{C\}, \{D\}$$

$$\{A, B\}, \{A, C\}, \{A, D\}, \{B, C\}, \{B, D\}, \{C, D\}$$

$$\{A, B, C\}, \{A, B, D\}, \{A, C, D\}, \{B, C, D\}$$

$$\{A, B, C, D\}.$$

7. An experiment in which two coins are thrown and the results classified as: 0 heads; 1 head; 2 heads. (See Exercise 4A, Question 1(c).)

9. (i) In the round-table problem sense of rotation matters, because the hostess' right- and left-sides differ; but there is no 'head of the table'.
   (a) 6!; (b) 6! − 2 × 5!.
   (ii) In the necklace problem sense can be reversed by turning the necklace over.
   (a) 360; (b) 240.

Exercise 7D (p. 221)

1. (i) $1 + 6\xi + 15\xi^2 + 20\xi^3 + 15\xi^4$,

   $1 + 7\xi + 21\xi^2 + 35\xi^3 + 35\xi^4$,

   $1 + 3\xi + 6\xi^2 + 10\xi^3 + 15\xi^4$,

   $1 + 4\xi + 10\xi^2 + 20\xi^3 + 35\xi^4$,

   $\frac{64}{729} + \frac{64}{243}\xi + \frac{80}{243}\xi^2 + \frac{160}{729}\xi^3 + \frac{20}{243}\xi^4$,

   $\frac{128}{2187} + \frac{448}{2187}\xi + \frac{224}{729}\xi^2 + \frac{560}{2187}\xi^3 + \frac{280}{2187}\xi^4$,

   $\frac{8}{27}\xi^3 + \frac{8}{27}\xi^4 + \frac{16}{81}\xi^5 + \frac{80}{729}\xi^6 + \frac{40}{729}\xi^7$,

   $\frac{16}{81}\xi^4 + \frac{64}{243}\xi^5 + \frac{160}{729}\xi^6 + \frac{320}{2187}\xi^7 + \frac{560}{6561}\xi^8$.

   (ii) $28 = \binom{8}{2} = \binom{8}{6} = \binom{-3}{6} = \binom{-7}{2}$.

   (iii) $1001 = 7 \times 11 \times 13$,

   $\binom{14}{6} = \left(\frac{14-6+1}{6}\right)\binom{14}{5}$; $\binom{14}{5} = \left(\frac{14-5+1}{5}\right)\binom{14}{4}$,

   $\binom{14}{4} = 2 \times 1001$.

   hence the numbers are 2002, 4004, 6006.

   (iv) $\binom{p}{r} = \frac{p}{1.2.3.....r}(p-1)(p-2)\ldots(p-r+1)$

   but none of 1, 2, 3, ..., r divides p, since it is prime and $\binom{p}{r}$ is an integer, so $\binom{p}{r}$ has a factor p.

   Even for $1 < r < p-1$ the converse is false; for $\binom{8}{3} = 56 = 8 \times 7$.

   (v) With slide-rule or 4-figure tables we get $3 \cdot 92 \times 10^6$.

   (vi) $\binom{22}{10} = \left(\frac{22-10+1}{10}\right)\binom{22}{9} = 646{,}646$,

   $\binom{22}{11} = \left(\frac{22-11+1}{11}\right)\binom{22}{10} = 705{,}432$,

   $\binom{23}{13} = \binom{23}{10} = \binom{22}{9} + \binom{22}{10} = 1{,}144{,}066$.

(vii) $$_{100}C_{21} = \tfrac{80}{21} \times 5 \cdot 36 \times 10^{20} = 2 \cdot 04 \times 10^{21}.$$

Now $$_{101}C_{20} = {}_{100}C_{19} + {}_{100}C_{20},$$

but $$_{100}C_{19} = \tfrac{20}{81} \times 5 \cdot 36 \times 10^{20} = 1 \cdot 32 \times 10^{20},$$

hence $$_{101}C_{20} = 6 \cdot 68 \times 10^{20}.$$

3. For each value of N the upper row has $\theta = \tfrac{1}{2}$, the lower has $\theta = \tfrac{1}{10}$.

N \ r	0	1	2	3	4	5	6	7, 8, 9, 10
2	0·250	0·500	0·250					
	0·810	0·180	0·010					
3	0·125	0·375	0·375	0·125				
	0·729	0·243	0·027	0·001				
5	0·031	0·156	0·312	0·312	0·156	0·031		
	0·590	0·328	0·073	0·008	0·000	0·000		
10	0·001	0·010	0·044	0·117	0·205	0·246	0·205	Symmetrically
	0·349	0·387	0·194	0·057	0·011	0·001	0·000	Zero.

Graphs similar to those required can be found in Chapter 8, Figure 8.1.

5. (i) Read the upper line for $N = 10$ in the answer to Question 3.
(ii) For $d \geqslant 1$ the probability of being right $\geqslant 0 \cdot 65$;
for $d \geqslant 3$ the probability of being wrong $\leqslant 2\%$, nearly.
(iii) For $N \leqslant 3$ the probability of being right $\geqslant 94\%$;
for $N \leqslant 2$ the probability of being wrong $\leqslant 1\%$, nearly.
(iv) $0 \cdot 234$; $0 \cdot 410$; $0 \cdot 246$.
The most likely split is $6:4$.

7. (i) Take course B; the ratio of the probabilities is $150:169$.
(ii) Yes. The ratio is now $216:169$.

9. (i) The probabilities are $0 \cdot 312$ and $0 \cdot 252$. The first is more likely.
Comment: The feats of getting 3 successes with probability $\tfrac{2}{4}$ and of getting 2 successes with probability $\tfrac{3}{4}$ are *not* the same (for four throws the probabilities are $0 \cdot 250$, $0 \cdot 211$), and they are not made the same by manipulating the tails of a probability diagram in this way. To think otherwise was the Chevalier de Méré's error; see Exercise 3 M, Question 40.
(ii) The probabilities are now $0 \cdot 145$ and $0 \cdot 114$. The first is, still, more likely.
Comment: The difficulty is altered by doubling the number of successes required and doubling the number of throws offered. Even if the 'tails' are not taken into account the probabilities are altered to $0 \cdot 109$, $0 \cdot 086$.

11. 58%. Use $(1 - \tfrac{1}{100})^{143} \simeq e^{-1 \cdot 43}$.

13. $0 \cdot 00010$; $0 \cdot 0017$; $0 \cdot 012$; $0 \cdot 055$.
Atkins' firm is working at a probability level of about 5 per cent (allowing about four weeks to the month). It installs a stand for 6 umbrellas and is distressed on about 5·5 per cent of days, when there are 7 or more umbrellas brought.

15. 0·104.

17. (i) 27%. (ii) No. 12 butterflies from the given type of habitat would show as few or fewer with spots on about 11 per cent of occasions.

19. Take pr (a mark exceeds the median) = $\frac{1}{2}$ and suppose that the trials are independent although there are only 300 candidates.

(i) Required probability = 14%. *Comment:* The performance of these pupils is not outstanding, about $\frac{1}{7}$ of all random selections of 8 of the pupils would show results as good or better than these.

(ii) Use table of Question 3 to find pr (7, 8, 9, 10 or 0, 1, 2, 3 pupils exceeding the median) = 34%. *Comment:* the results are not at all extreme.

21.

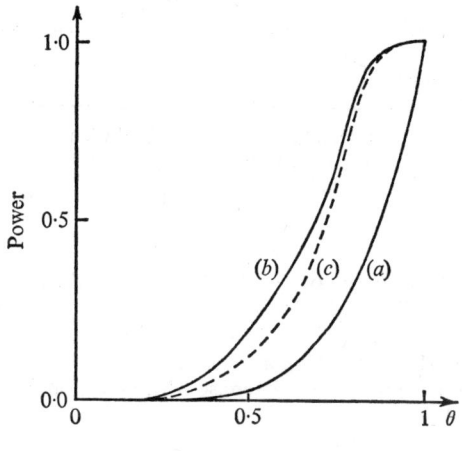

Fig. S.28

(i) In this case common-sense tells you you must not take too hard a line, and test (*a*) is thus least suitable.

You wish to avoid rejecting the dinghy when it is sound; that is: to avoid accepting the null hypothesis when it is false. You wish to avoid Type II error. Choose test (*b*), since it minimizes $1-p(\theta)$ for $\theta > \frac{1}{2}$. The snag is that when $\theta \leqslant \frac{1}{2}$, $p(\theta)$ is fairly large (until $\theta < 0.3$), and to reject the null hypothesis when $\theta < \frac{1}{2}$ is to commit a Type I error, that is to 'appear a mug'. (*c*) avoids this.

(ii) In this case common-sense tells you you must take a hard line, and test (*a*) seems best.

You wish to avoid accepting the dinghy when it is unsound; that is: to avoid rejecting the null hypothesis when it is true. You wish to avoid Type I error. Choose test (*a*), since it minimizes $p(\theta)$ for $\theta < \frac{1}{2}$.

The snag is that when $\theta > \frac{1}{2}$, $1-p(\theta)$ is fairly large, and to accept the null hypothesis when it is false is to commit a Type II error, that is to commit the more subtle error of looking every gift horse in the mouth. (*c*) avoids this.

(iii)

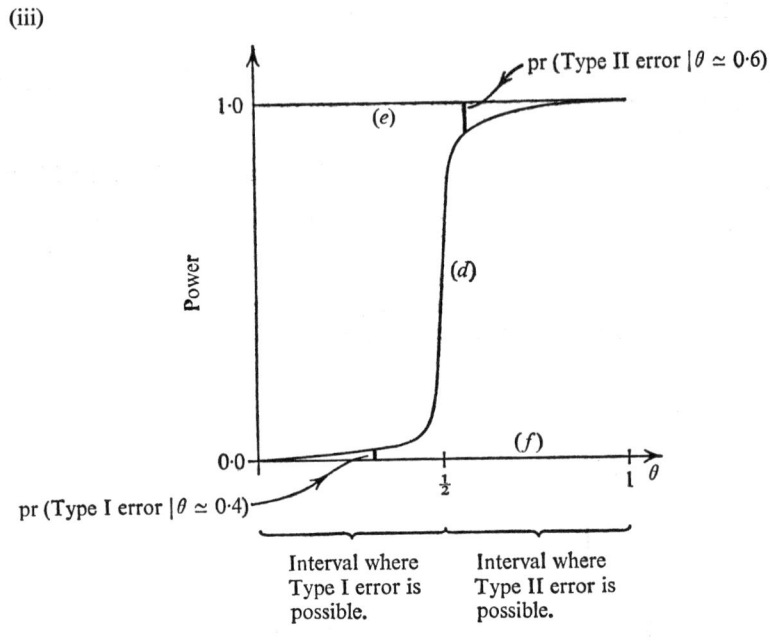

Fig. S.29

(c) gets nearer (d) than do (a) or (b), by using more throws. The diagram shows two typical probabilities of relevance.

(e) is liable to Type I error—being a 'mug'.

(f) is liable to Type II error—being a 'cynic'.

[If the reader has not already done Exercise 3 M, Question 28, he will find it helpful to do so now.]

The power functions are given by

$$p_a(\theta) = \theta^5; \quad p_b(\theta) = \theta^5 + 5\theta^4(1-\theta); \quad p_c(\theta) = \theta^6 + 6\theta^5(1-\theta).$$

Their graphs are shown in the figures.

23. $(1-\theta)^{10} + 10\theta(1-\theta)^{19}$. Plot the graph.

25. (i) The set of outcomes in which we obtain 100 heads in 200 throws contains as a *proper* subset the set of outcomes in which we obtain 50 on the first 100 throws and 50 on the next 100. Events defined by the latter (such as obtaining 50 heads on each of two batches of throws of 100 coins) thus have smaller probabilities than do events defined by the former.

(ii) 0·0796. (iii) 0·0804. (iv) ½.

(v) To suppose that we would have answers related in this way comes from arguing that pr (50 of each) is equal to pr (1 of each | 49 of each) × pr (49 of each), but the argument of part (i) reminds us that the equal split of the first 98 coins *and* of the last 2 is only one of many ways of achieving an equal split of 100; thus:

pr (50 of each) > pr (1 of each | 49 of each) × pr (49 of each) = ½ × 0·0804.

27. (a) 0·37%; (b) 1·99%.

29. (i) White flowers of genotype AA cannot produce blue flowers.
pr (a white flower is of genotype aA) $= 2\theta(1-\theta)/(1-\theta^2) = 2\theta/(1+\theta)$,
white flowers of genotype aA crossed with blue flowers produce white flowers
with probability $\frac{1}{2}$.

In experiment X, pr (blue flowers result) $= \dfrac{2\theta}{1+\theta} \times \frac{1}{2}$, hence $\phi = \theta/(1+\theta)$.

In experiment Y the only cross producing blue flowers is aA $\times$ aA,
for this cross, pr (blue flowers result) $= \frac{1}{4}$;

thus in experiment Y, pr (blue flowers result) $= \left(\dfrac{2\theta}{1+\theta}\right)^2 \times \frac{1}{4} = \phi^2$.

(ii) Two independent experiments are being carried with binomial probability
functions.

$$\psi = \text{required probability} = \binom{N}{r}(1-\phi)^{N-r}\phi^r . \binom{N}{s}(1-\phi^2)^{N-s}\phi^{2s}.$$

(iii) $\dfrac{d}{d\phi}(\log \psi) = \dfrac{r+2s}{\phi} - \dfrac{N-r}{1-\phi} - \dfrac{(N-s)\,2\phi}{1-\phi^2}$;

$\dfrac{d}{d\phi}(\log \psi) = 0$ when $3N\phi^2 + (N-r)\,\phi - (r+2s) = 0$.

The resulting value of ϕ with $0 < \phi < 1$ will *maximize* ψ, since $\psi = 0$ for
$\theta = 0, 1$ and $\psi > 0$ for $0 < \phi < 1$.
(iv) Maximizing values are $\phi = \frac{2}{3}$, $\theta = \frac{2}{7}$.
Expected value of r is $72 \times \frac{2}{3} = 16$.
Expected value of s is $72 \times (\frac{2}{3})^2 = 3\frac{5}{9}$.
[These are the average-values expected over a long run.]

31. We have pr (r successes)/pr ($r-1$) successes $= \dfrac{N-r+1}{r} \cdot \dfrac{\theta}{1-\theta}$.

Solving formally $\dfrac{N-r+1}{r} \cdot \dfrac{\theta}{1-\theta} = 1$ we get $r = (N+1)\theta$.

If $(N+1)\,\theta$ is integral we get a pair of equiprobable numbers of successes (for
example $\theta = \frac{1}{3}$, $N = 5$ leads to $r = 2$; and 1 and 2 successes are equiprobable).
The cases $(N+1)\,\theta$ not integral are still left to the reader!

Exercise 7E (p. 229)

1. No answers supplied.
It is very important for the reader to devise checks and counter-checks for
numerical work of this sort: for instance $\Sigma x\,\text{fr}(X = x)$ and Σx are both known
without the details of fr $(X = x)$, for each of $X = R, U, V$.

3. (i) The first may be either ....: probability $= 1$.
The next $(k-1)$ must be the same as the first: probability $= (\frac{1}{2})^{k-1}$.
The next must be the other ....: probability $= \frac{1}{2}$.
Result follows.
(ii) The previous result simplifies to $(\frac{1}{2})^k$

$$\text{pr (1st is of length } k \text{ and 2nd is of length } l-k) = (\tfrac{1}{2})^k . (\tfrac{1}{2})^{l-k}.$$

The minimum length at each end is 1 (*not* zero).

(iii) The previous result simplifies to $(l-1)(\frac{1}{2})^l$

$$0.0625 \ldots > 5\% \qquad 0.0351 \ldots \leqslant 5\%$$
$$0.0107 \ldots > 1\% \qquad 0.0058 \ldots \leqslant 1\%.$$

Referring to Exercise 1 A, Question 8, we now see that the groups of readings must be large enough for the probabilities to remain nearly constant.

Exercise 7 F (p. 231)

1. $\dfrac{4!}{2!\,1!\,1!\,0!\,0!}\left(\dfrac{1}{6}\right)^2\left(\dfrac{1}{3}\right)^1\left(\dfrac{1}{15}\right)^1\left(\dfrac{1}{60}\right)^0\left(\dfrac{25}{60}\right)^0 = \dfrac{1}{135}.$

$\frac{1}{135}$ is a fairly low probability. Examine whether the catch was really random. Was the fisherman using special lures for trout or throwing the pike and others back, and so on?

3. If the first sequence arose from Bernouilli trials with $\theta = \frac{1}{3}$, the probabilities of various values of U are given in the table below. The frequencies are also included.

u	pr $(U = u)$	fr $(U = u)$
0	$\frac{1}{3} = \frac{9}{27}$	8 (see note below)
1	$\frac{2}{3}.\frac{1}{3} = \frac{6}{27}$	20
2	$(\frac{2}{3})^2.\frac{1}{3} = \frac{4}{27}$	8
3 or more	$1-\left(\dfrac{9+6+4}{27}\right) = \dfrac{8}{27}$	14

[*Note:* If the sequence is Bernouillian the initial A can be taken as giving rise to an observation $u = 0$, since its predecessor would be immaterial.]

The probability of the observed frequencies of values of U occurring jointly is then calculated in Example 9 as 1.1×10^{-5}. Thus the first sequence is unlikely to have arisen from a sequence of Bernouilli trials with $\theta = \frac{1}{3}$.

Exercise 8 A (p. 236)

1. (i) $e^{-\frac{1}{2}} \simeq 0.6065$

with an obvious notation the probabilities are approximately

$$p_0 \simeq 0.61; \quad p_1 = \tfrac{1}{2}p_0 \simeq 0.30;$$
$$p_2 = \tfrac{1}{4}p_1 \simeq 0.08; \quad p_3 = \tfrac{1}{6}p_2 \simeq 0.01.$$

The remaining approximations are less than 0.005.

(ii) $\qquad\qquad\qquad e^{-2} \simeq 0.1353.$

The probabilities can again be calculated in succession; they are:

$$0.14, \quad 0.27, \quad 0.27, \quad 0.18, \quad 0.09, \quad 0.04, \quad 0.01.$$

The remainder are less than 0.005.

440

3.

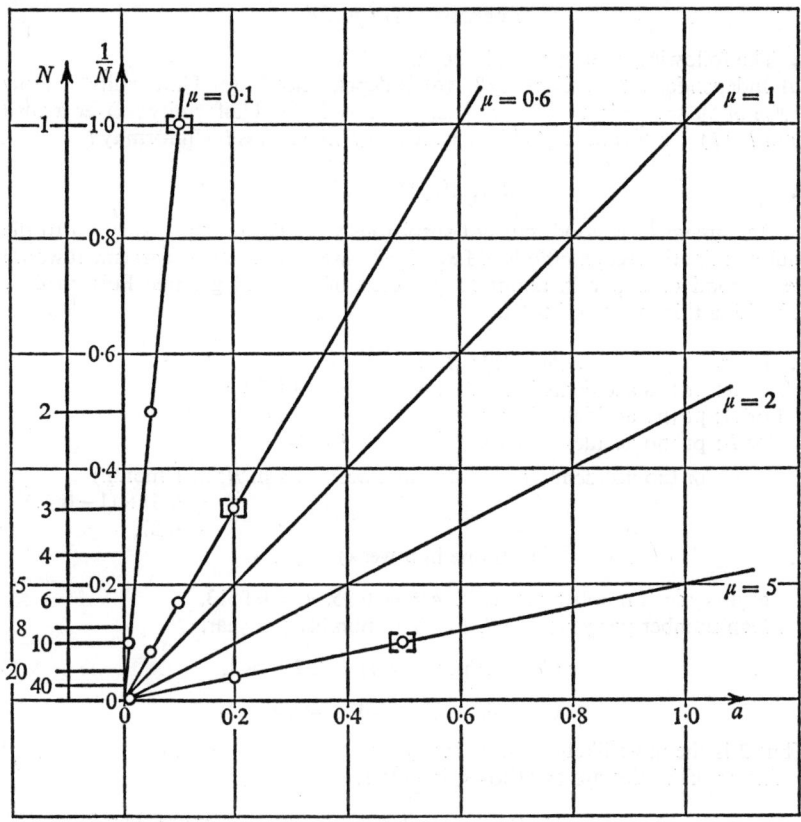

Fig. S.30

In the diagram, rather unsatisfactory approximations have their points marked within square brackets.

One might guess that a reasonable approximation would be obtained near (i) $N = 16$, $a = \frac{1}{8}$, or (ii) $N = 10$, $a = \frac{1}{5}$ or (iii) $N = 8$, $a = \frac{1}{4}$. The relevant probabilities are:

r	0	1	2	3	4	5	6
Approximations	0·14	0·27	0·27	0·18	0·09	0·04	0·01
(i)	0·12	0·27	0·29	0·19	0·09	0·03	0·01
(ii)	0·11	0·27	0·30	0·20	0·09	0·03	0·01
(iii)	0·10	0·27	0·31	0·21	0·09	0·02	0·00

Note that in all four cases pr $(R = 3)$/pr $(R = 2) = \frac{2}{3}$ *exactly.*

Note also that $\mu e^{-\mu}$ approximates rather better to $N\left(\dfrac{\mu}{N}\right)\left(1 - \dfrac{\mu}{N}\right)^{N-1}$ (for this range of values) than $e^{-\mu}$ approximates to $\left(1 - \dfrac{\mu}{N}\right)^{N}$.

441

Exercise 8 B (p. 241)

1. The following queries seem important.
(a) Independence? Uniformity? (b) Independence? (c) Uniformity? How is the book set up in type? (d) Independence? (e) Uniformity? Over a rush hour? (f) Uniformity? (g) Independence? Singleness? Uniformity?

3.   227, 212, 98, 31, 7, 1.

The agreement is good, but not suspiciously so. By an approximation to the multinomial theorem methods of Exercise 7F we can show that agreement would be as good or better in about 15 per cent of cases of genuine Poisson data exhibiting this mean and total.

5. The rates per month are $A:\frac{2}{3}$; $B:\frac{3}{4}$.
For B: pr (no accidents in 2 months) $= e^{-2\times\frac{3}{4}} \simeq 0\cdot22$
For A: pr (no accidents in 1 month) $= e^{-\frac{2}{3}} \simeq 0\cdot51$
For B: pr (no accidents in 1 month) $= e^{-\frac{3}{4}} \simeq 0\cdot48$

pr (no accidents in A, 1 or more accidents in B; in 1 month)
$$= 0\cdot51 \times (1 - 0\cdot48)$$
$$= 0\cdot27.$$

7.   pr (none in a week) $= \frac{19}{20}$.

Let μ = mean number per week. $e^{-\mu} = 0\cdot95$, $\mu = 0\cdot0513$.
Mean number per year = 2·7. Let N = number per year.

$$\text{pr}\,(N = 2)/\text{pr}\,(N = 1) = 2\cdot7/2 > 1,$$
$$\text{pr}\,(N = 3)/\text{pr}\,(N = 2) = 2\cdot7/3 < 1.$$

Thus 2 is the most likely number in a year.
The probabilities are as follows, in detail:

n	0	1	2	more than 2
pr $(N = n)$	0·07	0·19	0·25	0·50

The required probabilities are roughly $\frac{1}{4}$, $\frac{1}{4}$, $\frac{1}{2}$.
We assume (i) sightings are independent. This rules out false sightings triggered off; space ships parking in orbit.

(ii) Sightings are single. A convention could suitably ignore the extras in a multiple sighting.

(iii) Uniformity. No variations for seasons; or when Mars, say, or Venus is particularly near.

Since these conditions are likely *not* to be fulfilled by genuine spacecraft, a good Poisson fit to the sightings would suggest that they were *not* due to spacecraft at all, but to some other, and random, agent. This reminds us of the importance of constructing models even if it turns out that some of the models do *not* fit the data. Their interest is that of the dog which did *not* bark in the Sherlock Holmes story.

9. Let μ = mean number of strikes per day.

$$e^{-\mu} = \tfrac{134}{365} \Rightarrow \mu = 1.$$

We expect $365 \left(e^{-\mu} + \mu e^{-\mu} + \dfrac{\mu^2}{2} e^{-\mu} \right)$ days $= 134 \times 2\frac{1}{2}$ days to have 2 or fewer strikes.

We expect 30 days to have more than 2 strikes, and there to be 365 strikes per year.

11. (i) We check on 1 cm squares and on 2 cm squares, since a genuine Poisson pattern will show the proper frequencies on both. It would be easy to produce artificially a pattern with correct frequencies for 1 cm squares but failing to accumulate properly for 2 cm squares. The reader may like to try some other method of accumulation, for instance taking pairs of 1 cm squares diagonally opposite each other in each 2 cm square.

The tables show observed frequencies $f(x)$ and the mean frequencies $p(x)$ to be expected in a long run of Poisson patterns with the appropriate value of μ.

1st pattern: (a) 1 cm squares. $\mu = 0.525$, $80e^{-\mu} = 47.4$.

x	0	1	2	3	4 or more
$f(x)$	47	25	7	1	0
$p(x)$	47·4	24·9	6·5	1·1	0·1

(b) 2 cm squares: $\mu = 2.10$, $20e^{-\mu} = 2.4$.

x	0	1	2	3	4	5	6 or more
$f(x)$	2	5	6	4	2	1	0
$p(x)$	2·4	5·1	5·4	3·8	2·0	0·8	0·4

The agreements seem good; but perhaps suspiciously so; yet the approximate multinomial method (known as the χ^2 test of Goodness of Fit, which needs a continuous variate) shows that with a Poisson probability function having the appropriate value of μ a set of observed frequencies would show agreement as good as this in about 25 per cent of cases. If the agreement would only have been as good in say 1 per cent of cases we might have wondered whether the observations were somehow arranged to fit the theory.

2nd pattern: (a) 1 cm squares. $\mu = 0.275$, $80e^{-\mu} = 60.7$.

x	0	1	2	3 or more
$f(x)$	62	14	4	0
$p(x)$	60·7	16·7	2·3	0·3

(b) 2 cm squares. $\mu = 1.10$, $20e^{-\mu} = 6.7$.

x	0	1	2	3	4 or more
$f(x)$	7	4	9	0	0
$p(x)$	6·7	7·3	4·0	1·5	0·5

The second agreement may not seem good, but the χ^2 test, mentioned above, shows that with a Poisson pattern probability function having the appropriate value of μ a set of observed frequencies would show agreement as bad as this or worse in about 7 per cent of cases. We would not reject the Poisson hypothesis at the 5 per cent level of significance.

(ii) Keep your results for testing later when you study the χ^2 test. [Turn over.

(It may well be that your results show discrepancies from the Poisson frequencies so gross that you reject the Poisson hypothesis already. The usual fault is to have far too few squares showing zero frequency, this is because people confuse randomness with evenness of spread; evenness of spread is a product of *regularity*, not randomness!)

13. (i) $(\frac{1}{2})^5$.
   (ii) For new interval

$$\mu' = 5\mu, \ e^{-\mu'} = (e^{-\mu})^5 = (\frac{1}{2})^5,$$

$$\text{pr (1 flash in new interval)} = 5\mu e^{-5\mu} = 0.108.$$

   (iii) Because 'there remains a third way': to have *more* than 1 flash. The Binomial Theorem is concerned with dichotomy only.

15. Call the combined process C.

$$p_c(r, t) = \sum_{i=0}^{r} \frac{\mu_1^i t^i e^{-\mu_1 t}}{i!} \cdot \frac{\mu_2^{r-i} t^{r-i} e^{-\mu_2 t}}{(r-1)!}$$

$$= \frac{t^r e^{-(\mu_1+\mu_2)t}}{r!} \sum_{i=0}^{r} \frac{r!}{i! \, (r-i)!} \mu_1^i \mu_2^{r-i}$$

$$= \frac{t^r e^{-(\mu_1+\mu_2)t}}{r!} (\mu_1+\mu_2)^r$$

which is a Poissonian probability with mean $(\mu_1+\mu_2)$.

   (ii) The simulating processes were *nearly* Poissonian; though, since the finding of 33 at any instant excluded the finding of 77 at that instant, they were not independent of each other. The smaller the probability of finding the item, in this case one of a set of digit pairs, the better the approximation.

   (iii) The approximation breaks down because of the now serious lack of independence (compare Exercise 3 M, Question 65).

   (iv) By looking for a digit pair, say 33, and representing a minute by 100 pairs of random digits from the table, instead of by 1 pair.

17. (i) pr (at least the first event occurs in $0 \leqslant T < t$)

$$= \text{pr (not no events in } 0 \leqslant T < t)$$

$$= 1 - e^{-\lambda t}.$$

   (ii) pr (1st event is in $0 \leqslant T < t_0 + x$)

$$= \text{pr (1st event is in } 0 \leqslant T < t_0) + \text{pr (1st event is in } t_0 \leqslant T < t_0 + x)$$

since these two intervals are exclusive, and exhaust the intervals first mentioned.

Hence   pr (1st event is in $t_0 \leqslant T < t_0 + x$) $= \Phi(t_0 + x) - \Phi(t_0)$

$$= \int_{t_0}^{t_0+x} \lambda e^{-\lambda t} . dt.$$

(iii)

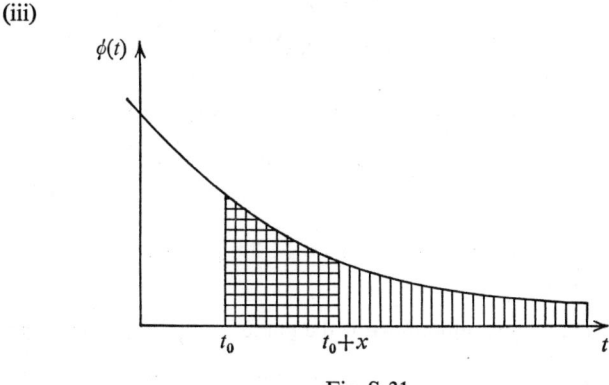

Fig. S.31

$\phi(t) = \lambda e^{-\lambda t}$

The horizontally striped area represents pr $(t_0 \leqslant t_1 < t_0 + x)$.
The vertically striped area represents pr $(t_0 \leqslant t_1)$

$$\text{pr} \, (t_0 \leqslant t_1 < t_0 + x | t_0 \leqslant t_1) = \frac{(1 - e^{-\lambda(t_0 + x)}) - (1 - e^{-\lambda t_0})}{e^{-\lambda t_0}}$$

$$= 1 - e^{-\lambda x}$$

$$= \text{pr} \, (0 \leqslant t_1 < x).$$

In words; the probability that the next event occurs in the next x seconds is independent of the initial instant, t_0. This may be thought of as a consequence of the fact that the parts of the graph of $\lambda e^{-\lambda t}$ to the right of any two ordinates differ only in vertical scale. It expresses the *uniformity* of the process.

(iv) The relevant data are:

x	Interval	Cumulative relative frequency	$\Phi(x)$
0	0–0	0	0
20	0–(20)	0·225	0·33
40	0–(40)	0·525	0·55
60	0–(60)	0·675	0·70
80	0–(80)	0·800	0·80
100	0–(100)	0·925	0·86
200	0–(200)	1·000	0·98

There seems to be a shortage of small values.

19. The situation described does not sound like a cycle (we notice 'long after', 'much sooner'). It sounds as if the article writer has misunderstood the descriptions of a Poisson process having mean about 1 event per 70 years.

If the situation is well modelled by a Poisson process then the next earthquake is more likely to be in the next year than in any *now-specified* later year; roughly speaking a specified later year (1990, say) for the *next* earthquake requires also the event, having probability less than 1, that no earthquake will occur until 1990. That *an* earthquake should occur in 1990 is as likely as that it will occur during the year from the time you read this, *if we have a Poisson process.*

445

Exercise 9A (p. 252)

1.
pr (five or six in 1 throw) = $\frac{1}{3}$; pr (no five or six in 6 throws) = $(\frac{2}{3})^6 = \frac{64}{729}$;

$$E\text{ (gain)} = -10 \times \frac{64}{729} + 1(1 - \frac{64}{729}) = \frac{25}{729}.$$

3. (i) Let x be the *forfeit*

$$0 = E\text{ (gain)} = -x \times \frac{1}{8} + 1 \times \frac{3}{8} + 3 \times \frac{3}{8} + 6 \times \frac{1}{8},$$

whence $x = 18$.

5. (i) The expected gain is unbounded if the game continues long enough. No entry fee can be large enough.

(ii) pr (Bank due to pay 2^{14} or more) = $(\frac{1}{2})^{15} + (\frac{1}{2})^{16} + (\frac{1}{2})^{17} + \dots$

$$= (\frac{1}{2})^{14}.$$

(iii) pr (Bank will not be broken in 256 games) = $\left(1 - \frac{1}{2^{14}}\right)^{256}$

$$\simeq 1 - \frac{256}{256 \times 64}$$

$$= \frac{63}{64}.$$

The terms of the binomial expansion are of alternate signs and decrease in magnitude so the error is less than the next term, $\dfrac{256 \cdot 255}{1 \cdot 2}\left(\dfrac{1}{256 \cdot 64}\right)^2$.

(iv) pr (Bank *will* be broken in 64 nights) = $1 - (1 - \frac{1}{64})^{64}$

$$\simeq 1 - e^{-1}$$

$$= 0 \cdot 63.$$

7. Expected scores $A: \frac{117}{32}$; $B: \frac{112}{32}$; $C: \frac{107}{32}$.

(i) No. We saw that C was the best choice.

(ii) No. We repeatedly choose C.

(iii) Yes. Expected values arise from totalled scores.

9. (i) The number of selections of 4 trumps and 9 non-trumps for a particular opponent to hold is $\binom{6}{4}\binom{20}{9}$.

These can be arranged in $\binom{6}{4}\binom{20}{9}$ 13! different-ordered-hands. The number of possible deals to this player, with order taken into account, is 26!/13!.

pr (he holds 4 trumps) = $\binom{6}{4}\binom{20}{9} 13! \Big/ \dfrac{26!}{13!}$

$$= \binom{13}{4}\binom{13}{2}\Big/\binom{26}{6}, \text{ on rearranging the factorials.}$$

But *either* opponent may hold 4 trumps, so the required probability is as given.

446

(ii) $\binom{26}{6}$ E (number of rounds)

$$= 3\binom{13}{3}\binom{13}{3}+4\times 2\binom{13}{2}\binom{13}{4}+5\times 2\binom{13}{1}\binom{13}{5}+6\times 2\binom{13}{0}\binom{13}{6},$$

whence E (number of rounds) $= \dfrac{3\times 5\times 41}{7\times 23} \simeq 3\cdot 8 < 4.$

11.

Remove from bag 1	Prob.	Remove from bag 2	Prob.	Final bag 1	Number of Red; Green	Prob.
G	$\frac{1}{3}$	R	$\frac{2}{3}$	RRR	3; 0	$\frac{2}{9}$
G	$\frac{1}{3}$	G	$\frac{1}{3}$	RRG	2; 1	$\frac{1}{9}$
R	$\frac{2}{3}$	R	$\frac{2}{3}$	RGR	2; 1	$\frac{4}{9}$
R	$\frac{2}{3}$	G	$\frac{1}{3}$	RGG	1; 2	$\frac{2}{9}$

E (number of Red) $= 2$; E (number of Green) $= 1$.

Remove from bag 1	Prob.	Remove from bag 2	Prob.	Final bag 1	Number of Red; Green	Prob.
G	$\frac{1}{3}$	R	$\frac{4}{7}$	RRR	3; 0	$\frac{4}{21}$
G	$\frac{1}{3}$	G	$\frac{3}{7}$	RRG	2; 1	$\frac{3}{21}$
R	$\frac{2}{3}$	R	$\frac{5}{7}$	RGR	2; 1	$\frac{10}{21}$
R	$\frac{2}{3}$	G	$\frac{2}{7}$	RGG	1; 2	$\frac{4}{21}$

E (number of Red) $= 2$; E (number of Green) $= 1$.

The reader should consider the following analogy; and the subsequent Question:

Two jars of different capacities holding mixtures of fluids in the same proportions can have contents poured from one to the other in any way and the proportions remain unchanged.

Does this analogy predict correct probability behaviour of bags holding differing proportions of Red and Green discs?

13. (ii) For each group:

$$\text{pr (1 analysis)} = (1-\theta)^k,$$

$$\text{pr }(1+k\text{ analyses}) = 1-(1-\theta)^k.$$

$$E\text{ (number of analyses)} = (1-\theta)^k+(1+k)\,(1-(1-\theta)^k)$$

$$= 1+k-k(1-\theta)^k.$$

For all N people:

$$E\text{ (number of analyses)} = \frac{N}{k}(1+k-k(1-\theta)^k).$$

The numbers predicted for the simulations given in part (i) are

$$40(1\cdot 25-0\cdot 656) \simeq 24$$

and

$$40(1\cdot 125-0\cdot 430) \simeq 28.$$

(iii) For small θ:

$$q \simeq N\left(\frac{1}{k}+k\theta\right),$$

which is minimized by $k = \theta^{-\frac{1}{2}}$, when $q \simeq 2N\theta^{\frac{1}{2}}$.
For $\theta = 0\cdot01$ the expected saving is 80 per cent.
   (iv) $\theta = 0\cdot2$ is too large for the approximation.
   We note the following values of $1+\dfrac{1}{k}-(0\cdot8)^k$:

$$[k = 1, \quad 1+1-0\cdot8 = 1\cdot20],$$
$$k = 2, \quad 1+0\cdot5-0\cdot64 = 0\cdot86,$$
$$k = 3, \quad 1+0\cdot33-0\cdot512 = 0\cdot82,$$
$$k = 4, \quad 1+0\cdot25-0\cdot41 = 0\cdot84.$$

The graphs of $1/x$ and $(0\cdot8)^x$ show that the quantity concerned has a single minimum, which is thus at $k = 3$ when k is integral. Percentage saving is then 18.
   (v) For $\theta = 0\cdot3$ the relevant values are

$$[k = 1, \quad 1+1-0\cdot7 = 1\cdot30],$$
$$k = 2, \quad 1+0\cdot5-0\cdot49 = 1\cdot01,$$
$$k = 3, \quad 1+0\cdot333-0\cdot343 = 0\cdot99,$$
$$k = 4, \quad 1+0\cdot25-0\cdot24 = 1\cdot01.$$

We have nearly reached the value of a for which the graphs of $1/x$ and of a^x touch.
   The following values of $(1-\theta)$ give minima for $k = 2, 3, 4, 5, \ldots,$

$$1-\theta = (\tfrac{1}{2})^{\frac{1}{2}}, \quad (\tfrac{1}{3})^{\frac{1}{2}}, \quad (\tfrac{1}{4})^{\frac{1}{2}}, \quad (\tfrac{1}{5})^{\frac{1}{2}}, \ldots$$

The smallest of these is $(\tfrac{1}{3})^{\frac{1}{2}}$, giving the required upper bound of θ as

$$1-(\tfrac{1}{3})^{\frac{1}{2}} \simeq 0\cdot307.$$

The explanation is that if θ is large enough then each group is likely to contain at least one member showing a reaction. This will entail testing all the group, individually, and the initial test will have been wasted. An increase, rather than a decrease in the number of tests is the result.

Exercise 9 B (p. 261)

1. (i) $E[cx] = \Sigma cx \operatorname{pr}(X = x) = c\Sigma x \operatorname{pr}(X = x) = c.E[x].$
    (ii) $E[x+a] = \Sigma(x+a)\operatorname{pr}(X = x) = \Sigma x \operatorname{pr}(X = x)+a\Sigma \operatorname{pr}(X = x)$
       $\qquad = E[x]+a.$
    (iii) A general linear function.
    (iv) $E[g(x)+f(x)] = \Sigma(g(x)\operatorname{pr}(X = x)+f(x)\operatorname{pr}(X = x))$
       $\qquad = E[g(x)]+E[f(x)].$
       Take $f(x) = g(x) = x$; put $\operatorname{pr}(X = x) = \frac{1}{2}$ for $x = -1, 1,$
       then $\qquad g(x)f(x) > 0, \quad$ so $E[g(x).f(x)] > 0,$
       but $\qquad E[g(x)] = E[f(x)] = 0.$

3. (i) Let $E[x] = \mu$, then $E[cx] = c\mu$ and $E[x+a] = \mu+a$.

$$\text{var } [cx] = E[(cx-c\mu)^2] = E[c^2(x-\mu)^2] = c^2E[(x-\mu)^2] = c^2 \text{ var } [x],$$

$$\text{var } [x+a] = E[((x+a)-(\mu+a))^2] = E[(x-\mu)^2] = \text{var } [x].$$

(ii) $$E[t] = E[(x-\mu)/\sigma] = \frac{1}{\sigma}E[x-\mu] = 0,$$

$$\text{var } [t] = \text{var } [(x-\mu)/\sigma] = \frac{1}{\sigma^2}\text{var } [x-\mu] = \frac{1}{\sigma^2}\text{var } [x] = 1.$$

5. Variance is $\frac{390}{8} \simeq 48 \cdot 8$.

7. The value of k is determinable from $\Sigma \text{ pr}(R = r) = 1$, but it need not be found.

(i) $\mu = 5$, $\sigma = (\frac{210}{36})^{\frac{1}{2}} \simeq 2 \cdot 4$.

(ii) $(\frac{24}{36})^2 \simeq 0 \cdot 44$.

9. (i) $\mu_0 = E[(x-\mu)^0] = E[1] = 1$,

$\mu_0' = E[x^0] = 1$.

(ii) $\mu_1 = E[x-\mu] = 0$,

$\mu_1' = E[x] = \mu$.

(iii) $\mu_2' = E[x^2] = \sigma^2+\mu^2$, from Question 2.

(iv) $\mu_3 = E[(x-\mu)^3] = E[x^3]-3\mu E[x^2]+3\mu^2 E[x]-\mu^3$

$= \mu_3'-3\mu_1'\mu_2'+2(\mu_1')^3$,

$\mu_4' = E[(x-\mu)^4] = E[x^4]-4\mu E[x^3]+6\mu^2 E[x^2]-4\mu^3 E[x]+\mu^4$

$= \mu_4'-4\mu_1'\mu_3'+6(\mu_1')^2\mu_2'-3(\mu_1')^4$.

11. (i) The sum of the probabilities is $\log 2+ \sum\limits_{k=1}^{\infty} 1/2^k k(k+1)$.

$$2\int_0^{\frac{1}{2}} -\log (1-t)\, dt = 2\int_0^{\frac{1}{2}} \sum\limits_{k=1}^{\infty} (t^k/k)\, .dt, \text{ using Taylor series}$$

$$= 2 \sum\limits_{k=1}^{\infty} \int_0^{\frac{1}{2}} (t^k/k)\, .dt, \text{ assuming the validity of}$$
$$\text{interchanging these operations}$$

$$= 2 \sum\limits_{k=1}^{\infty} (\tfrac{1}{2})^{k+1}/k(k+1)$$

$$= \sum\limits_{k=1}^{\infty} 1/2^k k(k+1).$$

$$2\int_0^{\frac{1}{2}} -\log (1-t)\, dt = 2\left[-t \log (1-t)\right]_0^{\frac{1}{2}} -2\int_0^{\frac{1}{2}} \frac{t}{1-t}.dt$$

$$= -\log \tfrac{1}{2}+2\int_0^{\frac{1}{2}} \left(1-\frac{1}{1-t}\right) dt$$

$$= -\log \tfrac{1}{2}+2[\tfrac{1}{2}+\log \tfrac{1}{2}]$$

$$= 1-\log 2.$$

The required equality follows.

(ii) $E(\text{gain}) = \sum_{1}^{\infty} 2^k/2^k k(k+1) = 1.$

Exercise 9C (p. 266)

1. (i) $0 \times q^5 + 1 \times 5q^4p + 2 \times 10q^3p^2 + 3 \times 10q^2p^3 + 4 \times 5qp^4 + 5 \times p^5$

$$= 5p(q^4 + 4q^3p + 6q^2p^2 + 4qp^3 + p^4)$$

$$= 5p(q+p)^4$$

$$= 5p.$$

(ii) $$\sum_{r=0}^{N} r\binom{N}{r} q^{N-r}p^r = \sum_{r=1}^{N} r\binom{N}{r} q^{N-r}p^r$$

$$= \sum_{r=1}^{N} N\binom{N-1}{r-1} q^{N-r}p^r$$

$$= Np \sum_{r=1}^{N} \binom{N-1}{r-1} q^{N-r}p^{r-1}$$

$$= Np \sum_{i=0}^{N-1} \binom{N-1}{i} q^{(N-1)-i}p^i$$

$$= Np.$$

3. Mean number of successes $= \frac{38}{30}$.

$k = 2$ gives $\theta = 0.3$ which is the available value nearest to $\frac{1}{4}(\frac{38}{30})$. With $\theta = 0.3$, the expected frequencies are:

$$30(\tfrac{7}{10})^4 = 7.2; \quad \tfrac{4}{1} \times \tfrac{3}{7} \times 7.2 = 12.4; \quad \tfrac{3}{2} \times \tfrac{3}{7} \times 12.4 = 7.9;$$

$$\tfrac{2}{3} \times \tfrac{3}{7} \times 7.9 = 2.3; \quad \tfrac{1}{4} \times \tfrac{3}{7} \times 2.3 = 0.2.$$

The 'expected' and observed variances are $4 \times \frac{3}{10} \times \frac{7}{10} = 0.84$; and 0.73.

5. Mean number of successes $= 0.4$.

$\theta = \frac{1}{10}$ gives $N = 4$ as the best value to fit $N\theta = 0.4$. With $N = 4$ the expected frequencies are

0 zeros: 26·2;   1 zero: 11·7;   2 zeros: 1·9;   more than 2 zeros: 0·2.

The 'expected' and observed variances are $4 \times \frac{1}{10} \times \frac{9}{10} = 0.36$; and 0.34.

7. (i) $m = \frac{38}{24} = 1.58 \simeq N\theta.$

$S^2 = (\frac{88}{24}) - (\frac{38}{24})^2 = 1.16 \simeq N\theta(1-\theta).$

$1 - \theta \simeq 0.734, \theta \simeq 0.266, N \simeq 5.94.$

Take $N = 6$, then $\theta \simeq 1.58/6 = 0.26.$

Try $k = 2$, then $\theta = \frac{3}{10}, N\theta = 1.8, N\theta(1-\theta) = 1.26.$

Try $k = 1$, then $\theta = \frac{2}{10}, N\theta = 1.2, N\theta(1-\theta) = 0.96.$

The best fit seems to be $N = 6, k = 2$.

(ii) The frequencies from the model would be

r	averaged fr (number of successes = r)	r	averaged fr (number of successes = r)
0	2·83	3	4·44
1	7·27	4	1·44
2	7·78	5 or more	0·26

9. (i) $\sigma^2 = \sum_{\text{all } x} (x-\mu)^2 . \text{pr} (X = x)$

$\qquad = \sum_{|x-\mu| \geqslant \lambda\sigma} (x-\mu)^2 \, \text{pr} (X = x) \quad + \quad \sum_{|x-\mu| < \lambda\sigma} (x-\mu)^2 \, \text{pr} (X = x)$

$\qquad \geqslant \sum_{|x-\mu| \geqslant \lambda\sigma} \lambda^2\sigma^2 \, \text{pr} (X = x) \quad + \quad 0$

$\qquad = \lambda^2\sigma^2 \, \text{pr} (|X-\mu| \geqslant \lambda\sigma)$, by the addition of probabilities of exclusive events.

The result follows.

(ii) Take $X = R$ then $\mu = N\theta$, $\sigma = (N\theta(1-\theta))^{\frac{1}{2}}$.

Take $\lambda = k.N^{\frac{1}{2}}\theta^{\frac{1}{2}}(1-\theta)^{-\frac{1}{2}}$.

The result follows.

11. Coefficient of variation $= \sigma/\mu = (N\theta(1-\theta))^{\frac{1}{2}}/N\theta \propto N^{-\frac{1}{2}}$.

13. (i) $H'(\xi) = \xi^k . G'(\xi) + k.\xi^{k-1}.G(\xi)$,

$\qquad H''(\xi) = \xi^k . G''(\xi) + 2k\xi^{k-1}G'(\xi) + k(k-1)\,\xi^{k-2}G(\xi)$.

The results follow, using $G(\xi) = 1$

(ii) H is the p.g.f. for a random variable Y, say, which exceeds by k the random variable X, say, defined by G.

The two results show that $E(Y) = E(X) + k$; var $(Y) =$ var (X).

(iii) The total number of trials to the kth success exceeds by k the (as so far defined) waiting time to the kth success; thus the power series for the p.g.f. of the latter is simpler by a factor ξ^k.

15. $E[\xi^r] = \Sigma\xi^r . \text{pr} (R = r)$ which leads to the p.g.f. for random variable R.

17. (i) If the reader has any difficulty in following these arguments he should expand the sums and consider the first few terms and a general term.

$\qquad E(R) = \sum_{r=0}^{\infty} r \, \text{pr} (R = r)$

$\qquad\qquad = \sum_{r=1}^{\infty} r \, \text{pr} (R = r)$

$\qquad\qquad = \sum_{r=1}^{\infty} r e^{-\mu}\mu^r/r!$

$\qquad\qquad = \mu e^{-\mu} \sum_{r=1}^{\infty} \mu^{r-1}/(r-1)!$

$\qquad\qquad = \mu e^{-\mu}.e^{\mu} = \mu$, as required.

$\qquad E(R^2) = \sum_{r=0}^{\infty} r^2 \, \text{pr} (R = r)$

$\qquad\qquad = \sum_{r=1}^{\infty} r^2 \, \text{pr} (R = r)$

$\qquad\qquad = \sum_{r=1}^{\infty} ((r-1)+1)\, e^{-\mu}\mu^r/(r-1)!$

$\qquad\qquad = \sum_{r=2}^{\infty} e^{-\mu}\mu^r/(r-2)! + \sum_{r=1}^{\infty} e^{-\mu}\mu^r/(r-1)!$

$\qquad\qquad = \mu^2 e^{-\mu} \sum_{r=2}^{\infty} \mu^{r-2}/(r-2)! + \mu e^{-\mu} . \sum_{r=1}^{\infty} \mu^{r-1}/(r-1)!$

$\qquad\qquad = \mu^2 + \mu$.

$$\text{var } (R) = E(R^2) - (E(R))^2$$
$$= (\mu^2 + \mu) - (\mu)^2$$
$$= \mu$$
$$= E(R).$$

(ii) $G(\xi) = e^{-\mu(1-\xi)}$,

$G'(\xi) = \mu e^{-\mu(1-\xi)}$; $E(R) = G'(1) = \mu$,

$G''(\xi) = \mu^2 e^{-\mu(1-\xi)}$; var $(R) = G''(1) + G'(1) - (G'(1))^2 = \mu^2 + \mu - (\mu)^2 = \mu$.

19. $m = 3.87$, and since the results are nearly Poissonian we nearly have

$$S^2 = 3.87 \quad \text{and} \quad 2S = 3.94, \quad m - 2S < 0, \quad m + 2S = 7.81.$$

The proportion in the interval is $2520/2608 = 96.5$ per cent.

21. (i) With the usual notation we have, for a binomial probability function

$$\sigma^2 = N\theta(1-\theta) < N\theta = \mu.$$

Thus the frequency diagram is 'underdispersed' relative to a Poissonian one and the *binomial* pattern exhibits regularity. A uniform frequency diagram is clearly 'overdispersed' and the *uniform function* pattern shows contagion.

(ii) Mean for each pattern is μ.

(a) Binomial: $B = N\theta(1-\theta)/N\theta = 1 - \theta = 1 - \dfrac{\mu}{N}$,

$$\phi = \frac{N-1}{N} = 1 - \frac{1}{N}.$$

(b) Uniform: $E[r] = \mu$, var $[r] = \frac{1}{12}((2\mu+1)^2 - 1)$, from Exercise 4F, Question 1 (iii)

$$= \mu(\mu+1)/3.$$

$$B = \tfrac{1}{3} + \tfrac{1}{3}\mu, \quad \phi = 2.$$

(c) Poissonian: $B = 1$, $\phi = 1$.

(iii) *If any simple correspondences exist* between the behaviours of B, ϕ and the type of pattern, the examples in (ii) suggest:

Contagion. $B > 1$ for patterns with sufficiently large μ (we need $\mu > 2$).
$\phi > 1$

Regularity. $B < 1$, and more markedly for larger μ/N.
$\phi < 1$, and more markedly for large N.

Convenience. ϕ seems more convenient, since it only needs the nearly empty squares to be counted; whereas B requires a complete survey and then some extensive calculations.

Effect of size of quadrat. The Binomial model suggests an orchard-type of plant-pattern with N vacancies for plants in each quadrat and a probability θ that any particular vacancy is occupied. If the quadrats are larger, μ, N will both be larger but μ/N will be unaffected.

For larger and larger quadrats we will have:

	B	ϕ
Binomial (and regularity generally?)	unaffected (< 1)	increased to nearly 1
Poisson (that is: random)	1	1
Uniform (and contagion generally?)	increased unboundedly	unaffected ($= 2$)

On these grounds B seems more satisfactory.

[A typical binomial situation has been indicated above; a typical uniform-frequency-function situation would arise from the following counts in successive quadrats along a line:

$$0, \quad 1, \quad 2, \quad 3, \quad 4, \quad 4, \quad 3, \quad 2, \quad 1, \quad 0.$$

This example may help to make clear the *clumping* (that is *non-uniformity* of plant spread) typically arising from a *uniform* frequency function. In these confusing circumstances it is probably better to use the adjective 'rectangular' to describe the frequency function.]

(iv) *Carex flacca.*

$$\bar{r} = 0{\cdot}725, \quad \text{var}\,[r] = 21{\cdot}3$$
$$B = 2{\cdot}9 > 1, \quad \phi = 2{\cdot}8 > 1.$$

This may exhibit contagion.

    Artificial data.

$$\bar{r} = 1, \quad \text{var}\,[r] = 1$$
$$B = 1, \quad \phi = 0.$$

The value of B suggests regularity(?) *or* randomness,   *or* contagion;
The value of ϕ suggests regularity   *or* non-randomness, *or* non-contagion.

(v) For the Welshmen: $\bar{r} = 1{\cdot}2$, var $[r] = 1{\cdot}43$, $N = 4$, $\mu/N = 0{\cdot}3$.

    Observed values: $B = 1{\cdot}2$, $\phi = 0{\cdot}67$.
    Binomial model values $B = 0{\cdot}7$, $\phi = 0{\cdot}75$.

A binomial model does not seem suitable, but we do not know how serious the discrepancies are. A binomial model is unlikely to be suitable on commonsense grounds since the Welshmen might well be in parties and their opponents in other parties as they left the ground. We would expect contagion, due to clumping.

23. (a) pr (a four and a five) $= \frac{1}{18}$.

    E(waiting times) $= 17, \; 2 \times 17, \; 3 \times 17, \; 4 \times 17$.
    Expected values of the numbers $= 18, 36, 54, 72$.

(b) The independence of the trials ensures that the record of successes has the same probability structure read forwards or backwards. We thus seek the expectation and variance of the waiting time to the first recess: the results are $(1-\theta)$ and $(1-\theta)/\theta^2$.

25. (i) For $l \geqslant 1$,

    pr $(L = l \,|\, \text{first result is a success}) = \theta^{l-1}(1-\theta)$,
    pr $(L = l \,|\, \text{first result is a failure}) = (1-\theta)^{l-1}\,\theta$,

so   pr $(L = l) = \theta \,.\, \theta^{l-1}(1-\theta) + (1-\theta)\,(1-\theta)^{l-1}\,\theta$,

            $= \theta^l(1-\theta) + (1-\theta)^l\,.\,\theta.$

(ii) $G_L(\xi) = (1-\theta) \sum_{l=1}^{\infty} \theta^l \xi^l + \theta \sum_{l=1}^{\infty} (1-\theta)^l . \xi^l$

$= (1-\theta) \left\{ \dfrac{1}{1-\theta\xi} - 1 \right\} + \theta \left\{ \dfrac{1}{1-(1-\theta)\xi} - 1 \right\}$

$= \dfrac{1-\theta}{1-\theta\xi} + \dfrac{\theta}{1-(1-\theta)\xi} - 1.$

(iii) $G'_L(\xi) = \dfrac{(1-\theta)\theta}{(1-\theta\xi)^2} + \dfrac{\theta(1-\theta)}{(1-(1-\theta)\xi)^2},$

$E(L) = G'_L(1) = \dfrac{\theta}{1-\theta} + \dfrac{1-\theta}{\theta}$

$= (1-\theta) \left(\dfrac{1-\theta}{\theta} + 1 \right) + \theta \left(\dfrac{\theta}{1-\theta} + 1 \right)$

$= $ pr (failure).{expected waiting time to first success $+ 1$}
$+ $ pr (success).{expected waiting time to first failure $+ 1$}

which is the required result.

This is the result we would anticipate if we knew that expectations behaved like averages, and could be averaged like averages.

It is less obvious why $E(L)$ is also expressible as the sum of the expected waiting time to the first success (namely $(1-\theta)/\theta$) and the expected waiting time to the first failure (namely $\theta/(1-\theta)$).

(iv) $\text{var}(L) = \dfrac{\theta}{(1-\theta)^2} + \dfrac{1-\theta}{\theta^2} - 2.$

27. (i) $\dbinom{k+r-1}{k-1} = \dbinom{k+r-1}{r} = (-1)^r \dbinom{-k}{r}$, see Chapter 7, Section 10.2,

$\text{pr}(R = r) = \dbinom{-k}{r} \theta^k (-(1-\theta))^r,$

$\Rightarrow G_R(\xi) = \theta^k (1-(1-\theta)\xi)^{-k}.$

(ii) In the model let each cyst result in, say, u mature cysts; then the p.g.f. for U is that for the waiting time to the first death in the chain. The independence of the k original cysts means that the p.g.f. for the total number of mature cysts is that for the waiting time to the kth death, which ends all the chains.

The required generating function for the model is thus the inverse binomial one obtained in part (i).

(iii) $E(\text{number of cysts}) = k(1-\theta)/\theta,$

$\text{var (number of cysts)} = k(1-\theta)/\theta^2.$

$B = 1/\theta; \phi = 1 + 1/k$ both exceed 1 and this suggests contagion, which fits the model well. Notice that for large quadrat or soil-specimen B is unaffected, but ϕ decreases towards 1.

(iv) No answers meaningful.

(v) We find $\bar{r} = 1.75$, $\text{var}[r] = 2.39$,

$$\theta \simeq \bar{r}/\text{var}[r] = 0.73, \quad 1-\theta \simeq 0.27,$$

$$k \simeq \dfrac{1.75 \times 0.73}{0.27} = 4.73.$$

Take $k = 5$, since the model uses a uniform spread of original cysts, and not a probabilistic spread with mean $= k$. Then we reapproximate to θ by $1.75 \simeq 5(1-\theta)/\theta$.

Whence $1/\theta = 1.35$ and $\theta = 0.74$.

We check that $\dfrac{k(1-\theta)}{\theta^2} = \dfrac{1.75}{0.74} = 2.38 \simeq 2.39$.

29. Material towards a discussion is available in the next section. The point raised in this question is an important one and the reader would do well to formulate some interpretation before proceeding; it is sure to help to clarify his ideas.

Exercise 9D (p. 281)

1. $(x_1 y_1^2 + x_1 y_2^2) + (x_2 y_1^2 + x_2 y_2^2) + (x_3 y_1^2 + x_3 y_2^2)$

$$= x_1(y_1^2 + y_2^2) + x_2(y_1^2 + y_2^2) + x_3(y_1^2 + y_2^2)$$

$$= (x_1 + x_2 + x_3)(y_1^2 + y_2^2).$$

3. (i) In the general result

$$\mathrm{var}\,[z] = E[z^2] - (E[z])^2$$

put $z = ax + by$, and note that $E[ax+by] = a\mu_x + b\mu_y$.

(ii) right-hand side

$$= E[a^2 x^2 + 2abxy + b^2 y^2] - (a^2\mu_x^2 + 2ab\mu_x\mu_y + b^2\mu_y^2)$$

$$= a^2\{E[x^2] - \mu_x^2\} + b^2\{E[x^2] - \mu_y^2\} + 2ab\{E[xy] - \mu_x\mu_y\}$$

$$= a^2\,\mathrm{var}\,[x] + b^2\,\mathrm{var}\,[y] + 0,$$

$E[xy] = \mu_x\mu_y$ requires X, Y to be independent.

5. If on a given day all the coaches have the same number of passengers, and all the passengers on all the coaches have the same appetite for buns then the number of passengers on a day is np and the number of buns is npb.

$$E[np] = \mu_N\mu_P; \quad E[npb] = \mu_N\mu_P\mu_B$$

provided N, P, B are independent, but nothing can be said about the variances.

If differing numbers of passengers arrive on each bus and they eat differing numbers of buns the problem is harder and will be deferred.

7. (a) $\qquad E(\text{total e.m.f./V}) = 4 \times E(\text{e.m.f./V}) = 90.4,$

$\qquad\qquad \mathrm{var}\,(\text{total e.m.f./V}) = 4 \times \mathrm{var}\,(\text{e.m.f./V}),$

so $\qquad\qquad \text{stad}\,(\text{total e.m.f./V}) = 2 \times \text{stad}\,(\text{e.m.f./V}),$

limits are (90.4 ± 0.12) volts.

(b) $\qquad E(\text{difference in diam./cm}) = \text{difference in } E(\text{diam./cm})$

$$= 0,$$

$\qquad \mathrm{var}\,(\text{difference in diam./cm}) = \text{sum of var (diam./cm)}$

$$= (0.006)^2 + (0.008)^2$$

$$= (0.010)^2,$$

so   stad (difference in diam./cm) = 0·01,

$$0·02 = 2 \times \text{stad (difference in diam./cm)}.$$

The probability function for the difference in diam./cm is symmetrical, so 50 per cent of the differences will be less than the mean, that is: negative. A further $\frac{1}{2}(100-95$ per cent), that is: $2\frac{1}{2}$ per cent, will have the bolts too small. Thus $47\frac{1}{2}$ per cent will be acceptable.

[It would be better industrial practice to make the mean diameter of bolts rather less than that of washers.]

9. (i) $E[(x-\mu_x)(y-\mu_y)] = E[xy-\mu_x y-\mu_y x+\mu_x \mu_y]$

$$= E[xy]-\mu_x E[y]-\mu_y E[x]+\mu_x \mu_y.$$

The result follows.

(ii)

x	y	xy	pr $(X = x, Y = y)$
-2	4	-8	$\frac{1}{5}$
-1	1	-1	$\frac{1}{5}$
0	0	0	$\frac{1}{5}$
1	1	1	$\frac{1}{5}$
2	4	8	$\frac{1}{5}$

$$E[xy] = (-8-1+0+1+8)/5 = 0; \ \mu_x = 0.$$

Hence cov $(X, Y) = 0$ but Y, X are not independent.

Exercise 9E (p. 285)

1. (i) $K(\xi)$ has a series expansion in powers of ξ with all coefficients positive and with $K(1) = 1$.

(ii) $K'(\xi) = G'(\xi) H(\xi)+G(\xi) H'(\xi); \quad K'(1) = G'(1)+H'(1),$

$$K''(1) = G''(1)+2G'(1) H'(1)+H''(1), \text{ using } G(1) = H(1) = 1.$$

$$\begin{aligned} K''(1)+K'(1)-(K'(1))^2 = \ & G''(1)+2G'(1) H'(1)+H''(1) \\ & +G'(1) \qquad\qquad\qquad +H'(1) \\ & -(G'(1))^2-2G'(1) H'(1)-(H'(1))^2, \end{aligned}$$

as required.

(iii) If G, H are the p.g.f. for independent random variables X, Y then K is the p.g.f. for the random variable $(X+Y)$.

The results of part (ii) state that

$$E(X+Y) = E(X)+E(Y) \quad \text{and} \quad \text{var } (X+Y) = \text{var } (X)+\text{var } (Y).$$

(iv) Write $G(\xi) = e^{-a(1-\xi)}, \quad H(\xi) = e^{-b(1-\xi)}$

for Poisson p.g.f. with parameters a, b.

The total number of events is given by the sum of these random variables, for which the p.g.f. is

$$G(\xi).H(\xi) \equiv e^{-a(1-\xi)}.e^{-b(1-\xi)}$$

$$= e^{-(a+b)(1-\xi)}$$

as required.

456

(v) T is the total number of multiples of 3 and of multiples of 2, counting 'six' once for each.

pr (throw is a multiple of 3 and a multiple of 2)

$$= \tfrac{1}{6} = \tfrac{1}{2} \times \tfrac{1}{3}$$

$$= \text{pr (throw is a multiple of 2).pr (throw is a multiple of 3).}$$

At each trial the results are independent so the total numbers of each will be values of independent random variables.

$G_X(\xi) = (\tfrac{2}{3} + \tfrac{1}{3}\xi)^{36}$	$G_Y(\xi) = (\tfrac{1}{2} + \tfrac{1}{2}\xi)^{36}$
$G_X'(\xi) = 12(\tfrac{2}{3} + \tfrac{1}{3}\xi)^{35}$	$G_Y'(\xi) = 18(\tfrac{1}{2} + \tfrac{1}{2}\xi)^{35}$
$G_X''(\xi) = 140(\tfrac{2}{3} + \tfrac{1}{3}\xi)^{34}$	$G_Y''(\xi) = 315(\tfrac{1}{2} + \tfrac{1}{2}\xi)^{34}$
$E(X) = 12$	$E(Y) = 18$
var $(X) = 140 + 12 - 144 = 8$	var $(Y) = 315 + 18 - 324 = 9$

$$E(T) = E(X) + E(Y) = 30,$$

$$\text{var } (T) = \text{var } (X) + \text{var } (Y) = 17, \quad \text{since } X, Y \text{ independent.}$$

$$G_T(\xi) = (\tfrac{2}{3} + \tfrac{1}{3}\xi)^{36} (\tfrac{1}{2} + \tfrac{1}{2}\xi)^{36}$$

$$= (\tfrac{1}{3} + \tfrac{1}{2}\xi + \tfrac{1}{6}\xi^2)^{36}$$

which is multinomial, but not binomial.

3. (i)
$$G_X(\xi) = (1 - \theta) + \theta\xi,$$

$$E(X) = G_X'(1) = \theta,$$

$$\text{var } (X) = G_X''(1) + G_X'(1) - (G_X'(1))^2 = \theta(1 - \theta).$$

(ii)
$$G_R(\xi) = ((1 - \theta) + \theta\xi)^N.$$

This is a binomial p.g.f. for a random variable having $E(R) = N\theta$

$$E(R) = N.E(X) = N\theta,$$

$$\text{var } (R) = N.\text{var } (X) = N\theta(1 - \theta).$$

These are the usual results, seen as consequences of the additive nature of the record of successes in the trials.

(iii) It can, since p.g.f. for waiting time to 1st success is $\theta(1 - (1 - \theta)\xi)^{-1}$ and p.g.f. for waiting time to kth success is $(\theta(1 - (1 - \theta)\xi)^{-1})^k$. Again the variables are additive.

(iv) The variances are also additive.

5. Let the winnings be X
$$E(X) = G'(1) = g, \quad \text{say.}$$
p.g.f. for profit is $\xi^{-g}G(\xi)$.

7. Let G be the p.g.f. for a random variable Y.
(i) H is the p.g.f. for the random variable $-Y$.
(ii) Let F be the p.g.f. for the random variable X.

$K(\xi) \equiv F(\xi) H(\xi)$ is the p.g.f. for the random variable $X + (-Y)$,

since X, Y are independent,

or $K(\xi) = F(\xi).G(\xi^{-1})$ is the p.g.f. for the random variable $X-Y$.

Then $K'(\xi) = F'(\xi).G(\xi^{-1}) - F(\xi).G'(\xi^{-1}).\xi^{-2}$

$K'(1) = F'(1) - G'(1)$

so $E(X-Y) = E(X) - E(Y).$

Furthermore

$K''(\xi) = F''(\xi).G(\xi^{-1}) - 2F'(\xi).G'(\xi^{-1})\ \xi^{-2}$

$+ F(\xi).G''(\xi^{-1}).\xi^{-4} + 2F(\xi).G'(\xi^{-1}).\xi^{-3},$

$K''(1) = F''(1) - 2F'(1)\ G'(1) + G''(1) + 2G'(1),$

so $K''(1) + K'(1) - (K'(1))^2$

$= F''(1) - 2F'(1)\ G'(1) + G''(1) + 2G'(1)$

$+ F'(1)\qquad\qquad\qquad - G'(1)$

$- (F'(1))^2 + 2F'(1)\ G'(1) - (G'(1))^2$

$= F''(1) + F'(1) - (F'(1))^2$

$+ G''(1) + G'(1) - (G'(1))^2$

and we have $\text{var}\,(X-Y) = \text{var}\,(X) + \text{var}\,(Y).$

Exercise 10 A (p. 289)

1. Yes.

3. No. pr ((r+1)th result is odd | rth result is odd) is estimated as zero, and so *not* equal to pr ((r+1)th result is odd). Thus successive events are not independent and cannot arise from independent trials.

In the table below, p_{ij} the entry in the ith row and jth column is pr (result is i | preceding result was j)

	1	2	3	4
1	0	$\frac{7}{17}$	0	$\frac{1}{3}$
2	$\frac{1}{2}$	0	$\frac{11}{20}$	0
3	0	$\frac{10}{17}$	0	$\frac{2}{3}$
4	$\frac{1}{2}$	0	$\frac{9}{20}$	0

5. (i) The occurrence of an accident reduces the chance of another following immediately; but a period free of accidents increases the chance of one occurring. This corresponds to taking warnings, and to becoming careless.

(ii) This has the opposite effect to (i). The occurrence of a case of the disease increases the chance of another and *vice-versa*. We have the possibility of an epidemic or the virtual extinction of the disease.

(iii) Random selection without replacement.

In (iv) the colour-chain is time-dependent Markovian, otherwise the chain is non-Markovian.

In (iii) and (iv) the number chain is Markovian.

Exercise 10 B (p. 299)

1. The diagrams show the unit square $OUBA$ and its transform $OU'B'A'$.
Eigen-vectors are marked: In (i) they are **OA**, **OU**; in (iii) they are **OV**, **OW**.
The numerical results are

(i) $\lambda = 3$; $\quad \dfrac{1}{2}\begin{pmatrix} \sqrt{2} \\ \sqrt{2} \end{pmatrix}$, $\begin{pmatrix} \frac{1}{2} \\ \frac{1}{2} \end{pmatrix}$, $\mathbf{OU} = \begin{pmatrix} 1 \\ 1 \end{pmatrix}$.

$\quad \lambda = 2$; $\quad \begin{pmatrix} 1 \\ 0 \end{pmatrix}$, $\begin{pmatrix} 1 \\ 0 \end{pmatrix}$, $\mathbf{OA} = \begin{pmatrix} 1 \\ 0 \end{pmatrix}$.

(iii) $\lambda = \pm\sqrt{6}$; $\quad \dfrac{1}{\sqrt{5}}\begin{pmatrix} \sqrt{3} \\ \pm\sqrt{2} \end{pmatrix}$, $\dfrac{1}{\sqrt{3}\pm\sqrt{2}}\begin{pmatrix} \sqrt{3} \\ \pm\sqrt{2} \end{pmatrix}$, $\mathbf{OV}, \mathbf{OW} = \begin{pmatrix} 1 \\ \pm\sqrt{\frac{2}{3}} \end{pmatrix}$.

(i)

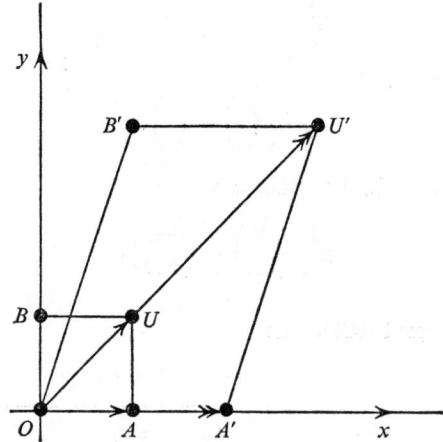

Fig. S. 32

(iii)

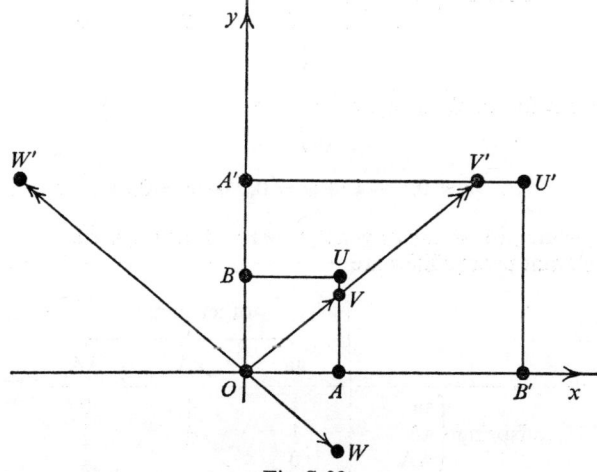

Fig. S. 33

3.
$$\mathbf{M}^4 \simeq 10^{-2} \begin{pmatrix} 23 & 22 & 22 \\ 20 & 20 & 20 \\ 57 & 58 & 58 \end{pmatrix}.$$

An eigen-vector is nearly
$$\begin{pmatrix} 0 \cdot 22 \\ 0 \cdot 20 \\ 0 \cdot 58 \end{pmatrix}.$$

5. The transition probabilities are:

	from	
	R	L
to $\begin{cases} R \\ L \end{cases}$	$\frac{2}{3}$	$\frac{1}{6}$
	$\frac{1}{3}$	$\frac{5}{6}$

The steady state vector is
$$\frac{1}{\frac{1}{3}+\frac{1}{6}} \begin{pmatrix} \frac{1}{6} \\ \frac{1}{3} \end{pmatrix} = \begin{pmatrix} \frac{1}{3} \\ \frac{2}{3} \end{pmatrix}.$$

Omitting the first result, the estimate is
$$\frac{1}{249} \begin{pmatrix} 88 \\ 161 \end{pmatrix} = \begin{pmatrix} 0 \cdot 35 \\ 0 \cdot 65 \end{pmatrix}.$$

7. (i) The relevant probabilities are

	other parent		
	aa	aA	AA
offspring $\begin{cases} aa \\ aA \\ AA \end{cases}$	1	$\frac{1}{2}$	0
	0	$\frac{1}{2}$	1
	0	0	0

The equations for an eigen-vector $\begin{pmatrix} u \\ v \\ w \end{pmatrix}$ are

$$\tfrac{1}{2}v = 0, \quad -\tfrac{1}{2}v + w = 0, \quad -w = 0,$$

giving steady-state probability of genotype-aa = 1, as required.
(ii) The relevant probabilities are

	other parent		
	aa	aA	AA
offspring $\begin{cases} aa \\ aA \\ AA \end{cases}$	$\frac{1}{2}$	$\frac{1}{4}$	0
	$\frac{1}{2}$	$\frac{1}{2}$	$\frac{1}{2}$
	0	$\frac{1}{4}$	$\frac{1}{2}$

We must satisfy

$$-\tfrac{1}{2}u+\tfrac{1}{4}v \quad\quad = 0,$$

$$\tfrac{1}{2}u-\tfrac{1}{2}v+\tfrac{1}{2}w = 0,$$

$$\tfrac{1}{4}v-\tfrac{1}{2}w = 0.$$

the probabilities are in the ratio $1:2:1$ as required.

9. No answers meaningful.

11. No answers meaningful.

13. The matrix arises from:

state at nth trial

state at $(n+1)$th trial
$$
\begin{array}{c|ccccc}
 & E_0 & E_1 & E_2 & E_3 & E_4 \\
\hline
E_0 & \tfrac{1}{3} & \tfrac{1}{3} & \tfrac{1}{3} & \tfrac{1}{3} & \tfrac{1}{3} \\
E_1 & \tfrac{2}{3} & 0 & 0 & 0 & 0 \\
E_2 & 0 & \tfrac{2}{3} & 0 & 0 & 0 \\
E_3 & 0 & 0 & \tfrac{2}{3} & 0 & 0 \\
E_4 & 0 & 0 & 0 & \tfrac{2}{3} & \tfrac{2}{3} \\
\end{array}
$$

The steady-state vector is
$$\frac{1}{81}\begin{pmatrix}27\\18\\12\\8\\16\end{pmatrix}\simeq\begin{pmatrix}0\cdot33\\0\cdot22\\0\cdot15\\0\cdot10\\0\cdot20\end{pmatrix}.$$

The relevance to Chapter 7. Success and failure have been interchanged. The length of run of successes here is r in state E_r; this corresponds to length of run of failures in Chapter 7. This in turn corresponds to the waiting time to the next success if *the simple (Bernouilli) sequence in Chapter 7 is read backwards*; this is permissible in a Bernouilli sequence since the trials are independent. The analyses are thus identical.

Exercise 10 C (p. 303)

1. $\mathbf{u}_1 = (1, 0, 0, 0, 0)'$; $\mathbf{u}_2 = (0, 0, 0, 0, 1)'$;

$\lambda\mathbf{u}_1+\mu\mathbf{u}_2$ for $\lambda+\mu = 1$ $(0\leqslant\lambda\leqslant1)$;

$\{A\}$, $\{B, C, D\}$, $\{E\}$.

[Note $\mathbf{u}'$ denotes the transpose of the matrix $\mathbf{u}$; that is to say the matrix with rows and columns of $\mathbf{u}$ interchanged.]

Exercise 10 D (p. 305)

1.

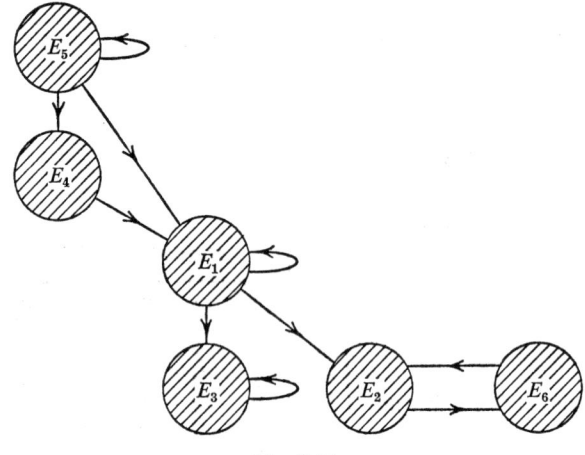

Fig. S. 34

	E_5	E_4	E_1	E_3	E_2	E_6
E_5	×	•	•	•	•	•
E_4	×	•	•	•	•	•
E_1	×	×	×	•	•	•
E_3	•	•	×	×	•	•
E_2	•	•	×	•	•	×
E_6	•	•	•	•	×	•

Transient classes: $\{E_5\}$, $\{E_4\}$, $\{E_1\}$;

closed classes: $\{E_3\}$, $\{E_2, E_6\}$;

absorbing state: E_3.

Exercise 10 E (p. 307)

1.

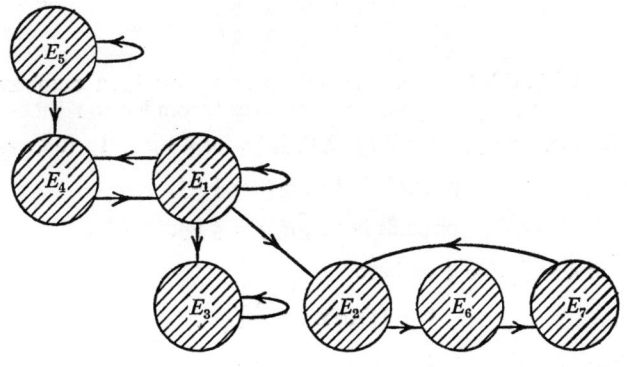

Fig. S.35

The diagram and table have been drawn up to resemble that for Exercise 10D as closely as possible.

	E_5	E_4	E_1	E_3	E_2	E_6	E_7
E_5	×	·	·	·	·	·	·
E_4	×	·	×	·	·	·	·
E_1	·	×	·	·	·	·	·
E_3	·	·	×	×	·	·	·
E_2	·	·	×	·	·	·	×
E_6	·	·	·	·	×	·	·
E_7	·	·	·	·	·	×	·

Transient classes: $\{E_5\}, \{E_4, E_1\}$;

closed classes: $\{E_3\}, \{E_2, E_6, E_7\}$;

absorbing state: E_3;

only state to which all are E_5.
    consequent:

463

3. (i) 1 class; $\mathbf{u} = (\frac{1}{3}, \frac{1}{6}, \frac{1}{3}, \frac{1}{6})'$.

(ii)
$$
\mathbf{N} = \begin{pmatrix} \frac{3}{4} & \frac{1}{2} & 0 & 0 \\ \frac{1}{4} & \frac{1}{2} & 0 & 0 \\ 0 & 0 & \frac{3}{4} & \frac{1}{2} \\ 0 & 0 & \frac{1}{4} & \frac{1}{2} \end{pmatrix};
$$

closed classes: $\{E_1, E_2\}$, $\{E_3, E_4\}$, where the states have been numbered corresponding to maxtrix columns read from left to right.

steady state vectors: $\lambda(\frac{2}{3}, \frac{1}{3}, 0, 0)' + \mu(0, 0, \frac{2}{3}, \frac{1}{3})'$ for $\lambda + \mu = 1$ $(0 \leqslant \lambda \leqslant 1)$.

(iii)
$$\text{pr (ends in } \{E_1, E_2\}) = \tfrac{1}{3} + \tfrac{1}{5},$$
$$\text{pr (ends in } \{E_3, E_4\}) = \tfrac{2}{5} + \tfrac{1}{5}.$$

(iv) From $\lambda = \frac{2}{5}$, $\mu = \frac{3}{5}$ we have
$$\mathbf{u} = (\tfrac{4}{15}, \tfrac{2}{15}, \tfrac{6}{15}, \tfrac{3}{15})'.$$

5. (i) Write
$$
\begin{pmatrix} 1 & 0 & \cdots & & 0 \\ 0 & \lambda_2^r & \cdots & & \\ \vdots & & & \vdots & \\ & & & & 0 \\ 0 & & \cdots & 0 & \lambda_n^r \end{pmatrix} = \mathbf{T}_r.
$$

Prove that: $\mathbf{T}_r \mathbf{T} = \mathbf{T}_{r+1}.$

$$\mathbf{M}^r = \mathbf{S}\mathbf{T}_r\mathbf{S}^{-1} \Rightarrow \mathbf{M}^{r+1} = \mathbf{M}^r.\mathbf{M} = (\mathbf{S}.\mathbf{T}_r\mathbf{S}^{-1}).(\mathbf{S}\mathbf{T}\mathbf{S}^{-1})$$
$$= \mathbf{S}\mathbf{T}_r\mathbf{T}\mathbf{S}^{-1}$$
$$= \mathbf{S}\mathbf{T}_{r+1}\mathbf{S}^{-1}$$

but $\mathbf{M}^1 = \mathbf{S}\mathbf{T}_1\mathbf{S}^{-1}$ so, the result is proved, by Induction.

(ii) $\lambda_1 = 1, \mathbf{u}_1 = \begin{pmatrix} \frac{3}{5} \\ \frac{2}{5} \end{pmatrix}; \quad \lambda_2 = -\frac{1}{4}, \mathbf{u}_2 = \begin{pmatrix} 1 \\ -1 \end{pmatrix};$

$$\begin{pmatrix} \frac{3}{5} & 1 \\ \frac{2}{5} & -1 \end{pmatrix} \begin{pmatrix} 1 & 0 \\ 0 & -\frac{1}{4} \end{pmatrix} \begin{pmatrix} 1 & 1 \\ \frac{2}{5} & -\frac{3}{5} \end{pmatrix} = \begin{pmatrix} \frac{1}{2} & \frac{3}{4} \\ \frac{1}{2} & \frac{1}{4} \end{pmatrix};$$

$$\mathbf{M}^6 = \begin{pmatrix} \dfrac{3}{5} + \dfrac{2}{10 \times 2^{11}} & \dfrac{3}{5} - \dfrac{3}{10 \times 2^{11}} \\ \dfrac{2}{5} - \dfrac{2}{10 \times 2^{11}} & \dfrac{2}{5} + \dfrac{3}{10 \times 2^{11}} \end{pmatrix}.$$

The limit is $\begin{pmatrix} \frac{3}{5} & \frac{3}{5} \\ \frac{2}{5} & \frac{2}{5} \end{pmatrix}.$

(iii) That the eigen-value 1 does not arise as a repeated root and that all other eigen-values are real and have modulus less than 1.

[Further query for the reader: we are looking for sufficient conditions, could we do without the *reality* of the other eigen-vectors?]

7. 1, ω, ω^2, where $\omega = (-1 + j\sqrt{3})/2$ and $j^2 = -1$.

Exercise 10F (p. 314)

1. (i) Closed classes: $\{A, B, C\}$ only;
transient classes: none.

(ii) Yes; M^4 shows this. $(\frac{2}{5}, \frac{2}{5}, \frac{1}{5})'$.

(iii) Matrix of times:

		from		
		A	B	C
	A	—	3	4
to	B	2	—	1
	C	6	4	—

Notice that
$$T_{CA} = T_{CB} + T_{BA},$$
$$T_{AC} = T_{AB} + T_{BC}.$$

Expected duration in each state:
$$2, \quad 1, \quad 1.$$

Notice that expected duration in $A = T_{BA}$;

$$\text{in } C = T_{BC}.$$

(iv) No answer meaningful.

3. $T_{AB} = \dfrac{3 - 3p + 3p^2}{p^3}$; $T_{AC} = \dfrac{3 - 2p + 3p^2}{p^3}$; $T_{AD} = \dfrac{3 - 2p + 4p^2}{p^3}$.

It does.
$$T_{BD} = p(1 + T_{BC}) + (1 - p)(1 + T_{BD})$$

and
$$T_{CD} = p.1 \qquad + (1 - p)(1 + T_{CD})$$

give
$$p(T_{BD} - T_{BC}) = 1$$

and
$$pT_{CD} \qquad = 1.$$

The result follows. [Compare Question 1 (iii).]

5. Let the probabilities be pr $(R \to L) = a$, pr $(L \to R) = b$.
Problem (B): We find the probabilities of lengths of run in R.
A typical sequence is $.LLLRRRRLR\ldots$; we define the part $.LLL\ldots$ as
giving 2 runs in R of zero length each.

$$\text{pr (the length of run in } R = 0) = 1 - b,$$

$$\text{pr (the length of run in } R = u) = b(1 - a)^{u-1}a \text{ for } u \geqslant 1.$$

Problem (C): Expected length of non-zero run in R

$$= \frac{1.ba + 2.b(1 - a)a + 3b(1 - a)^2 a + \ldots}{ba + b(1 - a)a + b(1 - a)^2 a + \ldots}$$

$$= \frac{ba(1 - (1 - a))^{-2}}{ba(1 - (1 - a))^{-1}} = a^{-1}.$$

Problem (*D*): To find the expected time to the next *R* from *L*, we find T_{RL} given by

$$T_{RL} = b.1 + (1-b)(1+T_{RL}),$$

whence $T_{RL} = b^{-1}$

= expected length of non-zero run in *L*.

Problem (*E*): The results for problems (*C*), (*D*) are not affected individually by the ratio $a:b$.

Comparison with tables in Chapter 7. Table 7.3 shows waiting times to the next 'success'.

R corresponds to 'success', waiting times are then lengths of run of *L*. We have $a = \frac{1}{3}; b = \frac{1}{6}$.

We find the mean of the non-zero waiting times is $\frac{161}{27} = 6.0$; the result expected symetrically from the solution to problem (*C*) is b^{-1}, that is 6.

To calculate the expected length of run of *R* including zeros we modify (*C*) to get

$$0(1-b) + 1.ba + 2.b(1-a)a + 3b(1-a)^2 a + \ldots = ba^{-1}.$$

The symmetrical value involving *L* is ab^{-1} which is 2.

From Table 7.3 the value is $\frac{161}{88} = 1.83$.

[See also Question 7.]

7. (i) For example:

$$T_{CA} = \tfrac{1}{2}(1+T_{CA}) + \tfrac{3}{10}(1+T_{CB}) + \tfrac{1}{5}.1,$$

$$T_{CB} = \tfrac{1}{5}(1+T_{CA}) + \tfrac{7}{10}(1+T_{CB}) + \tfrac{1}{10}.1,$$

$$T_{CC} = \tfrac{1}{10}(1+T_{CA}) + \tfrac{1}{2}(1+T_{CB}) + \tfrac{2}{5}.1.$$

In full we have

$$\begin{pmatrix} T_{AA} & T_{AB} & T_{AC} \\ T_{BA} & T_{BB} & T_{BC} \\ T_{CA} & T_{CB} & T_{CC} \end{pmatrix} = \begin{pmatrix} 1 & 1+T_{AB} & 1+T_{AC} \\ 1+T_{BA} & 1 & 1+T_{BC} \\ 1+T_{CA} & 1+T_{CB} & 1 \end{pmatrix} . \mathbf{M}.$$

The first three equations written solve to give:

$$T_{CA} = \tfrac{20}{3}, \quad T_{CB} = \tfrac{70}{9}, \quad T_{CC} = \tfrac{50}{9}.$$

and finally we have:

$$T_{AA} = \tfrac{50}{13}, \quad T_{BB} = \tfrac{25}{14}, \quad T_{CC} = \tfrac{50}{9}.$$

(ii) $T_{AA}:T_{BB}:T_{CC} = (0.26)^{-1}:(0.56)^{-1}:(0.18)^{-1}$

$$= u^{-1} \quad :v^{-1} \quad :w^{-1}$$

as required. (See Section 3.3.)

(iii) Difficulties arise over the zero-values of *u*, *v*. A suitable interpretation would be that the expected recurrence times were indefinitely large (because of the possibility of absorption at *C*).

(iv) Suppose we are finding the mean recurrence time for E_1. The numbers shown underneath the following sequence are the values that would be added to give the total of recurrence times as a step towards calculating the required mean:

$$E_3 \ E_4 \ E_1 \ E_2 \ E_3 \ E_2 \ E_4 \ E_1 \ E_2 \ E_1 \ E_1 \ E_1 \ E_3 \ E_4.$$

$$5 \qquad\qquad 2 \quad 1 \quad 1$$

It will be seen that these numbers total to $\Sigma r_i - c_1$, where $c_1 =$ number of terms up to and including first $E_1 +$ number of terms after last E_1.
c_i does not depend on the total length of sequence.

The observed mean will be obtained as $\dfrac{\Sigma r_i - c_1}{r_1 - 1}$.

For a sufficiently long sequence this is nearly $\dfrac{\Sigma r_i}{r_1} = u_1^{-1}$.

9.

$$
\begin{bmatrix}
q(0, t+1) \\
q(1, t+1) \\
q(2, t+1) \\
\vdots \\
q(n, t+1) \\
q(n+1, t+1) \\
\vdots \\
q(t+1, t+1)
\end{bmatrix}
=
\begin{bmatrix}
1-\lambda & 0 & 0 & & & & & \\
\lambda & 1-\lambda & 0 & & & & & \\
0 & \lambda & 1-\lambda & & & & & \\
& & & \ddots & & & & \\
& & & \lambda & 1-\lambda & 0 & & \\
& & & 0 & \lambda & 1-\lambda & & \\
& & & & & & \ddots & \\
& & & & & 1-\lambda & 0 \\
& & & & & 0 & \lambda & 1-\lambda
\end{bmatrix}
\begin{bmatrix}
q(0, t) \\
q(1, t) \\
q(2, t) \\
\vdots \\
q(n, t) \\
q(n+1, t) \\
\vdots \\
q(t, t) \\
0
\end{bmatrix}
$$

The matrix is shown square for simplicity, but note that the last column could be omitted with the last row of the probability vector for time t.

Further complications as to the size of the matrices would arise if we did not have $q(0, 0) = 1$, $q(n, t) = 0$, $(n > 0)$ which are the initial conditions specified.

Exercise 11 A (p. 321)

1. No answers can be meaningfully supplied.

3. (a) $f(x) = \frac{1}{2}$ for $x = 0, 1$.

(b) the even moments of higher orders; the population with the larger such parameters would tend to have the larger tails on the frequency diagram, and be less concentrated near the mean.

Exercise 11 B (p. 324)

1. (i) Let the mean be m, then the sample is $m - \frac{1}{2}r$, $m + \frac{1}{2}r$ and $S^2 =$ mean of $(\frac{1}{2}r)^2$ and $(\frac{1}{2}r)^2 = (\frac{1}{2}r)^2 = \frac{1}{4}r^2$.

(ii)

2nd member						
6	(7, 25)	(8, 16)	(9, 9)	(10, 4)	(11, 1)	(12, 0)
5	(6, 16)	(7, 9)	(8, 4)	(9, 1)	(10, 0)	(11, 1)
4	(5, 9)	(6, 4)	(7, 1)	(8, 0)	(9, 1)	(10, 4)
3	(4, 4)	(5, 1)	(6, 0)	(7, 1)	(8, 4)	(9, 9)
2	(3, 1)	(4, 0)	(5, 1)	(6, 4)	(7, 9)	(8, 16)
1	(2, 0)	(3, 1)	(4, 4)	(5, 9)	(6, 16)	(7, 25)
	1	2	3	4	5	6 1st member

(iii)

m	$36 \times$ probability	S^2	$36 \times$ probability
1	1	0	6
$1\frac{1}{2}$	2	$\frac{1}{4}$	10
2	3	1	8
$2\frac{1}{2}$	4	$2\frac{1}{4}$	6
3	5	4	4
$3\frac{1}{2}$	6	$6\frac{1}{4}$	2
4	5		
$4\frac{1}{2}$	4		
5	3		
$5\frac{1}{2}$	2		
6	1		

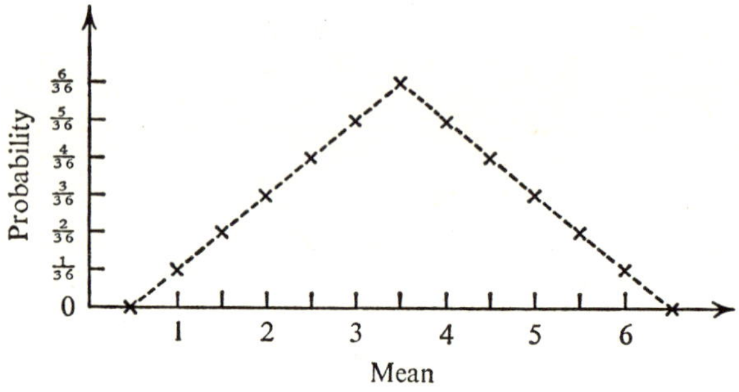

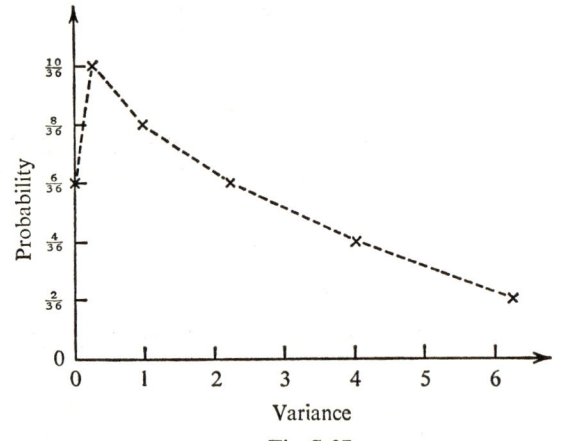

Fig. S.36

Fig. S.37

(iv)
$$E[m] = 3\tfrac{1}{2} \qquad \mathrm{var}\,[m] = \tfrac{35}{24};$$
$$E[v] = \tfrac{35}{24} \qquad \mathrm{var}\,[v] = \tfrac{1673}{576}.$$

(v) The pattern does not seem one of independence; the necessary symmetry seems lacking.
$$\rho[(m, v)] = \frac{E[mv] - E[m] \cdot E[v]}{\sqrt{(\mathrm{var}\,[m] \cdot \mathrm{var}\,[v])}} = \frac{\tfrac{735}{144} - \tfrac{7}{2} \times \tfrac{35}{24}}{\sqrt{(\tfrac{35}{24} \times \tfrac{1673}{576})}} = 0.$$

The value for ρ does not show that the variables are independent. See Exercise 9 D, Question 9. For instance pr $(m = 1|v = 0) \neq$ pr $(m = 1|v \neq 0)$.

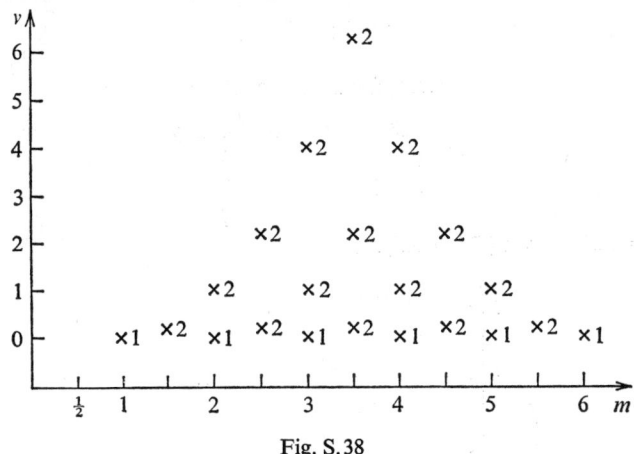

Fig. S.38

Exercise 11 C (p. 329)

1.

m	$p_3(m)$	$\phi_3(m)$
0	0·008	0·024
0·33	0·024	0·072
0·67	0·048	0·144
1·00	0·080	0·240
1·33	0·120	0·360
1·67	0·144	0·432
2·00	0·152	0·456
⋮	⋮	⋮
4·00	0·008	0·024

3. (i)

Sample size	1	2	3	4
Range	$1·5 \pm 1·5$	$1·5 \pm 1·0$	$1·5 \pm 0·83$	$1·5 \pm 0·75$

The necessary range decreases with increasing sample size, though less rapidly later.

(ii)

Sample size	1	2	3	4
$100 \times$ Probability	40	48	49·6	67·2

The probability increases with increasing sample size; this might be anticipated from part (i).

(iii)

Sample size	1	2	3	4
$100 \times$ Probability	20	52	68	77·6

For each sample-size (except $N = 1$) the probability is larger for intervals of the given length centred at 2·0 (the population's mean) than at 1·5.

[The anomaly at $N = 1$ is due to the discontinuity of $p_1(m)$.]

Exercise 11 D (p. 332)

1. $\sigma_2^2 = 1; \quad \sigma_3^2 = \frac{2}{3}.$

Exercise 11 E (p. 340)

1. (*a*)

S^2	Probability
0	$\frac{1}{8}$
$3a^2/4$	$\frac{1}{2}$
a^2	$\frac{3}{8}$

$E[S^2] = 3a^2/4; \quad \text{var}\,[S^2] = 3a^4/32.$

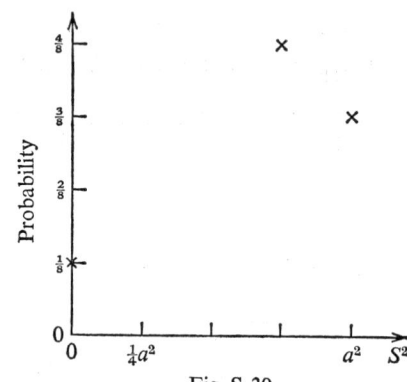

Fig. S. 39

(*b*)

S^2	Probability
0	$\frac{20}{64}$
a^2	$\frac{30}{64}$
$4a^2$	$\frac{12}{46}$
$9a^2$	$\frac{2}{64}$

$E[S^2] = 3a^2/2; \quad \text{var}\,[S^2] = 15a^4/4.$

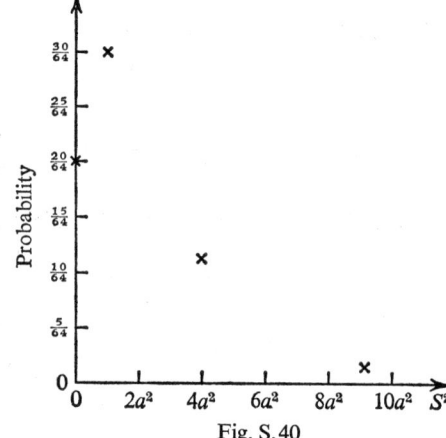

Fig. S. 40

Exercise 11 F (p. 342)

1. *For Section 6.2:* $\sigma^2 = \frac{1}{4}$, $\sigma^4 = \frac{1}{16}$, $\mu_4 = \frac{1}{16}$.

By the formula: for $N = 1$,   var $[S^2] = 0$;

$$N = 2, \quad \text{var } [S^2] = \tfrac{1}{8}.1.(1.\tfrac{1}{16}+1.\tfrac{1}{16}) = \tfrac{1}{64};$$

$$N = 3, \quad \text{var } [S^2] = \tfrac{1}{27}.2.(2.\tfrac{1}{16}+0) = \tfrac{1}{108}.$$

These agree with the results obtained by calculation for the probability functions for S^2 given in Section 6.2.

For Section 6.3: $\sigma^2 = a^2/2$; $\sigma^4 = a^4/4$; $\mu_4 = a^4/2$.

By the formula: for $N = 3$,   var $[S^2] = \tfrac{1}{27}.2(2.a^4/2-0) = \tfrac{2}{27}a^4$.

This agrees with the result calculated from the probability table.

Exercise 12 A (p. 359)

1. For $\rho = 1$, $E[\text{Bayesian estimates}] = \frac{5}{6} \neq \rho$, so the estimator is biased.

3. (Values in the table are all to be divided by 1024.)

(i)

ρ						
1	0	0	0	0	0	1024
$\frac{3}{4}$	1	15	90	270	405	243
$\frac{1}{2}$	32	160	320	320	160	32
$\frac{1}{4}$	243	405	270	90	15	1
0	1024	0	0	0	0	0
	0	$\frac{1}{5}$	$\frac{2}{5}$	$\frac{3}{5}$	$\frac{4}{5}$	$\frac{5}{5}$, r

(ii)

r	$\bar{\rho}$	r	$\bar{\rho}$
0	0·0	0·6	0·5
0·2	0·25	0·8	0·75
0·4	0·5	1·0	1·0

5. (i)

$$E[w] = r\frac{(1-\theta)}{\theta} = \frac{r}{\theta}-r.$$

Hence

$$E\left[\frac{w}{r}+1\right] = \frac{1}{\theta}.$$

An unbiased estimator for $1/\theta$ is $w/r+1$.

This unbiased estimator may be written as (total number of throws)/r, it can then be compared with the unbiased estimator r/(total number of throws) for θ, in the problem of Question 4, *where, however, r plays a different role.*

(ii) $\dfrac{d}{d\theta}L(\theta; w) = \dbinom{r+w-1}{r-1}(-w(1-\theta)^{w-1}\theta^r+(1-\theta)^w r\theta^{r-1})$

$\quad = 0$ for $\theta = r/(w+r)$.

The maximum likelihood estimator for θ is $r/($total number of throws$)$.

For $r = 1$, the estimator is $1/(w+1)$ and we investigate the bias in this simple case.

$$E\left[\frac{1}{w+1}\right] = \tfrac{1}{1}.\theta + \tfrac{1}{2}(1-\theta)\,\theta + \tfrac{1}{3}.(1-\theta)^2\,\theta + \dots$$

$$= \frac{\theta}{1-\theta}\{-\log(1-(1-\theta))\} = \frac{\theta}{1-\theta}\log\left(\frac{1}{\theta}\right) \neq \theta.$$

Thus, in this case the estimator is biased. The differing role of r is clear.

7. (i)
$$E[s] = 1.\theta + 2(1-\theta).\theta + 3.(1-\theta)\,\theta^2 + \dots$$

$$= \theta\,\frac{1}{(1-(1-\theta))^2} = \frac{1}{\theta}.$$

$$E\left[\frac{1}{s}\right] = \frac{\theta}{1-\theta}\log\left(\frac{1}{\theta}\right)$$

follows from the proof for $E[1/(w+1)]$ in Question 5.

(ii) We have a Poisson process with $\mu = 1$, from which $\theta = e^{-1}$.

(iii) $E\left[\dfrac{1}{s}\right] = \dfrac{e^{-1}}{1-e^{-1}}.1 = \dfrac{1}{e-1} = \dfrac{1}{E[s]-1} = \dfrac{1}{E[s-1]}.$

(iv) We cannot *always* have s an unbiased estimator for ϕ, say

$$and \qquad \frac{1}{s} \text{ an unbiased estimator for } 1/\phi.$$

For if we did, then we would always have:

$$E\left[\frac{1}{s}\right] = \frac{1^i}{\phi} = \frac{1}{E[s]},$$

whereas we do not have this in part (iii).

9. (*a*) Maximum likelihood estimator $= r/($total number of throws$)$ for both columns.

Estimates are $\frac{50}{120}$ and $\frac{48}{122}$, that is $0\cdot417$ and $0\cdot393$.

[These results are obtained from Questions 4(ii) and 5(ii).]

(*b*) Unbiased estimator $= r/N$ for first column (notation of Question 4(i))
$= (k-1)/(r-1)$ for second column (notation of Question 6(iii)).

Estimates for θ are thus $\frac{50}{120}$ and $\frac{47}{121}$, that is $0\cdot417$ and $0\cdot389$.

Unbiased estimator for $1/\theta = (w+r)/r$ for second column (notation of Question 5(i)).

Estimate for $1/\theta$ is thus $\frac{122}{48} = 2\cdot54$.

11. (i) $A1$: Working from 16 as origin:

$$\hat{\mu} = 16\tfrac{4}{12} = 16\cdot33; \quad \widehat{\sigma^2} = \frac{1}{11}\left(18 - \frac{4^2}{12}\right) = 1\cdot52.$$

$A2$: Working from 16 as origin:

$$\hat{\mu} = 15\tfrac{11}{16} = 15\cdot69; \quad \widehat{\sigma^2} = \tfrac{1}{15}(147 - \tfrac{25}{16}) = 9\cdot70.$$

$B1$: Working from 110 as origin:

$$\hat{\mu} = 111; \ \widehat{\sigma^2} = \tfrac{1}{7}(416 - \tfrac{64}{8}) = 58\cdot3.$$

$B3$: Working from 110 as origin:

$$\hat{\mu} = 107\cdot5; \ \ \widehat{\sigma^2} = \tfrac{1}{5}(303 - \tfrac{225}{6}) = 53\cdot1;$$

$$9\cdot70/1\cdot52 = 6\cdot4; \ \ 58\cdot3/53\cdot1 = 1\cdot1.$$

[Where there is no risk of ambiguity we may follow the custom of using 'hatted' symbols to denote *estimates* rather than *estimating functions* or *estimators*.]

(ii) For $A1$: $\dfrac{2}{1\cdot52} = 1\cdot32$;   for $A2$: $\dfrac{9\cdot70}{8\cdot25} = 1\cdot18$.

     For $B1$: $\dfrac{100}{58\cdot3} = 1\cdot72$;   for $B2$: $\dfrac{90}{53\cdot1} = 1\cdot69$.

13. It is not meaningful to quote answers.

Exercise 12 B (p. 385)

1. (i) By forming a frequency table and calculating: var $[t] = \tfrac{203}{108}$.
By using var $[m] = \tfrac{1}{3}$ var $[1, 2, 3, 4, 5, 6]$; var $[m] = \tfrac{35}{36} = \tfrac{135}{108}$.
By forming a frequency table and calculating: var $[u] = 92\tfrac{3}{4}/108$.
In increasing order of efficiency we have: t, m, u.

(ii) Each member of the sample is an unbiased estimator of the mean of the population, but a and c are not *independent* estimators, as the following table of values of $216 \times$ probability of a given pair shows:

	6	30	24	18	12	6	1
	5	24	18	12	6	1	
values of c	4	18	12	6	1		
	3	12	6	1			
	2	6	1				
	1	1					
		1	2	3	4	5	6
				values of a			

The lack of independence prevents us having any general result about the variance of $\tfrac{1}{2}(a+c)$.

(iii) The reader should construct a table, like that in part (ii), to show that u, t are not independent random variables.

3. In each case we can use

$$\hat{\mu} = m, \ \ \ \widehat{\sigma^2} = \tfrac{20}{19}S^2, \ \ \text{estimated var}\,[\hat{\mu}] = \tfrac{1}{19}S^2.$$

It is interesting to see whether $\bar{\mu}$ falls in the range $\hat{\mu} + 2\sqrt{\text{est. var}\,[\hat{\mu}]}$, and this test range is also shown in the table below.

Sample	m	S^2	$\hat{\mu}$	$\widehat{\sigma^2}$	Est. var $[\hat{\mu}]$	Range
(i)	15·4	16·34	15·4	17·2	0·86	13·6–17·2
(ii)	4·75	3·89	4·75	4·10	0·20	3·9– 5·6
(iii)	36·85	140·1	36·85	147·5	7·4	31·4–42·3

Sample (i) is from a rectangular population so we could more efficiently use the mean of the extremes to estimate μ as 15·5. We have not developed a method here for estimating the variance of this estimator.

5. (i) $$\text{pr}\,(R = r) = 1/\beta.$$

(ii) $$E[r] = \sum_{1}^{\beta} r/\beta = (\beta+1)/2,$$

$$E[r^2] = \sum_{1}^{\beta} r^2/\beta = (\beta+1)\,(2\beta+1)/6,$$

$$\text{var}\,[r] = E[r^2] - (E[r])^2 = (\beta^2-1)/12.$$

(iii) $t = 2r-1$, $\text{var}\,[t] = (\beta^2-1)/3$.

(iv) $E[\text{mean of sample of } N \text{ values of } t] = E[t] = \beta$, so the estimator is unbiased.
$$\text{var}\,[m] = \frac{1}{N}\text{var}\,[t] = (\beta^2-1)/3N.$$

(v) In Example 2 we have $N = 2$; the numbers r are the values of X, Y, say x, y; the values t are $2x-1, 2y-1$; $m = x+y-1$.
Thus m corresponds to the value of the statistic $X+ Y-1$ in the example.
The variance is $(\beta^2-1)/6$ from part (iv) of the present question and from the calculations in Example 2.

7. *First point:* multiplying by $(N+1)/N$ to remove the bias produces an increase in variance by a factor $((N+1)/N)^2$. *Second point:* if the bias is negligible for large enough N, then the reduction in variance is even more so negligible.

9.

	The statistic	The parameter
(i)	$\left(\dfrac{N}{N-1}\right) S^2$	σ^2
(ii)	S^2	σ^2
(iii)	A single random member of the sample	μ
(iv)	The largest member of the sample	μ

11.

Estimates of mean	37·6	38·8	36·4	34·6
Max. likelihood estimates of variance	240·6	81·8	118·6	110·8
Unbiased estimates of variance	300·8	102·2	148·3	137·3

We recollect that the samples were drawn from a population with mean 40 and variance 100.

13. The necessary results are quoted in Section 6.1.
It follows that, in Question 11, we cannot obtain an unbiased estimate of the standard deviation by square rooting the first estimate of variance, but also that we can obtain a maximum likelihood estimate of the standard deviation by square rooting the second estimate. We shall see later, however, that the square root of an *unbiased* estimate of variance is what enters into many of the further methods.

15. $E[a_1 x_1 + a_2 x_2] = (a_1 + a_2)\,E[x]$, since $E[x_1] = E[x_2] = E[x]$.
The estimator is unbiased $\Leftrightarrow a_1 + a_2 = 1$.

Now the variance of the estimator, say v, is given by

$$\frac{a_1^2}{u_1^2}+\frac{a_2^2}{u_2^2} = v.$$

(i) *Method* 1. Write $a_2 = 1-a_1$ in the expression for v, and minimize the resulting quadratic in a_1.

Method 2. In the (a_1, a_2)-plane, choices of (a_1, a_2) for fixed v lie on the ellipse,

$$\frac{a_1^2}{u_1^2}+\frac{a_2^2}{u_2^2} = v$$

and those for unbiased estimators lie in the line

$$a_1+a_2 = 1.$$

We show that for suitable choice of v the tangent to ellipse at

$$\left(\frac{u_1^2}{u_1^2+u_2^2}, \frac{u_2^2}{u_1^2+u_2^2}\right) \quad \text{is} \quad a_1+a_2 = 1.$$

The tangent is in fact $\dfrac{a_1 \cdot \dfrac{u_1^2}{u_1^2+u_2^2}}{u_1^2}+\dfrac{a_2 \cdot \dfrac{u_2^2}{u_1^2+u_2^2}}{u_2^2} = v$

and this is $\qquad a_1 \qquad +a_2 \qquad = 1,$

provided $\qquad\qquad v = \dfrac{1}{u_1^2+u_2^2}.$

Thus the minimum variance possible is $\dfrac{1}{u_1^2+u_2^2}$ and this is obtained when

$$a_r = u_r^2/(u_1^2+u_2^2).$$

(ii) Method 2 can be applied in a four-dimensional (a_1, a_2, a_3, a_4)-space to show that the minimum variance will be $1 \Big/ \sum_1^4 u_r^2$ and will be obtained by the set of choices $a_r = u_r^2/\Sigma u_r^2$.

(iii) The values of u_r^2 are $(\frac{100}{8})^2$, $(\frac{100}{5})^2$, $(\frac{100}{10})^2$, $(\frac{100}{4})^2$; their sum is $1281 \cdot 25$.

The values of a_r are

$$0 \cdot 122, \quad 0 \cdot 312, \quad 0 \cdot 078, \quad 0 \cdot 488.$$

The unbiased estimate

$$= (0 \cdot 122 \times 2 \cdot 03)+(0 \cdot 312 \times 2 \cdot 20)+(0 \cdot 078 \times 2 \cdot 25)+(0 \cdot 488 \times 2 \cdot 12)$$
$$= 2 \cdot 144.$$

The standard deviation of the population of such estimates is estimated as $1/\sqrt{(1281 \cdot 25)}$, that is $0 \cdot 028$.

17. $\operatorname{var}[t_\lambda] = \lambda^2 \operatorname{var}[x]+(1-\lambda)^2 \operatorname{var}[y]$

$$= \frac{\lambda^2\sigma^2}{N_1}+\frac{(1-\lambda)^2\,\sigma^2}{N_2}$$

$$= \frac{\sigma^2}{N_1N_2}\left\{(N_1+N_2)\left(\lambda-\frac{N_1}{N_1+N_2}\right)^2+N_1-\frac{N_1^2}{N_1+N_2}\right\}$$

$$\geqslant \sigma^2/(N_1+N_2), \text{ the minimum value being obtained from } \lambda = \frac{N_1}{N_1+N_2}.$$

19. (i)
$$\text{var}\left(\tfrac{1}{2}(T_1+T_2)\right) = \tfrac{1}{4}\sigma_1^2 + \tfrac{1}{4}\sigma_2^2;$$
$$\tfrac{1}{4}\sigma_1^2 + \tfrac{1}{4}\sigma_2^2 < \sigma_1^2 \quad \Leftrightarrow \quad \sigma_2^2 < 3\sigma_1^2.$$

(ii) If the variations in the values of the statistics are due to random variation in the data and not to systematic errors in your work, then you may take the variance of your statistic as σ^2/N_0 and that of your friend's statistic as σ^2/N.

Part (i) then shows that you need

$$\frac{\sigma^2}{N} < \frac{3\sigma^2}{N_0}$$

that is
$$N > \tfrac{1}{3}N_0.$$

21. (i)
$$E[S_r^2] = \frac{N_r - 1}{N_r}\sigma^2 \quad (r = 1, 2).$$

(ii)
$$E[\lambda S_1^2 + \mu S_2^2] = \left\{ \lambda \frac{(N_1 - 1)}{N_1} + \mu \frac{(N_2 - 1)}{N_2} \right\} \sigma^2.$$

We need
$$\lambda \frac{(N_1 - 1)}{N_1} + \mu \frac{(N_2 - 1)}{N_2} = 1.$$

(iii) $\lambda_0 \dfrac{(N_1 - 1)}{N_1} + \mu_0 \dfrac{(N_2 - 1)}{N_2} = \dfrac{N_1 - 1}{N_1 + N_2 - 2}\left\{ 1 + \dfrac{\delta}{N_1 - 1} \right\} + \dfrac{N_2 - 1}{N_1 + N_2 - 2}\left\{ 1 - \dfrac{\delta}{N_2 - 1} \right\}$

$$= 1 + \frac{\delta}{N_1 + N_2 - 2}\{1 - 1\}.$$

(iv) $\text{var}\left[\lambda_0 S_1^2 + \mu_0 S_2^2\right]$

$$= \frac{2\sigma^4(N_1 - 1)}{(N_1 + N_2 - 2)^2}\left\{ 1 + \frac{2\delta}{N_1 - 1} + \frac{\delta^2}{(N_1 - 1)^2} \right\}$$

$$+ \frac{2\sigma^4(N_2 - 1)}{(N_1 + N_2 - 2)^2}\left\{ 1 - \frac{2\delta}{N_2 - 1} + \frac{\delta^2}{(N_2 - 1)^2} \right\}$$

$$= \frac{2\sigma^4}{(N_1 + N_2 - 2)}\left\{ 1 + \frac{\delta^2}{(N_1 - 1)(N_2 - 1)} \right\}.$$

(v) By taking $\delta = 0$ the given results follow.

(vi)
$$S_i^2 = \frac{1}{N_i}q_i - \left(\frac{p_i}{N_i}\right)^2, \quad \text{by standard result.}$$

$$\Rightarrow N_i S_i^2 = q_i - p_i^2/N, \quad \text{as required.}$$

23. (i) With the notation of Question 21 we have:

i	p_i	q_i	p_i^2/N_i
1	21	81	$73\tfrac{1}{2}$
2	27	139	$121\tfrac{1}{2}$
3	33	203	$181\tfrac{1}{2}$
4	24	118	96
		541	$472\tfrac{1}{2}$

Unbiased estimate of variance $= \dfrac{\Sigma(q_i - p_i^2/N_i)}{(\Sigma N_i) - 4}$

$$= (541 - 472\tfrac{1}{2})/20$$

$$= 3\cdot425.$$

(ii) The means are $3\cdot5$, $4\cdot5$, $5\cdot5$, 4.

For these, $q = 78\cdot75$, $p = 17\cdot5$.

Unbiased estimate of variance of means $= (78\cdot75 - 76\cdot5625)/3$

$$= 0\cdot73.$$

25. In the notation of Question 24, and using results from Question 23,

$$T = \Sigma p_i = 105, \quad S = 541, \quad C = 472\tfrac{1}{2}.$$

Estimated variance between classes $= \tfrac{1}{3}(472\tfrac{1}{2} - \tfrac{1}{24}(105)^2)$

$$= 4\cdot375.$$

Estimated variance within classes $= \tfrac{1}{20}(541 - 472\tfrac{1}{2})$

$$= 3\cdot425.$$

Estimated overall variance $= \tfrac{1}{23}(541 - \tfrac{1}{24}(105)^2)$

$$= 3\cdot549.$$

$$\frac{\text{Estimated variance between classes}}{\text{Estimated variance within classes}} = \frac{4\cdot375}{3\cdot425} = 1\cdot28.$$

27. (i) There are $\dbinom{N}{x}$ ways of selecting which x of the N chosen discs shall have $R = b$

pr (a given disc has $R = b$) $= 1/\beta$,

pr (a given disc has $R < b$) $= (b-1)/\beta$.

Required probability $= \dbinom{N}{x}\left(\dfrac{1}{\beta}\right)^x\left(\dfrac{b-1}{\beta}\right)^{N-x} = p_x$,   say.

(ii) pr (largest value of R is b)
= pr (1 disc has $R = b$ and all the rest have $R < b$,
    or 2 discs have $R = b$ and all the rest have $R < b$,
    or 3 discs have $R = b$ and all the rest have $R < b$,
    or ......,
    or $(N-1)$ discs have $R = b$ and one has $R < b$,
    or N discs have $R = b$ and none have $R < b$)

$$= \sum_{x=1}^{N} p_x, \text{ since the events are exclusive}$$

$$= \left(\sum_{x=0}^{N} p_x\right) - p_0.$$

$$= \left(\frac{b}{\beta}\right)^N\left(\frac{b-1}{b} + \frac{1}{b}\right)^N - p_0 \quad \text{[doing the summation by Binomial Theorem]}$$

$$= \left(\frac{b}{\beta}\right)^N - \left(\frac{b-1}{\beta}\right)^N.$$

For $b = 1$ we require all N discs to have $R = 1$ and the probability of this is $(1/\beta)^N$, which is also given by the above formula.

Example 2 would take the N-dimensional cube with side b and remove an N-dimensional cube of side $b-1$, dividing the result by the volume of the N-dimensional cube of side β. The result would arise as $(b^N - (b-1)^N)/\beta^N$.

(iii) To make $\left(\dfrac{b}{\beta}\right)^N \left(1 - \left(\dfrac{b-1}{b}\right)^N\right)$ a maximum we must reduce β as far as possible. The least possible value of β is b since there must be at least b discs if the number b is to occur. Thus $\beta = b$ maximizes the likelihood and b is the maximum likelihood estimator of β.

(iv) $E[b] = \sum\limits_{i=1}^{i=\beta} i \left(\left(\dfrac{i}{\beta}\right)^N - \left(\dfrac{i-1}{\beta}\right)^N\right)$

$\qquad = \beta^{-N} \cdot \sum\limits_{i=1}^{i=\beta} [\{i^{N+1} - (i-1)^{N+1}\} - (i-1)^N],\quad$ after rearrangement

$\qquad = \beta^{-N} \cdot \left[\{\beta^{N+1}\} - \sum\limits_{i=1}^{i=\beta} (i-1)^N\right]$

$\qquad = \beta^{-N} \left[\{\beta^{N+1}\} + \beta^N - \sum\limits_{i=0}^{i=\beta} i^N\right].$

But $\sum\limits_{0}^{\beta} i^N = \dfrac{\beta^{N+1}}{N+1} + \text{polynomial in } \beta \text{ of degree } N$

so $\qquad E[b] = \beta^{-N}\left[\dfrac{N}{N+1} \cdot \beta^{N+1} + \text{polynomial in } \beta \text{ of degree } N\right]$

$\qquad\qquad \sim \dfrac{N}{N+1} \cdot \beta, \quad \text{for large } \beta.$

For Example 2 we make the mappings $N \to 2$, $\beta \to N$, $b \to z$ and $E(Z) = \frac{2}{3}N$ as required.

(v) $E[u] = \dfrac{N+1}{N} E[b] \sim \beta, \quad \text{for large } \beta$

so u is a nearly unbiased estimator of β, for large β.

$E[b^2] = \sum\limits_{i=1}^{i=\beta} i^2 \left(\left(\dfrac{i}{\beta}\right)^N - \left(\dfrac{i-1}{\beta}\right)^N\right)$

$\qquad = \beta^{-N} \sum\limits_{i=1}^{i=\beta} [\{i^{N+2} - (i-1)^{N+2}\} - (2i-1)(i-1)^N]$

$\qquad = \beta^{-N} \cdot \left[\{\beta^{N+2}\} - \sum\limits_{i=1}^{i=\beta} (2(i-1)^{N+1} + (i-1)^N)\right]$

$\qquad = \beta^{-N} \cdot \left[\{\beta^{N+2}\} + 2 \cdot \beta^{N+1} + \beta^N - \sum\limits_{i=0}^{i=\beta} \cdot (2i^{N+1} + i^N)\right]$

$\qquad = \beta^{-N} \left[\beta^{N+2} - \dfrac{2}{N+2} \beta^{N+2} + \text{polynomial in } \beta \text{ of degree } (N+1)\right]$

$\qquad \sim \dfrac{N}{N+2} \cdot \beta^2, \quad \text{for large } \beta.$

Hence $\text{var}[b] \sim \dfrac{N}{N+2}\cdot\beta^2 - \left(\dfrac{N}{N+1}\cdot\beta\right)^2$,   for large β,

and $\text{var}[u] = \left(\dfrac{N+1}{N}\right)^2 \text{var}[b]$

$\qquad\qquad\qquad \sim \left(\dfrac{(N+1)^2}{N(N+2)}-1\right)\cdot\beta^2$,   for large β,

$\qquad\qquad\qquad = \beta^2/N(N+2)$.

Since $\text{var}[u] \to 0$ as N, (the sample size), $\to 0$.
We have that u is a consistent estimator.

(vi) $\text{var}[u]/\text{var}[m] = \dfrac{\beta^2}{N(N+2)}\cdot\dfrac{3N}{\beta^2-1} \to 0$,   as $N \to \infty$,

u is preferable to m as an estimator.
(vii) No answers supplied.

29. (i) $E[r] = N\theta, \quad E[r/N] = \theta.$

$\quad \text{var}[r] = N\theta(1-\theta), \quad \text{var}[r/N] = \theta(1-\theta)/N.$

(ii) $\text{pr}\left(\left|\dfrac{r}{N}-\theta\right| \leqslant \epsilon\right) = \text{pr}\left(\left|\dfrac{r}{N}-\theta\right| \leqslant \epsilon\sqrt{\left(\dfrac{N}{\theta(1-\theta)}\right)}\sqrt{\left(\dfrac{\theta(1-\theta)}{N}\right)}\right)$

$\qquad\qquad\qquad\qquad > 1 - \left(\dfrac{1}{\epsilon}\sqrt{\left(\dfrac{\theta(1-\theta)}{N}\right)}\right)^2$

$\qquad\qquad\qquad\qquad = 1 - \dfrac{\theta(1-\theta)}{N\epsilon^2}$

$\qquad\qquad\qquad\qquad \geqslant 1 - \dfrac{1}{4N\epsilon^2}$

since $\theta(1-\theta) \leqslant \frac{1}{4}$ for all θ.

(iii) Take $\epsilon = 0\cdot01, \quad 1-\dfrac{1}{4N\epsilon^2} \geqslant 95\%,$

then $\dfrac{10^4}{4N} \leqslant \frac{1}{20}$ and $N \geqslant 5 \times 10^4.$

(iv) Take $\epsilon = 0\cdot432 - 0\cdot410 = 0\cdot022;$

$\qquad\qquad\qquad N = 2,848; \ \theta = 0\cdot410;$

$\dfrac{\theta(1-\theta)}{N\epsilon^2} = \dfrac{0\cdot410 \times 0\cdot590}{2848 \times (0\cdot022)^2} \simeq 17\%.$

Since there is about a 17 per cent chance that any observed discrepancy would exceed the one observed at the time, we have no cause for alarm.

(v) $\text{pr}\left\{\left|\dfrac{r}{N}-0\cdot432\right| \leqslant (0\cdot432-0\cdot320)\right\} > 1 - \dfrac{0\cdot432 \times 0\cdot568}{37 \times 32 \times (0\cdot112)^2},$

$\quad \text{pr}\left\{\left|\dfrac{r}{N}-0\cdot432\right| > 0\cdot432-0\cdot320\right\} < 0\cdot01653,$

$$\text{pr} \left\{0\cdot432 - \frac{r}{N} > 0\cdot432 - 0\cdot320\right\} < 0\cdot01653, \text{ a fortiori,}$$

$$\text{pr} \left\{\frac{r}{N} < 0\cdot320\right\} < 0\cdot01653,$$

$$\text{pr} \left\{\frac{r}{N} \geqslant 0\cdot320\right\} > 98\cdot34\%.$$

If the true mean for brass pins was as observed for plastic, then in about 1 experiment in about $\dfrac{1}{0\cdot01653}$, say in about 1 experiment in 60 *or more*, would we expect to observe as *few* brass pins fall point-down.

(vi) Although the value of ϵ is smaller the value of N is doubled, so that $(4N\epsilon^2)^{-1}$ is much reduced.

31. (i)

q	pr $(Q = q)$	θ_q	$\frac{1}{4}(q-1)$
1	$(1-\theta)^2$	0	0
3	$2\theta(1-\theta)$	$\frac{1}{2}$	$\frac{1}{2}$
5	θ^2	1	1

$$E[q] = 1 \times (1-\theta)^2 + 3 \times 2\theta(1-\theta) + 5 \times \theta^2 = 1 + 4\theta,$$

$$E[q^2] = 1^2 \times (1-\theta)^2 + 9 \times 2\theta(1-\theta) + 25 \times \theta^2 = 1 + 16\theta + 8\theta^2,$$

$$\text{var}\,[q] = (1 + 16\theta + 8\theta^2) - (1 + 4\theta)^2 = 8\theta - 8\theta^2.$$

The required estimator is $\frac{1}{4}(q-1)$ and the variance is $\frac{1}{16}(8\theta - 8\theta^2)$.

The unbiased estimator from Question 30 part (iv) had variance $\frac{1}{16}(1 + 8\theta - 9\theta^2)$, which exceeds that above for $0 \leqslant \theta < 1$.

Thus the third opening merely increases the variance of the estimator and so 'only confuses the issue'.

(ii) The maximum likelihood estimates are given at the head of this solution, they are the same as the unbiased estimate obtained in part (i).

(iii) With the notation of Question 30 part (ii) we have

n	pr $(N = n)$	$\theta_n \equiv$ max. likelihood estimate of θ
0	$\frac{1}{2}(1-\theta)^2$	0
2	$\frac{1}{2}(1-\theta^2)$	0
4	$(1-\theta)\theta$	$\frac{1}{2}$
5	θ^2	1

Thus
$$E[\theta_n] = \frac{1}{2}.(1-\theta)\theta + 1.\theta^2 = \frac{1}{2}(\theta + \theta^2);$$

for $0 < \theta < 1$, $\quad |E[\theta_n] - \theta| = \dfrac{\theta - \theta^2}{2} \leqslant \frac{1}{8}$,

but
$$\frac{|E[\theta_n] - \theta|}{\theta} \to 50\%, \text{ as } \theta \to 0.$$

INDEX